Methods of Solving Solid Geometry Problem

Ellina Grigorieva

Methods of Solving Solid Geometry Problems

Birkhäuser

Ellina Grigorieva
Texas Woman's University
Denton, TX, USA

ISBN 978-3-031-86968-6 ISBN 978-3-031-86969-3 (eBook)
https://doi.org/10.1007/978-3-031-86969-3

Mathematics Subject Classification: 51-01

This book is published under the imprint Birkhäuser, www.birkhauser-science.com by the registered company Springer Nature Switzerland AG
The registered company address is: Gewerbestrasse 11, 6330 Cham, Switzerland

In memory of my wonderful parents, ***Natali and Valery Grigoriev****, who sincerely believed in me and encouraged me in writing my books.*

To my university mentor and scientific advisor, ***Prof. Nikolai Fedorovich Stepanov****. Without your help and brilliant mind, my career as a scientist would not be possible.*

To my beautiful daughter ***Sasha****. Many thanks for your advice to publish with Birkhauser!*

Preface

Stereometry is a part of geometry that studies the properties of solid geometric bodies and spatial geometric figures, not only visible and known. It also allows us to model the structure of objects invisible to the eye, i.e., micro-objects, such as molecules, viruses, crystals, etc. Stereometry is an excellent tool for studying the macro- and micro-world.

In stereometry, as well as in planimetry, the properties of geometric figures are established by proving the corresponding theorems; however, the starting point is the properties of the basic geometric figures, expressed by axioms.

There are various geometry books that emphasize the pedagogical aspect of the subject, dedicated to the preparation of future teachers of mathematics ([39], [38]), books with standard problems in geometry and their solutions ([41], [40], [37]) or designed to prepare for the SAT or ACT. However, there are hardly any books on solving complex problems of three-dimensional geometry (stereometry), especially books for self-study to prepare for mathematical Olympiads, and to develop skills and abilities without help from an instructor or a tutor.

This book complements my four other problem-solving books ([2]–[5]) published by Birkhauser. It will help readers master their skills in solving solid geometry problems and to develop visualization, intuition, and ability to conduct their own proofs. The distinguishing feature of this book is that, along with the traditional components such as a short theoretical section, a problem-solving section, and a homework section for each chapter (containing 25–50 self-study problems with solutions and/or answers), the book provides detailed solutions to all 226 presented problems, as well as for the most intricate problems in the homework sections. These solutions can be used for solving other problems, and even for those that seem to be not connected with the given problem. This allows the student to independently advance in solving the problem directed by professional advice. Math teachers can find a readily made set of well-organized problems for all levels of knowledge.

The book is written in a friendly manner that encourages the reader to think and be creative through questions, comments, and advice. This book resembles a well-organized, self-paced, distance-learning course. The lack of an instructor is compensated by competent advice, which helps students solve complex problems on their own with the help of detailed solutions, ideas, or hints. Acquirement of such experience by students is easy due to another important feature of the book—the material is presented in a manner of a spiral pattern, where each topic is often repeated several times and each time at a more complex level than the previous. This fosters the better understanding of the material already covered and allows to gradually approach complex, olympiad-level problems. Many problems are solved in two or three ways; moreover, variants of some problems are presented in different chapters of the book, helping the reader to see the connection between different topics of geometry and therefore improving the learning outcomes.

Stereometry can be used to study existing chemical compounds and predict the properties of molecules, crystals, and can even provide some insight into the synthesis of new chemicals. In this book, I briefly show the various applications of stereometry to chemistry, molecular structures, and crystallography.

Although most problems can be understood by high-school students, the book uses vector-coordinate methods of analytic geometry and calculus, so will certainly be of interest to undergraduate and graduate-level students, taking a course in History of Mathematics or Euclidean Geometry. If a problem has a historic meaning, then its' original solution, as well as another one using modern methods of mathematics are included. I consider these instructions and commentary useful in the sense of getting acquainted with the historical course of the development of techniques and methods for solving problems practiced by people throughout time. In addition, depending on the importance of the questions raised, historical information about the origin of the methods and information about the activities of a particular mathematician, is included.

The goal of this book is:

1. To develop skills, spatial imagination, and intuition for solving problems in a historical context
2. To connect solid geometry and the sciences to enhance cognitive and research abilities
3. To be a source in preparation for math Olympiads or other similar contests.

Geometry, especially solid geometry, requires imagination, knowledge of the basic axioms and theorems, and the ability to apply them where necessary. Usually when students are studying the History of Mathematics, they already know much of higher mathematics. Thus, I find it useful to look for different solutions to ancient problems and motivate students to apply not only the methods of Euclid but also the methods of modern geometry, mathematical analysis, and analytic geometry. Therefore, they did not just learn how Thales or Archimedes solved the problem, but they saw how the same problem could be solved, for example, using vector-coordinate methods or applying a derivative or integral. This way each class session concludes with a fruitful discussion of the various proposed solution methods and their comparison.

Usually, in order to encourage everyone in the class, I start with a simple problem for everyone in the class to solve, and then we treat it as a research topic, developing from it to a real research journey. Below is one of the ideas of such teaching "tricks" that I also try to use in my book:

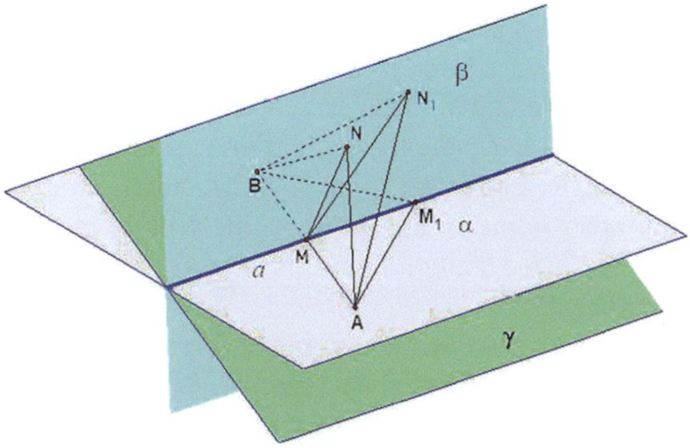

Fig. 1 Plane β is the locus of all points equidistant from the ends of segment *AB*.

Imagine that we have two points *A* and *B* on a line and try to find a point that would be equidistant from the ends of the segment *AB*. It is clear, that most of you will answer this question like this: It is very simple—this is the midpoint of the *AB* segment.

We denote the midpoint of the segment by *M*, and next, we turn not to a straight line, but to a plane containing this segment, and suppose that we want to find all points on the plane that are equidistant from the ends of the segment *AB*. Let's draw a perpendicular bisector to the given segment *AB*. Any point lying on this perpendicular bisector will be equidistant from the ends of the segment *AB*.

Now imagine that we have points *A* and *B* in space, and we want to find all possible points equidistant from the ends of the segment *AB*. Well then it probably comes to mind that this is a plane perpendicular to the segment *AB* and passing through its midpoint *M*. If we take any point of this plane, for example, point *N*, then the distance from this point to the ends of the segment will be the same. Moreover, this point will be the vertex of the isosceles triangle *ANB*, in which the segment *NM* is the median, altitude and bisector of ΔANB (See Fig. 1).

For any other point N_1 of this plane, there will be a new isosceles triangle, the distances from N_1 to points *A* and *B* will be equal. Of course, the distances from different points N_i of the plane β to the ends of a given segment will be different from point to point, but the distances from N_i to points *A* and *B* will be equal to each other ($N_iA = N_iB$). Plane β is called the **locus** of points equidistant from the ends of segment *AB*.

Now let's look again at the segment *AB* in space and think about how many planes can be constructed so that the distances from point *A* and point *B* to such a plane (perpendiculars dropped from each point to the plane) be equal. This task is different from the previous one. Of course, plane β in Fig. 1 satisfies this condition. However, in this book you will prove that there are infinitely many other planes (e.g., the green plane γ); all these planes must pass through the midpoint of the given segment, but perpendicularity to the segment is not required.

Next, let us solve a different problem: Given three points *A*, *B*, *C* that do not lie on the same line. Construct the locus of points equidistant from each of the three points.

Recalling lessons from plane geometry, we can find the center O of the circumscribed circle of triangle ABC. This point is located at the distance of the radius of the circumscribed circle from each of the three points. Is it possible to find other points in space? You can draw a straight line through point O perpendicular to the plane of the triangle. Let's take some point M on this straight line, then $MA = MB = MC$. Any other point on this straight line will be also equidistant from the vertices of $\triangle ABC$. Such a straight line is the locus of all points equidistant from the vertices of the triangle!

Now let's pose another problem: Draw a plane such that the distance to this plane from each of the three given points A, B, and C is the same. Think: should this plane pass through the center of the circumcircle? Not at all!

What is changing? Here we want to find the plane equidistant from all three points. It is clear, that all three points lie on one plane, so what happens? Probably, the plane of three points is itself equidistant from all points and, in principle, the distance from this plane to each point is the same and is equal to zero. But this is, so to speak, a trivial solution, which is not very interesting...

Is there another plane? Or maybe several planes? How do we find this plane?

Can we, for example, draw any plane through the midpoint of the segment AB? Well then it will be a plane equidistant from the ends of the segment AB (like the planes β or γ in Fig. 1), that is, the question arises: what to do with the third given point C? The plane must be equidistant from all three points. The answer to this question is also given in this book and is based on the standard laws of Euclidean geometry.

The most interesting question arises when we have not three, but four points.

Let's imagine that we have four points on a plane. Then, the problem may not have a solution, because, as we will find out in this book, the solution is unique (or not unique) only if the fourth point satisfies certain conditions—i.e., if for three points such a plane can always be found and it will be not even unique, then for four points, if they all lie on a plane, there may not be a solution to the problem. However, if the four points are non-coplanar and located in space, then the problem becomes very interesting and the methods of constructing a plane lying between four points, which is equidistant from all these four points, are given in this book. For example, an equidistant plane can be constructed between one point and the three other points because any three points form a plane.

What if the plane should go so that two given points lie above the plane, and two below the plane? This book explains how to construct a plane between four points so that the points are not just randomly placed around it, but are at equal distances from the plane. Is such a solution even possible? It turns out that such a plane exists, and an explanation of how to solve this problem is discussed in detail in Chap. 1 of this book.

Additionally, Chap. 1: "Methods of Solving Stereometry Problems" will refresh your memory of basic axioms and theorems of solid geometry, as well as will use construction and vector-coordinate approach, and compare these two basic methods. It will help the reader solve some interesting problems, as:

- *The little boy sawed off the legs of the stool and hid the sawn off pieces. It turned out that the stool still stands on the floor, but it is crooked and it is impossible to sit on it. Only 3 sawn-off pieces were found. What was the length of the fourth missing piece if all the sawn off pieces were of unequal length?*
- *Prove, that if in a tetrahedron the two pairs of opposite edges are orthogonal, then the third pair of opposite edges is also orthogonal.*
- *How many parts can N planes divide space into? Every three planes intersect at one point and no four planes have a common point.*
- *How many planes are there equidistant from four given points in space if the points do not lie in the same plane?*

The problems of Chap. 1 can be solved with either construction or vector-coordinate method, some using only construction method or only vector -coordinate method, and some (like the last one in the list) preferably with the construction method.

There is a section in the book dedicated to polyhedral angles and their properties. By going through this section the reader will practice a lot in proofs of important theorems and solve problems like these:

- *Prove that the sum of plane angels of a convex polyhedral angle is less than* 2π.
- *Is it possible to intersect an arbitrary trihedral angle with a plane so that the cross-section contains a regular triangle?*

Chapter 1 should help students who are struggling in Calculus 3 course to recognize the importance and the meaning of the scalar product, cross-product, or scalar-vector product through their applications in solving solid geometry problems. It teaches you to think geometrically. For example, if you are asked to find the distance from a point to a plane, it could be easier to find the height of a pyramid instead, especially if you can visualize one.

In Chap. 2: "Polyhedra and Their Properties," the reader will learn about convex and non-convex polyhedra, about pyramids and prisms, regular polyhedra and their properties. The chapter starts from photos of the greatest Egyptian Cheops pyramid, teaches the reader how to make an informative sketch of any pyramid and then gives a nice proof of the formula for a frustum of a pyramid based on homothetic transformations. The reader will learn why, for example, there are only five regular polyhedra and why ancient Greeks didn't just know all of them but also studied properties of semi-regular polyhedra. There will be a section on the Euler's convex polyhedron formula, introduction, and derivation of χ number.

One section of Chap. 2 is devoted to a regular tetrahedron and its special properties. In this chapter, the reader can see that solid geometry is important for modeling not only macro world but also the micro world. It also demonstrates how a regular tetrahedron can model the chemical structure and the sp^3-hybridization of methane and other molecules. Some natural or synthesized (man-made) molecules shaped like regular or semi-regular polyhedra are discussed, with explanations of their structure based on chemistry and stereometry. There is a section devoted to special properties of fullerene C_{60}, a molecule in the shape of a truncated icosahedron synthesized in 1991. By solving the corresponding problem with the use of the Euler formula, we will prove why any polyhedral model of fullerene C_{2n} consists of only regular pentagons and hexagons and has exactly twelve pentagonal faces.

The reader will also learn that some viruses have a structure or can be modeled as regular polyhedra. Such modeling helps study viruses and understand how, for example, to make a vaccine or neutralize a virus by modifying or changing its chemical bonds. In addition, the reader will learn how to construct a regular polyhedron, for example, by inscribing one into the other, and define a dual polyhedron.

The reader will see the advantage of using the unfolding of a polyhedron and become familiar with Alexandrov's theorem. One important section of this chapter will be learning how to solve complex problems using unfolding. This action turns a three-dimensional problem into a two-dimensional one and helps find relationships between angles, distances in a three-dimensional figure, looking at its unfolded analogue on a plane. Some interesting problems of this chapter are:

- *Pyramid DABC has the lateral edges of the same length a. The plane angles of the trihedral angle with vertex D are* α, β *and* γ. *Consider all planes which pass the vertex A of the pyramid and cut other two lateral edges at points M and K. Find the minimum value of the perimeter of the cross-section.*
- *All faces of a tetrahedron are equal acute scalene triangles with sides a, b, and c. Find the volume of the tetrahedron.*
- *Find all convex polyhedra with six faces.*

In Chap. 2 the reader will prove Pythagorean theorem in space, solve several optimization and physics problems:

- *All plane angles at vertex D of tetrahedron ABCD are right. Some point K of the face ABC is located at distances of a, b and c from the other faces BCD, ACD and ABD, respectively. Find the minimum value of the volume of this tetrahedron.*
- *A cube-shaped vessel filled with liquid is rotated* 30° *relative to the base. Obviously, some liquid is dropped out. Find what part of the original volume of the liquid remained in the vessel.*

The reader will continue learning about dihedral angles by solving problems on finding dihedral angles in dodecahedron and icosahedron and learn a lot about regular and semi-regular polyhedra by solving challenging problems as follows:

- *Inscribe a regular dodecahedron in the regular icosahedron with edge of a. Find the edge of the inscribed dodecahedron.*

Use vector-coordinate method to solve a problem like this one:

- *Construct a rhombic dodecahedron, a polyhedron with* 12 *faces,* 14 *vertices, and* 24 *edges.*

Additionally, Chap. 2 has many problems on inscribed pyramids and prisms, inscribed and circumscribed regular polyhedra. The last section of Chap. 2 is a collection of 40 problems involving polyhedra with answers, hints, and solutions.

Chapter 3: "Cross-sections of Polyhedra" is mainly devoted to the basic concepts of construction of cross-sections of different polyhedra, such as method of trace or method of projection. (Sects. 3.1–3.3) In Sect. 3.4, the reader will learn different methods of finding the area of a cross-section so as important theorems that help to find such areas using the area of the projection of the cross-section. This chapter is certainly the best in mastering the construction method and in developing spatial imagination. This chapter explains how to find the ratio of the volumes of figures in which a cutting plane divides a cube, pyramid, or other polyhedron. It will demonstrate how some useful auxiliary constructions can help solve Olympiad type problems like these:

- *Given the tetrahedron ABCS whose edges AB and CS have lengths a and b respectively. The distance between the skew lines AB and CS is d, and the angle between them is ω. Tetrahedron ABCS is divided into two solids by plane ε, parallel to lines AB and CS. The ratio of the distances of ε from AB and CS is equal to t. Compute the ratio of the volumes of the two obtained solids.*
- *Three non-collinear points M, K, and N are given. Construct a triangle ABC, such that $M \in AB$, $N \in AC$ and $K \in BC$, but the points M, N, K are not vertices of ΔMNK.*

This chapter is recommended for all college students taking a Geometry or History of Math course. The chapter will help the reader grasp different concepts of solid geometry by solving the same problem in different ways, including construction, vector-coordinate method, or their combination.

Additionally, the reader will learn that if a secant plane goes through the center of a cube, then it divides the cube into two equal parts. A cross-section of a cube by a plane passing through its center is always a polygon symmetric with respect to the center of the cube. The area of the orthogonal projection of a polygon onto a plane is equal to the product of its area and the cosine of the angle between the plane of the polygon and the projection plane.

In Chap. 3 the reader will see applications of solid geometry to construction of shadows made by the object under different light sources, such as floor or ceiling lamp or shadows caused by the rays of the sun. Additionally, the reader will learn how solid geometry can help solve some Olympiad-type plane geometry problems.

Chapter 4: "Cylinders, Cones, Spheres, and More" is devoted to cylinders, cones, spheres, figures of revolution and their combinations. There are many problems on inscribed and circumscribed figures with curvature. The reader will review how most important formulas were obtained more than 2000 years ago by the greatest ancient scientist—Archimedes and learn how to find volumes of solid figures using Cavalieri's principle without the use of integrals or trigonometric substitution. There are several challenging word problems like one about a mathematician who ordered a gold wedding ring and when the ring appeared to be too small, ordered to increase its size without paying extra money for the gold, because mathematically there was no need to add gold to the ring.

Some other interesting problems are

- *A cube is inscribed in a cone so that one of its edges lies on the diameter of the base of the cone, and the vertices of the cube that do not belong to this edge lie on the lateral surface of the cone. If the center of the cube lies at the height of the cone, find the ratio of the volumes of the cone and cube.*
- *Find the largest convex polyhedron with five vertices that can be inscribed in a sphere of radius R.*

This chapter continues exploring problems on finding volume of polyhedra, but now using new theorem about a sphere inscribed in a polyhedron. We will derive the formula for the volume of some "exotic" spatial figures.

There is a section dedicated to rotating figures, which may be of interest to both high school and college students. Two Pappus-Guldin theorems and their connection with Cavalieri's principle are shown. Using an example of finding the volume of a paraboloid inscribed into a cylinder, and then solving problems about such an interesting figure as a torus, the reader will be introduced to some methods of Calculus that clearly follow from the work of the greatest mathematicians of the past.

In all chapters, the college math is connected with elementary geometry studied at middle and high school, so it can help students or math instructors improve their problem-solving skills and develop deeper knowledge of geometry and see its connection with Calculus and Linear Algebra. Throughout the book, some historical background will be given showing a connection between modern methods of geometry and the methods known by Ancient Greeks, Babylonians, or Egyptians.

This book can be helpful in two ways:

- For self-education for those who cannot afford a personal tutor, but who want to do well in math classes, and to be prepared for any exam or competition.
- For math teachers and college professors who would like to use it as an extra source in the classroom.

On the Importance of Drawing a Good Sketch by Hand

Most geometry problems included in math competitions do not contain diagrams. It is usually a word problem indicating some kind of interaction and correspondence between the elements in the given figures. For example, the figure itself is described and some parameters of this figure are given. You either need to construct something, or find some measurement of angle or a distance, or you need to prove a statement related to the problem. You need to do the drawing yourself and this is where the difficulty may arise. That is why you need to develop the skills to create a good drawing, or at least such a drawing that will not interfere with the solution of the problem but will help solve it. It is very important to learn how draw three-dimensional objects by hand, with the use of a ruler and compass only. This will help you in constructing various intersections of volumetric figures or, for example, cross-sections of solid figures by different planes.

When I solve a geometry problem, I make some freehand sketches trying to understand the condition. There are about 400 drawings in this book. These figures are not perfect, they are made to help the reader to understand my solutions. However, this does not mean that a sketch is provided in advance. When I give a solution in the book, I draw a diagram to help the reader figure out what and how to do and in what sequence. The skills of construction of independent drawings by hand in accordance with the condition of a problem are very necessary. I hope that their fundamentals will be understood by the readers of this book. Additionally, making a good sketch by hand without the use of technology, following the condition of a problem often helps to decide on the method that will lead to the best solution for the problem. Even if the coordinates of the points in the problem statement are given, sketching often helps to choose the best and most beautiful solution using coordinates or without them.

The greatest mathematician of all time, **Leonard Euler**, said the following about the role of drawing in solving problems: My pencil is even sharper than my head. One of the greatest mathematicians of the 20th century, **Andrey Kolmogorov**, repeatedly said that many of his discoveries were brought to life by an unexpectedly emerging geometric picture with the help of a drawing. One can introduce a concept, interpret a definition, illustrate a formula, show hidden patterns and connections between concepts, visualize difficult-to-perceive provisions of theory and abstraction, trace the course of reasoning, combine several facts into a single visual image, and reveal a hint for solving problems.

How to Use My Book?

Here are my suggestions about how to use this book:

- Read the corresponding section and try to solve the problems without looking at the solution.
- If you find any question or section too difficult, skip it and go to another one. Later you may come back and try to understand it. Different people react differently to the same question and the reaction sometimes is not related to intelligence or education.

In case of failure, you need to carefully re-read the condition of the problem again, understand whether the solution method is chosen correctly, check for errors in the calculations. Therefore, even if now something does not work out, seems boring, unnecessary, unimportant, etc.—it is still important not to give up and interest yourself, find motivation and continue trying. Read the solution provided in the book after you have completed your own solution or when you feel like you are just absolutely stuck. Think about similar problems which you could solve using the same or similar approach. Find a similar problem from the homework section. Create your own problem and write it down alone with your original solution. Now, you have made this powerful method your own and will be able to use it where needed.

I hope that upon finishing this book you will love geometry and its language as much as I do. Good luck and my best wishes to you!

PS: Please send your thoughts, comments and suggestions to the following email: egrigorieva@twu.edu.

Ellina Grigorieva, Ph.D.
Professor of Mathematics
Texas Woman's University
Denton, TX, USA

Acknowledgements

Over the three years I have been working on this book, I have received feedback from friends and colleagues at Lomonosov Moscow State University, for which I will always be indebted.

I would like to express my gratitude to Dr. Paul Deignan for his contribution to the final formatting and preparation of the manuscript, and for his insightful comments on its style and content.

I appreciate the extremely helpful suggestions from the reviewers, which have made this book better. I really appreciate the production team for their help in improving my graphics. My special thanks to the Editor Christopher Tominich at Birkhauser, who has always been encouraging, helpful, and positive.

Contents

Chapter 1
Methods of Solving Stereometry Problems

There are different methods for solving stereometry problems. Two of the most popular are the so-called construction method and the vector-coordinate method. Each is useful in its own way. The most universal is the construction method. It can help solve almost any problem in stereometry of those given in the Mathematics Olympiad. However, to use the construction method, one must have a good spatial imagination and be able to apply an impeccable knowledge and understanding of the basic theorems of stereometry related to the mutual arrangement of lines and planes in space. The construction method is emphasized in this book. The vector-coordinate method avoids the necessity of working out spatial relations in finding the distances (from a point to a plane, between parallel planes, between skew lines) or all types of angles (between lines, between a line and a plane, between planes). Practice shows that students quickly master the vector-coordinate method, since when using it, one must adhere to a simple general algorithm: calculate the coordinates of the desired points and apply the appropriate formula, focusing on the computational part.

However, it would be naive to think that learning just one method would help you to tackle any problem. As a problem creator and USAMO grader for several years, I have witnessed how sometimes the best and brightest students get drowned in an ocean of complex calculations using the vector-coordinate method when a good sketch combined with the construction method is most expedient. Often the construction method allows for a much faster and more elegant solution. Here is an example of a problem where the introduction of coordinates can complicate the problem so much so that its solution becomes unexpectedly long and incomprehensible.

There are four different points in space that do not lie in the same plane. Draw a plane so that two points lie above this plane and two other points lie below this plane. In this case, this plane must be equidistant from each of the four given points.

Anyone who starts to solve this problem by entering some arbitrary coordinates of the four points and then trying to somehow find the conditions on the plane so that it satisfies the original problem will likely get confused before reaching a solution. Of course, such formulas can be obtained; they exist, but they are absolutely unnecessary.

E. Grigorieva, *Methods of Solving Solid Geometry Problems*,
https://doi.org/10.1007/978-3-031-86969-3_1

I believe that it is not a good idea to try to remember them or even derive them (unless the points are given by specific coordinates) because it is much easier to simply use construction rules and the laws of Euclidean geometry. For example, the solution to this problem using the construction method will be given in this chapter in the section "Planes and Lines in Space."

In my opinion, there is no golden rule and only problem-solving experience can help you to choose the best method. Usually the coordinate method works well when points or other objects are given by their coordinates or when the introduction of a coordinate system is straightforward. If a problem concerns evaluating volume, the ratio of the volumes obtained as a result of the plane's cutting a solid figure, or if you need to evaluate the area of the cross section obtained, in all these and similar cases the construction method might work much better and even faster than the coordinate method.

Let us start from solving one interesting Olympiad-type problem about a boy and a handyman (his grandfather).

Problem 1 The little boy sawed off all the legs of a square stool and hid the four sawn-off pieces. It turned out that the lengths of all the pieces were different and afterward the stool still stood on the floor, albeit obliquely but touching the floor with all four ends of the legs. Grandfather decided to repair the stool but found only three pieces of $8,\ 9$, and 10 cm long. How long can the fourth piece be?

Solution Method 1. (Construction method) Suppose that before cutting the legs of the stool, the boy made notches on each leg as shown in Fig. 1.1. In this case, the initial state of the stool is described by points $A,\ B,\ C,\ D$, and the positions of the notches are described by points $A_1,\ B_1,\ C_1,\ D_1$. Assume that the lengths of the sawn pieces will be described by segments $AA_1 = a,\ BB_1 = b,\ CC_1 = c,\ DD_1 = d$.

Since the stool is standing, touching the floor with four legs, the points $A_1,\ B_1,\ C_1$, and D_1 lie in the same plane. Hence, lines A_1C_1 and D_1B_1 intersect and are not skew lines. Let $\alpha = \{A_1, B_1, C_1, D_1\}$. The stool is a square shape, and all its legs are parallel; this means the planes ABA_1B_1 and CDC_1D_1 are parallel. Hence, plane α must cut these two parallel planes by parallel lines so we can state:

$$A_1B_1 \| C_1D_1.$$

By analogy, the planes DD_1A_1A and CC_1B_1B are also parallel and also must be cut by a plane α by parallel lines,

$$D_1A_1 \| C_1B_1.$$

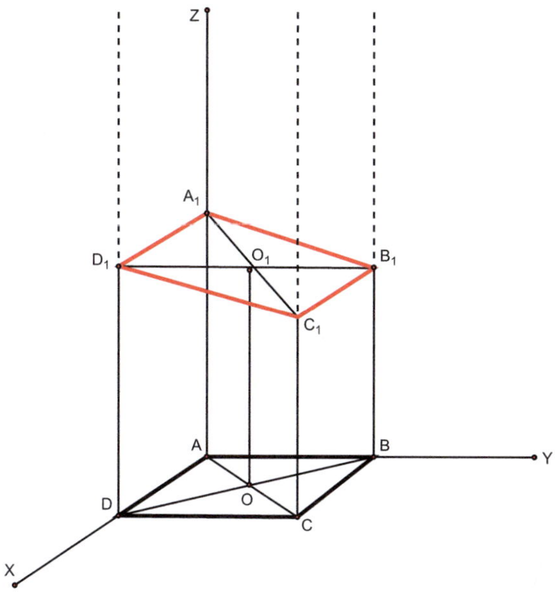

Fig. 1.1 Legs of the stool in a coordinate system

Therefore the quadrilateral $A_1B_1C_1D_1$ is a parallelogram and its diagonals intersect at the common midpoint O_1. Let O be the center of square $ABCD$. Note that the segment OO_1 is the midline of both the trapezoid ACC_1A_1 and the trapezoid BDD_1B_1, which means that

$$\frac{AA_1 + CC_1}{2} = OO_1 = \frac{BB_1 + DD_1}{2}$$

and can be rewritten as $a + c = b + d$.

Now let's go over the possible lengths of the sawn-off parts:

$$8 + x = 9 + 10$$
$$9 + x = 8 + 10$$
$$10 + x = 8 + 9.$$

Therefore $x = 7,\ x = 9,\ x = 11$. Since the lengths of all the pieces are different, x cannot be equal to 9, and the only options of 7 and 11 remain.

Method 2. (Vector-Coordinate method) Let us relate to the same sketch (Fig. 1.1). The initial state of the stool is described by points $A,\ B,\ C,\ D$, and the positions of the notches are described by points $A_1,\ B_1,\ C_1,\ D_1$. Then the lengths of the sawn pieces will be described by segments $AA_1,\ BB_1,\ CC_1,\ DD_1$. Without loss of generality, we can assume that original and new coordinates of the stool's feet are

$$\begin{aligned}
A(0,0,0) &\to A_1(0,0,a) \to AA_1 = a\\
B(0,1,0) &\to B_1(0,1,b) \to BB_1 = b\\
C(1,1,0) &\to C_1(1,1,c) \to CC_1 = c\\
D(1,0,0) &\to D_1(1,0,d) \to DD_1 = d.
\end{aligned}$$

The stool with cut-off ends still stands on the floor, so A_1, C_1, D_1 and B_1 are coplanar (they belong to the same plane). Hence, the three vectors $\overrightarrow{A_1C_1}$, $\overrightarrow{A_1B_1}$, and $\overrightarrow{A_1D_1}$ are dependent and their scalar triple product must be equal to zero. We have

$$\overrightarrow{A_1C_1} = (1, 1, c - a)$$

$$\overrightarrow{A_1B_1} = (0, 1, b - a)$$

$$\overrightarrow{A_1D_1} = (1, 0, d - a).$$

The condition that the three vectors $\vec{u}$, $\vec{v}$, $\vec{w}$ are coplanar, is equivalent to the determinant of the matrix being zero, the rows of which are the vectors:

$$\begin{vmatrix} u_1 & u_2 & u_3\\ v_1 & v_2 & v_3\\ w_1 & w_2 & w_3 \end{vmatrix} = 0.$$

Substituting vectors $\overrightarrow{A_1C_1}$, $\overrightarrow{A_1B_1}$, and $\overrightarrow{A_1D_1}$, in the determinant,

$$\begin{vmatrix} 1 & 1 & c-a\\ 0 & 1 & b-a\\ 1 & 0 & d-a \end{vmatrix}$$

and by equating it to zero, we obtain $d - a + b - a - (c - a) = 0$, that can simplified as $d + b = a + c$. Here (a, b, c, d) is the list of all measurements that allow this stool to stand. Three of the measurements are 8, 9, and 10, and the measurement of the missing piece that we denote by x. Trying different combinations in the formula above, we have three equations to solve

$$\begin{aligned}
8 + x = 9 + 10 &\Rightarrow \boxed{x = 7}\\
9 + x = 8 + 10 &\Rightarrow x = 9\\
10 + x = 8 + 9 &\Rightarrow \boxed{x = 11}.
\end{aligned}$$

We selected two solutions (7 and 11) for the remaining piece since all the pieces must be different.

Answer. The lost piece can be 7 or 11 cm long.

Remark. It is interesting to emphasize that in both solutions the key point was to notice that the four points (four ends of the stool) must belong to the same plane!

Obviously, this book will not deal with problems that are purely computational in terms of distance from a point to a line or distance from a point to a plane. As a rule, there will be more complex problems related to finding the distance between two skew lines of a polyhedron or, for example, the distance from a tangent point on the sphere to some opposite face of the figure in which the sphere is inscribed. On the other hand, a good choice of the coordinate system and an accurate sketch make it possible to significantly simplify the calculations. Therefore, the knowledge that will be offered in this chapter is of a general cognitive nature, but both methods will be used throughout the text to solve interesting problems and will often be compared or promoted, showing their advantages and disadvantages.

1.1 Basic Definitions, Axioms, and Theorems

The basic figures in space are a point, a line, and a plane.

A **point** has zero dimension. You can define a point as a point you just drew with a pencil on a piece of paper, or as a point drawn with chalk on a blackboard. A point can be defined as the intersection of the ceiling and two adjacent walls, but you can also say that this point defines the vertex of a trihedral angle formed by the planes of two adjacent walls and the ceiling. In geometry, each point usually has a name. Points are denoted by capital letters of the English alphabet A, B, C, etc. You can also imagine a point in some frame of reference or coordinate system.

A **line** contains infinitely many points. Your pencil can be thought of as a line segment; just imagine continuing this line from both ends to infinity. A line can also be defined as the intersection of two planes or surfaces. In a coordinate system, a line can be defined by its direction and a point on this line. Lines are usually denoted by lowercase letters of the English alphabet, for example, line a, b, c, etc.

A **plane**, like a line, is infinite. If we think of a plane, then in our daily life and at home we see many flat objects. For example, the table we work at is flat, and the surface of the computer is flat, and, say, the ceiling is also flat, as well as the floor we walk on is flat. All these are objects that define a certain infinite plane in three-dimensional space. An infinite plane can be defined by specifying certain figures on the plane, for example, three points that do not lie on a straight line, or we can define a plane as containing some plane figure, such as a triangle or a square. In most cases, we will draw planes as parallelograms, although, of course, a parallelogram is a finite figure, and a plane is infinite. For example, to draw a 3D cube on a piece of paper, we often draw its front and back faces as equal squares, and the rest of the faces as parallelograms. In sketches, we will show only part of the plane, but think of it as having unlimited expanse in all directions. Planes are usually denoted by the Greek letters $\alpha,\ \beta,\ \gamma$.

Plane and solid geometries are very connected. Most rules of plane geometry work for solid geometry. However, by going from a two-dimensional to three-dimensional space, we will see new terms and rules. For example, one of the basic things of solid geometry such as skew lines simply do not exist in plane geometry. In planimetry, we have only one plane on which all the figures under consideration are located, however, there are infinitely many planes in stereometry. In this regard, the formulation of some axioms of planimetry as axioms of stereometry is required:

- P1. Of the three collinear points, one and only one lies between the other two.
- P2. Each segment has a certain length greater than zero. The length of a segment is equal to the sum of the lengths of the parts into which it is divided by any of its points.
- P3. On any half-line from its starting point, one and only one segment of a given length can be constructed.
- P4. Each angle has a certain degree measure greater than zero. The straight angle is 180°. The degree measure of an angle is equal to the sum of the degree measures of the angles into which it is divided by any ray passing between its sides.
- P5. On a plane, through a given point not lying on a given line, at most one line can be drawn parallel to the given one.
- P6. From a half-line, on the plane containing it, in a given half-plane one and only one angle with a given degree measure less than 180 degrees can be constructed.
- P7. A straight line belonging to a plane divides this plane into two half-planes.
- P8. Whatever the triangle is, in a given plane, in a given position relative to a given half-line in this plane, there is a triangle equal to it.

With the expansion of the space in which geometric objects are embedded in solid geometry, we introduce a more expansive set of axioms for the basic properties of lines and planes [13].

Axiom 1 Whatever the line is, there are points that belong to this line and points that do not belong to it. Through any two points you can draw a line and only one.

Axiom 2 Whatever a plane is, there exist points that belong to this plane and the points that do not belong to it.

This fact is almost obvious. For example, a butterfly sitting somewhere other than on your desk, such as on a wall, may represent a point that does not belong to the plane of your desk.

Axiom 3 If two different planes have a common point, then they intersect along a straight line passing through this point.

This axiom states that if two different planes have a common point, then there is a line belonging to each of these planes. Moreover, a common point of two planes belongs to this line.

Axiom 4 If two different lines have a common point, then only one plane can be drawn through these two lines.

This axiom means that if two different lines a and b have a common point C, then there is a plane containing lines a and b and the plane with this property is unique. We understand that two different lines with a common point intersect at the point that can be written as $a \cap b = C$. No matter how two intersecting lines are located in space, there will always be a plane that will pass so that both of these lines lie on it. It is always possible to construct a plane passing through any two intersecting lines and in which these two intersecting lines will lie.

Let's find out what defines a unique plane in space. A plane, and only one, can be drawn given the following combination of points or figures.

Theorem 1 *Through a line and a point not lying on it, one can draw a plane and only one.*

Proof Let AB be a line and C a point not lying on it. Draw a line through points A and C. The lines AB and AC are different because point C does not lie on the line AB. If we draw the plane through the lines AB and AC, then plane α (by Axiom 4) passes through the line AB and the point C.

Let us prove that the plane α passing through the line and the point is unique. Assume a contradiction and say that there is another plane β passing through the same line and the same point; then, according to the Axiom 3, the planes α and β intersect along a straight line, and this line must contain points $A,\ B$ and C. However, these points do not lie on the same straight line! We obtained a contradiction and thereby proved the theorem. □

Remark. If a point lies on a line, then the principle of uniqueness is violated. There will be infinitely many planes passing through a given line.

Theorem 2 *If two points of a line belong to a plane, then the entire line belongs to that plane.*

Proof Let a be a given line and α be a given plane. According to Axiom 1, there exists a point A that does not lie on the line a. Let us draw a plane β through the line a and the point A.

(1) If the plane β coincides with the plane α, then the plane α contains the straight line a, which is stated by the theorem.
(2) If the plane β is different from the plane α, then these planes intersect along some line b containing two points of the line a. By Axiom 3, the line b coincides with a.

Therefore, the line a lies in the plane α. □

Now, let us consider the existence of a plane passing through three given points.

Theorem 3 *Through three points that do not lie on the same straight line, one can draw a plane and only one.*

Proof Let $A,\ B,\ C$ be three given points not lying on the same straight line. Draw lines AB and AC. They are different, since the points $A,\ B,\ C$ do not lie on the same line. According to Axiom 4, through lines AB and AC one can draw a plane α; this plane contains points A, B, and C. To prove that the plane α passing through these three points is unique, note that by Theorem 2 the plane contains the lines AB and AC. According to Axiom 4, such a plane is unique. □

Remark. It is important to know that if three points lie on the same line, or, in other words, if three given points are collinear, then an infinite number of planes will pass through these three points. In this case, the principle of uniqueness is violated.

Theorem 4 *There is a unique plane passing through two parallel lines.*

The proof of this theorem follows from Theorem 1. All four cases can be demonstrated in one drawing. From plane geometry, we know that there is a unique triangle that passes through three distinct points which do not lie on the same line. Thus, in Fig. 1.2 triangle ABC is unique. Line AB intersects with line BC at point B, and it defines the same. Using side $AC(m)$ through point B, we can draw line n parallel to line AC, such that $n \parallel m$. Further, by drawing a line through point C parallel to side

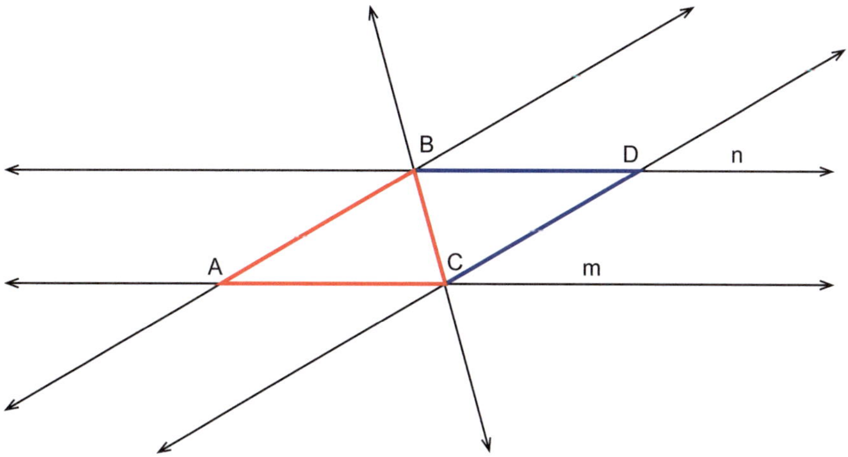

Fig. 1.2 Making a plane

AB of the triangle, we intersect this line with line n, obtaining point D; therefore parallelogram $ABDC$ is also in the same plane.

Problem 2 Prove that two different planes can be drawn through a given line.

Proof Let a be a given line. According to Axiom 1, there is a point A that does not lie on this line, and, according to Theorem 1, a plane can be drawn through the line a and the point A. Let's call it α. According to Axiom 2, there exists a point B that does not lie in the plane α. Let's draw a plane β through line a and point B (see Fig. 1.3).

The planes α and β are different since point B lies in plane β but does not lie in plane α. □

Problem 3 Four distinct points do not belong to one plane. Is it possible that any three of these points lie on the same line?

Solution If three (i.e., A, B, and C) of the four given points belong to the same line, for example, line a, then the fourth point D does not lie on this line. Then, according to Theorem 1, a line and a point that does not lie on it form a unique plane. Hence, all four points are on that plane. Consequently, the described case is impossible. If four points do not lie in the same plane, then no three of them can belong to the same line.

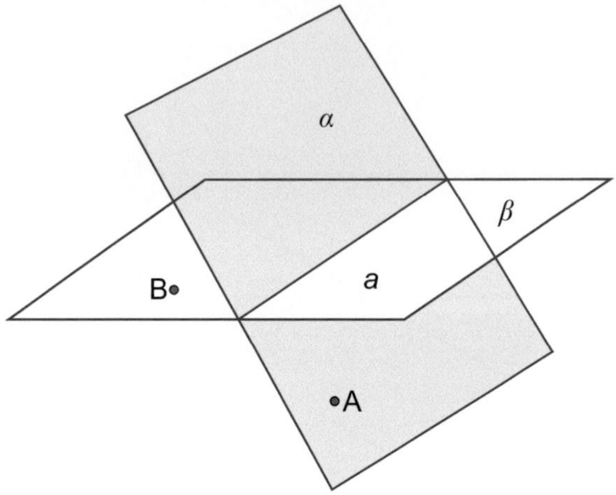

Fig. 1.3 Two planes through the same line

Since this book is not a textbook on Euclidean geometry but rather a guide to solving problems, we simply cannot present here all relevant theorems and formulas. We will give some of them as needed to solve problems, consolidate skills, and demonstrate approaches to solving problems.

1.2 Vectors, Coordinates, and Related Problems

In this book, we will solve problems using different methods, using different approaches, such as analytic geometry methods applied to points, lines, planes, and their positions relative to each other. Let us start by reviewing vectors.

A **vector** is a directed line segment. For example, in Fig. 1.4, we can see several vectors. First, let us recognize vector $\vec{u} = \overrightarrow{EG}$ shown in red. Points E and G mark the initial and the terminal points of the vector, respectively. The magnitude (length) of vector $\vec{u} = \overrightarrow{EG}$ is denoted by $\|\vec{u}\| = \left\|\overrightarrow{EG}\right\|$. The unit vector in the same direction can be obtained by dividing the given vector by its magnitude that can be expressed as follows:

$$\vec{e_u} = \frac{\vec{u}}{\|\vec{u}\|} = \frac{\overrightarrow{EG}}{\left\|\overrightarrow{EG}\right\|}. \tag{1.1}$$

Any vector can be placed anywhere in space and start from any point. For example, in the same sketch, we translated vector $\vec{u} = \overrightarrow{EG}$ obtaining the same vector in a different location denoted as $\vec{u} = \overrightarrow{OA}$. We can add and subtract vectors or multiply a

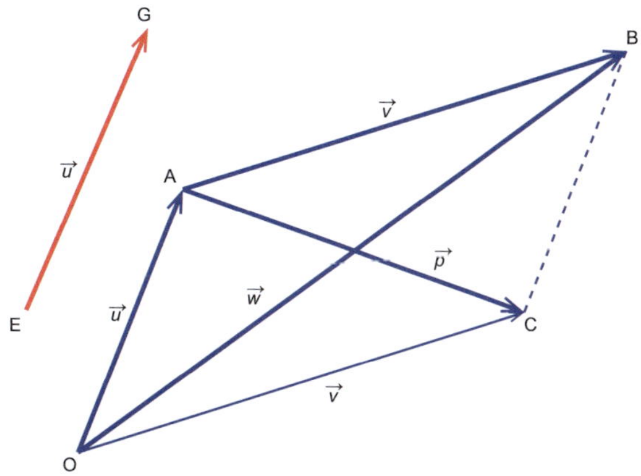

Fig. 1.4 Vectors in space

vector by a scalar. In Fig. 1.4, we added two vectors, $\vec{u} = \overrightarrow{OA}$ and $\vec{v} = \overrightarrow{OC} = \overrightarrow{AB}$ obtaining a new vector $\vec{w} = \overrightarrow{OB}$, $\vec{u} + \vec{v} = \vec{w}$.

In our sketch, we did it in two different ways:

1. By the **triangle rule**: We constructed vector $\vec{v} = \overrightarrow{AB}$ from the end point of vector $\vec{u} = \overrightarrow{OA}$ and connected points O and B to obtain vector $\vec{w} = \overrightarrow{OB}$ as the sum of two vectors.
2. By the **parallelogram rule**: We started both vectors at point O and then constructed a parallelogram $(AOBC)$. Its diagonal OB determines the resulting vector $\vec{w} = \overrightarrow{OB}$.

In Fig. 1.4, we showed the differences of two vectors:

$$\overrightarrow{AB} = \overrightarrow{OB} - \overrightarrow{OA} \tag{1.2}$$

or as

$$\vec{v} = \vec{w} - \vec{u}$$

and also

$$\overrightarrow{AC} = \overrightarrow{OC} - \overrightarrow{OA}. \tag{1.3}$$

In our sketch, vector $\overrightarrow{AC} = \vec{p}$, so the equation above can be written as $\vec{p} = \vec{v} - \vec{u}$.

Formulas (1.2) and (1.3) are very important for us. Please recall that any vector in space can be always written as a difference of two other vectors that both start at the same point. You subtract a vector that ends at the ending point of the given vector and vector that ends at the starting point of the given vector. The starting point of both vectors in the difference is the same but can be chosen anywhere in space and selected for our convenience!

Multiplying a vector by a positive scalar (constant) will change the vector magnitude and will keep its direction. If we multiply vector by a negative scalar, then the new vector will be in an opposite direction and with a different magnitude. Using the same Fig. 1.4 we can imagine several vectors. For example, the opposite of vector $\vec{u} = \overrightarrow{EG}$ would be $-\vec{u} = -\overrightarrow{EG} = \overrightarrow{GE}$. Thus, denoting vectors by two letters helps to work with vectors in either direction.

A **scalar product** (or dot product) of two vectors $\vec{a}$ and $\vec{b}$ can be expressed using

$$\vec{a} \cdot \vec{b} = (\vec{a}, \vec{b}) = \|\vec{a}\| \cdot \left\|\vec{b}\right\| \cdot \cos\varphi. \tag{1.4}$$

Angle φ is the angle between two vectors. Geometrically, we need to make sure that both vectors start at the same point. The scalar product is a commutative operation, so

$$\left(\vec{a}, \vec{b}\right) = \left(\vec{b}, \vec{a}\right). \tag{1.5}$$

For any two scalars λ_1 and λ_2 and three vectors, the following is true:

$$(\lambda_1 \vec{a} + \lambda_2 \vec{b}, \vec{c}) = \lambda_1(\vec{a}, \vec{c}) + \lambda_2(\vec{b}, \vec{c}). \tag{1.6}$$

If two vectors are orthogonal (lie on perpendicular lines), then the right side of (1.4) equals zero, hence, their scalar product is equal to zero:

$$\left(\vec{a}, \vec{b}\right) = 0. \tag{1.7}$$

Considering a scalar product of two identical vectors (by itself), we can express the magnitude of a vector using a self-scalar product as

$$\|\vec{a}\| = \sqrt{(\vec{a}, \vec{a})}. \tag{1.8}$$

Let us solve some problems right away.

Problem 4 Prove that for any four points $A,\ B,\ C$, and D the following is true:

$$\overrightarrow{AB}\cdot\overrightarrow{CD}+\overrightarrow{BC}\cdot\overrightarrow{AD}+\overrightarrow{CA}\cdot\overrightarrow{BD}=0.$$

Proof Let us replace all vectors by the vectors started at the same point A (you can use formula (1.2) as an example):

$$\overrightarrow{AB}\cdot\left(\overrightarrow{AD}-\overrightarrow{AC}\right)+\left(\overrightarrow{AC}-\overrightarrow{AB}\right)\cdot\overrightarrow{AD}-\overrightarrow{AC}\cdot\left(\overrightarrow{AD}-\overrightarrow{AB}\right).$$

Applying (1.5) and (1.6), we can simplify this expression as

$$\overrightarrow{AB}\cdot\overrightarrow{AD}-\overrightarrow{AB}\cdot\overrightarrow{AC}+\overrightarrow{AC}\cdot\overrightarrow{AD}-\overrightarrow{AB}\cdot\overrightarrow{AD}-\overrightarrow{AC}\cdot\overrightarrow{AD}+\overrightarrow{AC}\cdot\overrightarrow{AB}=0.$$

Each term above has its opposite. Therefore the expression is equal to zero. □

Problem 5 Prove that the sums of the squares of the opposite edges in a tetrahedron are equal if and only if the third pair of the opposite edges is orthogonal.

Proof Let us sketch a tetrahedron $ABCD$ (Fig. 1.5) and introduce vectors

$$\vec{a}=\overrightarrow{AB},\ \vec{b}=\overrightarrow{BC},\ \vec{c}=\overrightarrow{CD}. \tag{1.9}$$

In the tetrahedron $ABCD$, the following pairs of the edges are opposite: BD and AC, AD and BC, and AB and DC. By the condition of the problem, without loss of generality, assume that the following is true for the first two pairs of the opposite edges:

$$AC^2+BD^2=BC^2+AD^2. \tag{1.10}$$

Now we need to prove that if (1.10) holds, then $AB\perp DC$. Consider the corresponding vectors and express them in terms of $\vec{a}$, $\vec{b}$, and $\vec{c}$.

$$\overrightarrow{AC}=\overrightarrow{AB}+\overrightarrow{BC},\ \overrightarrow{BD}=\overrightarrow{BC}+\overrightarrow{CD},\ \overrightarrow{AD}=\overrightarrow{AB}+\overrightarrow{BC}+\overrightarrow{CD}. \tag{1.11}$$

Because the scalar product by itself is equal to the square of the magnitude and substituting (1.11) and (1.9) into (1.10), we obtain

$$(\vec{a}+\vec{b},\vec{a}+\vec{b})+(\vec{b}+\vec{c},\vec{b}+\vec{c})=(\vec{b},\vec{b})+(\vec{a}+\vec{b}+\vec{c},\vec{a}+\vec{b}+\vec{c}). \tag{1.12}$$

Let us apply properties of the scalar product and expand the formulas on the left and on the right of equation (1.12). Let us also omit the vector symbol:

$$\begin{aligned}&\|a\|^2+\|b\|^2+2(a,b)+\|b\|^2+\|c\|^2+2(b,c)\\&=\|b\|^2+\|a\|^2+\|b\|^2+\|c\|^2+2(a,b)+2(b,c)+2(a,c).\end{aligned} \tag{1.13}$$

Since the right side of equation (1.13) differs from the left side only by one term (the scalar product of vectors $\vec{a}$ and $\vec{c}$), then this scalar product must be zero, $(\vec{a},\vec{c})=0$. This means that vectors $\vec{a}=\overrightarrow{AB}$ and $\vec{c}=\overrightarrow{CD}$ are orthogonal as is the third pair of the opposite sides in the tetrahedron. Therefore, in order for (1.10) to be true, the third pair of opposite edges of a tetrahedron must be orthogonal. □

Problem 6 In a tetrahedron, two pairs of the opposite edges are perpendicular. Prove that the third pair of opposite edges must be also perpendicular.

Proof In order to prove this statement, we can use, for example, Fig. 1.5. Let $AB \perp DC$ and $DA \perp BC$. We need to prove that $AC \perp DB$.

Note that if $AB \perp DC$, then $(\overrightarrow{AB},\overrightarrow{DC})=0$. If $AD \perp BC$, then $(\overrightarrow{AD},\overrightarrow{BC})=0$. Consider the sum of the scalar products of all three pairs of opposite edges and assume that it is equal to some value of p:

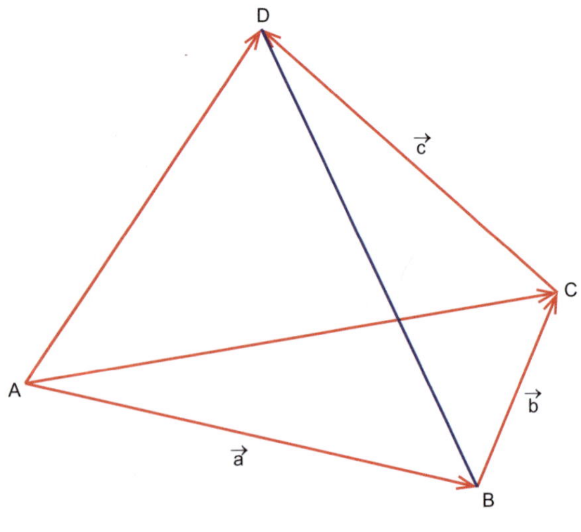

Fig. 1.5 The equal sums of the squares of two opposite edges in a tetrahedron

$$(\overrightarrow{AB}, \overrightarrow{DC}) + (\overrightarrow{AD}, \overrightarrow{BC}) + (\overrightarrow{AC}, \overrightarrow{DB}) = p. \tag{1.14}$$

On one hand, by the condition of the problem, we can state that

$$p = (\overrightarrow{AC}, \overrightarrow{DB}). \tag{1.15}$$

On the other hand, we can expand the left side of (1.14), and rewrite it as follows:

$$\begin{aligned}
&(\overrightarrow{AB}, \overrightarrow{AC} - \overrightarrow{AD}) + (\overrightarrow{AD}, \overrightarrow{AC} - \overrightarrow{AB}) + (\overrightarrow{AC}, \overrightarrow{AB} - \overrightarrow{AD}) \\
&= (\overrightarrow{AB}, \overrightarrow{AC}) - (\overrightarrow{AB}, \overrightarrow{AD}) + (\overrightarrow{AD}, \overrightarrow{AC}) - (\overrightarrow{AD}, \overrightarrow{AB}) + (\overrightarrow{AC}, \overrightarrow{AB}) - (\overrightarrow{AC}, \overrightarrow{AD}) \\
&= 2\left[(\overrightarrow{AB}, \overrightarrow{AC}) - (\overrightarrow{AB}, \overrightarrow{AD})\right] \\
&= 2 \cdot (\overrightarrow{AB}, \overrightarrow{AC} - \overrightarrow{AD}) \\
&= 2(\overrightarrow{AB}, \overrightarrow{DC}) = 0.
\end{aligned} \tag{1.16}$$

The scalar product in (1.16) is equal to zero, because by the condition of the problem $AB \perp DC$. Hence, the right side (p) in (1.15) is also zero. Then, the scalar product of vectors AC and DB is zero. Therefore $AC \perp BD$. □

Now, let us solve a problem concerning vectors of any dimensionality.

Problem 7 Vectors $\vec{m}$, $\vec{n}$, and $\vec{p}$ are unit vectors, such that $(\vec{p}, \vec{n}) = (\vec{p}, \vec{m}) = 0$ and the angle between vectors $\vec{n}$ and $\vec{m}$ is 60°. Find the angle between vectors $\vec{a}$ and $\vec{b}$, when

$$\vec{a} = 5\vec{n} - \vec{m} - \vec{p}, \ \vec{b} = \vec{n} + 2\vec{m} + 3\vec{p}.$$

Solution Denote the angle by φ, then cosine of this angle can be found as

$$\cos\varphi = \frac{(a, b)}{\sqrt{(a, a)}\sqrt{(b, b)}}. \tag{1.17}$$

We can find the numerator of the fraction in (1.17):

$$(\vec{a}, \vec{b}) = (5\vec{n} - \vec{m} - \vec{p}, \vec{n} + 2\vec{m} + 3\vec{p}).$$

The right-hand side of which can be rewritten as

$$\begin{aligned}
5(n, n) + 10(n, m) + 15(n, p) - (m, n) - 2(m, m) - 3(m, p) - (p, n) \\
- 2(p, m) - 3(p, p) = 9(n, m).
\end{aligned}$$

Evaluating the scalar product of vectors m and n, we obtain

$$(\vec{a}, \vec{b}) = 9(\vec{n}, \vec{m}) = 9 \cdot 1 \cdot 1 \cdot \cos 60^\circ = 4.5.$$

Next, we will evaluate the denominator:

$$(\vec{a}, \vec{a}) = (5\vec{n} - \vec{m} - \vec{p}, 5\vec{n} - \vec{m} - \vec{p}) = 25 - 5(\vec{n}, \vec{m}) + 1 - 5(\vec{n}, \vec{m}) + 1 = 27 - 10 \cdot 0.5 = 22$$

$$(\vec{b}, \vec{b}) = (\vec{n} + 2\vec{m} + 3\vec{p}, \vec{n} + 2\vec{m} + 3\vec{p}) = 16;$$

$$\begin{aligned}\|b\|^2 &= (\vec{n} + 2\vec{m} + 3\vec{p},\ \vec{n} + 2\vec{m} + 3\vec{p}) \\ &= (\vec{n}, \vec{n}) + 2(\vec{n}, \vec{m}) + 3(\vec{n}, \vec{p}) + 2(\vec{m}, \vec{n}) \\ &+ 4(\vec{m}, \vec{m}) + 6(\vec{m}, \vec{p}) + 3(\vec{p},\ \vec{n}) + 6(\vec{p},\ \vec{m}) + 9(\vec{p}, \vec{p}) \\ &= 1 + 2 \cdot \tfrac{1}{2} + 2 \cdot \tfrac{1}{2} + 4 + 9 = 16.\end{aligned}$$

Finally,

$$\cos\varphi = \frac{\frac{9}{2}}{\sqrt{22} \cdot 4} = \frac{9\sqrt{22}}{176}.$$

Answer. $\cos\varphi = \frac{9\sqrt{22}}{176}$.

Formulas (1.2)–(1.8) work for any space and do not depend on a coordinate system. Often in solid geometry we want to introduce a coordinate system. Some formulas listed below may help you to solve stereometry problems using the coordinate method.

1. **Coordinate form of a point and a vector**. In a three-dimensional coordinate system, a point A can be given by its coordinates, for example, as

   $$A(x_1, y_1, z_1).$$

   A vector (directed segment) is also defined by three coordinates. For example, a vector OA that starts at the origin and ends at point A will have the same coordinates as point A above that can be written as

   $$\overrightarrow{a} = \overrightarrow{OA} = (x_1, y_1, z_1).$$

   There are infinitely many other vectors in space that are equal to $\overrightarrow{a}$ and can be obtained by translation.

2. **Magnitude of a vector**. The magnitude of vector $\overrightarrow{a}$ is calculated by the formula:

   $$\|\overrightarrow{a}\| = \sqrt{x_1^2 + y_1^2 + z_1^2}. \tag{1.18}$$

3. **Distance between two points in space**. Suppose that there is another vector, $\overrightarrow{OB}$, with the coordinates

$$\overrightarrow{OB} = (x_2, y_2, z_2).$$

The vector AB has the following coordinates:

$$\overrightarrow{AB} = (x_2 - x_1, y_2 - y_1, z_2 - z_1).$$

The distance AB is equal to the magnitude of $\overrightarrow{AB}$:

$$d_{AB} = \sqrt{(x_2 - x_1)^2 + (y_2 - y_1)^2 + (z_2 - z_1)^2}.$$

4. **Coordinates of the midpoint of a segment**. If we want to determine the coordinates of point C, the midpoint of segment AB, let us connect the center of the coordinate system with points A, B, and C. The following is true:

$$\overrightarrow{OC} = \frac{1}{2} \cdot \left(\overrightarrow{OA} + \overrightarrow{OB}\right). \tag{1.19}$$

The coordinates of the midpoint can be expressed as

$$\overrightarrow{OC} = \left(\frac{x_1 + x_2}{2}, \frac{y_1 + y_2}{2}, \frac{z_1 + z_2}{2}\right). \tag{1.20}$$

Knowledge of this will be very helpful in case when a plane or straight line is given by the midpoints of some diagonals or edges of a polyhedron.

5. **Scalar product of two vectors**. A scalar product of two vectors $\overrightarrow{a} = (x_1, y_1, z_1)$ and $\overrightarrow{b} = (x_2, y_2, z_2)$ can be evaluated as

$$(\overrightarrow{a}, \overrightarrow{b}) = x_1 \cdot x_2 + y_1 \cdot y_2 + z_1 \cdot z_2. \tag{1.21}$$

The angle φ between two vectors can be evaluated as

$$\cos\varphi = \frac{x_1 \cdot x_2 + y_1 \cdot y_2 + z_1 \cdot z_2}{\sqrt{x_1^2 + y_1^2 + z_1^2}\sqrt{x_2^2 + y_2^2 + z_2^2}}. \tag{1.22}$$

6. **Cross product of two vectors**. A cross product or vector product of two vectors is another vector:

$$\overrightarrow{n} = \overrightarrow{a} \times \overrightarrow{b} = \begin{vmatrix} \overrightarrow{i} & \overrightarrow{j} & \overrightarrow{k} \\ x_1 & y_1 & z_1 \\ x_2 & y_2 & z_2 \end{vmatrix}. \tag{1.23}$$

This vector is perpendicular to the plane formed by the two given vectors. The magnitude of the cross-product vector is equal to the area of the parallelogram formed by vectors $\vec{a}$ and $\vec{b}$ and can be evaluated as follows:

$$\|\vec{n}\| = \|\vec{a}\| \cdot \|\vec{b}\| \cdot \sin\alpha, \tag{1.24}$$

where α is the angle between two vectors ($\vec{a}$ and $\vec{b}$).

7. **Scalar triple product**. A scalar triple product of three vectors (also commonly known as a *mixed product* or *a box product*) is the scalar product of a vector with the cross product of two other vectors. If $\vec{a}$, $\vec{b}$, $\vec{c}$ are three given vectors, then their scalar triple product is

$$\vec{a} \cdot \left(\vec{b} \times \vec{c}\right). \tag{1.25}$$

The magnitude of the scalar triple product gives the volume of the parallelepiped, where the three vectors represent adjacent sides of the parallelepiped. The volume of a triangular pyramid constructed on three vectors is equal to $1/6$ of the mixed vector product's magnitude. For example, if we know the coordinates of all four points shown in Fig. 1.6, we can calculate the volume of the pyramid $ABDC$ as $1/6$ of the volume obtained by the formula (1.25).

The scalar triple product (1.25) (as a scalar product) satisfies the commutative property (1.5):

$$\left(\vec{a} \times \vec{b}\right) \cdot \vec{c} = \vec{a} \cdot \left(\vec{b} \times \vec{c}\right). \tag{1.26}$$

It can be evaluated as

$$(\vec{a} \times \vec{b}) \cdot \vec{c} = \begin{vmatrix} a_1 & a_2 & a_3 \\ b_1 & b_2 & b_3 \\ c_1 & c_2 & c_3 \end{vmatrix} = a_1 \cdot \begin{vmatrix} b_2 & b_3 \\ c_2 & c_3 \end{vmatrix} - a_2 \cdot \begin{vmatrix} b_1 & b_3 \\ c_1 & c_3 \end{vmatrix} + a_3 \cdot \begin{vmatrix} b_1 & b_2 \\ c_1 & c_2 \end{vmatrix}. \tag{1.27}$$

8. **Collinear vectors** can be represented as the scalar multiple of one another. If $\vec{a}$ and $\vec{b}$ are two collinear vectors, then $\vec{a} = \lambda \vec{b}$, $\lambda \in R$ where λ is a scalar number. Collinear vectors are on the same line in space. Using vectors we can define collinear point as it is shown in Fig. 1.7. For example, let points A, B, and C lie on the same line. Without loss of generality we can assume that point C is between points A and B. Assume that $\overrightarrow{AC} = k\overrightarrow{AB}$, where $0 < k < 1$. The following is true for three **collinear points** A, B, and C:

$$\overrightarrow{OC} = (1 - k) \cdot \overrightarrow{OA} + k \cdot \overrightarrow{OB}, \tag{1.28}$$

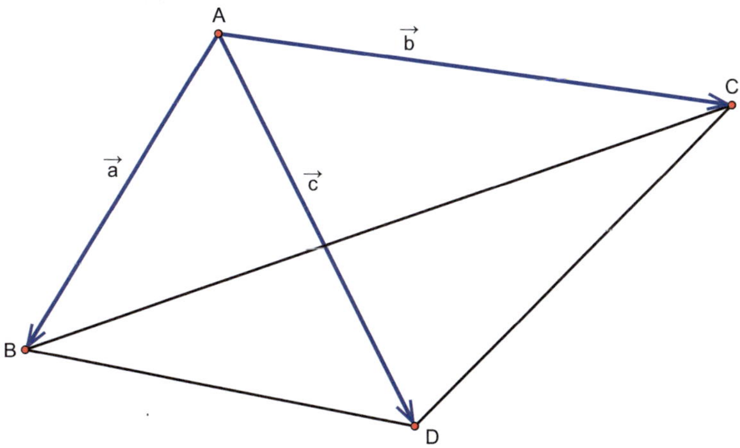

Fig. 1.6 Application of the mixed vector product

Fig. 1.7 Collinear points

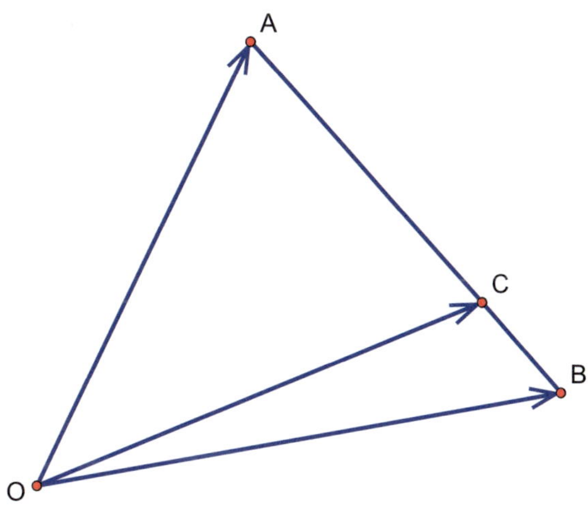

where O is any point in space.

Remark. You will prove this statement by solving one of the problems below.

9. **Coplanar vectors** are vectors located in the same plane. Coplanarity can be used for both points and vectors, and they are coplanar points and coplanar vectors. Coplanar points lie in the same plane. Coplanar vectors can be represented as the linear combination of one another. If $\overrightarrow{u}$, $\overrightarrow{v}$, $\overrightarrow{w}$ are three coplanar vectors, then each of them can be represented as linear combination of the other two vectors. For example,

$$\overrightarrow{u} = m \cdot \overrightarrow{v} + n \cdot \overrightarrow{w}, \tag{1.29}$$

where m and n are scalar numbers.

If three vectors $\vec{u}$, $\vec{v}$, $\vec{w}$ are linearly independent, then they are not coplanar. In such case, any other vector from three-dimensional space can be written as their linear combination:

$$\vec{s} = m \cdot \vec{u} + n \cdot \vec{v} + k \cdot \vec{w}. \tag{1.30}$$

In (1.30) vectors $(\vec{u}, \vec{v}, \vec{w})$ are also called **basis**. In many problems, it is convenient to choose an orthogonal basis, however, this formula can be used for any three independent vectors.

As we remember from linear algebra or calculus, a cross product is a vector that is orthogonal (perpendicular) to two other linearly independent vectors. So, if we consider the scalar triple product of three vectors, then this will be a kind of test for their coplanarity. If the scalar triple product is zero, then the three vectors are coplanar and lie in the same plane. If this value is not zero, then the three vectors are linearly independent in space, span three-dimensional space, and can be chosen as its basis.

We know that the determinant of a square matrix is equal to zero if its rows are dependent. Assume that vector $\vec{c}$ (for example, in (1.27)) is a linear combination of two other vectors ($\vec{a}$ and $\vec{b}$), which means that all three vectors lie in the same plane, then

$$\vec{c} = m \cdot \vec{a} + n \cdot \vec{b}, \quad \vec{c} = (ma_1 + nb_1,\ ma_2 + nb_2, ma_3 + nb_3).$$

Let us use (1.25) and show that the scalar triple product of three coplanar vectors is equal to zero:

$$\begin{aligned}(\vec{a} \times \vec{b}) \cdot \vec{c} &= \left(\left(\vec{a} \times \vec{b}\right), m\vec{a} + n\vec{b}\right) \\ &= m\left(\left(\vec{a} \times \vec{b}\right), \vec{a}\right) + n\left(\left(\vec{a} \times \vec{b}\right), \vec{b}\right) = 0.\end{aligned}$$

Other important formulas, such as those for a line or a plane in space, will be given and discussed in appropriate sections of this and other chapters.

Let us solve some problems.

Problem 8 Given three collinear points A, B, C, such that point C is somewhere between A and B. Prove that

$$\overrightarrow{OC} = k \cdot \overrightarrow{OA} + (1 - k) \cdot \overrightarrow{OB}, \quad 0 < k < 1.$$

Proof Let us use Fig. 1.7. Using the properties of vector's summation we can state that

$$\overrightarrow{AC} + \overrightarrow{CB} = \overrightarrow{AB}.$$

Assume that $\overrightarrow{AC} = k \cdot \overrightarrow{AB}$, then $\overrightarrow{CB} = (1-k)\overrightarrow{AB}$. Next, using properties of vectors we can write

$$\overrightarrow{OC} = \overrightarrow{OA} + \overrightarrow{AC}$$

or

$$\overrightarrow{OC} = \overrightarrow{OA} + k \cdot \overrightarrow{AB}.$$

Because

$$\overrightarrow{AB} = \overrightarrow{OB} - \overrightarrow{OA},$$

then

$$\overrightarrow{OC} = \overrightarrow{OA} + k(\overrightarrow{OB} - \overrightarrow{OA}).$$

After simplification, we have $\overrightarrow{OC} = k\overrightarrow{OB} + (1-k)\overrightarrow{OA}$. □

Remark. By solving this problem, we proved formula (1.28) for three points lying on the same line.

Problem 9 Given a vector $\overrightarrow{a} = (1, 2, 3)$. Find collinear to it vector that starts at point $A(1, 1, 1)$ and ends on the plane XOY.

Solution Let the vector end at point B so the third coordinate of point B must be 0 and hence $B = (x, y, 0)$. The coordinates of vector $\overrightarrow{AB}$ are

$$\overrightarrow{AB} = (x-1, y-1, 0-1) = (x-1, y-1, -1).$$

Since vectors $\overrightarrow{a}$ and $\overrightarrow{AB}$ are collinear, then their coordinates are proportional:

$$\frac{x-1}{1} = \frac{y-1}{2} = \frac{-1}{3}.$$

From this we find the coordinates of point B as $x = \frac{2}{3},\ y = \frac{1}{3},\ z = 0$.

Answer. $\overrightarrow{AB} = (-1/3, -2/3, -1)$.

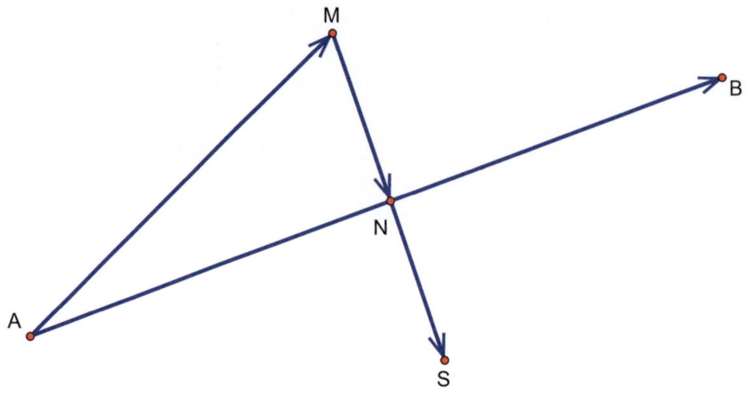

Fig. 1.8 Point symmetric to the given point with respect to a line

Problem 10 Given three non-collinear points in space, A, B, and M. Find the point symmetric to point M with respect to the line passing points A and B.

Solution Let us make a little sketch of the situation shown in Fig. 1.8. This will help us to find the idea of solving the problem.

We understand that if point S is symmetric to point M with respect to the line AB, then $MS \perp AB$ and $MN = NS$, where $N \in (AB)$. So using vector notation:

$$\overrightarrow{MN} = \overrightarrow{AN} - \overrightarrow{AM}.$$

Vectors $\overrightarrow{AN}$ and $\overrightarrow{AB}$ are collinear (points $A,\ N,\ B$ belong to the same line), then we can write it as

$$\overrightarrow{AN} = t \cdot \overrightarrow{AB}.$$

Now, we can substitute this into expression for $\overrightarrow{MN}$:

$$\overrightarrow{MN} = t \cdot \overrightarrow{AB} - \overrightarrow{AM}.$$

Since $MN \perp AB$, then the scalar (dot) product of the corresponding vectors is equal to zero:

$$\begin{aligned}
&(\overrightarrow{MN}, \overrightarrow{AB}) = 0 \\
&(t \cdot \overrightarrow{AB} - \overrightarrow{AM}, \overrightarrow{AB}) = 0 \\
&t(AB, AB) = (AM, AB) \\
&t = \tfrac{(AM,AB)}{(AB,AB)}.
\end{aligned}$$

Vector MN can be rewritten as

$$\overrightarrow{MN} = \frac{(\overrightarrow{AM}, \overrightarrow{AB})}{(\overrightarrow{AB}, \overrightarrow{AB})} \cdot \overrightarrow{AB} - \overrightarrow{AM}. \tag{1.31}$$

Equation (1.31) can be used to find the distance from a point to a line. For this, we just need to find the magnitude of vector MN:

$$d(M, l) = \sqrt{\overrightarrow{MN}, \overrightarrow{MN}}.$$

Finally, we can find the image of point M under reflection over line l. In general point S in a vector form can be written as

$$S = M + 2\overrightarrow{MN}$$

or

$$S = M + 2 \cdot \left(\frac{(AM, AB)}{(AB, AB)} \cdot AB - AM \right). \tag{1.32}$$

Most people don't remember complex formulas. It is much better to understand what needs to be calculated and have a plan for solving a particular or similar problem. Below are some basic steps:

1. Sketch the problem and place all points on the sketch.
2. Mark and evaluate vectors $\overrightarrow{AB}$ and $\overrightarrow{AM}$ using coordinates of points $A,\ B,\ M$.
3. Calculate scalar products (AM, AB) and (AB, AB).
4. Evaluate parameter $t = \frac{(AM,AB)}{(AB,AB)}$.
5. Find vector $\overrightarrow{MN}$ perpendicular to the given line (AB) using (1.31).
6. Use formula (1.32) to find the coordinates of point S symmetric to point M.

Problem 11 Given four points $A = (0, 1, -1),\ B = (1, -1, 2),\ C = (3, 1, 0),\ D = (2, -3, 1)$. Are they on the same plane? If yes, then find the equation of the plane.

Solution Let us make three vectors starting at one point, i.e., A. Using these four points we can make three vectors that start at A and end at each of the three other points:

$$\overrightarrow{AB} = (1, -2, 3),\ \overrightarrow{AC} = (3, 0, 1),\ \overrightarrow{AD} = (2, -4, 2),$$

and evaluate the determinant, in which the coordinates of three vectors are the rows:

$$\begin{vmatrix} 1 & -2 & 3 \\ 3 & 0 & 1 \\ 2 & -4 & 2 \end{vmatrix} = -24 \neq 0.$$

Answer. The vectors are linearly independent and the points are not coplanar.

Problem 12 Find the maximum angle between vectors $\overrightarrow{u} = (a, b, c)$ and $\overrightarrow{v} = (b, c, a)$.

Solution The given vectors are different but they have the same magnitude. The cosine of the angle between the two vectors can be found using formula (1.22):

$$\cos\varphi = \frac{ab + bc + ac}{a^2 + b^2 + c^2}.$$

The following operation will help to simplify the numerator of the fraction:

$$2ab + 2bc + 2ac = (a + b + c)^2 - a^2 - b^2 - c^2.$$

So the expression for the cosine can be rewritten as

$$\cos\varphi = \frac{1}{2} \cdot \frac{(a + b + c)^2}{a^2 + b^2 + c^2} - \frac{1}{2} \cdot \frac{a^2 + b^2 + c^2}{a^2 + b^2 + c^2},$$

or after simplification as

$$\cos\varphi = \frac{1}{2} \cdot \frac{(a + b + c)^2}{a^2 + b^2 + c^2} - \frac{1}{2}.$$

Let us inspect the right-hand side of this formula. The first term is always nonnegative so the right-hand side is always greater or equal to $-\frac{1}{2}$. It is equal to $-\frac{1}{2}$, if the first term is zero and then $\varphi = 120°$ and the angle takes its maximum possible value (please recall the fact that $\cos\varphi$ is a monotonically decreasing function of $\varphi \in (0, \pi)$). Such case occurs if $a + b + c = 0$. Since the given vectors are nonzero, it is clear that some of the components must be negative. There are an infinite number of pairs of $\vec{u} = (a, b, c)$ and $\vec{v} = (b, c, a)$ that form a $120°$ angle. For example, $\vec{u} = (2023, 1, -2024)$ and $\vec{v} = (1, -2024, 2023)$, etc.

Answer. The maximum possible angle is 120°.

Problem 13 Prove that for any real numbers $x_1,\ x_2,\ x_3,\ y_1,\ y_2$, and y_3:

$$\begin{aligned}&\left((x_2y_3-x_3y_2)^2+(x_1y_3-x_3y_1)^2+(x_1y_2-x_2y_1)^2\right)^{\frac{1}{2}}\\&\le\left(x_1^2+x_2^2+x_3^2\right)^{\frac{1}{2}}\cdot\left(y_1^2+y_2^2+y_3^2\right)^{\frac{1}{2}}\end{aligned}.$$

Find the condition for which the inequality becomes an equality.

Proof Consider two vectors in $\Re^3$: $\overrightarrow{u}=(x_1,x_2,x_3)$ and $\overrightarrow{v}=(y_1,y_2,y_3)$. Then on the right-hand side of the given inequality we have the product of their magnitudes. Let the third vector $\overrightarrow{w}$ be their cross product, such that

$$\overrightarrow{w}=\overrightarrow{u}\times\overrightarrow{v}=\begin{vmatrix}\overrightarrow{i}&\overrightarrow{j}&\overrightarrow{k}\\x_1&x_2&x_3\\y_1&y_2&y_3\end{vmatrix}=\begin{vmatrix}x_2&x_3\\y_2&y_3\end{vmatrix}\cdot\overrightarrow{i}-\begin{vmatrix}x_1&x_3\\y_1&y_3\end{vmatrix}\cdot\overrightarrow{j}+\begin{vmatrix}x_1&x_2\\y_1&y_2\end{vmatrix}\cdot\overrightarrow{k}. \tag{1.33}$$

If we evaluate its magnitude, using formula (1.24), then on the left we will obtain

$$\|w\|=\|u\|\cdot\|v\|\cdot\sin\alpha\le\|u\|\cdot\|v\|=\left(x_1^2+x_2^2+x_3^2\right)^{\frac{1}{2}}\cdot\left(y_1^2+y_2^2+y_3^2\right)^{\frac{1}{2}}.$$

This is an expression for the area of a parallelogram formed by two vectors. On the other hand, if we evaluate the magnitude of the vector $\overrightarrow{w}$ by finding its coordinates using the right-hand side of (1.33), we get the left-hand side of the inequality, which we need to prove.

We see that the equal sign in the original inequality will appear if the vectors $\overrightarrow{u}$ and $\overrightarrow{v}$ are orthogonal, $\alpha=90°$, so the following condition (zero of the scalar product) must be satisfied

$$x_1y_1+x_2y_2+x_3y_3=0.$$

□

Problem 14 The pyramid $ABCDE$ has a parallelogram $ABCD$ as its base. Point E is the apex of the pyramid. Point F is the midpoint of the edge DE and the point G divides the edge BE in the ratio $3\div1$, starting from the vertex. The point H is the intersection of the line EC and the plane α containing the points $A,\ F$, and G. Find the ratio in which the point H divides the edge EC.

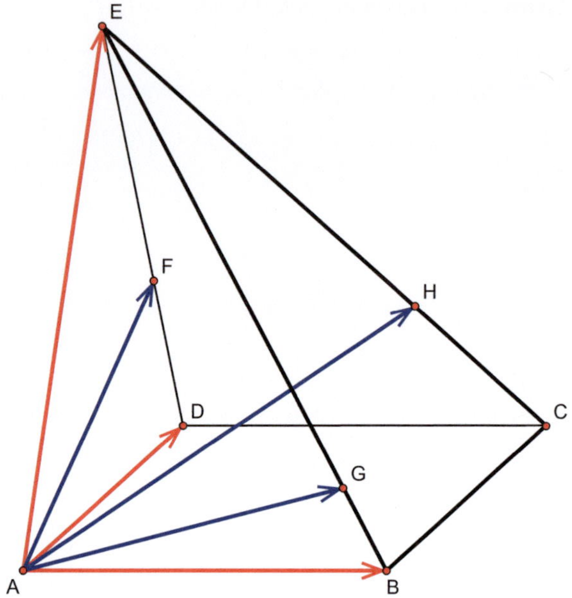

Fig. 1.9 Pyramid $ABCDE$ and point H

Solution Let us make a sketch of a pyramid regarding to the condition of the problem and place points F and G on edges ED and EB, respectively. Let's also place point H on the side EC. We do not know its location. However, we need to find the ratio $EH \div HC$. In our pyramid $ABCDE$, we can choose and mark independent (non-coplanar) vectors $\overrightarrow{AE}$, $\overrightarrow{AD}$, and $\overrightarrow{AB}$ as the basis. (They are shown in red in Fig. 1.9.)

We also showed in our sketch (blue), the vectors $\overrightarrow{AF}$, $\overrightarrow{AG}$, and $\overrightarrow{AH}$. By the condition of the problem, $H = EC \cap AFG, \quad (AFG) = \alpha$. Since point H belongs to the plane α and vectors $\overrightarrow{AF}, \overrightarrow{AG}$ are independent in this plane, then the vector $\overrightarrow{AH}$ can be written as their linear combination:

$$\overrightarrow{AH} = m \cdot \overrightarrow{AF} + n \cdot \overrightarrow{AG}, \tag{1.34}$$

where scalars m and n are to be determined. By the condition of the problem, we know that $EG = 0.75EB$ and $GB = 0.25EB$, and that $EF = 0.5ED$. Because point F is collinear to points E and D, and points $E, \ G, \ B$ are also collinear, we can then use formula (1.28) and rewrite (1.34) as

$$\overrightarrow{AH} = m(0.5 \cdot \overrightarrow{AD} + 0.5 \cdot \overrightarrow{AE}) + n(0.25 \cdot \overrightarrow{AE} + 0.75 \cdot \overrightarrow{AB}).$$

After collecting like terms it can be rewritten as

$$\overrightarrow{AH} = 0.75n \cdot \overrightarrow{AB} + 0.5\,m \cdot \overrightarrow{AD} + (0.5\,m + 0.25n) \cdot \overrightarrow{AE}. \tag{1.35}$$

On the other hand, point H is collinear to points E and C, so we can write $\overrightarrow{AH}$ again using (1.28) as

$$\overrightarrow{AH} = (1-k) \cdot \overrightarrow{AE} + k \cdot \overrightarrow{AC},$$

where k is an unknown parameter. Vector $\overrightarrow{AC}$ is the diagonal of parallelogram $ABCD$, so

$$\overrightarrow{AC} = \overrightarrow{AD} + \overrightarrow{AB}.$$

Now we can substitute this into the previous equation:

$$\overrightarrow{AH} = (1-k) \cdot \overrightarrow{AE} + k \cdot (\overrightarrow{AD} + \overrightarrow{AB}),$$

which can be simplified as

$$\overrightarrow{AH} = (1-k) \cdot \overrightarrow{AE} + k \cdot \overrightarrow{AD} + k \cdot \overrightarrow{AB}. \tag{1.36}$$

Since the left sides of (1.35) and (1.36) are the same, we can equate the right-hand sides by terms obtaining a system of three equations in three variables:

$$\begin{cases} 0.75n = k \\ 0.5\,m = k \\ 0.5\,m + 0.25n = 1 - k \end{cases}.$$

Solving this system we obtain that $n = \frac{4k}{3}$ and $m = 2k$, so

$$k + \frac{k}{3} = 1 - k, \quad k = \frac{3}{7}; \quad 1 - k = \frac{4}{7}.$$

Since $\overrightarrow{EH} = k \cdot \overrightarrow{EC}$, then $\overrightarrow{EH} = \frac{3}{7} \cdot \overrightarrow{EC}$, from which we can find that

$$EH \div HC = 3 \div 4.$$

Answer. Point H divides the edge AC in ratio 3 : 4.

Remark. In Chap. 3, we will study various methods for constructing cross sections of polyhedra. Here, using the vector-coordinate approach, we simply solved the problem of cutting a pyramid with a plane without constructing the cross section.

1.3 Planes in Space

The location of two planes in space may be summarized by Fig. 1.10.

The following theorems are very important in study of solid geometry. Some of the theorems are listed here without proof because either they can be easily proven with the axioms or from ideas of those theorems with proofs.

Theorem 5 (Intersecting planes) *If two different planes have a common point, then they have a common line on which all common points of these planes lie. If two planes have one common point, they have another common point. These planes intersect by the line containing this point (Fig. 1.11).*

This very important theorem is based on Axioms 3. It will be used many times in construction of cross sections and in solving other problems.

Definition. Two different planes are **parallel** if they do not intersect.

Theorem 6 (Parallelism of two planes) *If two intersecting lines of the same plane are, respectively, parallel to two intersecting lines of another plane, then these planes are parallel.*

Proof Let α and β be two planes and lines a_1 and a_2 be the intersecting lines in plane α and b_1 and b_2 be intersecting and, respectively, parallel to these lines in plane β. Assume that α is not parallel to plane β. Then two planes must intersect by some line c that can be written as $\alpha \cap \beta = c$. Since $a_1 \| b_1$, then $a_1 \| \beta$. Since $a_2 \| b_2$, then

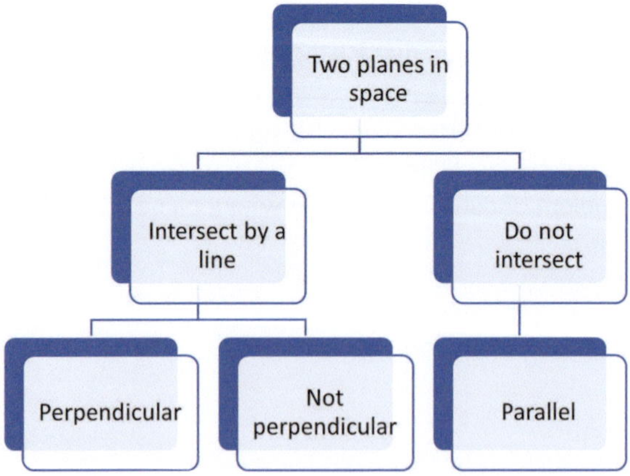

Fig. 1.10 Two planes in space

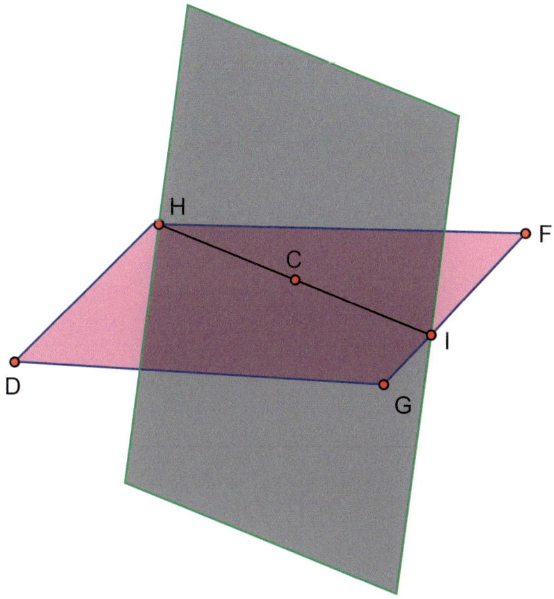

Fig. 1.11 Intersecting planes

$a_2 \| \beta$. Because $c \in \alpha$ then either a_1 or a_2 will intersect with c. However, $c \in \beta$, so, line a_2, for example, that intersects with line c must also intersect plane β where c is located. This contradicts the fact that $a_2 \| \beta$. Therefore, planes α and β must be parallel. □

Theorem 7 (Perpendicularity of two planes) *A plane that passes through a line perpendicular to another plane is also perpendicular to this plane.*

Remark. That is, to prove the perpendicularity of two planes, one can prove that one of the planes passes through a line perpendicular to the other plane.

In the following section, we will introduce a dihedral angle. We can also define two perpendicular planes, if the corresponding dihedral angle is the right angle (90°).

Theorem 8 *If a given plane is perpendicular to another plane, then the given plane will be parallel to all lines perpendicular to the second plane.*

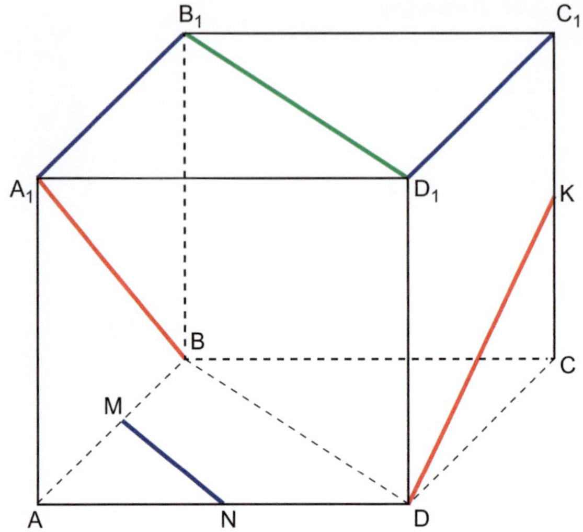

Fig. 1.12 Lines and planes in a cube

For illustration we can consider a cube sketched in Fig. 1.12. We can see the plane $\alpha = AA_1D_1D$ is perpendicular to plane $\beta = ABCD$ that can be written as $\alpha \perp \beta$. Line $l = (BB_1)$ is perpendicular to plane β, then plane α is parallel to line l $(\alpha \| l)$.

Theorem 9 *Let two perpendicular planes intersect by a line and one line of the first plane be perpendicular to the line of intersection of two planes. Then this line is perpendicular to any line in the second plane.*

In Fig. 1.13, plane $DHFG$ is perpendicular to plane HJI, and line JC is perpendicular to HI, then line JC is perpendicular to CK. (Line CK is some line in plane $DMFG$, then by this theorem $JC \perp CK$.)

Question. Consider Theorem 9 and associated with it Fig. 1.13. Is $CK \perp HI$?

Answer. In general, the answer is "no." CK can be any line in plane $DHFG$ and we know only that $JC \perp CK$. Therefore CK does not have to be perpendicular to HI.

Theorem 10 *If two parallel planes are intersected by a third plane, then the lines of intersection are parallel.*

Proof Let us sketch two parallel planes $\alpha \| \beta$. Assume that there is the third plane γ that intersects both the given parallel planes, $\alpha \cap \gamma = a$ and $\beta \cap \gamma = b$. In Fig. 1.14,

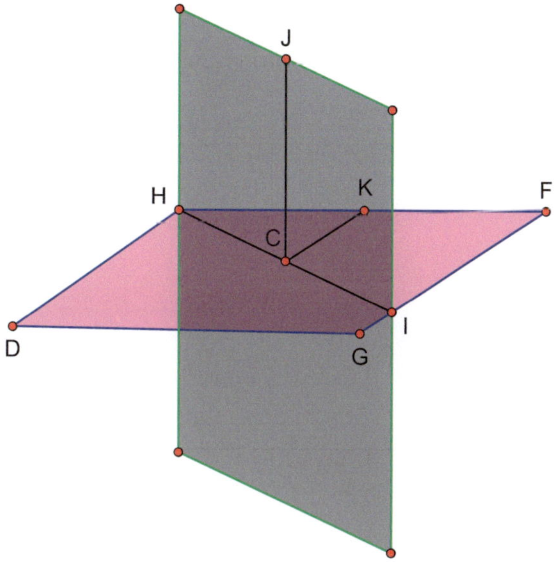

Fig. 1.13 Perpendicular planes

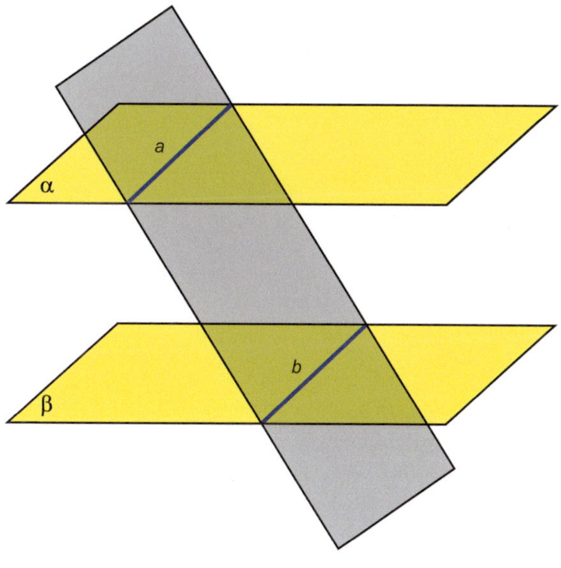

Fig. 1.14 Parallel planes intersected by a third plane

plane γ is shown in gray and the parallel planes in yellow. Because the lines of intersection belong to one plane (γ), then they are either parallel or intersect in that plane. Since lines a and b belong to parallel planes, then they cannot have common points. Therefore, they must be parallel, $a \| b$. □

Let us consider a problem regarding location of three planes.

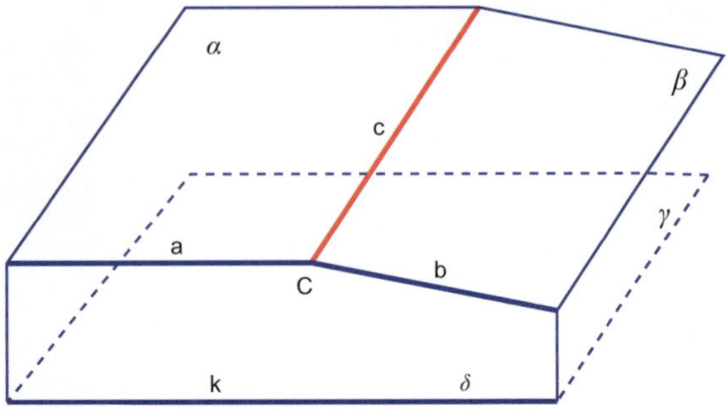

Fig. 1.15 Assume that planes α and β intersect

Problem 15 Each of the two planes α and β is parallel to a third plane γ. Can the planes α and β intersect?

Solution Suppose that the planes α and β intersect along some line c. Take in the plane γ a line k not parallel to line c and also take a point C on line c. Draw the plane δ through the line k and point C (Fig. 1.15). This plane intersects the planes α and β along the lines a and b. The lines a and k are parallel as the lines of intersection of the parallel planes α and γ with the plane δ, and in the same way the lines b and k are parallel (as the lines of intersection of the parallel planes β and γ with the plane δ). Thus, in the plane δ we have obtained different intersecting lines a and b that both parallel to the line k, which contradicts the **axiom of parallel lines**. The resulting contradiction proves that the planes α and β cannot intersect.

1.3.1 Several Planes. Space Partitioning

So far we have talked about how two planes can be positioned in space, and also discussed a little about the position of the three planes relative to each other, but not in general. In this section, we will try to understand how three or more planes can divide space and into how many parts. It is clear that one plane (and, as we know, it is infinite) divides space into two parts. If, for example, we take a point that does not belong to the plane, then this point can be located both above the plane and below the plane, that is, either in one or the other part of space.

Next, let's assume that we have two planes. If the planes are parallel to each other, then they divide the space into three parts. But what happens if our planes intersect? If the planes intersect, then it does not matter if they intersect at a right angle or at some other angle—the space will be divided into four parts. We can visualize four parts of the space formed by two planes in Fig. 1.11. I think we can easily imagine the case of three parallel planes—three parallel planes divide space into four parts. What happens if, for example, we have two parallel planes and a third plane intersecting

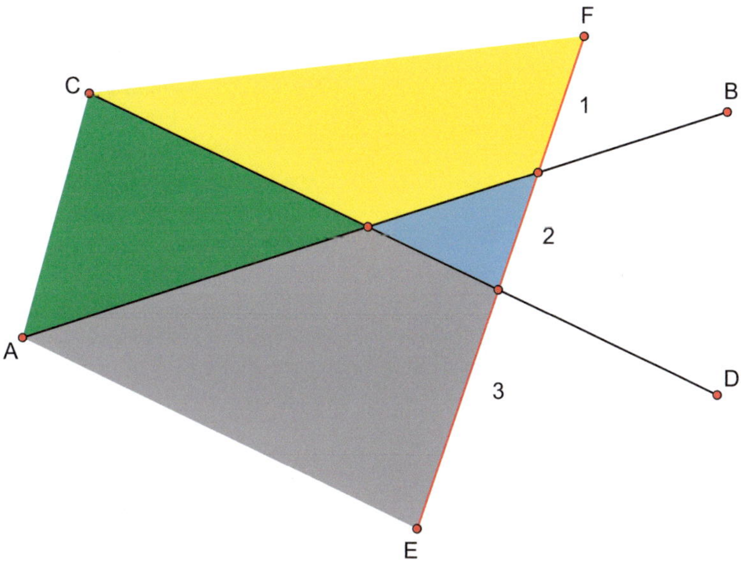

Fig. 1.16 The third line EF makes three additional regions in the plane

those two parallel planes? This case can be observed in Fig. 1.14. There are six spatial regions.

Now suppose that of the three planes, every two intersect. In this case, it is convenient to consider how our planes intersect the auxiliary, fourth plane. Then we can represent three pairwise intersecting lines in our flat drawing in the form of a diagram of pairwise intersecting planes. You can use the sketch in Fig. 1.16 where we see that the third line adds three additional areas to the original four formed by the two intersecting lines. In the case of three planes, our space is divided into seven parts. If three planes intersect at one point, like a coordinate system in three-dimensional space, then the space is divided into eight regions. This can also be seen on a plane projection—three lines intersecting at one point divide the plane into eight parts.

Before solving space partitioning problems, we prove some relevant theorems.

Theorem 11 *n straight lines can divide a plane into at most $\frac{n^2+n+2}{2}$ regions.*

Proof Let us prove this statement using mathematical induction.

1. We can see that the statement is true for $n = 1$. One line divides a plane into two regions and by the formula $\frac{1^2+1+2}{2} = 2$.
2. Suppose that the statement is true for $n = k$, which means that k lines form at most $\frac{k^2+k+2}{2}$ regions in one plane.

3. Let us draw an additional $(k+1)^{st}$ line, such that this line is not parallel to any of the given k lines and not passing through a point of intersection of any pair of the k given lines. Then this line will make $k+1$ additional regions added to those formed by the given k lines. (For example, in Fig. 1.16, we have two intersecting lines $L_1 = AB$ and $L_2 = CD$ that divide the plane into 4 regions. The third line $L_3 = EF$ shown in red color adds 3 additional regions to those four, so it divides the plane into $4+3=7$ regions.) For $(k+1)$ lines, we have

$$\frac{k^2+k+2}{2}+k+1=\frac{k^2+k+2+2k+2}{2}=\frac{(k^2+2k+1)+(k+1)+1}{2}$$
$$=\frac{(k+1)^2+(k+1)+1}{2},$$

which completes the proof.

□

Problem 16 How many parts can n planes divide space into? Every three planes intersect at one point, and no four planes have a common point.

Solution Consider n planes satisfying the condition. Assume that these planes divide space into a_n parts and see how this number changes when drawing $(n+1)^{st}$ plane. In general, this additional plane is intersected by other planes along n straight lines and so is divided by them into $\frac{n^2+n+2}{2}$ regions (pieces). Each of these pieces splits one of the old parts of space into two, which means we have the formula:

$$a_{n+1}=a_n+\frac{n^2+n+2}{2}.$$

This recursive formula allows us to obtain the general formula for the problem. Let us rewrite it for the nth term:

$$a_n=a_{n-1}+\frac{n^2-n+2}{2}.$$

We know that one plane divides space into two parts, then $a_1=2$. If we have no planes in space, then $a_0=1$. Let's multiply both sides of the last formula by 2, and then write down n consecutive terms of the recursion:

$$\begin{aligned}2a_1 &= 2\cdot 1 + \left(1^2 - 1 + 2\right)\\ 2a_2 &= 2a_1 + \left(2^2 - 2 + 2\right)\\ 2a_3 &= 2a_2 + \left(3^2 - 3 + 2\right)\\ 2a_4 &= 2a_3 + \left(4^2 - 4 + 2\right)\\ &\dots\\ 2a_{n-1} &= 2a_{n-2} + \left((n-1)^2 - (n-1) + 2\right)\\ 2a_n &= 2a_{n-1} + \left(n^2 - n + 2\right).\end{aligned}$$

Next, let us add the left-hand sides and the right-hand sides separately. We notice that the same terms on the left and on the right will be cancelled. The left-hand side will have only a_n and the terms on the right-hand side can be combined in the known sums of the sum of the first n squares minus the sum of the first n natural numbers plus number 2 repeated $n+1$ times:

$$2a_n = 2\cdot(n+1) + \frac{n(n+1)(2n+1)}{6} - \frac{n(n+1)}{2}.$$

Dividing both sides by 2 and factoring $(n+1)$ on the right, we obtain

$$a_n = \frac{(n+1)(12 + n(2n+1) - 3n)}{12} = \frac{(n+1)(n^2 - n + 6)}{6}.$$

This formula also can be rewritten as

$$a_n = \frac{n^3 + 5n + 6}{6}. \tag{1.37}$$

Answer. n planes can divide space into $\frac{n^3+5n+6}{6}$ parts.

Theorem 12 *n planes can divide space into at most $\frac{n^3+5n+6}{6}$ parts.*

Proof Let us prove this statement using mathematical induction.

1. We can see that the statement is true for $n = 1$. One plane divides space into two parts and by the formula $\frac{1^3+5\cdot 1+6}{6} = 2$.
2. Suppose that the statement is true for $n = k$, which means that k planes can split space into at most $\frac{k^3+5k+6}{6}$ parts.
3. Let us draw the $(k+1)^{st}$ plane such that it is not parallel to any of the k planes and does not pass through the line of intersection of any pair of the k planes or a point of intersection of any three of the k planes. This $(k+1)^{st}$ additional plane

will intersect with the k planes by k lines that divide the $(k+1)^{st}$ plane into at most $\frac{k^2+k+2}{2}$ regions. Thus the $(k+1)$ plane will make at most

$$\frac{k^3+5k+6}{6}+\frac{k^2+k+2}{2}=\frac{k^3+5k+6+3k^2+3k+1+5}{6}=\frac{(k+1)^3+5(k+1)+6}{6}$$

parts.

□

Remark. This theorem gives the maximum number of parts into which space can be divided by n intersecting planes. However, you may need to find how many spatial regions there are if all planes intersect at one point, or if only a few planes intersect at one point, and some plane intersects in a straight line with another plane or runs parallel to that plane, and so on. In such cases, counting parts is unique and this can even be a rather interesting task that develops spatial and combinatorial skills.

Problem 17 In how many parts is space divided by

(a) three planes that intersect at one point?

(b) four planes, such that no four planes intersect at one point?

(c) four planes intersecting at one point?

Solution

(a) Let's consider three planes intersecting at one point. For example, we can imagine a coordinate system in three-dimensional space. It is clear that three such planes divide the space into eight parts. If we look at this coordinate system parallel to one of the planes, for example, the XOY plane, we will see two mutually perpendicular lines that divide the plane into four regions. You can visualize that the space will be divided into twice as many parts as the number of plane regions, that is, eight.

(b) Now let's imagine that in space there is another, fourth plane that intersects each of the three given planes, but does not pass through the point of intersection of the three planes, O (the origin of the coordinate system). We have two possible cases:

1. The fourth plane can be perpendicular to one of the given coordinate planes, i.e., plane XOY and intersect with each of the three given planes. We see that it is as if a third straight line intersects two perpendicular straight lines. That is, to the four already known areas on the plane, three more areas are added that this straight line forms. In fact, Fig. 1.16 can be useful where the red line shows a projection of the fourth plane onto the XOY plane. Since we got seven regions on the plane,

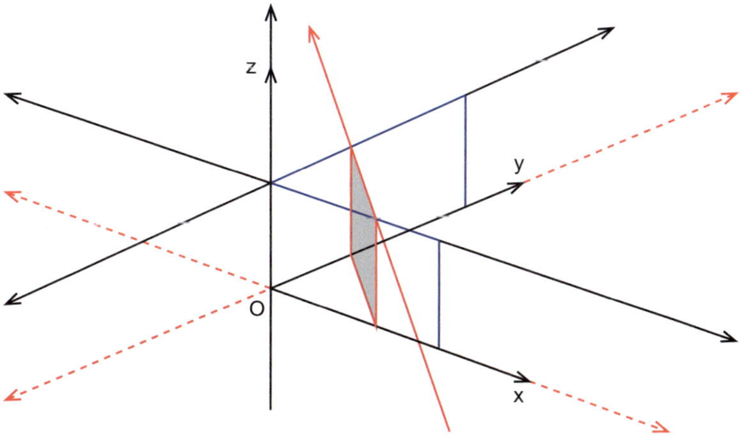

Fig. 1.17 The fourth plane intersects the three planes

Fig. 1.18 Tetrahedron as intersection of four planes (blue, pink, gray, and yellow)

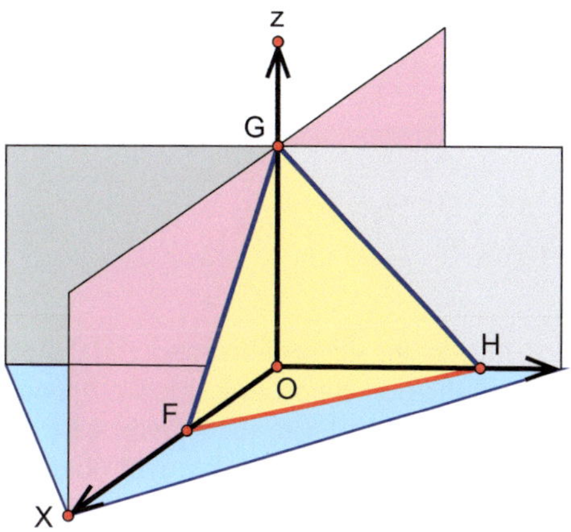

then multiplying by two, we get 14 parts, the maximum possible number in this case (Fig. 1.17).

2. Let us now consider a situation where four planes intersect so that every three planes have a common point. Therefore, the four planes form a tetrahedron (Fig. 1.18). The four planes divide the space into the maximum possible, 15, as confirmed by the formula (1.37) ($a_4 = \frac{4^3+5\cdot 4+6}{6} = 15$).

(c) Consider now four planes α, β, γ, and δ intersecting at one point, O. It is interesting to look into construction of four planes passing through the same point from a different point of view. For example, in the following chapter, we will discuss properties of an octahedron. Each face of an octahedron is a triangle, opposite faces are parallel and it looks like a bi-pyramid with a square base (Fig. 1.19).

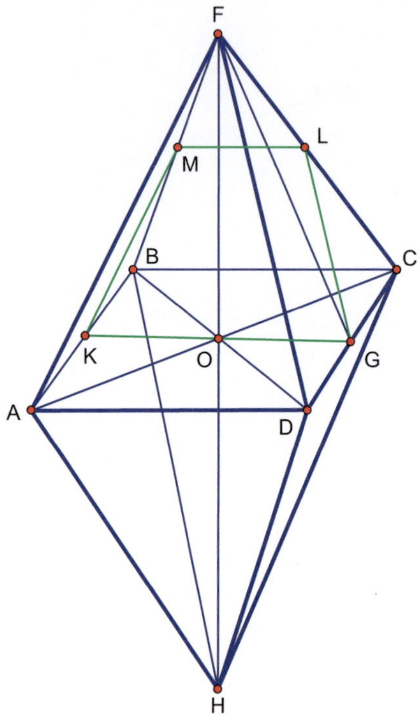

Fig. 1.19 Regular octahedron: demonstration of three and four intersecting planes

Consider the plane of the base, $ABCD$ and two other planes, each of which passes through the line connecting the two opposite vertices of the octahedron (F and H) that is perpendicular to the square $ABCD$ and one of the diagonals of the square (planes AFC and BFD). We will obtain three planes intersecting at one point O (that is also the point of intersection of the diagonals of the square). By connecting point O with the vertices of each triangular face of the octahedron, we observe eight triangular pyramids and eight trihedral angles. The space is divided into eight parts.

If a plane is drawn through O parallel to one of the faces of the octahedron (i.e., plane $KMLG$ shown in green color is parallel to planes AFD and HBC), then it will be the fourth plane going through center O. This plane will not intersect two faces of the octahedron to which it is parallel, but it will divide each of other six planes, consequently such a plane will divide each of the six trihedral angles. Therefore, the entire space will be divided into 14 parts ($14 = 8 + 6$).

Answer. (a) 8, (b) 15, and (c) 14 parts.

Problem 18 Into how many parts is space divided by (a) three planes? (b) four planes that do not intersect at one point? (c) five planes passing through one point (no three planes have a common line)?

Solution

(a) Three parallel planes divide space into four parts. Let us look at Fig. 1.14. We can see that three planes, two of which are parallel, divide the space into six parts. However, if no planes are parallel, then the space will be divided into eight parts.

(b) Four planes divide space into 14 parts. Let's prove it. Three planes $\alpha,\ \beta,\ \gamma$ divide space into eight parts. As we draw the fourth δ plane, it intersects with the previous three along three lines passing through their common point. These lines divide the δ plane into six parts. Therefore, of those eight parts into which space was divided by three planes, the fourth plane δ intersects six parts and divides each of them into two parts. Thus, six more parts are added, total $8 + 6 = 14$.

(c) Five planes divide the space into 22 parts. Similarly, the fifth plane, intersecting with the previous ones along four straight lines, adds $4 \cdot 2 = 8$ parts, and they become $14 + 8 = 22$.

Answer. (a) 4, 6, or 8 parts; (b) 14 parts; (c) 22 parts.

Remark. We're going to come back to this topic of partitioning later in this chapter when we learn more about polyhedral angles.

1.3.2 Different Forms of a Plane in Space

A plane (i.e., α) can be expressed algebraically as

$$ax + by + cz = d, \tag{1.38}$$

where $a,\ b,\ c,\ d$ are real numbers. Here vector $\overrightarrow{n} = (a, b, c)$ is perpendicular to the plane and can be written as $\overrightarrow{n} \perp \alpha$. If the parameter $d = 0$, then the plane goes through the origin, and if $d \neq 0$, then it does not.

In books on analytic geometry, a plane is usually given by the following formula:

$$ax + by + cz + d_1 = 0, \tag{1.39}$$

where all terms are on one side. I use the plane equation (1.39) in a slightly different form because, in my opinion, this form (1.38) makes it easier to geometrically interpret the solutions of a system of two or three linear equations in three variables in terms of two or three planes in space. The same plane (α) containing point $M_0 = (x_0, y_0, z_0)$ has the equation:

$$a(x - x_0) + b(y - y_0) + c(z - z_0) = 0. \tag{1.40}$$

Problem 19 Find the equation of a plane that contains points $A(0, 1, 0)$, $B(1, 0, 0)$, and $C(1, 1, 1)$.

Solution Let us solve the system of three linear equations:

$$\begin{cases} a \cdot 0 + b \cdot 1 + c \cdot 0 = d \\ a \cdot 1 + b \cdot 0 + c \cdot 0 = d \\ a \cdot 1 + b \cdot 1 + c \cdot 1 = d \end{cases}$$

From this system we obtain that $a = b = d$ and $c = -d$, then $dx + dy - dz = d$. Dividing by $d \neq 0$, we obtain $x + y - z = 1$.

Answer. $x + y - z = 1$.

Now let us discuss the location of two or three planes in space. We know that two planes are parallel, if they do not have common points. In general, two different planes can be written as

$$\begin{array}{ll} \alpha : & a_1x + b_1y + c_1z = d_1, \quad \vec{n}_1 = (a_1, b_1, c_1) \\ \beta : & a_2x + b_2y + c_2z = d_2, \quad \vec{n}_2 = (a_2, b_2, c_2) \end{array}.$$

1. If two planes are parallel, $\alpha \parallel \beta$, then vectors $\overrightarrow{n_1}, \overrightarrow{n_2}$ must be collinear, so $\overrightarrow{n_1} = k \cdot \overrightarrow{n_2},\ k \in R$, which also indicates that the coefficients of the variables are proportional. Thus, the planes, $x + 3y - 5z = 1$ and $2x + 6y - 10z = 5$ are parallel.
2. Two planes will intersect by a line, if vectors $\vec{n}_1,\ \vec{n}_2$ are not collinear. This can be written as $\alpha \cap \beta = a$. For example, plane $x + 3y - 5z = 1$ will intersect another $x - 3y - z = 3$ by the line $x = 2 + 3z$ that can be obtained by adding the left and the right sides of these two equations together and after simplification.
3. Two planes are perpendicular to each other if and only if their normal vectors are also perpendicular (orthogonal), which means that their scalar product equals zero, $(n_1, n_2) = a_1 \cdot a_2 + b_1 \cdot b_2 + c_1 \cdot c_2 = 0$.

For example, we can find for what value of the parameter k , the two given planes, $x + 2y + z = 1$ and $x - y + k \cdot z = 2$, are perpendicular to each other $(\overrightarrow{n_1}, \overrightarrow{n_2}) = 1 \cdot 1 + 2 \cdot (-1) + 1 \cdot k = 0 \Rightarrow k = 1$. In all other cases when $k \neq 1$ the two planes will intersect but never be perpendicular!

Let us discuss the behavior of the three planes,

$$\begin{cases} a_1x + b_1y + c_1z = d_1, & \alpha, \quad \overrightarrow{n_1} = (a_1, b_1, c_1) \\ a_2x + b_2y + c_2z = d_2, & \beta, \quad \overrightarrow{n_2} = (a_2, b_2, c_2) \\ a_3x + b_3y + c_3z = d_3, & \gamma, \quad \overrightarrow{n_3} = (a_3, b_3, c_3) \end{cases}$$

The following simple cases are possible:

1. If $a_1 = a_2 = a_3,\ b_1 = b_2 = b_3$, and $c_1 = c_2 = c_3$, but $d_1 \neq d_2 \neq d_3$, then all three planes go parallel to each other.
2. If $a_1 = a_2 = a_3,\ b_1 = b_2 = b_3$, and $c_1 = c_2 = c_3$ but $d_1 = d_2 \neq d_3$, then two planes coincide that we have only two parallel planes.
3. If $a_1 = a_2 \neq a_3,\ b_1 = b_2 \neq b_3$, and $c_1 = c_2 \neq c_3$ but $d_1 \neq d_2 \neq d_3$, then the first two planes are parallel to each other and both are intersected by the third plane. In this case, the system is inconsistent because three planes do not have common points, but each of the two parallel planes is intersected by the third line and the lines of the intersection are parallel.
4. Three planes can intersect at one point, like the planes in the corner of your room where the ceiling and two walls meet. Algebraically, this occurs when the rank of the coefficient matrix of the above system is 3, which is equivalent to the determinant of the coefficient matrix being nonzero:

$$\det A = \begin{vmatrix} a_1 & b_1 & c_1 \\ a_2 & b_2 & c_2 \\ a_3 & b_3 & c_3 \end{vmatrix} \neq 0.$$

All other cases that can result from system of three linear equations if we think of each equation as a plane in space R^3

$$\begin{cases} a_1x + b_1y + c_1z = d_1 \\ a_2x + b_2y + c_2z = d_2 \\ a_3x + b_3y + c_3z = d_3 \end{cases} \tag{1.41}$$

can be obtained from the theory of solutions for system of linear equations and explained based on the ranks of the coefficient matrix A

$$A = \begin{bmatrix} a_1 & b_1 & c_1 \\ a_2 & b_2 & c_2 \\ a_3 & b_3 & c_3 \end{bmatrix} \tag{1.42}$$

and the augmented matrix A^b(that is obtained from matrix A^b by adding an additional column of the constant terms)

$$A^b = \begin{bmatrix} a_1 & b_1 & c_1 & d_1 \\ a_2 & b_2 & c_2 & d_2 \\ a_3 & b_3 & c_3 & d_3 \end{bmatrix}. \tag{1.43}$$

As you recall, rank of a matrix is the number of independent rows or columns. For any rectangular matrix $A_{n\times m}$, its rank

$$r(A) \leq \min(n, m). \tag{1.44}$$

Hence, the maximum rank of the augmented matrix $A^b_{3\times 4}$ is 3. We have the cases:

1. **Case 1**. $r(A) = r(A^b) = 3$, then the system (1.41) is consistent. Moreover, its solution is unique. Geometrically we have the case when three different planes intersect at one point, for example, point $M(x_0, y_0, z_0)$. It can be visualized as intersections of the three orthogonal planes at the center of the coordinate system, as the vertex of a trihedral angle (we will learn about such angles later in this chapter) or simply by looking at the corner of your room where the planes of two walls intersect with the ceiling plane. This solution (x_0, y_0, z_0) can be obtained by different methods, such as Gaussian elimination, use of the inverse of the coefficient matrix or, for example, Cramer's formulas. You can do it by hand or analytically. At least one of such examples will be shown in this chapter.

2. **Case 2**. $r(A) = r(A^b) = 2$, then the system (1.41) is also consistent, and its solution exists. There are infinite number of the solutions. In this case, geometrically we have two planes that after simplification coincide, so they represent just one plane that intersects with the 3rd plane. Algebraically, since rank of the matrices is 2 and less than the number of the variables (3) by 1, we have one degree of freedom or one parameter. Since two planes intersect by one line, then as the result from solving such a system, we will obtain an equation of a line in a parametric form.

3. **Case 3**. $r(A) = 2 \neq r(A^b) = 3$. The system (1.41) is inconsistent. No solution. This means that three planes represented the equations of the system but do not have common points in the way that two planes go parallel and each of them is intersected by the third plane. The lines of intersections are parallel. This is an interesting case represented by Theorem 12 and that will be seen many times in this book when we construct a cross section of a polyhedron by a plane.

4. **Case 4**. $r(A) = 1 \neq r(A^b) = 2$. The system (1.41) is inconsistent, and all planes go parallel to each other.

5. **Case 5**. $r(A) = 1 \neq r(A^b) = 1$. The system is consistent, it has infinite number of solutions. Three planes eventually are reduced to only one on which all possible solutions lie. In general, solution to system (1.41) represents parametric form of

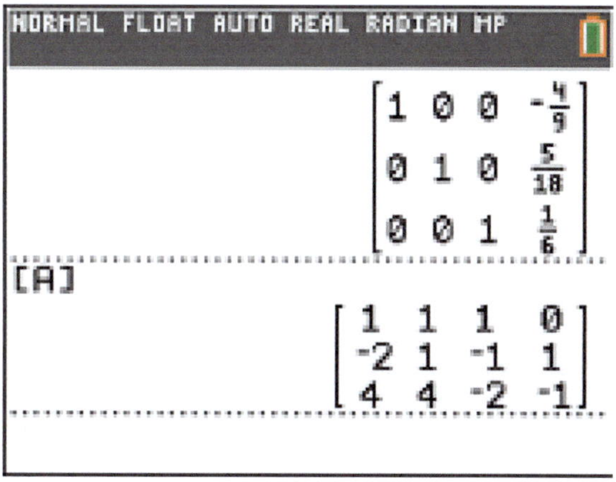

Fig. 1.20 A unique solution in the form of a screenshot of a graphing calculator

a plane. Since the ranks of the matrices equal 1 that is two less than the number of the variables, we have two degrees of freedom, which indicate the presence of two independent parameters in our solution.

Let's look at problems that illustrate some of these cases.

Problem 20 (Case 1) Show that planes $x + y + z = 0$, $-2x + y - z = 1$, and $4x + 4y - 2z = -1$ intersect at one (unique) point. Find its coordinates.

Solution After the row reduction, we can see that the coefficient matrix became the identity matrix and has rank 3, then the augmented matrix also has rank 3, then the solution is unique. Therefore, all three planes intersect at one point, $x_0 = -\frac{4}{9},\ y_0 = \frac{5}{18},\ z_0 = \frac{1}{6}$ (see also Fig. 1.20, the screenshot of a graphing calculator).

$$\begin{cases} x + y + z = 0 \\ -2x + y - z = 1 \\ 4x + 4y - 2z = -1 \end{cases}$$

$$A^b = \begin{bmatrix} 1 & 1 & 1 & 0 \\ -2 & 1 & -1 & 1 \\ 4 & 4 & -2 & -1 \end{bmatrix} \Rightarrow \begin{bmatrix} 1\ 0\ 0 \vdots & -\frac{4}{9} \\ 0\ 1\ 0 \vdots & \frac{5}{18} \\ 0\ 0\ 1 \vdots & \frac{1}{6} \end{bmatrix} \Rightarrow x = -\frac{4}{9},\ y = \frac{5}{18},\ z = \frac{1}{6}.$$

Answer. All three planes intersect at one point with the following coordinates: $x = -\frac{4}{9},\ y = \frac{5}{18},\ z = \frac{1}{6}$.

Problem 21 (Case 2.) Consider three planes that make the system:

$$\begin{cases} x + y + z = 0 \\ 2x + 2y + 2z = 0 \\ 4x + 4y - 2z = -1. \end{cases}$$

Explain their location with respect to each other.

Solution We can see that this system differs from the system in the previous problem by the second equation which is a multiple of the first equation. Hence, we actually have only two planes that can intersect. This can be seen because the normal vectors of these planes are different. Solving the first and the last equations together and using the matrix approach we obtain the reduced form of the augmented matrix:

$$A^b = \begin{bmatrix} 1 & 1 & 1 & 0 \\ 2 & 2 & 2 & 0 \\ 4 & 4 & -2 & -1 \end{bmatrix} \Rightarrow \begin{bmatrix} 1\,1\,0 & \vdots & -\frac{1}{6} \\ 0\,0\,1 & \vdots & \frac{1}{6} \\ 0\,0\,0 & \vdots & 0 \end{bmatrix} \Rightarrow r(A) = r(A^b) = 2.$$

The simplified system is

$$\begin{cases} x + y = -\frac{1}{6} \\ \quad\;\; z = \frac{1}{6}. \end{cases}$$

Here we have one degree of freedom so one parameter or one independent variable. It cannot be variable z because $z = \frac{1}{6}$. Assume that $y = t,\ t \in R$. From the first equation we obtain that $x = -\ y - \frac{1}{6} = -t - \frac{1}{6}$. The solution to the problem (the line in a parametric form) can be written as

$$\begin{bmatrix} x \\ y \\ z \end{bmatrix} = \begin{bmatrix} -t - \frac{1}{6} \\ t \\ \frac{1}{6} \end{bmatrix}$$

or after factoring parameter t:

$$\begin{bmatrix} x \\ y \\ z \end{bmatrix} = \begin{bmatrix} -t - \frac{1}{6} \\ t \\ \frac{1}{6} \end{bmatrix} = \begin{bmatrix} -\frac{1}{6} \\ 0 \\ \frac{1}{6} \end{bmatrix} + t \cdot \begin{bmatrix} -1 \\ 1 \\ 0 \end{bmatrix}, \quad t = 0, \pm 1, \pm 2, \ldots \tag{1.45}$$

Equation (1.45) is the line in a vector form. Here vector $\overrightarrow{p} = \begin{bmatrix} -\frac{1}{6} \\ 0 \\ \frac{1}{6} \end{bmatrix}$ is the position of a point on the line and vector $\overrightarrow{v} = \begin{bmatrix} -1 \\ 1 \\ 0 \end{bmatrix}$ is its direction.

Students sometimes find it difficult to rewrite a solution obtained in vector form, for example, given by (1.45), in another form. We can rewrite (1.45) as

$$\left[\begin{array}{l} x = -\frac{1}{6} - t \\ y = t \\ z = 6. \end{array}\right. \tag{1.46}$$

By solving the first equation of (1.46) for t we get $t = -x - \frac{1}{6} = \frac{6x+1}{-6}$, then the line of intersection of two planes has another familiar representation:

$$\frac{6x+1}{-6} = \frac{y}{1} = t, \quad z = \frac{1}{6}. \tag{1.47}$$

What kind of line is the line given by formula (1.47)? Because $z = \frac{1}{6}$, this line is in the plane $z = \frac{1}{6}$ that goes parallel to XOY coordinate plane above it by $\frac{1}{6}$ units. (It is shown in Fig. 1.21 in red.) Next, think about the line $y = -x - \frac{1}{6}$ in the plane XOY and move it up by $\frac{1}{6}$ units; this will be the answer in a graphical form. Obviously, $\frac{1}{6}$ units are much less than one unit, so when we sketch two planes, the line of their intersection will look like it lies in the XOY plane (shown as gold in Fig. 1.21).

Problem 22 (Case 3). Find how the planes $x + y + z = 0$, $x + y + z = 10$, and $4x + 4y - 2z = -1$ are located relative to each other.

Solution Just by looking at the equations we can state that the first two planes go parallel to each other (they differ by only the right-hand sides). This will result in no solutions to the system:

$$\begin{cases} x + y + z = 0 \\ x + y + z = 10 \\ 4x + 4y - 2z = -1 \end{cases} \tag{1.48}$$

Geometrically, each of the two parallel planes will intersect with the third plane. The lines of intersection will be parallel as it is stated by Theorem 10 and shown in Fig. 1.22. Note that soon we will use this property in our constructions of cross sections. If we do not have a geometric consideration, we can try to find the ranks of the coefficient and augmented matrices:

Fig. 1.21 The two blue planes intersect along the red line, which is parallel to the XOY plane

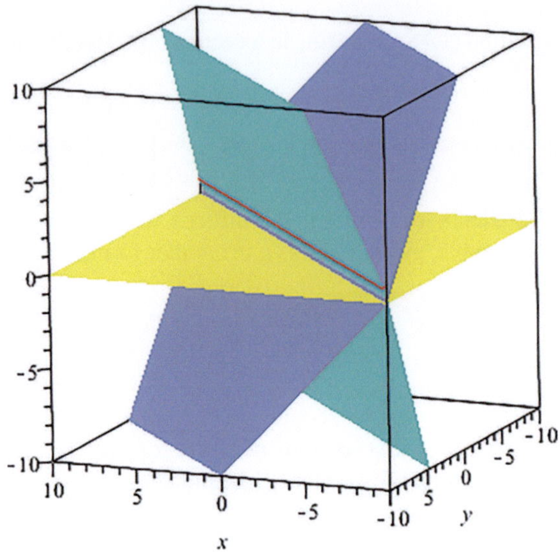

$$A^b = \begin{bmatrix} 1 & 1 & 1 & 0 \\ 1 & 1 & 1 & 10 \\ 4 & 4 & -2 & -1 \end{bmatrix} \Rightarrow \begin{bmatrix} 1 & 1 & 0 & \vdots & 0 \\ 0 & 0 & 1 & \vdots & 0 \\ 0 & 0 & 0 & \vdots & 1 \end{bmatrix} \Rightarrow r(A) = 2 \neq r(A^b) = 3.$$

Algebraically this is Case 3 and therefore the system (1.48) is inconsistent.

Answer. Two parallel planes are intersected by the third plane.

1.3.3 *Other Forms of a Plane in 3D Space*

1. A plane defined by the segments it cuts off on the coordinate axes:

$$\frac{x}{a} + \frac{y}{b} + \frac{z}{c} = 1. \tag{1.49}$$

2. A plane passing through three given points $M(x_1, y_1, z_1)$, $N(x_2, y_2, z_2)$, and $K(x_3, y_3, z_3)$ can be defined as

$$\begin{vmatrix} x - x_1 & y - y_1 & z - z_1 \\ x_2 - x_1 & y_2 - y_1 & z_2 - z_1 \\ x_3 - x_1 & y_3 - y_1 & z_3 - z_1 \end{vmatrix} = 0. \tag{1.50}$$

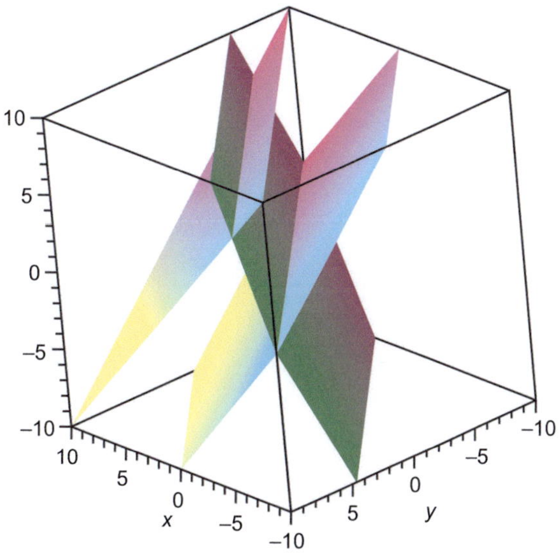

Fig. 1.22 Two parallel planes intersected by a third plane

3. Finding a plane that goes through the origin and contains vectors $w_1 = (x_1, y_1, z_1)$, $w_2 = (x_2, y_2, z_2)$, then its normal vector $n = (a, b, c)$ can be found as vector product:

$$\vec{n} = \vec{w_1} \times \vec{w_2} = \begin{vmatrix} \vec{i} & \vec{j} & \vec{k} \\ x_1 & y_1 & z_1 \\ x_2 & y_2 & z_2 \end{vmatrix} \tag{1.51}$$

and then the plane has the equation $ax + by + cz = 0$, while

$$ax + by + cz = d, \quad d \neq 0 \tag{1.52}$$

is the plane parallel to it.

Remark. Formula (1.51) can be also used when we need to find the equation of a plane that contains three given points, $M(x_{01}, y_{01}, z_{01})$, $N(x_{02}, y_{02}, z_{02})$, and $K(x_{03}, y_{03}, z_{03})$. We can make two vectors:

$$\begin{aligned} \overrightarrow{MN} &= (x_{02} - x_{01}, y_{02} - y_{01}, z_{02} - z_{01}) = \vec{w_1} \\ \overrightarrow{MK} &= (x_{03} - x_{01}, y_{03} - y_{01}, z_{03} - z_{01}) = \vec{w_2} \end{aligned} \tag{1.53}$$

and then apply formula (1.51) again to find the plane passing through the origin and later substitute coordinates of one of the given points into (1.52) to find the value of d and the equation of the plane.

Problem 23 Find the equation of the plane passing through the origin and containing the vectors $w_1 = \vec{i} + 2\vec{j} - \vec{k}$, $w_2 = -2\vec{i} - 5\vec{j}$.

Solution Using (1.51), we have the formula for the normal vector:

$$\vec{n} = \vec{w_1} \times \vec{w_2} = \begin{vmatrix} \vec{i} & \vec{j} & \vec{k} \\ 1 & 2 & -1 \\ -2 & -5 & 0 \end{vmatrix} = -5\vec{i} + 2\vec{j} - \vec{k}.$$

The plane equation is $5x - 2y + z = 0$.

Answer. $5x - 2y + z = 0$.

Problem 24 Rectangular parallelepiped $ABCDA_1B_1C_1D_1$ has edges $AB = a$, $AD = b$, and $AA_1 = c$. Find the angle between planes BB_1D and ABC_1.

Solution This problem can be solved using the vector-coordinate method and formula (1.50). Introduce the coordinate system centered at point A, with the X-axis in direction of vector AB, the Y-axis in direction of vector AD, and Z-axis in direction of AA_1. The vertices of the box have the following coordinates:

$$A(0,0,0),\ B((a,0,0),\ C(a,b,0)\ D(0,b,0)\ B_1(a,0,c),\ C_1(a,b,c),\ D_1(0,b,c).$$

Let us find the equations for both planes using (1.50), then vectors normal to each planes and finally using the scalar product of these two vectors, we will find the angle between two planes.

Let us find the equation of plane BB_1D. Substituting coordinates of points $B(a,0,0) = (x_1, y_1, z_1)$, $B_1(a,0,c) = (x_2, y_2, z_2)$, and $D(0,b,0) = (x_3, y_3, z_3)$ in (1.50), we obtain

$$\begin{vmatrix} (x-a) & y & z \\ 0 & 0 & c \\ -a & b & 0 \end{vmatrix} = 0. \tag{1.54}$$

Evaluating the determinant we get

$$acy + bcx + abc = 0.$$

Dividing both sides by c, we can rewrite it as the equation for plane BB_1D:

$$bx + ay + 0 \cdot z = -ab.$$

Hence, the vector, normal to this plane, can be written as

$$\vec{n}_1 = (b, a, 0). \tag{1.55}$$

Next, we will find the equation of the second plane, ABC_1. Substituting coordinates of points $A(0, 0, 0) = (x_1, y_1, z_1)$, $B(a, 0, 0) = (x_2, y_2, z_2)$, and $C_1(a, b, c) = (x_3, y_3, z_3)$ in (1.50), we obtain

$$\begin{vmatrix} x & y & z \\ a & 0 & 0 \\ a & b & c \end{vmatrix} = 0. \tag{1.56}$$

After evaluating determinant (1.56) we get

$$abz - yac = 0.$$

This can be rewritten as the equation for plane ABC_1:

$$0 \cdot x + cy - bz = 0.$$

Its normal vector is

$$\vec{n}_2 = (0, c, -b). \tag{1.57}$$

The angle φ between two normal vectors (1.55) and (1.57) is the required angle between two planes. It can be found using the scalar product formula (1.22):

$$\cos\varphi = \frac{ac}{\sqrt{a^2 + b^2}\sqrt{b^2 + c^2}}. \tag{1.58}$$

Answer. $\varphi = \arccos \frac{ac}{\sqrt{a^2+b^2}\sqrt{b^2+c^2}}$

Remark. It is interesting that if the parallelepiped is a cube with side a, then for any value of a, formula (1.58) gives the same value of cosine, $\cos\varphi = 1/2$. Then for any cube the angle between planes ABC_1 and BB_1D will be always $\varphi = 60°$.

Problem 25 Find the condition on parameters $b_i,\ i = 1, 2, 3$, for which the system.

$$\begin{cases} x + 2y + 3z = b_1 \\ 4x + 5y + 6z = b_2 \\ 7x + 8y + 9z = b_3 \end{cases}$$

is consistent.

Solution The system is consistent if the ranks of the coefficient and augmented matrices are equal. Since the coefficient matrix is entirely included in the augmented matrix, we will use the Gaussian elimination row reduction strategy:

$$\begin{bmatrix} 1\,2\,3 & \vdots & b_1 \\ 4\,5\,6 & \vdots & b_2 \\ 7\,8\,9 & \vdots & b_3 \end{bmatrix} \Rightarrow \begin{bmatrix} 1 & 2 & 3 & \vdots & b_1 \\ 0 & -3 & -6 & \vdots & -4b_1 + b_2 \\ 0 & -6 & -12 & \vdots & -7b_1 + b_3 \end{bmatrix} \Rightarrow \begin{bmatrix} 1 & 2 & 3 & \vdots & b_1 \\ 0 & -3 & -6 & \vdots & -4b_1 + b_2 \\ 0 & 0 & 0 & \vdots & (b_1 - 2b_2 + b_3) \end{bmatrix}.$$

After the row reduction we see that the rank of the coefficient matrix equals 2. In order for the system to be consistent, the rank of the augmented matrix must also be 2. This depends only on the very last entry of the matrix, $(b_1 - 2b_2 + b_3)$. If it equals zero, then the system is consistent. Therefore the necessary condition is

$$b_1 - 2b_2 + b_3 = 0. \tag{1.59}$$

Formula (1.59) determines the locus of all the points given by a triple (b_1, b_2, b_3) that belong to the plane $x - 2y + z = 0$. Otherwise, the given system is inconsistent.

Answer. $b_1 - 2b_2 + b_3 = 0$.

Remark. Sometimes it is difficult to understand what exactly we got by solving a particular problem. Let's look at some numerical examples and find at least one triple that satisfies (1.59). For example, the triple $(1, 1, 1)$ is fine. It means that the system

$$\begin{cases} x + 2y + 3z = 1 \\ 4x + 5y + 6z = 1 \\ 7x + 8y + 9z = 1 \end{cases}$$

is consistent and that $r(A) = r(A^b) = 2$. After Gaussian elimination, the augmented matrix has the form:

$$A^b = \begin{bmatrix} 1\;0\;-1 & \vdots & -1 \\ 0\;1\;\;\;2 & \vdots & 1 \\ 0\;0\;\;\;0 & \vdots & 0 \end{bmatrix}.$$

We now understand that the system represents two intersecting planes. Originally, we had three variables. Subtracting the number, 2, of the independent rows, we obtain one degree of freedom—the two planes intersect in a line. Working backward we can obtain the system of two linear equations:

$$\begin{cases} x - z = -1 \\ y + 2z = 1 \end{cases} \Rightarrow$$

$$x = t - 1,\;\; y = -2t + 1,\;\; z = t \in R. \tag{1.60}$$

This line can be also written in the following form:

$$\begin{bmatrix} x \\ y \\ z \end{bmatrix} = \begin{bmatrix} -1 + t \\ 1 + 2t \\ t \end{bmatrix} = \begin{bmatrix} -1 \\ 1 \\ 0 \end{bmatrix} + t \cdot \begin{bmatrix} 1 \\ 2 \\ 1 \end{bmatrix} = p + t \cdot v.$$

Expressing parameter t from each coordinate in (1.60) we obtain the canonical form of the same line: $t = x + 1,\;\; t = \frac{y-1}{-2},\;\; t = z$ or

$$t = \boxed{x + 1 = \frac{y - 1}{-2} = z}.$$

Problem 26 Find the volume of a solid figure concluded between planes $\alpha:\;\; x + y + z = 1$ and $\beta:\;\; x + 2y + 3z = 6$ and the three planes of the coordinate system.

Solution The first plane intersects all the axes at values of 1. The second plane intersects the X-axis at 6, the Y-axis at 3, and Z-axis at 2. The volume of the figure obtained is the difference of the volumes of two pyramids shown in Fig. 1.23.

$$V = \frac{1}{3} \cdot \frac{1}{2}(2 \cdot 3 \cdot 6 - 1 \cdot 1 \cdot 1) = \frac{35}{6}.$$

Answer. $V = 35/6$.

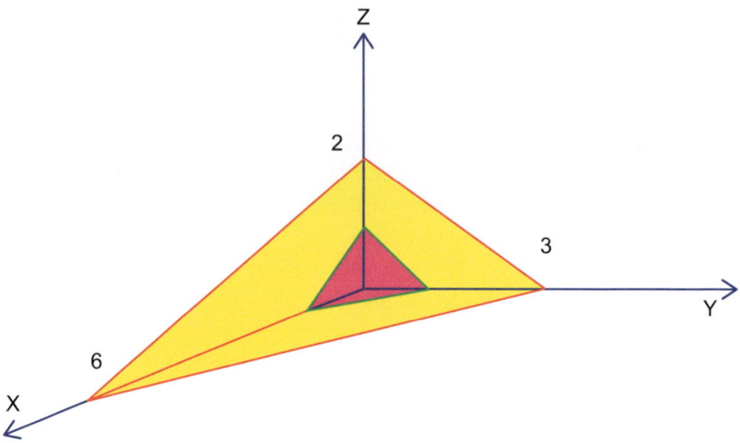

Fig. 1.23 A solid body enclosed between two planes and a coordinate system

1.4 Planes and Lines in Space

Definition 1 A **line is parallel to plane** if they do not intersect.

Definition 2 A **line is perpendicular to plane** if it is perpendicular to any line in the plane.

Obviously, that if a line is perpendicular to a given plane, then it is perpendicular to any line drawn in the plane through the point of intersection.

Definition 3 If a line that intersects a plane, but is not perpendicular to it, then it is called an **oblique** or **slant** line.

Thus, a line can belong to a plane (if it has with the plane at least two different points), intersect a plane, be perpendicular to a plane, or be parallel to a plane (do not intersect that plane). These can be summarized by the diagram given in Fig. 1.24.

Example 1. Any line drawn on one wall of a room and the plane of the opposite wall are parallel. Any line on the wall of a room, except for lines parallel to the line of intersection of two adjacent walls of the room, will intersect the plane of the adjacent wall of the room.

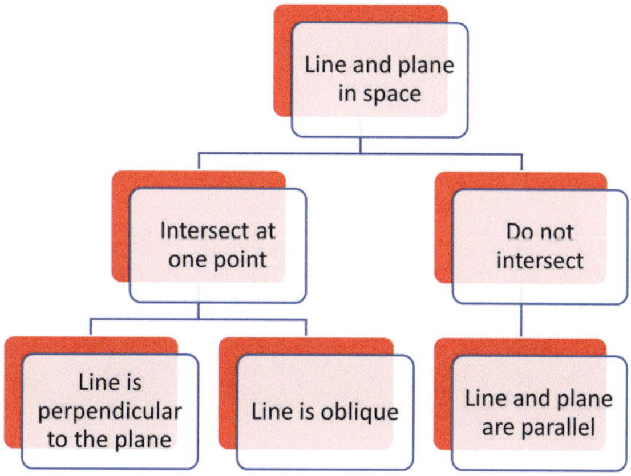

Fig. 1.24 Line and plane in space

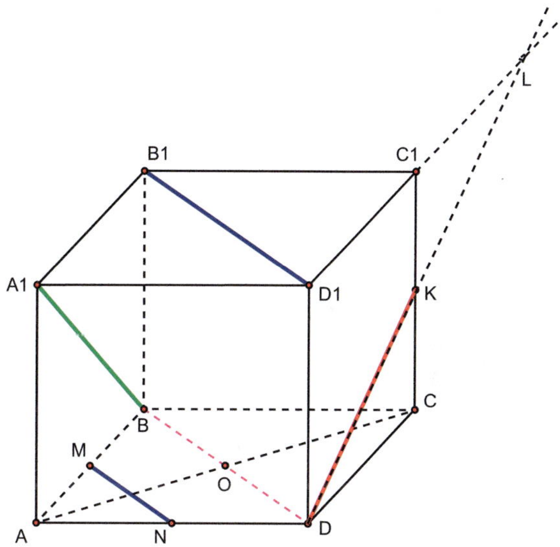

Fig. 1.25 Lines and planes in a cube

Example 2. Line A_1B is parallel to the cube face DD_1C_1C or line B_1D_1 is parallel to the face $ABCD$. The straight line DK intersects the adjacent face $A_1B_1C_1D_1$ at point L (see Fig. 1.25); this point is common to the planes $A_1B_1C_1D_1$ and DD_1C_1C. Also $L = D_1C_1 \cap DK$.

- The projection of a point onto a plane is the base of the perpendicular dropped from a given point to a given plane.
- The projection of an oblique onto a plane is a straight line consisting of the projections of all points of an oblique onto a given plane.
- The angle between an inclined (oblique) line and a plane is the angle between the inclined line and its projection.

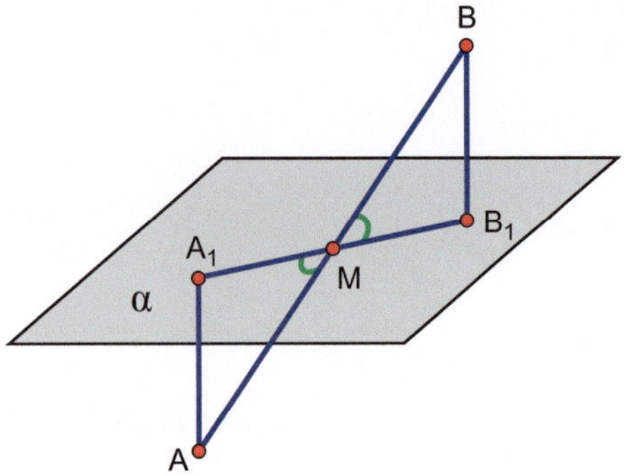

Fig. 1.26 Distances to the plane from the ends of the segment

Theorem 13 *Two different perpendiculars to the same plane are parallel.*

Lemma 1 *The distance from each of the two points located on opposite sides of a plane to this plane is the same if and only if the plane passes through the midpoint of the segment connecting these two points.*

Proof To prove this statement, we need to show that if the points are on different sides of the plane, then the distance from each of them to the plane will be the same if the plane passes through the middle of the segment connecting the points. Let's use the sketch in Fig. 1.26.

Assume that the line passing through points A and B intersects plane α at point M and makes angle φ with the plane α and that M is the midpoint of segment AB. Let us drop perpendiculars AA_1 and BB_1 from points A and B to the plane α, respectively. It follows from Theorem 13 that $AA_1 \| BB_1$. Next, we can state that points A, A_1, M, B_1 and B belong to the same plane perpendicular to plane α. From the right triangles $\triangle AA_1M$ and $\triangle BB_1M$, then $AA_1 = BB_1 = \frac{AB}{2} \cdot \sin\varphi$ which proves the lemma. □

Let us solve some problems.

Problem 27 Find the locus of all points equidistant from the points $A(1, 1, 1)$ and $B(3, -1, -3)$. Find how this plane intersects the coordinate axes.

Solution Any point of the plane that passes through the midpoint of segment AB and perpendicular to this segment will be equidistant from both of the given points. First, we can find the coordinates of the midpoint M of segment AB:

$$M = \left(\frac{1+3}{2}, \frac{1-1}{2}, \frac{1-3}{2}\right) = (2, 0, -1).$$

Second, the plane must be orthogonal to segment AB and so $\overrightarrow{AB}$ is its normal vector:

$$\vec{n} = \overrightarrow{AB} = (3-1, -1-1, -3-1) = (2, -2, -4).$$

Third, the equation of the plane perpendicular to AB can be written as $2x - 2y - 4z = d$. Since this plane must go through M, parameter d is found by substituting the coordinates of M into the equation of the plane, $2 \cdot 2 - 2 \cdot (-2) - 4 \cdot (-4) = d, \ \ d = 8$. So the equation of the plane is $x - y - 2z = 4$. Finally, we want to find how this plane intersect the coordinate axes. This means that two other coordinates are equal to zero, so it intersects the X-axis at $P = (4, 0, 0)$, the Y-axis at $Q = (0, -4, 0)$, and the Z-axis at $R = (0, 0, -2)$.

Answer. $x - y - 2z - 4 = 0$. $(4, 0, 0)$, $(0, -4, 0)$, and $(0, 0, -2)$.

Problem 28 Given three non-collinear points $A, \ B$, and C, find the geometric position (locus) of all points equidistant from each of the given points.

Solution Recalling lessons from plane geometry, we can find the center O of the circumscribed circle of the triangle ABC. This point is located at the distance of the radius of the circumscribed circle from each of the three points. Is it possible to find other points in space? You can draw a straight line through point O perpendicular to the plane of the triangle and then taking any point of this straight line (it will be like the top of a pyramid) the distance from it to each vertex of triangle ABC will be the same. Such a straight line is the locus of all points equidistant from the vertices of the triangle. Let's take some point M on this straight line, then $MA = MB = MC$. Any other point on this straight line will be also equidistant from the vertices of $\triangle ABC$.

Answer. The locus is the line passing the circumcenter of $\triangle ABC$ and perpendicular to plane ABC.

Problem 29 Draw a plane such that the distance to this plane from each of the three given points A, B, and C is the same.

Solution

Case 1. If the three given points are collinear, then any plane parallel to the line containing three points will satisfy the condition.

Case 2. If points A, B, and C are not on the same line, then they belong to a plane of triangle ABC. Let points D, E, and F be the midpoints of segments AB, BC, and AC, respectively. By connecting these points we have the midline segments: DE, EF, and DF. A plane passing a midline segment of $\triangle ABC$ and perpendicular to plane ABC is equidistant from the points. There are three such planes.

Question. Do you think there are other planes that meet the requirements for this problem? Should an equidistant plane passing the midline segment of a triangle always be perpendicular to the plane of a triangle?

By solving the next problem, you will show that through each middle segment of a triangle, an infinite number of planes equidistant from each vertex can be constructed. Also, that these planes may not be orthogonal to the plane of the triangle.

Problem 30 Prove that any plane passing through the midline segment of a triangle is equidistant from all three vertices of the triangle.

Proof Sketch triangle ABC and draw the middle segment EF such that $EF \parallel BC$ and $AE = EB, \quad AF = FC$ (Fig. 1.27). Draw a plane passing through this midline segment and denote it as α. The plane α intersects the plane of triangle ABC along a line containing the midline segment EF. Now construct perpendiculars from points $A, \ B$, and C to this plane and prove that the lengths of these perpendiculars are equal:

$$\left.\begin{array}{l} AB \cap \alpha = E \\ AA_1 \perp \alpha \\ BB_1 \perp \alpha \\ AE = EB \end{array}\right\} \Rightarrow \Delta AA_1E \cong BB_1E \Rightarrow AA_1 = BB_1.$$

Similarly,

$$\left.\begin{array}{l} AC \cap \alpha = F \\ AA_1 \perp \alpha \\ CC_1 \perp \alpha \\ AF = FC \end{array}\right\} \Rightarrow \Delta AA_1F \cong CC_1F \Rightarrow AA_1 = CC_1.$$

Please note that the triangles AA_1E and AA_1F are not equal, but they have a common side AA_1. Therefore $AA_1 = BB_1 = CC_1$. □

The following problem was offered at one Hungarian Math Olympiad.

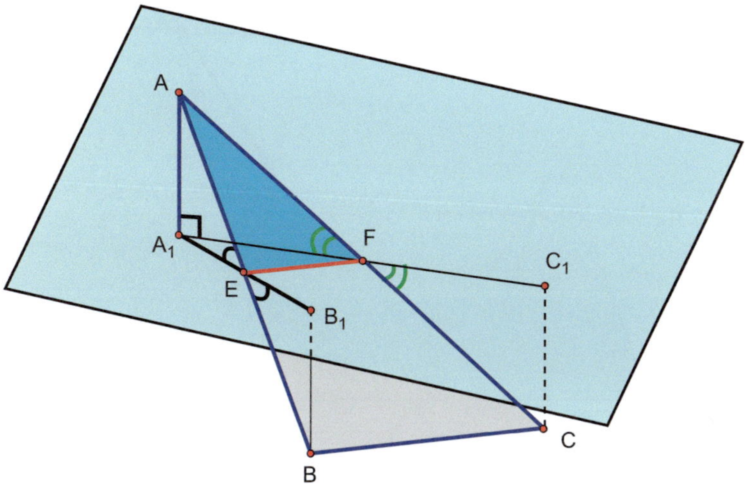

Fig. 1.27 Distances to the plane from the vertices of triangle ABC

Problem 31 Given four distinct points A, B, C, and D, draw a plane so that points A and C are on one side of the plane and points B and D are on the other so that the distance from each of the four points to this plane is the same.

Solution
Case 1. All four points are non-coplanar.

Method 1. Let us place the four points in our sketch shown in Fig. 1.28. Connect points A and B and find the midpoint E of segment AB and also connect points A and D and mark the midpoint of segment AD by F. In triangle ABD, $EF \| BD$, $EF = \frac{1}{2}BD$. Connect points B and C with midpoint H and also points D and C and with midpoint G. Triangles $\triangle CBD$ and $\triangle ABD$ have a common side and so we can state that

$$EF \| HG \| BD.$$

Next, by connecting points E and H and also points F and G, we obtain the parallelogram $EFGH$ of plane, say α. Points A and C lie above α with B and D below. Applying Lemma 1, we can state that α is equidistant from each of four points (A , B, C, and D.)

Method 2. In general, if four points are not in the same plane, then we can connect them and make a tetrahedron like two different cases shown in Fig. 1.29. It follows from Lemma 1 that, if the plane passes through the midpoint of the segment connecting two points on opposite sides of the plane, then the distance from each point to the plane will be the same. Let M be the midpoint of AB, N be the midpoint of BC,

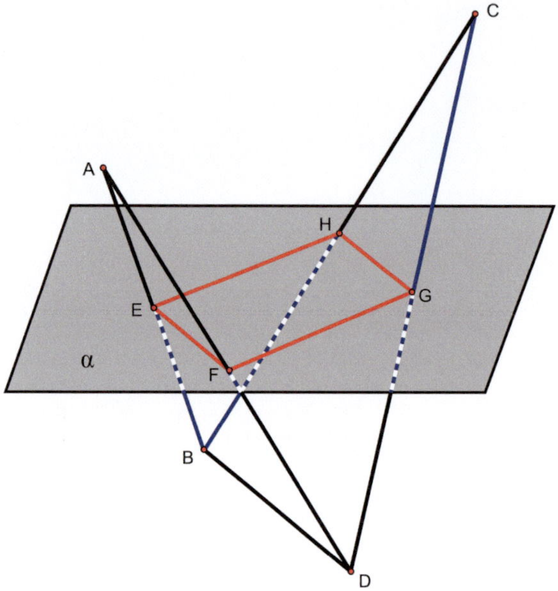

Fig. 1.28 Analysis: construction of an equidistant plane

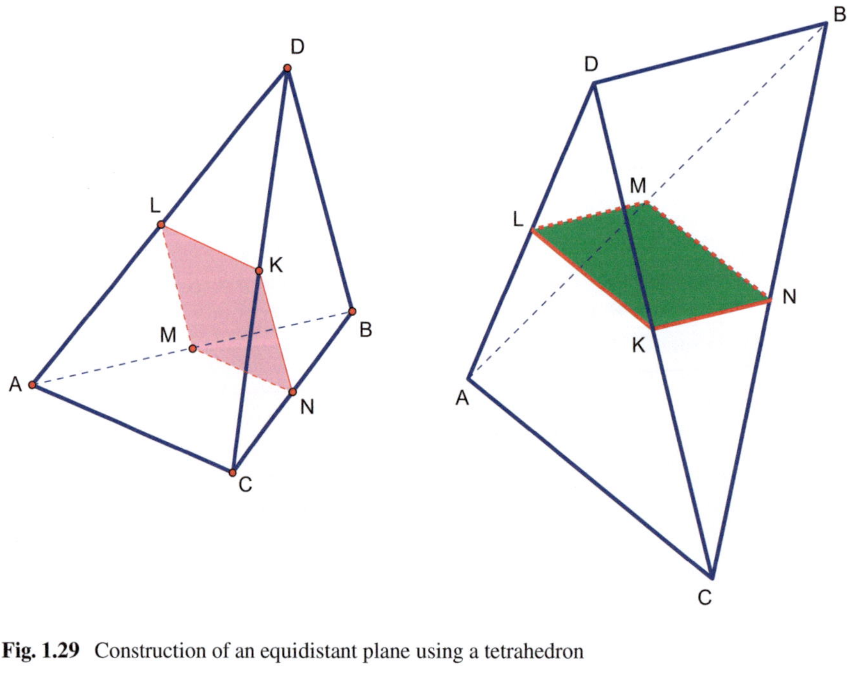

Fig. 1.29 Construction of an equidistant plane using a tetrahedron

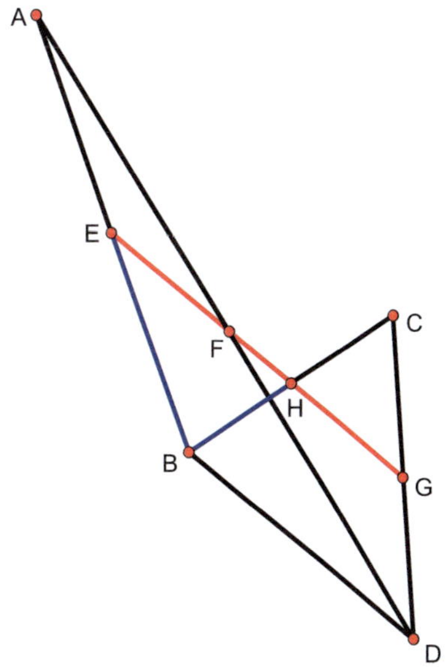

Fig. 1.30 The midpoints E, F, H, G belong to one line

and K—the midpoint of DC. Connecting points M and N and then N and K we obtain the segments MN and NK, the traces of the plane on faces ABC and CBD, respectively. It is clear that $MN \| AC$ and that $KN \| DB$. Next, we can draw a segment $KL \| MN$ and connect points L and M obtaining $LM \| KN$. The parallelogram $MNKL$ is the cross section of tetrahedron $ABCD$ by plane α.

Let us prove that plane α is equidistant from A , B, C, and D. The bases of the perpendiculars dropped from the vertices A, B, C, and D to the α plane will be denoted by the points A_1, B_1, C_1, and D_1, respectively. However, we will not show this in the drawing so as not to clutter up the sketch. It follows from Lemma 1 that $AA_1 = BB_1$ (since $AM = MB$), also $BB_1 = CC_1$ (since $CN = NB$), and we can state that $AA_1 = BB_1 = CC_1$. However, $CC_1 = DD_1$ (since K is the midpoint of DC). Therefore, $AA_1 = BB_1 = CC_1 = DD_1$, which proves that the requested plane α has been constructed.

Case 2. Assume that we are given four points A, B, C, and D. Suppose that we connect A and B and found the midpoint of AB as E. Then we connect A and D and marked the midpoint of segment AD as F. Next, we connect B and C and denote H as the midpoint of segment BC. Assume that points E, F, and H appear on the same line. Let us prove that this case is possible only if the four given points are coplanar.

If we connect points E and F, then the line EF belongs to the plane of triangle ABD (Fig. 1.30). Denote this plane by α, so $EF \in \alpha$. On the other hand, point H is the midpoint of BC, and $H \in (EF)$, so $(BC) \cap (EF) = H$. Hence, $H \in \alpha$ and since $C \in (BH)$, then $C \in \alpha$. Therefore, the solution in this case is possible only if all four points belong to the same plane α. Then also G will be the midpoint of segment CD. If the line EH intersects the segment CD at its midpoint G, any plane that goes through four points $E,\ F,\ H$, and G will be a solution. Otherwise, the problem does not have a solution for four points belonging to the same plane.

Problem 32 How many planes are there equidistant from four given points in space if the points do not lie in the same plane?

Solution To draw a plane that is at the same distance from each of the four given points, let's think about how a plane can be located in relation to four non-coplanar points. Because any plane bisects space, one possibility is for one point to be above the plane (three points below) and the other is that two points are above the plane (two below). Recall the important statement that we proved as Lemma 1: The points A and B are on opposite sides of the plane at equal distances from it if and only if this plane intersects segment AB at its midpoint. Consider the cases:

Case 1. One of the points lies in a different part of space from the others. This results in four planes (since instead of point A you can take any of the points $B,\ C$, or D, which gives three more planes, i.e., four in total). Let's consider a situation where point A lies on the other side of plane β formed by points $B,\ C$, and D. Find the midpoints of the segments AB, AC, and AD. The plane passing through the midpoints $E,\ F,\ G$ of these segments will be equidistant from all four given points $A,\ B,\ C$, and D (Fig. 1.31). Because any three distinct points lie on a plane, then there are three other similar constructions possible. The equidistant planes can go between point B and plane ADC, between point C and plane ABD, or between point D and plane ABC.

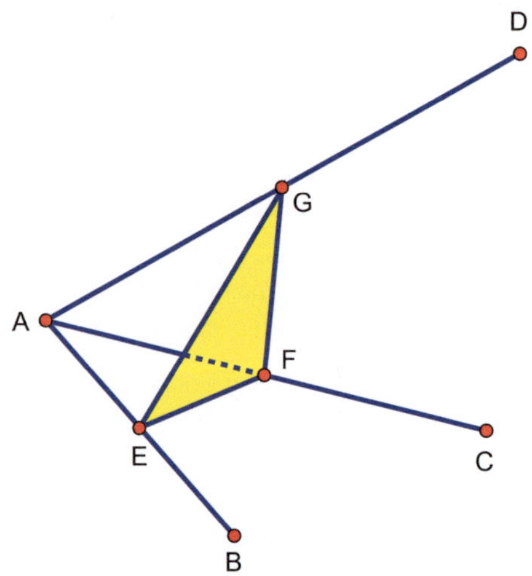

Fig. 1.31 The midpoints $E,\ F,\ G$ form a plane equidistant from the given four points

Case 2. The points are located in pairs on opposite sides of the plane. Let, for example, in one half-space with point A there be point B and in the other spatial region—points C and D. The plane equidistant from points A, B, C, and D must pass through points K, L, and M, the midpoints of segments AC, AD, and BD, respectively. Points K, L, and M do not lie on the same line, otherwise, all points A, B, C, and D would be in the plane containing this line and point A (please explain why) which contradicts the condition. Therefore, the plane containing points K, L, and M is unique and the distances from it to points A, B, C, and D are equal by construction.

Note that point C or D could also lie in the same half-space with point A, which gives two more planes. In fact, the case where points A and C are above the plane and B and D are below was considered in the previous problem and illustrated in Fig. 1.28. There are a total of three such planes. Between the two cases, there are altogether seven planes in space that can be drawn equidistant from four given points.

Answer. Seven possible planes can be constructed.

Theorem 14 *Two planes perpendicular to the same line are parallel.*

Theorem 15 *A perpendicular line to one of two mutually perpendicular planes passing through the line of intersection of the planes lies in the second plane.*

Corollary 1 *If each of two non-parallel planes is perpendicular to a third plane, then the line of intersection of the two planes is perpendicular to this third plane.*

Theorem 16 *If one of the two parallel lines is perpendicular to a plane, then the other line is also perpendicular to that plane.*

Theorem 17 (Parallelism of a straight line and a plane) *If a straight line that does not lie in a given plane is parallel to any straight line from this plane, then it is parallel to the plane.*

Theorem 18 (Perpendicularity of a straight line to a plane) *If a straight line is perpendicular to two intersecting straight lines of a plane, then it is perpendicular to the plane.*

Proof Let line a be perpendicular to the intersecting lines p and q located in plane α (Fig. 1.32). We will prove that $a \perp \alpha$. Assume m is any line in plane α. Let us show that $a \perp m$. Create on each of these lines the corresponding vectors $\overrightarrow{a}$, $\overrightarrow{p}$, $\overrightarrow{q}$, and $\overrightarrow{m}$. Since the lines a and p are perpendicular, then the following scalar product is zero:

$$\overrightarrow{a} \cdot \overrightarrow{p} = 0.$$

Similarly, we have that

$$\overrightarrow{a} \cdot \overrightarrow{q} = 0.$$

Vector $\overrightarrow{m}$ is coplanar on α with vectors $\overrightarrow{p}$ and $\overrightarrow{q}$. Because the lines p and q intersect, vectors $\overrightarrow{p}$ and $\overrightarrow{q}$ are non-collinear (linearly independent).

Next, we can rewrite vector $\overrightarrow{m}$ as linear combination of vectors $\overrightarrow{p}$ and $\overrightarrow{q}$ as

$$\overrightarrow{m} = x\overrightarrow{p} + y\overrightarrow{q}, \ x, y, \in R.$$

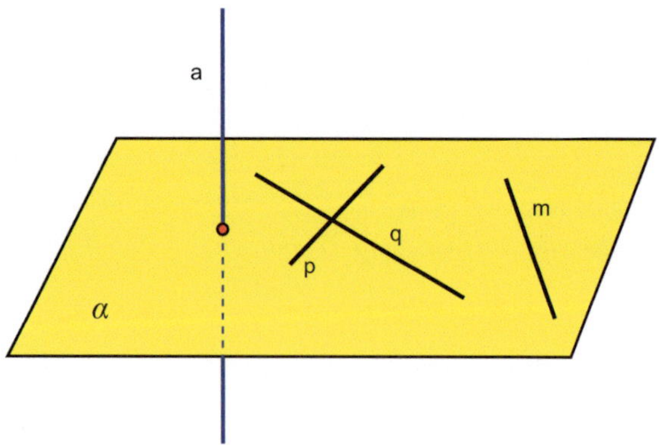

Fig. 1.32 Line is perpendicular to a plane

Let us now consider the scalar product of vectors $\overrightarrow{a}$ and $\overrightarrow{m}$ and simplify it using distributive law:

$$\overrightarrow{a} \cdot \overrightarrow{m} = \overrightarrow{a}(x\overrightarrow{p} + y\overrightarrow{q}) = x(\overrightarrow{a} \cdot \overrightarrow{p}) + y(\overrightarrow{a} \cdot \overrightarrow{q}) = 0.$$

Hence, the vectors $\overrightarrow{a}$ and $\overrightarrow{m}$ are orthogonal, and then line $a \perp m$. Therefore line a is perpendicular to any line in plane α and so it is perpendicular to this plane α. □

Remark. Before you start solving problems, pay attention to the importance of Theorem 18 about the perpendicularity of a line and a plane: A line is perpendicular to a plane if it is perpendicular to two intersecting lines of this plane. Therefore, the requirement that lines lying in a plane and a line perpendicular to these lines pass through one common point is superfluous.

Problem 33 Given a tetrahedron $DABC$ with vertex D such that $\angle ADC = \angle ADB$ and $DB = DC$. If point M is the midpoint of edge BC, prove that line segment BC is perpendicular to plane DAM and that plane DAM is the perpendicular bisector of BC.

Proof 1. In $\triangle\ DBC$ we have $DC = DB$ and M is the midpoint of edge BC, so triangle BDC is an isosceles and $DM \perp BC$.
2. Since triangles DAB and DAC have common side AD, then $\triangle\ DAB = \triangle\ DAC$ by two sides and the angle between them, and then $AB = AC$.
3. Next, triangle ABC is isosceles, then $AM \perp BC$.
4. $DM \cap AM = M$ and define one unique plane $\alpha = DMA$. If BC is perpendicular to two intersecting lines in plane α, then by Theorem 18 we can state that $BC \perp \alpha$.
5. Let β be the perpendicular bisector of segment BC. Then any point in plane β is equidistant from points B and C. Points $D,\ A,\ M$ are all equidistant from B and C. Therefore $\beta = \alpha$. □

Problem 34 Pyramid $ABCDE$ has a rectangle with sides $AB = 3$ and $AD = 4$ as its base. The edge EA is perpendicular to the base of the pyramid, $EA = 2$. A plane α passes the midpoint K of edge EC and is perpendicular to edge EC. Assume that the plane intersects sides AB and AD at points N and M, respectively. Find the length of MN.

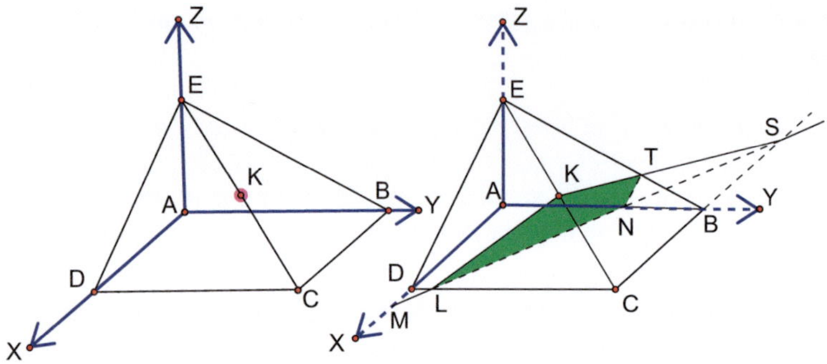

Fig. 1.33 The plane intersects the given pyramid perpendicular to its lateral edge

Solution
(1) Because edge $EA \perp ABCD$, then it is convenient to introduce a rectangular coordinate system with the center at point A as it is shown in Fig. 1.33. In this system, vertices of the pyramid and the given midpoint will have coordinates:

$$E(0,0,2),\ \ C(3,4,0),\ \ B(3,0,0),\ \ D(0,4,0),\ K(1.5,2,1),$$

where the coordinates of point K were found using the coordinates of the ends of the segment, points E and C.
(2) Plane $\alpha \perp EC$, then vector $\overrightarrow{EC} = (3, 4, -2)$ is $\overrightarrow{n}$, the normal vector of α. Hence, plane α has the following equation:

$$\alpha:\ 3x + 4y - 2z = d,$$

where d can be easily found since we know that α goes through point K. Substituting coordinates of points K into the equation of the plane above, we obtain $3 \cdot 1.5 + 4 \cdot 2 - 2 \cdot 1 = 10.5 = d$. The equation of plane α is $3x + 4y - 2z = 10.5$ or $6x + 8y - 4z = 21$. Since point $M \in AD$, we can get its coordinate by replacing $y = z = 0$, so $x = \frac{21}{6} = \frac{7}{2}$. At this point, M plane α cuts the X-axis. Replacing $x = z = 0$, we obtain coordinates of point N, $y = \frac{21}{8}$. At this point, the plane cuts the Y-axis. By the Pythagorean theorem applied to $\triangle MAN$, we have the distance MN:

$$MN = \sqrt{\frac{49 \cdot 16 + 441}{64}} = \frac{\sqrt{1225}}{8} = \frac{35}{8}.$$

Answer. $MN = 35/8$.

Remark. We solved this problem without constructing the cross section of the pyramid with a plane passing through the point K and perpendicular to the edge EC.

A sketch of the cross section inside the pyramid is shown on the right in Fig. 1.33. However, we did not exactly follow the rules of construction, which will be explained in Chap. 3. Although the cross section is correct, we drew it by approximately placing the points M and N on X- and Y-axes, by knowing their coordinates. We then connected points M and N inside the base of the pyramid and marked point L. This gives us two sides of the cross-sectional polygon: sides LK and LN. Then we got a point S belonging to two adjacent planes $ABCD$ and EBC and the perpendicular plane. By connecting the points K and S, we got the point T on the edge EB and two other sides of the quadrilateral $LKTN$.

However, we can construct this section without "cheating." This will be explained in the following chapters. Note that the shape of this cross section was not required, since the answer to the questions (the distance between two points) could be found without knowing how to construct the cross section.

Theorem 19 (Property of perpendicularity of a line to a plane) *If a line is perpendicular to a plane, then it is perpendicular to all lines from this plane that pass through the point of intersection of the line and plane.*

Proof Let the line a be perpendicular to the lines b and c lying in plane α and intersecting at point A (Fig. 1.34). Assume that line a passes through the point A. Let us prove that a is perpendicular to plane α.

Draw an arbitrary line p through the point A in plane α and show that p is perpendicular to line a. Draw another line (i.e., CB) in α that does not pass through point A and intersects lines $b,\ c,\ p$ at points $B,\ C,\ P$, respectively. On line a from point A in different directions on α take equal segments AA_1 and AA_2. Triangle A_1CA_2 is isosceles, since segment AC is both, its height (by the conditions of the theorem) and its median (by construction). Similarly, we get that the triangle A_1BA_2

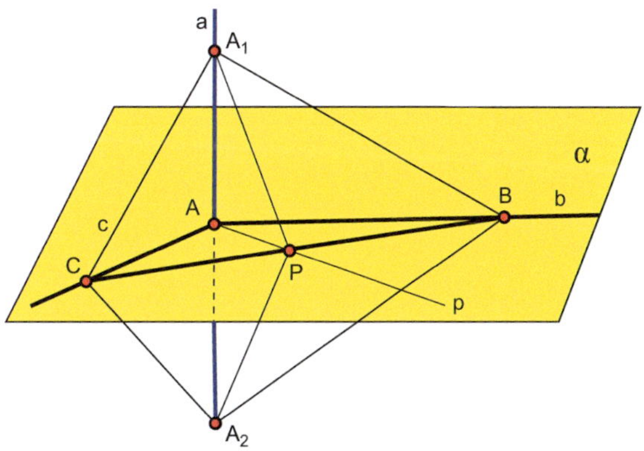

Fig. 1.34 Proof that the line a is perpendicular to the plane α

is also isosceles. Triangle A_1BC is equal to triangle A_2BC (by three sides), therefore the angles A_1BP and A_2BP are equal. Triangles A_1BP and A_2BP are equal (by two sides and the angle between them). Hence, $A_1P = A_2P$ and so triangle A_1PA_2 is isosceles, so its median is also its altitude ($AP \perp a$), i.e., p is perpendicular to line a. By the condition, line a is perpendicular to the plane α which proves the theorem. □

Theorem 20 (On three perpendiculars) *A line in the plane is perpendicular to a slant (oblique) line if and only if it is perpendicular to the projection of this slant line onto the same plane. The line drawn in the plane through the foot of the slant and perpendicular to its projection is perpendicular to the slant itself.*

Let us use Fig. 1.35 and rephrase Theorem 20 as "If a line (i.e., (DE)) in a plane α drawn through the foot of the slant line C is perpendicular to its orthogonal projection BC on the plane ($BC \perp DE$), then it is perpendicular to the slant AC." The converse of the statement is also true, "If a line in the plane is perpendicular to the slant, then it is perpendicular to the projection of the slant onto the plane."

Remark. Theorem 20 can be reformulated as "If the projection of the oblique line onto a plane is perpendicular to the line lying in this plane, then the oblique itself is perpendicular to this line." Therefore, one should not require that the oblique line referred to in the three perpendicular theorem and the straight line lying in the plane pass through a common point.

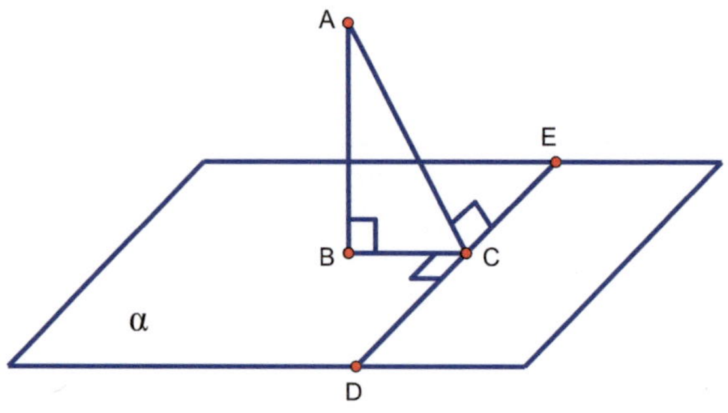

Fig. 1.35 Theorem on three perpendiculars

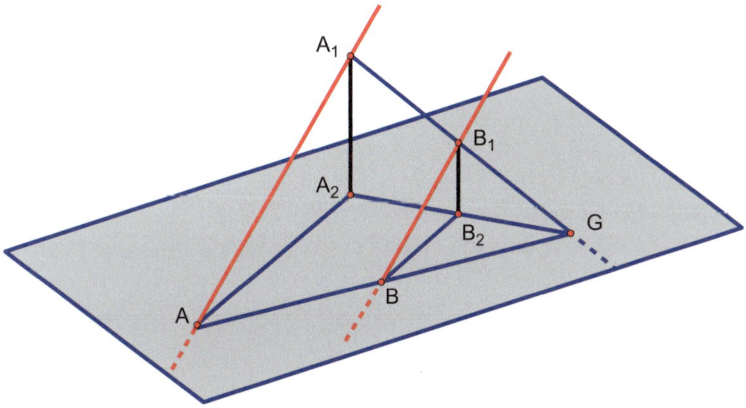

Fig. 1.36 Two parallel lines intersect a plane

Problem 35 Line a is parallel to line b. Line a intersects plane α at point A and line b at point B. From point $A_1 \in a$ and point $B_1 \in b$, such that $A_1B_1 \nparallel AB$, the corresponding perpendiculars A_1A_2 and B_1B_2 are dropped to plane α. Prove that

$$\frac{AA_1}{BB_1} = \frac{A_1A_2}{B_1B_2}.$$

Proof Sketch a plane and two parallel lines that intersect the plane at points A and B (Fig. 1.36). Next, mark the points A_1 and B_1 on line a and b, respectively. In our sketch, it is important to locate points A_1 and B_1 so that the mentally connected points A_1 with B_1 and A with B would not form parallel lines, but instead intersect somewhere in the plane of our sketch. Any two parallel lines always belong to one plane, say plane β. The condition of the problem can be written as

$$\left.\begin{array}{l} a \cap \alpha = A \\ b \cap \alpha = B \\ a \parallel b \end{array}\right\} \Rightarrow \alpha \cap \beta = (AB).$$

Next, let us drop perpendiculars from points A_1 and B_1 onto plane α and denote the feet of the perpendiculars regarding the condition of the problem by A_2 and B_2. We know that two lines perpendicular to the same plane are parallel:

$$\left.\begin{array}{l} A_1A_2 \perp \alpha \\ B_1B_2 \perp \alpha \end{array}\right\} \Rightarrow A_1A_2 \parallel B_1B_2.$$

Denoting the plane formed by parallel lines B_1B_2 and A_1A_2 by γ, $\gamma \perp \alpha$. Since $AA_1 \neq BB_1$, A_2B_2 is not parallel to AB. (We have already planned this in our sketch.) So,

$$AB \cap A_2B_2 = G \in \alpha.$$

Because $A_1A_2 \perp \alpha$ and $B_1B_2 \perp \alpha$, each is perpendicular to any line in the plane, i.e., they are perpendicular to AA_2 and to BB_2. Therefore, $\triangle AA_1A_2 \approx \triangle BB_1B_2$ (similar right triangles) and hence the ratio of their corresponding sides is the same,

$$\frac{AA_1}{BB_1} = \frac{A_1A_2}{B_1B_2}.$$

□

1.5 Intersecting, Perpendicular, Parallel, and Skew Lines

Definition 4 Two lines in space are called **parallel** if they lie in one plane and do not intersect.

Definition 5 Two lines that do not intersect and do not belong to one plane are called **skew** lines.

In space, two lines can be in the following positions:

1. They intersect. (They lie in the same plane. In other words, there will be a plane that will be located in space so that both of these lines will lie inside this plane = the plane will pass through both lines.)
2. They are parallel. (There is also a plane that will pass through both of these lines.)
3. They are skew lines, i.e., do not intersect and not parallel. (There will no longer be a plane in which both lines will be located. In another way, the skew lines do not lie in the same plane.) Not intersecting and not parallel lines are skew lines.

All the cases can be seen in Fig. 1.37.

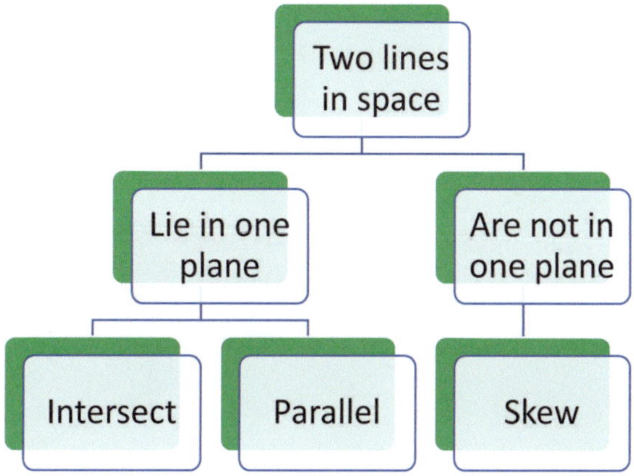

Fig. 1.37 Two lines in space

Theorem 21 *If one of two parallel lines is perpendicular to a third line, then the other line is also perpendicular to this line.*

Theorem 22 *If two lines are parallel to a third, then they are parallel to each other.*

Feature of skew lines. If one of the two lines (for example, line $n = AB$) lies in a certain plane, (for example, in the plane of parallelogram $DEFG$) and the other line m intersects this plane at point C that does not lie on the first line, then the lines n and m are skew lines (Fig. 1.38). Why do we say that lines n and m do not intersect? The line through point C goes through the plane (its segment below the purple plane is shown as a dashed line). On the other hand, line $n = AB$ lies on the plane $DEFG$ and since point C does not belong to line AB, then n and m will never intersect.

Two lines are skew if they belong to parallel planes but are not parallel to one another. Let's modify the drawing of Fig. 1.38 and show that we can always draw a plane containing one of two skew lines. In Fig. 1.39 line n is parallel to a plane containing another skew line, m. We can draw a line HI parallel to AB through point C, forming a plane DFE, and then draw a new unique plane β through the two intersecting lines m and HI. The straight line n is parallel to the plane β and does not intersect any line of this plane. Consequently, n will be skew to any line of plane β that is not parallel to n, such as line m.

Using an example of a cube, $ABCDA_1B_1C_1D_1$, we can demonstrate pairs of skew lines defined by the edges and other segments of the cube. Assume that M, N, K are the midpoints of AB, AD, and C_1C, respectively (Fig. 1.12). Hence, MN is a midline segment of triangle ABD, then $MN \| BD$. Since $B_1D_1 \| BD$, then by the ***transitivity*** property $MN \| B_1D_1$. Consider parallel faces, planes $ABCD$ and $A_1B_1C_1D_1$ and

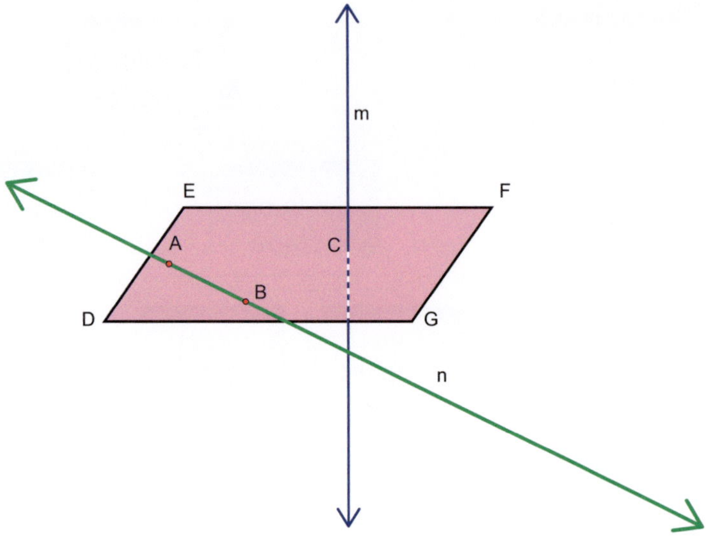

Fig. 1.38 Skew lines

Fig. 1.39 Construction of two skew lines

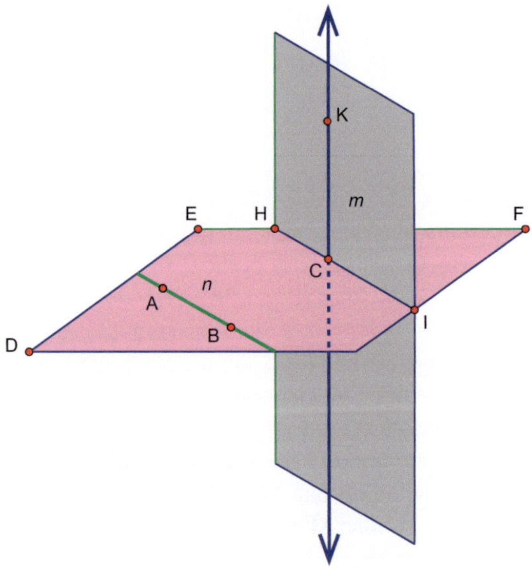

let's find pairs of skew lines there. The pairs (MN and D_1C_1) and (MN and A_1B_1) are skew lines because they lie in parallel planes and are not parallel. Now, consider parallel faces AA_1B_1B and DD_1C_1C. Here the lines A_1B and DK are skew. Also, if we connected D and C_1, then we could state that A_1B is skew to DC_1 as well.

We can notice that the MN line is skew in respect to the DK line. Let's focus on these two lines. MN lies on the lower face of the cube $ABCD$ and the line KD intersects this plane at point D. MN and DK are skew lines. If you mentally continue the line DK, you can imagine how it intersects the bottom plane of the cube and remains below this plane, while the line MN never leaves this plane. These two skew lines behave similarly to those shown in Fig. 1.38.

Definition 6 The **angle between two skew lines** is the angle formed between two lines parallel to the given skew lines and passing through the same point.

Thus, in Fig. 1.39, the angle between skew lines n and m is $\angle KCH$. In Fig. 1.12, the angle between the lines MN and DK is the angle $\angle KDB$ between the diagonal of the base of the cube BD and segment DK ($BD \| MN$). Let us find this angle by solving the following problem.

Problem 36 Given a cube $ABCDA_1B_1C_1D_1$ with side of 1. Points M, N, and K are the midpoints of edges AB, AD, and CC_1, respectively. Find the angle between lines MN and KD.

Solution Sketch a cube satisfying the condition of the problem (Fig. 1.40). Since MN is the midline segment of triangle ABD, $MN \| BD$. Hence, finding the angle between two skew lines is equivalent to evaluating the angle KDB. Consider the isosceles triangle BKD, in which $BD = \sqrt{2}$, $BK = KD$, and can be found using the Pythagorean theorem applied to the right triangle DCK, $KD = \sqrt{\frac{1}{4} + 1} = \frac{\sqrt{5}}{2}$. In ΔKBD let O be the point of intersection of the diagonals of the base $ABCD$, then O is the midpoint of BD and $KO \perp BD$. We can find cosine of the angle $\varphi = \angle KDO$:

$$\cos\varphi = \frac{OD}{KD} = \frac{\sqrt{2}}{2} \div \frac{\sqrt{5}}{2} = \frac{\sqrt{10}}{5}.$$

Therefore, $\varphi = \arccos\frac{\sqrt{10}}{5}$.

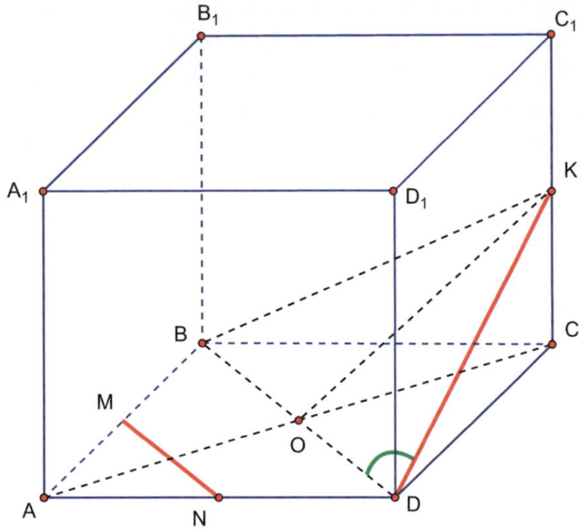

Fig. 1.40 The angle between two skew lines in a cube

Answer. $\varphi = \arccos \frac{\sqrt{10}}{5} \approx 51°$.

Definition 7 The **common perpendicular** of two skew lines is the segment perpendicular to both lines, the ends of which lie on these lines. Such a segment always exists and it is unique.

Definition 8 The **distance between two skew lines** is the length of their common perpendicular. Note that this is also the distance between two parallel planes containing these skew lines.

Finding the distance between two skew lines is sometimes quite a difficult task so it will be discussed more extensively later in this chapter. In the case of the cube shown in Fig. 1.12, some of these distances can be easily found. For example, if we consider two skew lines, $A_1B \in AA_1B_1B$ and $DK \in DD_1C_1C$, because the corresponding faces are parallel, the distance between these particular skew lines can be found as the distance between two parallel faces of the cube and so equal to the side of the cube, $AD = a$.

In the same cube shown in Fig. 1.12, we can find many lines that intersect. For example, both DK and D_1C_1 lie in the plane DD_1C_1C and are not parallel, so they will intersect at some point of this plane. The point of intersection L can be found by continuing these lines as it is shown in Fig. 1.25. Here $DK \cap D_1C_1 = L$. Note, that point L belongs to both $A_1BC_1D_1$ and DD_1C_1C planes.

1.5.1 *Other Forms of a Line in Space*

Let us now use analytic geometry to describe a line or lines in space.
Any nonzero vector, collinear to the given line, is called its direction vector. Assume that line $l \in R^3$, point $M(x_0, y_0, z_0) \in l$, and the direction vector is $\overrightarrow{a} = (q, r, s)$. Then the line l can be written in **parametric form** or its equivalent **vector form** as

$$\begin{cases} x = x_0 + qt \\ y = y_0 + rt \\ z = z_0 + st \end{cases} \Leftrightarrow \begin{bmatrix} x \\ y \\ z \end{bmatrix} = \begin{bmatrix} x_0 \\ y_0 \\ z_0 \end{bmatrix} + t \cdot \begin{bmatrix} q \\ r \\ s \end{bmatrix}. \tag{1.61}$$

In Eq. (1.61), the first vector represents the position vector and the second vector represents the direction of the line. We can eliminate parameter t from (1.61) by solving each of the equations for the parameter t and then equating them:

$$\frac{x - x_0}{q} = \frac{y - y_0}{r} = \frac{z - z_0}{s}. \tag{1.62}$$

Formula (1.62) is the **canonic form of a line**.

If we are given two distinct points $M_0(x_0, y_0, z_0),\ M_1(x_1, y_1, z_1) \in l$ that belong to a line, then the vector connecting two points is the direction vector; it is collinear to line l. We have

$$\overrightarrow{M_0M_1} \parallel l,\ \ \overrightarrow{M_0M_1} = (x_1 - x_0, y_1 - y_0, z_1 - z_0).$$

Then (1.62) can be rewritten as the equation of a straight line passing through two given points:

$$\frac{x - x_0}{x_1 - x_0} = \frac{y - y_0}{y_1 - y_0} = \frac{z - z_0}{z_1 - z_0}. \tag{1.63}$$

We will use (1.63) if we know coordinates of two points on the line.

Problem 37 Find the point symmetric to $M = (1, 1, 1)$:
(a) with respect to plane α given by $x + 4y + 3z + 5 = 0$;
(b) with respect to the line given by point $A = (3, 4, 1)$ and point $B = (1, 6, -7)$.

Solution

(a) First, we can find the equation of the line $l \perp \alpha$, $M \in L$. Let's find the equation of this line:

$$l: \frac{x-1}{1} = \frac{y-1}{4} = \frac{z-1}{3}, \tag{1.64}$$

where the direction vector of the line l is collinear to the normal vector to the plane, and the line passes point M. We can rewrite it as

$$\begin{aligned} x-1 &= t, \quad x = t+1 \\ y-1 &= 4t, \quad y = 4t+1 \\ z-1 &= 3t, \quad z = 3t+1. \end{aligned} \tag{1.65}$$

Second, we know that line l is perpendicular to plane α and intersects it at one point. Denote this point by M_0 and find its coordinates by substituting coordinates from (1.65) into equation of the plane $x + 4y + 3z + 5 = 0$,

$$(1+t) + 4(4t+1) + 3(1+3t) = -5,$$

solving which we get $t = -0.5$. Substituting this value of the t parameter back into (1.65), we find the coordinates of point M_0:

$$M_0 = (0.5, -1, -0.5).$$

If point M is reflected about the plane α, its image is point M_1, the coordinates of which we are asked to find. Both points M and M_1 are equidistant from plane α. Moreover, point M_0 is the midpoint of segment MM_1:

$$\begin{aligned} x_0 &= \tfrac{x(M)+x(M_1)}{2} \\ y_0 &= \tfrac{y(M)+y(M_1)}{2} \\ z_0 &= \tfrac{z(M)+z(M_1)}{2}. \end{aligned}$$

We know these coordinates so we can find the coordinates of point M_1 that is symmetric to M as

$$\begin{aligned} x(M_1) &= 2x(M_0) - x(M) = 2 \cdot 0.5 - 1 = 0 \\ y(M_1) &= 2y(M_0) - y(M) = 2 \cdot (-2) - 1 = -3 \\ z(M_1) &= 2z(M_0) - z(M) = 2 \cdot (-0.5) - 1 = -2. \end{aligned}$$

Therefore we found point $M_1 = (0, -3, -2)$ symmetric to point $M = (1, 1, 1)$ with respect to plane α.

(b) Now let us find point S symmetric to point M with respect to the line that goes through points A and B. If point S is symmetric to the given point M, then it lies on the perpendicular to line AB. Moreover, $MN = NS$, where N is the midpoint of

MS. It is clear that as soon as we find the coordinates of point N, we will find the coordinates of point S using, for example, (1.19) and referring to Fig. 1.8:

1. Evaluate vectors $\overrightarrow{AB} = (-2, 2, -8)$ and $\overrightarrow{AM} = (-2, -3, 0)$.
2. Calculate the scalar products $(AB, AM) = (-2)(-2) + 2(-3) + (-8) \cdot 0 = -2$.
3. Calculate scalar product $(AB, AB) = 72$.
4. Find parameter $t = \frac{-2}{72} = -\frac{1}{36}$.
5. Find vector $\overrightarrow{MN} = t \cdot \overrightarrow{AB} - \overrightarrow{AM} = (\frac{35}{18}, \frac{55}{18}, \frac{2}{9})$.
6. Find coordinates of the symmetric point S, $S = (1, 1, 1) + 2(\frac{35}{18}, \frac{55}{18}, \frac{2}{9})$.

$$S = \left(\frac{44}{9}, \frac{64}{9}, \frac{13}{9}\right).$$

Answer. (a) $M_1 = (0, -3, -2)$, (b) $S = (\frac{44}{9}, \frac{64}{9}, \frac{13}{9})$.

1.5.2 Line as the Intersection of Two Planes

If two planes intersect in space, then they intersect in a straight line, so one can imagine that a line can also be given by a system of two linear equations in three variables representing the intersecting planes:

$$\begin{cases} a_1x + b_1y + c_1z = d_1 & (\alpha) \\ a_2x + b_2y + c_2z = d_2 & (\beta)\,. \end{cases} \tag{1.66}$$

However, we know that planes do not always intersect by a line. They can coincide or be parallel to each other. Let us solve the following problem and find the condition by which two planes will always intersect in a line.

Problem 38 Prove that any line in space can be given by system of equations (1.66) if either $\frac{a_1}{a_2} \neq \frac{b_1}{b_2}$ or if $\frac{b_1}{b_2} \neq \frac{c_1}{c_2}$ or if $\frac{a_1}{a_2} \neq \frac{c_1}{c_2}$.

Proof Without loss of generality, we can prove only the case any equations of (1.66) with aforementioned restrictions on the parameters, such as $\frac{a_1}{a_2} \neq \frac{b_1}{b_2}$ describe a line in space. Let us multiply the first equation of system (1.66) by $-a_2$:

$$-a_2a_1x - a_2b_1y - a_2c_1z = -a_2d_1$$

and the second equation by a_1,

$$a_1a_2x + a_1b_2y + a_1c_2z = a_1d_2.$$

Next, after we add the left- and right-hand sides of the two equations and factor out y and z, we obtain

$$(a_1b_2 - a_2b_1) \cdot y + (a_1c_2 - a_2c_1) \cdot z = a_1d_2 - a_2d_1. \tag{1.67}$$

We need to consider four cases:

1. The coefficient of y is equal to 0 and the coefficient of z is also equal to zero. Assume that the right-hand side of (1.67) is not equal to zero, then we would get that $0 = a_1d_2 - a_2d_1 \neq 0$ and the system is inconsistent. Therefore the two planes are parallel.
2. The coefficient of y is equal to 0 and the coefficient of z is also equal to zero, so as $a_1d_2 - a_2d_1 = 0$. In this case, the two planes coincide.
3. The coefficient of y is equal to 0, but the coefficient of z is not, so in (1.67) will not have the first term, but can be solved for z:

$$y = \frac{a_1d_2 - a_2d_1}{a_1c_2 - a_2c_1}. \tag{1.68}$$

4. The coefficient of y is not equal to 0, then $\frac{a_1}{a_2} \neq \frac{b_1}{b_2}$. Hence, for any coefficient of z, (1.67) will give us an equation of the line as the intersection of two planes. We can see that in this case two planes given by (1.66) intersect with each other, if $\frac{a_1}{a_2} \neq \frac{b_1}{b_2}$.

□

The problem we just solved helps us better understand how to describe two planes in space from the point of view of analytical geometry.

Theorem 23 *Let two planes be defined by their normal vectors:* $\overrightarrow{n_1} = (a_1, b_1, c_1)$ *and* $\overrightarrow{n_2} = (a_2, b_2, c_2)$. *If the planes intersect in a line, then it is necessary and sufficient for the following strict inequality to hold:*

$$\begin{vmatrix} b_1, c_1 \\ b_2, c_2 \end{vmatrix}^2 + \begin{vmatrix} a_1, c_1 \\ a_2, c_2 \end{vmatrix}^2 + \begin{vmatrix} a_1, b_1 \\ a_2, b_2 \end{vmatrix}^2 > 0. \tag{1.69}$$

Proof Suppose that the two planes α and β given by formula (1.66) intersect in line l. Suppose also that line l is defined by its direction vector $\overrightarrow{a} = (q, r, s)$. From Theorem 19, we know that if a line is perpendicular to a plane, then it is perpendicular to any line lying on this plane. Therefore, the normal vectors of both planes will be orthogonal to this direction vector and the corresponding dot products will be zero:

$$\begin{aligned} (\overrightarrow{n_1}, \overrightarrow{a}) = a_1q + b_1r + c_1s = 0 \\ (\overrightarrow{n_2}, \overrightarrow{a}) = a_2q + b_2r + c_2s = 0. \end{aligned} \tag{1.70}$$

It is obvious that the determinant with two equal rows will be equal to zero:

$$\begin{vmatrix} a_1 & b_1 & c_1 \\ a_1 & b_1 & c_1 \\ a_2 & b_2 & c_2 \end{vmatrix} = 0.$$

Let us expand it by the first row. We obtain

$$\begin{vmatrix} a_1 & b_1 & c_1 \\ a_1 & b_1 & c_1 \\ a_2 & b_2 & c_2 \end{vmatrix} = a_1 \cdot \begin{vmatrix} b_1 & c_1 \\ b_2 & c_2 \end{vmatrix} - b_1 \cdot \begin{vmatrix} a_1 & c_1 \\ a_2 & c_2 \end{vmatrix} + c_1 \cdot \begin{vmatrix} a_1 & b_1 \\ a_2 & b_2 \end{vmatrix} = 0. \tag{1.71}$$

By analogy the determinant will be also zero (here we repeat a row),

$$\begin{vmatrix} a_2 & b_2 & c_2 \\ a_1 & b_1 & c_1 \\ a_2 & b_2 & c_2 \end{vmatrix} = 0.$$

Again we will expand it by the first row:

$$\begin{vmatrix} a_2 & b_2 & c_2 \\ a_1 & b_1 & c_1 \\ a_2 & b_2 & c_2 \end{vmatrix} = a_2 \cdot \begin{vmatrix} b_1 & c_1 \\ b_2 & c_2 \end{vmatrix} - b_2 \cdot \begin{vmatrix} a_1 & c_1 \\ a_2 & c_2 \end{vmatrix} + c_2 \cdot \begin{vmatrix} a_1 & b_1 \\ a_2 & b_2 \end{vmatrix} = 0. \tag{1.72}$$

Inspecting and comparing the right-hand sides of (1.70), (1.71), and (1.72) we can see that the right-hand sides of (1.71) and (1.72) represent the scalar products of the normal vectors of the planes and the direction vector of the line of intersection of two planes. This can be written as

$$\overrightarrow{a} = (q, r, s) = \left(\begin{vmatrix} b_1 & c_1 \\ b_2 & c_2 \end{vmatrix}, -\begin{vmatrix} a_1 & c_1 \\ a_2 & c_2 \end{vmatrix}, \begin{vmatrix} a_1 & b_1 \\ a_2 & b_2 \end{vmatrix} \right). \tag{1.73}$$

It is clear that if all the components of this vector are equal to zero, then the planes α and β do not intersect. Hence, we can state that if two planes intersect in a line, then it is necessary and sufficient that the following condition is valid:

$$\| \vec{a} \| \neq 0, \quad q^2 + r^2 + s^2 \neq 0. \tag{1.74}$$

We know that if two planes are parallel $\alpha \parallel \beta$ or if $\alpha \equiv \beta$, then their normal vectors are collinear that can be written as

$$a_2 = \lambda a_1, \quad b_2 = \lambda b_1, \quad c_2 = \lambda c_2. \tag{1.75}$$

If we substitute relationships (1.75) into (1.73), then each component of the direction vector will become zero. In order that two planes given by (1.66) intersect, at least one component of the direction vector must be nonzero. We can generalize this as

$$q^2 + r^2 + s^2 > 0.$$

or

$$\begin{vmatrix} b_1 & c_1 \\ b_2 & c_2 \end{vmatrix}^2 + \begin{vmatrix} a_1 & c_1 \\ a_2 & c_2 \end{vmatrix}^2 + \begin{vmatrix} a_1 & b_1 \\ a_2 & b_2 \end{vmatrix}^2 > 0. \tag{1.76}$$

□

Condition (1.76) states that in order for two planes to intersect in a line, at least one of the components of the direction vector (1.73) cannot be zero. Assume that $\begin{vmatrix} a_1 & b_1 \\ a_2 & b_2 \end{vmatrix} \neq 0$. Then we can solve the system (1.66). Let point $M_0(x_0, y_0, z_0) \in l$ (the line of intersection), then using (1.62) we can find the formula for this line as

$$\frac{x - x_0}{\begin{vmatrix} b_1 & c_1 \\ b_2 & c_2 \end{vmatrix}} = \frac{y - y_0}{-\begin{vmatrix} a_1 & c_1 \\ a_2 & c_2 \end{vmatrix}} = \frac{z - z_0}{\begin{vmatrix} a_1 & b_1 \\ a_2 & b_2 \end{vmatrix}}. \tag{1.77}$$

Remark. Formula (1.77) can be derived if we suppose that the line of intersection of two planes is given by a position and direction vector. The direction vector must be perpendicular to the plane formed by the two normals of the intersecting planes, and therefore it can be found using the vector product of the normals:

$$\vec{a} = \vec{n_1} \times \vec{n_2} = \begin{vmatrix} \vec{i} & \vec{j} & \vec{k} \\ a_1 & b_1 & c_1 \\ a_2 & b_2 & c_2 \end{vmatrix} = (b_1c_2 - b_2c_1)\vec{i} - (a_1c_2 - a_2c_1)\vec{j} + (a_1b_2 - a_2b_1)\vec{k}.$$

Next, we can always find the position vector (the coordinates of point M_0 belonging to the intersection line). To do this, we can set one variable in the system (1.66) equal to 0, i.e., $z = z_0 = 0$, and then find the other two coordinates (x_0 and y_0) of the point by solving a system of two linear equations in two variables. Thus, we obtain the equation of the line of intersection of two planes in the form (1.61), which can immediately be rewritten as (1.77).

It is clear that if the two planes intersect, then the line along which they intersect is unique. However, if you are given a straight line and you need to find which two planes can intersect along this straight line, then there are infinitely many such planes. From (1.77), it follows that we have three degrees of freedom in choosing the parameters of vectors, normal to two intersecting planes.

Problem 39 Find two planes intersecting along the line,

$$\frac{x-1}{1}=\frac{y-2}{-1}=\frac{z-3}{5}.$$

Solution Assume that two intersecting planes can be described by system (1.66). Comparing the given equation of the line with formula (1.77), we can find certain relationships on parameters $a_1,\ b_1,\ c_1,\ a_2,\ b_2,\ c_2$ of the system:

$$\begin{vmatrix} b_1 & c_1 \\ b_2 & c_2 \end{vmatrix}=1,\quad \begin{vmatrix} a_1 & c_1 \\ a_2 & c_2 \end{vmatrix}=-1,\quad \begin{vmatrix} a_1 & b_1 \\ a_2 & b_2 \end{vmatrix}=5. \tag{1.78}$$

Here we can select any values of three parameters of the first determinant, i.e., $b_1=1,\ b_2=1,\ c_1=1$ and evaluate $c_2=2$ to satisfy the condition of the first determinant:

$$\begin{vmatrix} b_1 & c_1 \\ b_2 & c_2 \end{vmatrix}=\begin{vmatrix} 1 & 1 \\ 1 & 2 \end{vmatrix}=1\cdot 2-1\cdot 1=1.$$

The second and the third determinants give us the condition for the two other parameters:

$$\begin{vmatrix} a_1 & 1 \\ a_2 & 2 \end{vmatrix}=2a_1-a_2=1$$

and

$$\begin{vmatrix} a_1 & 1 \\ a_2 & 1 \end{vmatrix}=a_1-a_2=5.$$

Solving the system,

$$\begin{cases} 2a_1-a_2=1 \\ a_1-a_2=5 \end{cases}$$

we obtain $a_1=-4,\ a_2=-9$. Now system (1.66) can be written as

$$\begin{cases} -4\cdot x+1\cdot y+1\cdot z=d_1 \\ -9\cdot x+1\cdot y+2\cdot z=d_2. \end{cases}$$

From the equation of the given line, we have that $x = 1,\ y = 2,\ z = 3$ satisfy this system, so $d_1 = 1$ and $d_2 = -1$. Multiplying both equations of the last system by -1, we get the equations of two planes intersecting along the given line:

$$\begin{cases} 4x - y - z = -1 \\ 9x - y - 2z = 1. \end{cases}$$

With different values of the three parameters of any of the three determinants in (1.78), we obtain other sets of planes that satisfy the conditions of the problem.

Note. Another way to solve this problem without using the formula (1.77) is given as a solution to a similar Problem 43 of the homework section.

Answer. Planes $4x - y - 1 + 1 = 0$ and $9x - y - 2z - 1 = 0$ intersect by the given line.

1.6 Some Important Formulas of Analytic Geometry

In this section, we will recall or derive some formulas from analytic geometry that often help us to find distances between two lines or between a point and a plane. They also can be used to find an angle between two lines or planes. Some problems will be solved by several methods so you can see their advantage and disadvantage. Often one problem can be solved by a combination of vector-coordinate methods and basic rules of construction and theorems of solid geometry.

1.6.1 Distance Between a Point and a Plane

We will give the formula without proof (it can be found in any book in Analytic Geometry or Calculus III). Assume that we want to find the distance between point $M(x_0, y_0, z_0)$ and the plane α that is described in the orthogonal coordinate system as $ax + by + cz = d$, then the distance $\rho(M, \alpha)$ from M to α is

$$\rho(M, \alpha) = \frac{|ax_0 + by_0 + cz_0 - d|}{\sqrt{a^2 + b^2 + c^2}}, \tag{1.79}$$

where $a,\ b,\ c$ are coordinates of the normal vector (perpendicular) to plane α.

Problem 40 Given a cube $ABCDA_1B_1C_1D_1$ with edge of 1 unit. Find the distance from vertex B_1 to the plane going through points $E,\ F,\ K$. It is known that $AE \div AA_1 = 1 \div 3$, $E \in AD_1$, point K is the midpoint of the main diagonal DB_1, $F \in BC$ and $BF \div BC = 1 \div 4$.

Solution This problem can be solved without construction. This method is based on introducing the coordinates system and then applying formula (1.79) directly. Let's place the center of coordinate system at point $B = (0, 0, 0)$. Let vector BA be in the direction of the X-axis, vector BC of the Y-axis, and vector BB_1 in the direction of the Z-axis. In this system, our points will have the following coordinates:

$$E = (1, 0, \frac{1}{3}),\ B_1 = (0, 0, 1),\ \ F = (0, \frac{1}{4}, 0),\ \ K = (0.5, 0.5, 0.5).$$

Next, we will find the equation of the plane α that passes through points E, K, F using formula (1.50):

$$\begin{vmatrix} x-1 & y & z-\frac{1}{3} \\ -1 & \frac{1}{4} & -\frac{1}{3} \\ -\frac{1}{2} & \frac{1}{6} & \frac{1}{6} \end{vmatrix} = 0.$$

After evaluating the determinant and simplification we have the equation for α:

$$5x + 8y - 9z = 2. \tag{1.80}$$

Substituting the information from (1.80) and coordinates of B_1 into (1.79), we have the distance from B_1 to α:

$$\rho(B_1, \alpha) = \frac{|-9 \cdot 1 - 2|}{\sqrt{5^2 + 8^2 + (-9)^2}} = \frac{11}{\sqrt{170}}.$$

Answer. $\frac{11}{\sqrt{170}}$.

Problem 41 Points A and B are located in the same half-space relative to the given plane α and do not lie in it. Construct in plane α a point M so that the sum of the distances from which to points A and B is minimal.

Solution Let us sketch the described situation and place two points above plane α (Fig. 1.41). Assume that there is some point M on plane α. We want

$$AM + BM \to \min. \tag{1.81}$$

Construct point A_1 symmetric to point A with respect to the plane α. Let $AA_1 \perp \alpha$, $AA_1 \cap \alpha = C$, and $AC = CA_1$. Since the point M lies on plane α, its image will coincide with itself. For the sum of the distances, we can state that $AM + MB = A_1M + MB$. We know that the length of the broken line $A_1 - M - B$ will be minimal if all three points A_1, M, and B lie on the same line. This means that the point

M that satisfies the conditions of the problem must be the intersection point of the line A_1B and the plane α. This gives us an idea of the solution:

1. Construct the image of point A relative to the plane of symmetry α, $A_1 = S_\alpha(A)$.
2. Draw the plane β through the line AA_1 and point B.
3. Denote the line of intersection of the plane α and β by c, $\alpha \cap \beta = c$.
4. Point M is the point of intersection of line A_1B and line c, $A_1B \cap c = M$. The desired point M is constructed.

Next, let us show that only point M constructed as described minimizes (1.81). Assume there exists another point $F \in \alpha$, such that

$$AF + FB < AM + MB.$$

Point F can be seen in Fig. 1.41 so that

$$AF + FB = A_1F + FB.$$

Consider triangle $\triangle A_1FB$. One of its sides is

$$A_1B = A_1M + MB.$$

Moreover, using the triangular inequality for the sides of $\triangle A_1FB$ we can state that

$$A_1M + MB = A_1B < A_1F + FB.$$

Therefore, the point M in plane α that minimizes the sum of distances is unique.

Fig. 1.41 The sum of the distances from point M to A and to B is minimal

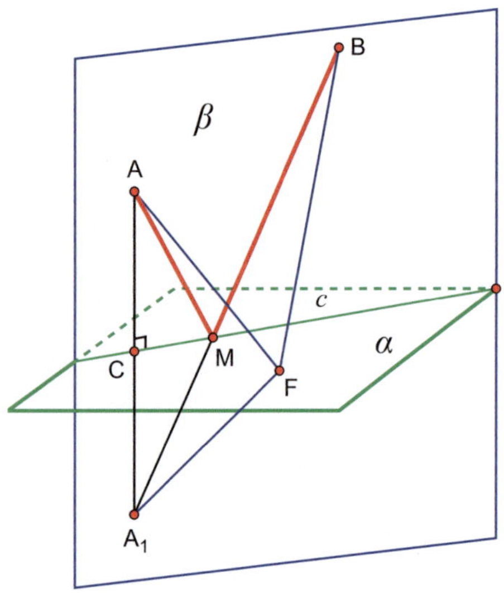

Problem 42 Find the equation of the plane that is equidistant from the points: $A(1, 1, 0)$, $B(3, 3, 4)$, $C(9, 3, 6)$, and $D(3, 1, -10)$.

Solution

Method 1. This method is based on the direct use of distance formula (1.79).

Assume that some plane $ax + by + cz = d$ satisfies the condition of the problem. Then the distances from each point to this plane must be equal. Because the denominators for each such quantity are equal, then the following must be true:

$$(a + b - d)^2 = (3a + 3b + 4c - d)^2 = (9a + 3b + 6c - d)^2 = (3a + b - 10c - d)^2.$$

Instead of equating the absolute values, we equated their squares which does not change the number of the solutions but does make the task easier. Therefore, we need to solve six nonlinear equations in four variables in order to find four parameters of the plane α. The original equation can be written as

$$u^2 = v^2 = w^2 = t^2$$

and then rewritten as the system:

$$\begin{cases} u^2 = v^2 \\ u^2 = w^2 \\ u^2 = t^2 \\ v^2 = w^2 \\ v^2 = t^2 \\ w^2 = t^2 \end{cases} \Leftrightarrow \begin{cases} (u - v)(u + v) = 0 \\ (u - w)(u + w) = 0 \\ (u - t)(u + t) = 0 \\ (v - w)(v + w) = 0 \\ (v - t)(v + t) = 0 \\ (w - t)(w + t) = 0 \end{cases} \Leftrightarrow \begin{cases} u = v, \quad \boxed{u = -v} \\ \boxed{u = w}, \quad u = -w \\ u = z, \quad \boxed{u = -t} \\ v = w, \quad v = -w \\ v = t, \quad v = -t \\ w = t, \quad \boxed{w = -t} \end{cases}.$$

We broke each equation using the difference of square formula from one quadratic into two linear cases. On one hand, this idea simplifies the solution. On the other hand, we have to consider 36 possible combinations. One of them gives us the system whose solution is the answer for the equidistant plane. Substituting actual expressions for u, v, w and t and using only four equations placed in the frame, we obtain the homogeneous system of four linear equations in four variables:

$$\begin{cases} a + b - d = -3a - 3b - 4c + d \\ a + b - d = 9a + 3b + 6c - d \\ a + b - d = -3a - b + 10c + d \\ 9a + 3b + 6c - d = -3a - b + 10c + d \end{cases} \Leftrightarrow \begin{cases} 2a + 2b + 2c - d = 0 \\ 4a + b + 3c = 0 \\ 2a + b - 5c - d = 0 \\ 6a + 2b - 2c - d = 0 \end{cases} \Leftrightarrow \left[\begin{array}{l} a = -\frac{d}{10} \\ b = \frac{7d}{10} \\ c = -\frac{d}{10} \end{array} \right..$$

The system has infinitely many solutions. Let us substitute a, b, c, and d in the equation of a plane,

$$-\frac{d}{10} \cdot x + \frac{7d}{10} \cdot y - \frac{d}{10} \cdot z = d.$$

After simplification ($d \neq 0$) we have an equation of the equidistant plane,

$$x - 7y + z = -10.$$

We obtained the solution but showed only a few lines and omitted many other. This approach is clearly inefficient in terms of time.

Method 2. This method is based on construction.

We solved a similar problem (Problem 32) where four points without coordinates were given: A, B, C, and D. The midpoints are: $M = (2, 2, 2)$ is the midpoint AB, $N = (2, 1, -5)$ is the midpoint of AD, and $K = (6, 3, 5)$ is the midpoint of BC. The plane, let's name it α, must go through these three points, so $\alpha = (MNK)$. Now we are going to find the equation of this plane using formula (1.50):

$$\begin{vmatrix} x-2 & y-2 & z-2 \\ 0 & -1 & -7 \\ 4 & 1 & 3 \end{vmatrix} = 0.$$

After calculating the determinant, we obtain the equation of the plane α in the form $ax + by + cz = d$, $x - 7y + z = -10$. We can see this method is much easier than the previous one and gives the same answer much faster. Next, let us check that all points are equidistant from plane α. For this, we need to use formula (1.79) and find the distance from each point to the plane. However, we can compare only the numerators of each point since the denominator is the same and is equal to

$$\sqrt{1^2 + (-7)^2 + 1^2} = \sqrt{52}.$$

The distance from point A to the plane is

$$\rho(A, \alpha) = \frac{|1 \cdot 1 - 7 \cdot 1 + 1 \cdot 0 + 10|}{\sqrt{52}} = \frac{4}{\sqrt{52}} = \frac{2}{\sqrt{13}}.$$

The distance from point B to plane α is

$$\rho(B, \alpha) = \frac{|1 \cdot 3 - 7 \cdot 3 + 1 \cdot 4 + 10|}{\sqrt{52}} = \frac{4}{\sqrt{52}} = \frac{2}{\sqrt{13}}.$$

The distance from point C to plane α is

$$\rho(C, \alpha) = \frac{|1 \cdot 9 - 7 \cdot 3 + 1 \cdot 6 + 10|}{\sqrt{52}} = \frac{9 - 21 + 16}{\sqrt{52}} = \frac{2}{\sqrt{13}}.$$

The distance from point D to plane α is

$$\rho(D,\alpha)=\frac{|1\cdot 3-7\cdot 1+1\cdot(-10)+10|}{\sqrt{52}}=\frac{4}{\sqrt{52}}=\frac{2}{\sqrt{13}}.$$

We can see that plane α is indeed equidistant from all four given points.

Answer. $x-7y+z=-10$.

1.6.2 Distance Between Two Skew Lines

The distance between the points at which the common perpendicular to the skew lines intersects these lines is called the distance between two skew lines. This problem also can be reduced to a problem of evaluating the distance from a point to a plane. Please consider Fig. 1.42 and compare the common perpendicular M_1M_2 of the skew lines l_1 and l_2 with any segment (for example, C_1C_2, the end of which lie on each of the skew lines.) Such a definition of the distance between the skew lines is natural since it is easy to show the distance between the points at which the common perpendicular to the skew lines intersects these lines is equal to the minimum of the lengths of all segments $M_1M_2,\ \ M_1\in l_1,\ \ M_2\in l_2$. For any choice of points C_1 and C_2, $\|M_1M_2\|<\|C_1C_2\|$.

Let us calculate the distance between two skew lines. Let two skew lines $l_1,\ l_2$ be given parametrically with the corresponding directed vectors $\overrightarrow{s_1}=(a_1,b_1,c_1),\ \ \overrightarrow{s_2}=(a_2,b_2,c_2)$. Now our further constructions can be illustrated in Fig. 1.43. Select a point on each of the skew lines, i.e., $P_1\in l_1,\ \ P_2\in l_2$. Through point P_2 let us draw a line parallel to l_1. Construct from point P_2 both directed segments, and denote the ends of the directed segments by Q and R, respectively, so $\overrightarrow{P_1Q}=\overrightarrow{s_1}$ and $\overrightarrow{P_1R}=\overrightarrow{s_2}$.

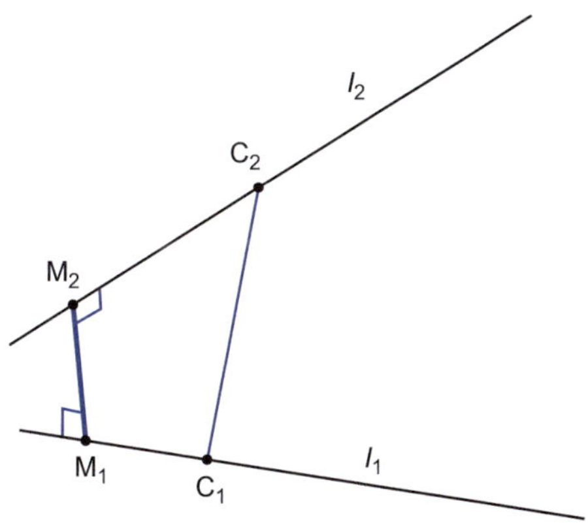

Fig. 1.42 Distance between two skew lines

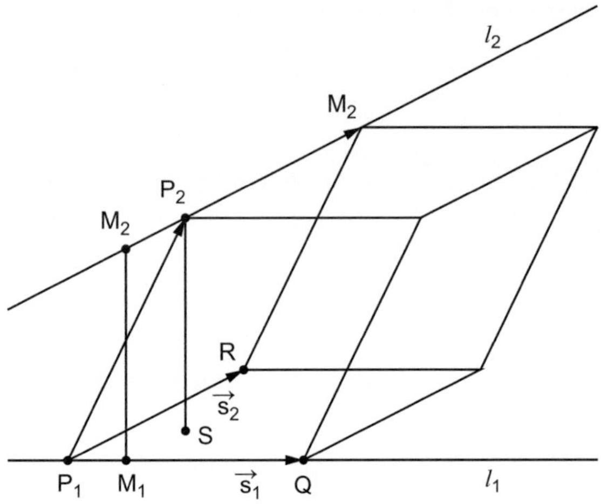

Fig. 1.43 Finding the distance between two skew lines using construction of a parallelepiped

Next, we construct a parallelepiped on the vectors $\overrightarrow{P_1Q},\ \overrightarrow{P_1R},\ \overrightarrow{P_1P_2}$. From the vertices of this parallelepiped, we drop the height to the plane of vectors $\overrightarrow{P_1Q},\ \overrightarrow{P_1R}$ (Fig. 1.43) and denote the foot of this height by S. In the same sketch, we show segment M_1M_2 as the common perpendicular of two skew lines, the length of which we need to evaluate. Regarding our construction, the line P_2S is perpendicular to the both skew lines ($P_2S \perp l_1,\ \ P_2S \perp l_2$) and therefore it is parallel and equal to the common perpendicular to these lines, M_1M_2. Therefore, the distance between the skew lines is equal to the length of the height, $h = P_2S$.

In order to find this distance, we recall that it is equal to the height of the parallelepiped and we know how to calculate the volume of a parallelepiped (see Chap. 2 in case you forgot or if you do not remember), $V = h \cdot B$. We can independently evaluate the volume, V using the magnitude of the scalar triple product and the area of the base B using the magnitude of the cross product of the direction vectors, s_1 and s_2. The height h and the required distance between two skew lines can be found as the quotient of these two quantities:

$$d(l_1, l_2) = \frac{\left\| \overrightarrow{P_1P_2}, (s_1 \times s_2) \right\|}{\|s_1 \times s_2\|}. \tag{1.82}$$

The importance of formula (1.82) cannot be overstated. I do not recommend using the formula for the mixed scalar product here. Most of us know how to evaluate a cross product and a scalar product, so, I recommend to follow the steps:

1. Find coordinates of one point on each given skew line and find the coordinates of vector P_1P_2. The points can have different names but we usually chose the simplest points.
2. By taking one more point on each skew line, i.e., similar to R and Q, find the corresponding direction vectors of each skew line, $P_1Q = s_1$ and $P_1R = s_2$.

3. Evaluate their cross product, the vector, $\overrightarrow{u} = \overrightarrow{P_1Q} \times \overrightarrow{P_1R}$.
4. Find the magnitude of the cross-product vector $\overrightarrow{u}$, $\|\overrightarrow{u}\|$.
5. Evaluate the scalar product of vectors $\overrightarrow{P_1P_2}$ and $\overrightarrow{u}$.
6. Find the absolute value of this scalar product.
7. Find the quotient of the quantities found in steps 6 and 4.

By solving the following problems in different methods, we will be able to compare the methods and understand the material better.

Problem 43 Given a cube $ABCDA'B'C'D$. Find the distance between lines AB' and BD and also find the angle between two skew lines.

Solution

Method 1. Since $DC' \parallel AB'$, then $\left(BDC'\right) \parallel AB'$, then the distance $\rho(AB', BD) = \rho(AB', BDC') = \rho(A, BDC')$. Let us introduce a coordinate system as it is shown in Fig. 1.44 and define the points $A(0, 0, 0)$, $B(0, 1, 0)$, $D(1, 0, 0)$, $C'(1, 1, 1)$, and $B'(0, 1, 0)$. Using these coordinates, we find the equation of the plane passing through points $B,\ D,\ C'$ to be $x + y - z = 1$. The distance between lines AB' and BD equals the distance from point A to plane BDC':

$$\rho(A, BDC') = \frac{|1 \cdot 0 + 1 \cdot 0 - 1 \cdot 0 - 1|}{\sqrt{1^2 + 1^2 + (-1)^2}} = \frac{1}{\sqrt{3}}.$$

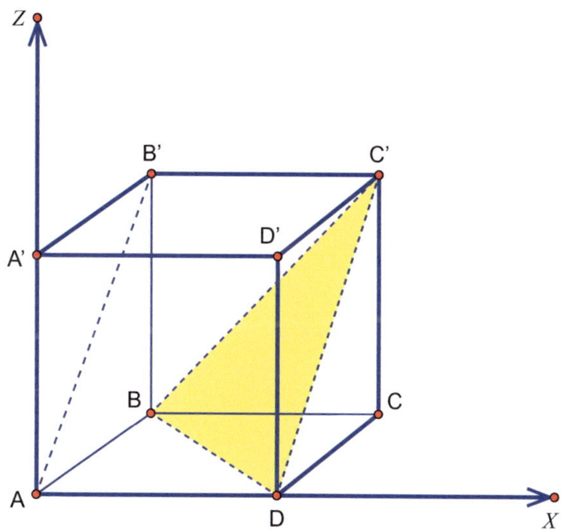

Fig. 1.44 Two skew lines in a cube

The angle (φ) between two lines is that of an equilateral triangle $C'DB$, and is therefore $\varphi = 60°$.

Remark. Here we applied knowledge of stereometry. By constructing the line parallel to edge AB' through vertex D, we obtained the plane DBC' parallel to the edge AB'. This allowed us to reduce the problem of finding the distance between two skew lines to that of finding the distance from a point on the line parallel to the plane and the plane.

Method 2. This is an application of formula (1.82) or, namely, the direct application of the coordinate method without construction of a line parallel to the edge AB' or any plane parallel in space. We need only two points on each line and to introduce the coordinate system. We can use the same coordinate system as it was selected in Method 1. We will first find the coordinates of the direction vectors for the skew lines, $\overrightarrow{BD} = \overrightarrow{s_1} = (1, -1, 0)$, $\overrightarrow{AB'} = \overrightarrow{s_2} = (1, 0, 1)$, and the coordinates of vector, $\overrightarrow{P_1P_2} = \overrightarrow{AB} = (0, 1, 0)$. Secondly, let us find the cross product of vectors $\overrightarrow{s_1}$ and $\overrightarrow{s_2}$:

$$\overrightarrow{u} = \overrightarrow{s_1} \times \overrightarrow{s_2} = \begin{vmatrix} \overrightarrow{i} & \overrightarrow{j} & \overrightarrow{k} \\ 1 & 0 & 1 \\ 1 & -1 & 0 \end{vmatrix}.$$

Thus, $\overrightarrow{u} = \overrightarrow{s_1} \times \overrightarrow{s_2} = (1, 1, -1)$ and its magnitude is $\|u\| = \sqrt{1^2 + 1^2 + (-1)^2} = \sqrt{3}$. The scalar product of vectors $\overrightarrow{AB}$ and $\overrightarrow{u}$ is $(\overrightarrow{AB}, \overrightarrow{u}) = 1 \cdot 0 + 1 \cdot 1 + (-1) \cdot 0 = 1$. Finally, after dividing 1 by $\sqrt{3}$ we obtain the distance between two skew lines, $d = \frac{1}{\sqrt{3}}$.

Method 3. (Purely geometric method). In order to find the distance from point A to plane BDC' we can use the volume of pyramid $ABDC'$ found in two different ways. As we know the volume of any pyramid equals $1/3$ of the product of height and base, $V = \frac{1}{3}h \cdot B$. First, consider the base as $\triangle BC'D$, then the height h is the perpendicular dropped form vertex A, i.e., the required distance. Second, consider the base of the same pyramid to be the $\triangle ABD$ and then the height dropped to this base from vertex C will be the edge of the cube, $CC' = 1$. Now equate the two volumes as

$$\frac{1}{3}h \cdot [\triangle BC'D] = \frac{1}{3} \cdot CC' \cdot [\triangle ABD].$$

The areas of both triangles can be easily found. Triangle $BC'D$ is an equilateral with side of $a = \sqrt{2}$. If an equilateral triangle has side a, then its area is $\frac{a^2\sqrt{3}}{4}$, so $[\triangle BC'D] = \frac{\sqrt{3}}{2}$. The area of the second triangle is half the area of the square with side 1, $[\triangle ABD] = \frac{1}{2}$. Finally, the unknown distance h is equal to

$$h = \frac{1}{2} \div \frac{\sqrt{3}}{2} = \frac{1}{\sqrt{3}}.$$

Answer. $d = \frac{1}{\sqrt{3}}$; $\varphi = 60°$.

1.6.3 Angle Between Two Lines

In order to find the angle between two lines, AB and CD, we determine the coordinates of their directed vectors. Let $\overrightarrow{AB}(x_1, y_1, z_1)$ and $\overrightarrow{CD}(x_2, y_2, z_2)$. Assume that φ is the angle between lines AB and CD. Because the scalar product of two vectors $\overrightarrow{u}$ and $\overrightarrow{v}$ is

$$\left(\overrightarrow{u}, \overrightarrow{v}\right) = \|u\| \|v\| \cdot \cos\varphi, \tag{1.83}$$

we can find the cosine of the angle between two vectors as

$$\cos\varphi = \frac{x_1x_2 + y_1y_2 + z_1z_2}{\sqrt{x_1^2 + y_1^2 + z_1^2}\sqrt{x_2^2 + y_2^2 + z_2^2}}. \tag{1.84}$$

The cosine can be positive, zero, or negative for an acute, right, and obtuse angles between vectors, respectively. However, we are looking for an acute angle between two lines, so if the scalar product of two vectors is negative, we will need to take its absolute value. This can be explained using formulas of trigonometry:

$$\cos(180° - \varphi) = -\cos\varphi.$$

Hence, if a scalar product is negative, we must take its absolute value, which algebraically means finding the value of the corresponding acute angle. The acute angle between two lines can be found as

$$\varphi = \arccos\left(\frac{|x_1x_2 + y_1y_2 + z_1z_2|}{\sqrt{x_1^2 + y_1^2 + z_1^2}\sqrt{x_2^2 + y_2^2 + z_2^2}}\right). \tag{1.85}$$

Angle between Two Skew Lines You can construct an angle between skew lines if you move them parallel so that they pass through one point. In Fig. 1.44, the angle $\angle BDC'$ is the angle between the two skew lines ($B'A$ and BD).

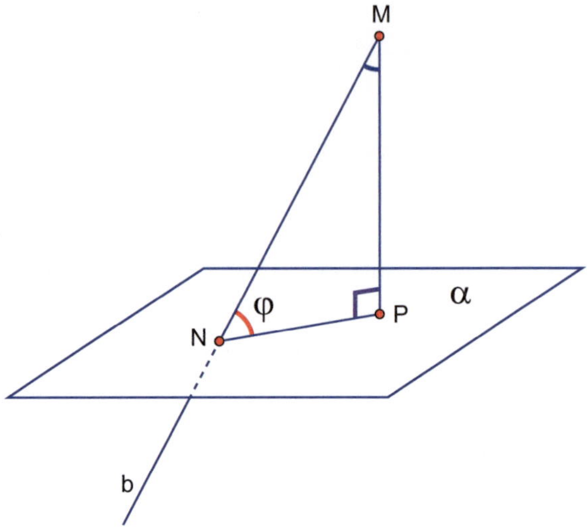

Fig. 1.45 Angle between line b and plane α

1.6.4 Angle Between a Line and a Plane

Let line b be given by its directed vector $\overrightarrow{MN} = (x_0, y_0, z_0)$ intersect plane α at point N (Fig. 1.45). Assume that the plane is given by $ax + by + cz = d$. If we drop a perpendicular to the plane α from point M and denote its orthogonal projection by point P, vector PM will be orthogonal to the plane, so it is collinear to the normal vector of the plane $\overrightarrow{n} = (a, b, c)$. From the right triangle $\triangle MPN$, we have that the angle between line b and plane α is $\varphi = \angle MNP$. However, using the scalar product of vectors MN and n and by taking the absolute value of this scalar product, we will evaluate not the angle MNP, but rather its complimentary angle $(90° - \varphi)$. From trigonometry we know $\cos(90° - \varphi) = \sin\varphi$. Hence, what we can find this way is the sin of the angle between line b and plane α:

$$\sin\varphi = \frac{|ax_0 + by_0 + cz_0|}{\sqrt{a^2 + b^2 + c^2} \cdot \sqrt{x_0^2 + y_0^2 + z_0^2}}. \tag{1.86}$$

In the following problem, we can see the advantage of thinking geometrically and of the construction method.

Problem 44 Given a cube $ABCDA'B'C'D'$ with side a, find the distance from vertex B to the plane passing the vertices A, B', and C.

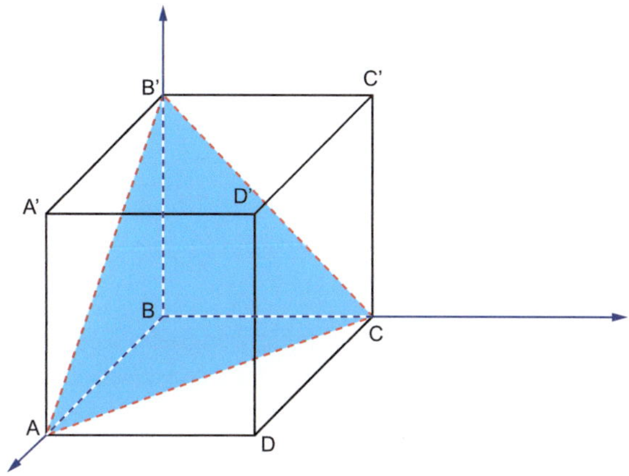

Fig. 1.46 Distance between a vertex of a cube and plane ABC

Solution

Method 1. The plane passing through given vertices cuts off the equilateral triangle inside the cube, each side of which is equal to $a\sqrt{2}$ (Fig. 1.46). The distance between point B and the triangle $AB'C$ equals the length of the perpendicular erected from vertex B to this plane. Of course, we could use analytical geometry, introduce coordinate system with the center at point B as it is shown in the Fig. 1.46, and then use formula (1.79). However, we can recognize a right, regular tetrahedron $ABCB'$ inside the cube. Its volume can be found in two different ways:

$$V = \frac{1}{3} \cdot BB' \cdot [\triangle\ ABC] = \frac{1}{3} \cdot h \cdot \left[\triangle\ AB'C\right],$$

where h is the height dropped to the base $AB'C$ from vertex B. The area of triangle ABC can be then found as

$$[ABC] = \frac{b^2\sqrt{3}}{4}, \quad b = a\sqrt{2}.$$

Hence, $[ABC] = a^2\sqrt{3}/2$.

On the other hand, triangle ABC is a right triangle, so its area is half the product of the legs. Both areas can be substituted in the formula:

$$a \cdot \frac{1}{2} \cdot a^2 = h \cdot \frac{a^2\sqrt{3}}{2}.$$

By reducing further common factors, we have the required distance as $h = \frac{a}{\sqrt{3}}$.

Answer. $h = \frac{a}{\sqrt{3}}$.

Method 2. Many readers probably like the solution and would like to explore the situation a little more. For example, we could find the same distance using methods of analytic geometry. Right? Let us study this case a little bit deeper and draw a plane parallel to the previous one, $AB'C$. For this we can simply take

$$A'D \parallel B'C$$
$$DC' \parallel AB'$$
$$A'C' \parallel AC.$$

We constructed another triangle, congruent to $\triangle\ AB'C$. Because the corresponding sides of two triangles are parallel by construction, the planes of the triangles are also parallel. This can be written as

$$\triangle\ AB'C \in \alpha, \quad \triangle\ DC'A' \in \beta, \quad \alpha \parallel \beta.$$

Let O_1 and O_2 be the centers of triangles $AB'C$ and $DC'A'$, respectively. To find these points, in each triangle, we can construct two medians and find the points of their intersection. In Fig. 1.47, in order not to clutter the drawing, we showed only one median in each triangle. Points M and N obviously lie in different planes:

$$M \in A'D, \quad N \in AC.$$

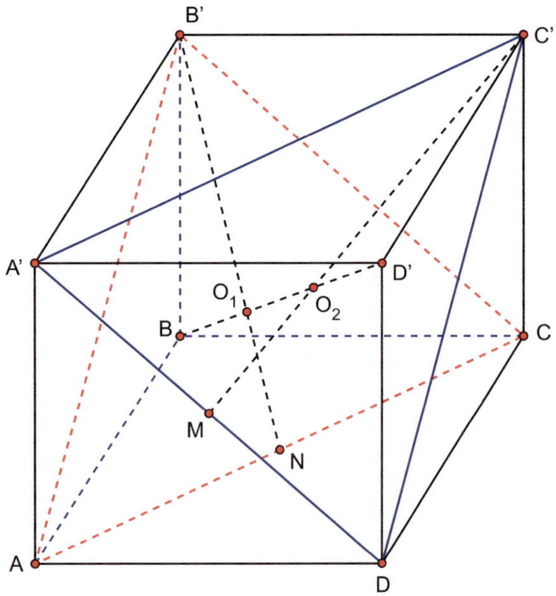

Fig. 1.47 Two parallel planes in a cube and cube's main diagonal

The regular pyramid $A'C'DD'$ is congruent to a regular pyramid $AB'CB$, and so their heights dropped from the corresponding vertices must be equal. Moreover, the foots of the heights must land onto orthocenters O_1 and O_2 of each triangle. Segment $O_1O_2 = d$ plays the role of the distance between two parallel planes. Next. let us connect the vertices D' and B of the cube. The main diagonal of the cube satisfies

$$BD' = BO_1 + O_1O_2 + O_2D'$$

and is divided by the points O_1 and O_2 into three equal parts:

$$d = O_1O_2 = BO_1 = O_2D' = \frac{BD'}{3} = \frac{a\sqrt{3}}{3} = \frac{a}{\sqrt{3}}.$$

As you recall this was the exact answer obtained by Method 1.

Right now, we will solve the following problem.

Problem 45 Consider the cube shown in Fig. 1.47. Prove that the main diagonal of the cube, BD', pierces the triangle ACB' (in plane α) at its orthocenter.

Proof Let us find the point of intersection of plane α with the line BD'. The coordinates of vector BD' are $\overrightarrow{BD'} = (1, 1, 1)$. The plane, α, that passes through points A, C, and B' is described by the equation $x + y + z = 1$. It is clear that $\overrightarrow{BD'}$ is a normal vector for plane α. On the other hand, vector BO_1 is also perpendicular to the same plane and has the same vertex, i.e.:

$$\left.\begin{array}{l}\overrightarrow{BD'} \perp ACB' \\ \overrightarrow{BO_1} \perp ACB'\end{array}\right\} \Rightarrow \overrightarrow{BO_1} = k\overrightarrow{BD'}, \ \ \overrightarrow{BO_1} \parallel \overrightarrow{BD'}.$$

If Q_1 is the orthocenter of triangle ACB', then

$$\overrightarrow{BO_1} = \frac{1}{3}\left(\overrightarrow{BB'} + \overrightarrow{BC} + \overrightarrow{BA}\right).$$

On the other hand, $\overrightarrow{BD'}$ can be expressed as

$$\overrightarrow{BD'} = \overrightarrow{BB'} + \overrightarrow{BC} + \overrightarrow{BA},$$

so $BO_1 = \frac{1}{3}BD'$. □

Problem 46 Given cube $ABCDA'B'C'D'$ and the following points that belong to the cube: K is a midpoint of edge AA', $M \in B'C'$, such that $B'M \div MC' = 2 \div 1$, and point N is on edge CD such that $DN \div NC = 1 \div 2$. Find the distance from point K to the plane that passes through points M, N and vertex B.

Solution Let us solve this problem using the analytic geometry approach. If the center of the coordinate system is vertex $B(0,0,0)$ and the X-axis coincides with vector BA, the Y-axis with vector BC, and the Z-axis with vector BB', then the given points will have the following coordinates:

$$B(0,0,0),\ M(0,2/3,1),\ N(2/3,1,0),\ K(1,0,1/2).$$

Since plane $\alpha = BMN$ goes through the origin, its equation has the form:

$$ax + by + cz = 0.$$

Substituting the coordinates of points M and N we obtain the system:

$$\begin{cases} \frac{2}{3}b + c = 0, \\ \frac{2}{3}a + b = 0, \\ a + \frac{c}{2} = 0. \end{cases} \tag{1.87}$$

Expressing c and a in terms of b from the first and the second equations of (1.87) we have $c = -3b/2$ and $a = -3b/2$. Substituting into the third equation of the system:

$$\frac{-3b}{2}x + by - \frac{2b}{3}z = 0.$$

Solving for parameter b, the equation of the plane can be rewritten as

$$9x - 6y + 4z = 0. \tag{1.88}$$

Now using formula (1.79), we can find the distance from point K to this plane as

$$\rho(K,\alpha) = \frac{9 \cdot 1 + 9(-6) \cdot 0 + 4 \cdot \frac{1}{2}}{\sqrt{9^2 + (-6)^2 + 4^2}} = \frac{11}{\sqrt{133}}.$$

Answer. $\frac{11}{\sqrt{133}}$.

Fig. 1.48 Distance KP to plane MBN

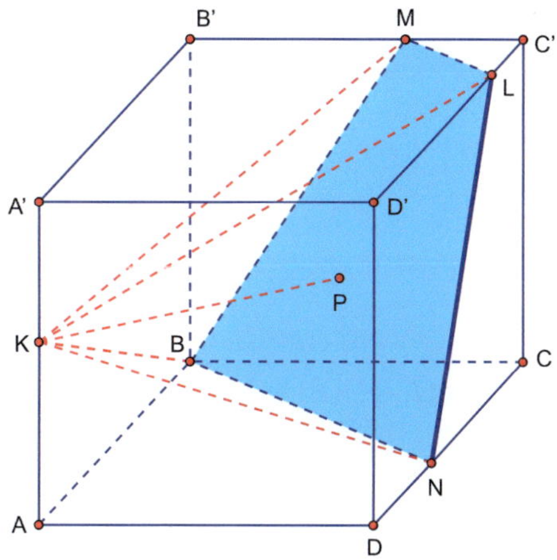

Remark. Note that we found the distance from a point to the plane without construction. However, a little construction is always helpful in solving problems.
Consider Fig. 1.48. We can construct cross section of the cube by plane α by connecting points B and N, then B and M and then drawing line through point M in plane $A'B'C'D'$ parallel to BN (property of parallel planes intersected by the third plane given by Theorem 10) and finally connecting points L and N. If you have difficulties to understand how to construct this cross section, then you may want to study Chap. 3 for detailed information on different methods and ideas. If we connect K with vertices M, N, B, and L, we will obtain a pyramid $KBMLN$. Now we can look, for example, for the volume of this pyramid. The volume can be found if we find the area of the trapezoid $MLNB$ and then use the formula for the volume of a pyramid.

1.6.5 Distance from a Point to a Line in Space

Lemma 2 *If $M_0(x_0,\ y_0,\ z_0)$ is a point in a coordinate system, $\overrightarrow{a} = (q, r, s)$ is a direction vector of line l, $M_1(x_1,\ y_1,\ z_1)$ is a point on line l, then the distance between point $M_0(x_0,\ y_0,\ z_0)$ and line l is*

$$\rho(M_0, l) = \frac{\left\|\overrightarrow{M_0M_1} \times \overrightarrow{a}\right\|}{\|\overrightarrow{a}\|}. \tag{1.89}$$

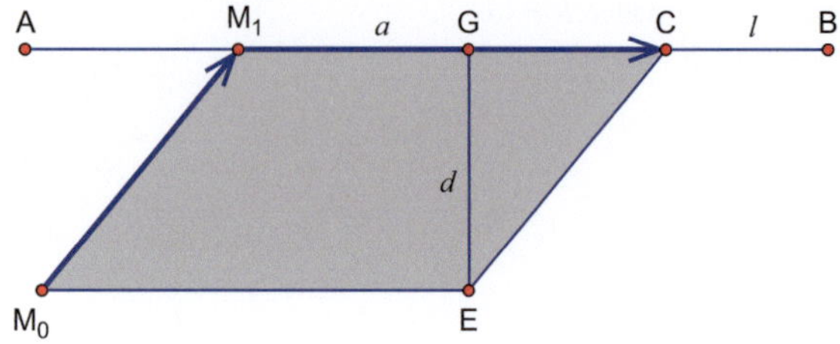

Fig. 1.49 Distance from a point to a line in space

Proof Let $l = (AB)$ be a line in space and $\overrightarrow{a} = (q, r, s)$ be its direction vector, and $M_1(x_1,\ y_1,\ z_1)$ be a point on the line. From the properties of the cross product, it is known that the magnitude of the cross product of vectors is equal to the area of a parallelogram constructed on these two vectors,

$$A = \left\| \overrightarrow{M_0 M_1} \times \overrightarrow{a} \right\|.$$

In Fig. 1.49, $\overrightarrow{a} = \overrightarrow{M_1 C}$. On the other hand, the parallelogram's $(M_0 M_1 C E)$ area is equal to the product of its side and height ($h = EG = d$) dropped to this side, $A = h \cdot a$. Having equated the areas, it is simple to obtain the formula (1.89) of the distance from a point to a line. □

Let us apply this formula and solve a problem.

Problem 47 Using Fig. 1.48, find the distance from point M to line BN.

Solution

Method 1. Let us introduce a coordinate system with center B and axes directed along vectors BA, BC, and BB'. We have, $M = (0, \frac{2}{3}, 1)$, $\overrightarrow{BN} = (\frac{2}{3}, 1, 0)$, $\overrightarrow{MB} = (0, -\frac{2}{3}, -1)$. Using formula (1.89), we need to calculate the cross product of these two vectors:

$$\begin{vmatrix} i & j & k \\ 0 & -\frac{2}{3} & -1 \\ \frac{2}{3} & 1 & 0 \end{vmatrix} = 1 \cdot \overrightarrow{i} - \frac{2}{3} \cdot \overrightarrow{j} + \frac{4}{9} \cdot \overrightarrow{k} = \left(1, -\frac{2}{3}, \frac{4}{9}\right).$$

Its magnitude is $\sqrt{1^2 + \left(-\frac{2}{3}\right)^2 + \left(\frac{4}{9}\right)^2} = \frac{\sqrt{133}}{9}$ and the magnitude of the direction vector is $\sqrt{\left(\frac{2}{3}\right)^2 + 1^2 + 0^2} = \frac{\sqrt{13}}{3}$. By dividing the two quantities we obtain the distance, $d = \rho(M, l) = \frac{\sqrt{133}}{9} \div \frac{\sqrt{13}}{3} = \frac{\sqrt{399}}{39}$.

Method 2. Let us find the distance MT differently. First, we can drop perpendicular MS from M to BC. Since MT is perpendicular to BN, MT is the distance between point M and line $l = BN$. Also, ST is perpendicular to BN by Theorem 20 on three perpendiculars. Angle MTS is the angle between the plane of the base ABC and plane α shaded by blue in Fig. 1.50. The equation of the blue plane is $9x - 6y + 4z = 0$, so its normal vector is $\overrightarrow{n_1} = (9, -6, 4)$. On the other hand, the base plane is the XOY plane and its normal vector is $\overrightarrow{n_2} = (0, 0, 1)$. The angle between the two planes is

$$\cos\varphi = \frac{(\overrightarrow{n_1}, \overrightarrow{n_2})}{\|\overrightarrow{n_1}\| \cdot \|\overrightarrow{n_2}\|} = \frac{9 \cdot 0 + (-6) \cdot 0 + 4 \cdot 1}{\sqrt{133} \cdot 1} = \frac{4}{\sqrt{133}}.$$

Next, we need to find the sine of the same angle, that is, $\sin\varphi = \sqrt{1 - \cos^2\varphi} = \frac{\sqrt{13}}{\sqrt{133}}$. Finally, from the right triangle MST we get $MT = \frac{MS}{\sin\varphi} = 1 \div \frac{\sqrt{13}}{\sqrt{133}} = \frac{\sqrt{399}}{39}$. This answer matches the answer obtained using the coordinate method.

Answer. $\frac{\sqrt{399}}{39}$.

Problem 48 Find the equation of a line parallel to the plane α given by the formula $x + y + z = 1$ and passing through the point $(4, 2, -3)$.

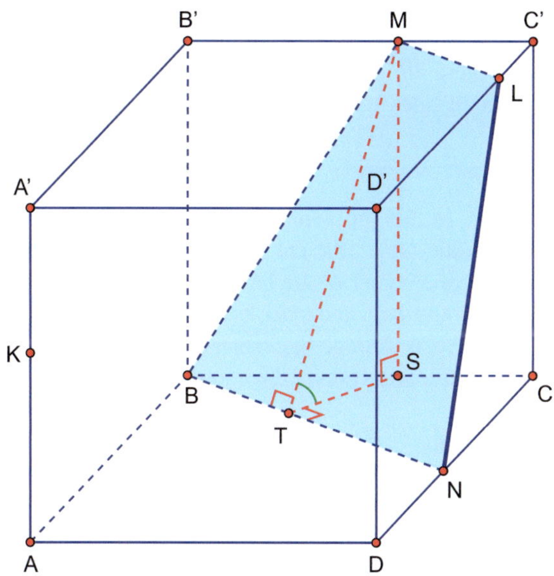

Fig. 1.50 Angle between two planes

Solution Let us find the normal vector to the plane α,

$$\overrightarrow{n_1} = (1, 1, 1).$$

Assume that the direction vector of the line has coordinates

$$\overrightarrow{a} = (n, m, p).$$

These two vectors must be orthogonal if the line is parallel to the plane. Then the scalar product of two vectors is equal to zero

$$(\overrightarrow{n_1}, \overrightarrow{a}) = n + m + p = 0,$$

where we have the infinite number of choices. For example, we can take $n = 5$, $m = -7$, then $p = 2$, and the direction vector will be

$$\overrightarrow{a} = (5, -7, 2).$$

Hence, the line will have the equation:

$$\frac{x-4}{5} = \frac{y-2}{-7} = \frac{z+3}{2}.$$

Answer. $\frac{x-4}{5} = \frac{y-2}{-7} = \frac{z+3}{2}$.

Problem 49 Given a regular hexagonal prism $ABCDEFA_1B_1C_1D_1E_1F_1$ with all sides and edges equal to 1. Find the angle between AB_1 and BF_1.

Solution In this problem, it is important to choose the coordinate system and its center in the most efficient way. For example, it is nice to place the center of the system at the center of the basic hexagon. Then find the midpoints of two parallel sides, AB and BD, and mark OY-axis. The OX-axis is through points F and C and the OZ-axis is through the centers of both hexagons as it is shown in Fig. 1.51.
In this coordinate system, the coordinates of the points are

$$A\left(-\frac{1}{2}, -\frac{\sqrt{3}}{2}, 0\right), \; B\left(\frac{1}{2}, -\frac{\sqrt{3}}{2}, 0\right), \; B_1\left(\frac{1}{2}, -\frac{\sqrt{3}}{2}, 1\right), \; F_1(-1, 0, 1).$$

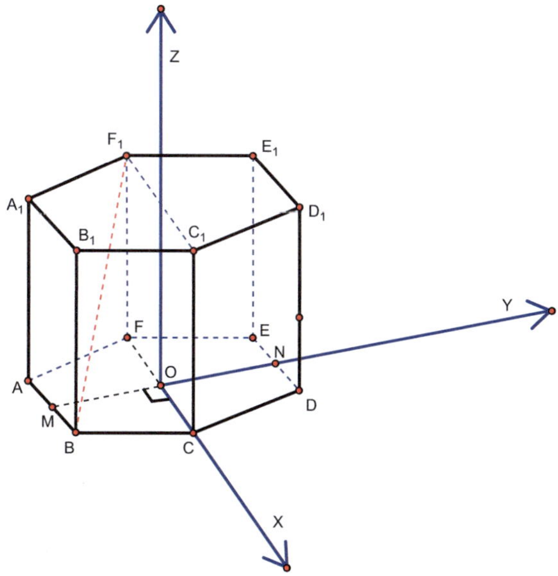

Fig. 1.51 Regular hexagonal prism

Next, we can find coordinates of the direction vectors for both lines:

$$\overrightarrow{AB_1} = (1, 0, 1), \ \overrightarrow{BF_1} = \left(-\frac{3}{2}, \frac{\sqrt{3}}{2}, 1\right).$$

Finally we can evaluate the cosine of the angle between two vectors using scalar product formula (1.85):

$$\varphi = \arccos \frac{\sqrt{2}}{8}.$$

Answer. $\varphi = \arccos \frac{\sqrt{2}}{8}$.

Note. Why did we use this orientation and location of the coordinate system? Is line FC really perpendicular to midline segment MN? The answer will come quickly if we consider the base of the prism, a regular hexagon $ABCDEF$ (Fig. 1.52). Connect the midpoints M and N and the vertexes F and C of the hexagon that became our X- and Y-axes directions, respectively. Triangle EOD is isosceles, then $ON \perp ED$. In $\triangle EOD \angle EON = 30°$. On the other hand, angle FOE is a central angle of a regular hexagon, then its measure is 60°, so

$$\measuredangle FON = \measuredangle FOE + \measuredangle EON = 90°.$$

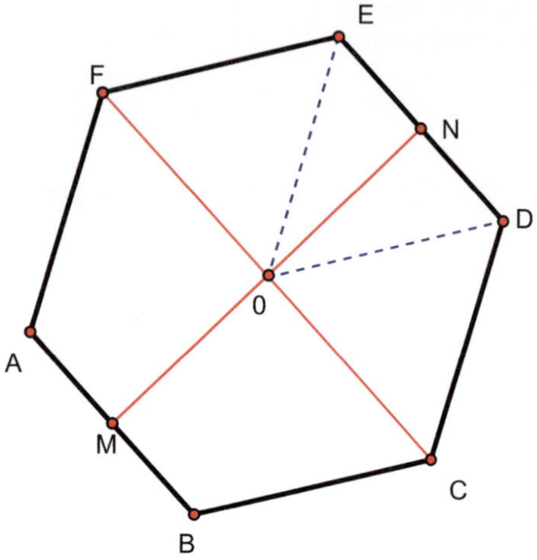

Fig. 1.52 Regular hexagon

1.6.6 Affine Transformation

A space transformation that maps each plane to a plane is an **affine transformation**. A straight line is a line of intersection of two planes. Therefore, according to the definition, an affine space transformation maps every line onto a line. The affine transformation of space preserves the parallelism of planes, the parallelism of lines, the parallelism of a line and a plane, and maps skew lines to skew lines. Each of these statements can be proven by contradiction. An important role in the study of affine transformations was played by the French mathematician and geometer **J. Darboux** (1842–1917).

In a three-dimensional space, affine transformation can be represented by the matrix formula:

$$X' = AX + X_0, \tag{1.90}$$

where A is a 3×3 nonsingular matrix (i.e., the determinant of matrix A is not zero, $\det A \neq 0$). If we want to stretch or compress a given figure without changing its shape (homothetic transformations), then matrix A can be written as $A = k \cdot I$. Here I is the identity matrix, and if $k > 0$, then the object is stretching and if $0 < k < 1$, then it is compressed equally in all directions. If we want to change the shape of a solid figure, we must start changing its stretching or compression factor in all directions, then the matrix A will have the form:

$$A = \begin{pmatrix} k_x & 0 & 0 \\ 0 & k_y & 0 \\ 0 & 0 & k_z \end{pmatrix}.$$

Using an affine transformation, we can easily obtain some relations for a cube and then obtain the corresponding formulas for a parallelepiped. Next, we will look at examples of solving problems using the method of affine transformations.

Problem 50 Given a rectangular parallelepiped $ABCDA_1B_1C_1D_1$ with edges $AB = a,\ AD = b,\ AA_1 = c$, find the distance from vertex B_1 to the plane going through points $E,\ F,\ K$. It is known that $AE \div AA_1 = 1 \div 3$, $E \in AD_1$, point K is the midpoint of the main diagonal DB_1, $F \in BC$ and $BF \div BC = 1 \div 4$. Evaluate the distance if $a = 1,\ b = 2,\ c = 3$.

Solution

Method 1. (Using affine transformations). We recall solving a very similar problem earlier in this chapter, but it was concerned about the cube with side 1. Using affine transformations we can stretch the cube in all three directions. For any point on a parallelepiped and cube, we have a one-to-one mapping. Assume that a hypothetical point on the unit cube has coordinates $P = (x, y, z)$. This point will have an image P_1 under the affine transformation by matrix A

$$A = \begin{pmatrix} a & 0 & 0 \\ 0 & b & 0 \\ 0 & 0 & c \end{pmatrix}$$

and will have coordinates $P = (x_1, y_1, z_1)$, where $x_1 = ax,\ y_1 = by,\ z_1 = cz$. As you remember, the equation of the plane α passing points $E,\ F,\ K$ on the unit cube was α: $5x + 8y - 9z = 2$. The equation of the corresponding plane α_1 in the given parallelepiped will be $5x_1 + 8y_1 - 9z_1 = 2$, for $x_1 = \frac{x}{a}$, $y_1 = \frac{y}{b}$, and $z_1 = \frac{z}{c}$, we have $5\frac{x}{a} + 8\frac{y}{b} - 9\frac{z}{c} = 2$, or after simplification we obtain $\alpha_1 : \frac{5}{a} \cdot x + \frac{8}{b} \cdot y - \frac{9}{c} \cdot z = 2$. We can also multiply both sides by abc to have a different form of the same plane, α_1: $5bcx + 8acy - 9abz = 2abc$. The distance from B_1 on the unit cube to plane α is

$$\rho(B_1, \alpha) = \frac{|-9 \cdot 1 - 2|}{\sqrt{5^2 + 8^2 + (-9)^2}} = \frac{11}{\sqrt{170}}.$$

The distance from point $B_1 = (0, 0, c)$ to plane α_1 is

$$\rho(B_1, \alpha_1) = \frac{|-\frac{9}{c} \cdot c - 2|}{\sqrt{\left(\frac{5}{a}\right)^2 + \left(\frac{8}{b}\right)^2 + \left(\frac{9}{c}\right)^2}} = \frac{11abc}{\sqrt{25b^2c^2 + 64a^2c^2 + 81a^2b^2}}.$$

Therefore, the distance from point B_1 to the plane containing points $E,\ F,\ K$ is

$$\rho(B_1, \alpha_1) = \frac{11abc}{\sqrt{25b^2c^2 + 64a^2c^2 + 81a^2b^2}}. \tag{1.91}$$

Method 2. This method is based on introducing the coordinates system and then applying formula (1.79) directly. Let's place the center of coordinate system at point $B = (0, 0, 0)$. Let vector BA be in the direction of the X-axis, vector BC of the Y-axis, and vector BB_1 in the direction of the Z-axis. In this system, our points will have the following coordinates: $E = (1, 0, \frac{1}{3}),\ B_1 = (0, 0, c),\ F = (0, \frac{b}{4}, 0)$, and $K = (\frac{a}{2}, \frac{b}{2}, \frac{c}{2})$. Next, we will find the equation of the plane α_1 that passes through points E, K, F using formula (1.50):

$$\begin{vmatrix} x - a & y & z - \frac{c}{3} \\ -a & \frac{b}{4} & -\frac{c}{3} \\ -\frac{a}{2} & \frac{b}{2} & \frac{c}{6} \end{vmatrix} = 0.$$

After evaluating the determinant and equating it to zero and after multiplying both sides by 24, we have

$$bc(x - a) + 4acy - 12ab\left(z - \frac{c}{3}\right) + 3ab\left(z - \frac{c}{3}\right) + 4acy + 4bc(x - a) = 0.$$

Next, after simplification we obtain the equation of the plane α_1:

$$5bcx + 8acy - 9abz = 2abc. \tag{1.92}$$

Using (1.92), coordinates of point B_1, and formula (1.79), we can find the required distance from B_1 to plane α_1:

$$\rho(B_1, \alpha_1) = \frac{|-9ab \cdot c - 2abc|}{\sqrt{25b^2c^2 + 64a^2b^2 + 81a^2b^2}} = \frac{11abc}{\sqrt{25b^2c^2 + 64a^2c^2 + 81a^2b^2}}. \tag{1.93}$$

For a particular case requested in the problem, we can simply replace $a = 1,\ b = 2,\ c = 3$ in formula (1.93) and obtain $d = \frac{11 \cdot 1 \cdot 2 \cdot 3}{\sqrt{1800}} = \frac{11\sqrt{2}}{10}$. If $a = b = c = 1$ we have the case of the unit cube and the distance obtained using (1.93) is equal to $\frac{11}{\sqrt{170}}$.

Both methods lead us to the same answer. However, the second method requires careful algebraic operations and it is easy to make mistakes when calculating the determinant with three parameters and fractions.

Answer. $\frac{11abc}{\sqrt{25b^2c^2+64a^2c^2+81a^2b^2}}$.

Problem 51 Prove that the diagonal AC' of the parallelepiped $ABCDA'B'C'D'$ passes the intersection points of the medians of triangles $A'BD$ and $CB'D'$ and is divided by these points into three equal segments.

Proof In this problem, using affine transformations, we will prove the equality of three segments. Let's check the affine properties of the figure and the conditions of the problem. The affine image of any parallelepiped can be a cube. The division of a segment in a given ratio is an affine property. Consider pyramid $C'CD'B'$ shown in Fig. 1.53. In this pyramid $C'B' = C'C = C'D'$ are the sides of the cube and $B'D' = D'C = CB'$ as the diagonals of the equal faces. The point O_2 is the centroid of face $CB'D'$ and is also the point of intersections of the bisectors, and is so the center of the circle inscribed into this equilateral triangle $CB'D'$. The segment $C'O_2$ is the height of the regular pyramid $C'CD'B'$. Assume that the cube side is 1, then $B'D' = D'C = CB' = \sqrt{2}$ and $B'O_2 = \frac{\sqrt{2}}{\sqrt{3}}$ is the radius of the circumscribed circle.

We can find $C'O_2$ from triangle $C'B'O_2$ as $C'O_2 = \sqrt{1 - (B'O_2)^2} = \frac{1}{\sqrt{3}}$. Similarly, in pyramid $ABDA'$, we can find $A'O_1 = \frac{\sqrt{2}}{\sqrt{3}}$. The diagonal of the cube is equal to $AC' = \sqrt{3}$. Next, we will evaluate the distance

$$O_1O_2 = AC' - \left(AO_1 + C'O_2\right) = \sqrt{3} - 2 \cdot \frac{\sqrt{3}}{3} = \frac{1}{\sqrt{3}}.$$

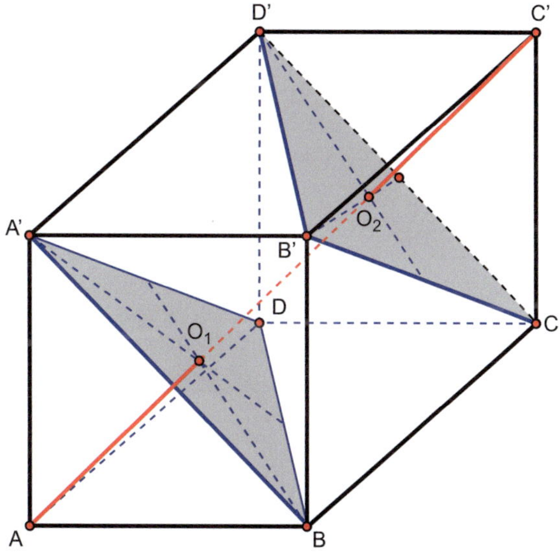

Fig. 1.53 Diagonal of the parallelepiped is divided into three equal parts

We can see that $AO_1 = O_1O_2 = C'O_2 = \frac{1}{\sqrt{3}}$. Therefore, points O_1, O_2 divide the diagonal into three equal parts. □

One difficulty in doing constructions is that the ratios of the segments can visually look correct, but just a little inaccuracy can change the entire cross section of the figure by a plane. For example, let us try to solve the following problem.

Problem 52 The base of the quadrangular pyramid $SABCD$ is the parallelogram $ABCD$. Points K, L, M belong to the edges SA, SB, SC, respectively, and divide these edges in ratios $2 \div 1$, $1 \div 2$, and $3 \div 1$, counting from the top of the pyramid. Does the plane KLM intersect the edge SD? If so, in what ratio does it divide this edge?

Solution We can construct a cross section as shown in Fig. 1.54, where the cross section does not intersect with the lateral edge of SD. However, the drawing is just our sketch. Even if you are absolutely sure of the correctness of your constructions, we need to complete this task with the appropriate proof. We will use the vector-coordinate method and the properties of affine transformations.

1. Let us introduce the following vectors based on the condition of the problem:

$$\overrightarrow{SK} = \frac{2}{3}\overrightarrow{SA}, \ \overrightarrow{SL} = \frac{1}{3}\overrightarrow{SB}, \ \overrightarrow{SM} = \frac{3}{4}\overrightarrow{SC}. \tag{1.94}$$

2. Next, we will express $\overrightarrow{SD}$ in terms of three other vectors.

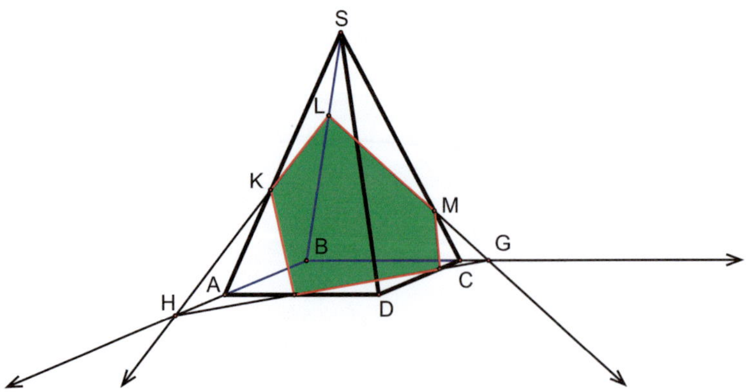

Fig. 1.54 Pentagonal cross section of a quadrangular pyramid

$$\begin{aligned}\overrightarrow{SD} &= \overrightarrow{SB} + \overrightarrow{BD} = \overrightarrow{SB} + \overrightarrow{BA} + \overrightarrow{BC} \\ &= \overrightarrow{SB} + \left(\overrightarrow{SA} - \overrightarrow{SB}\right) + \left(\overrightarrow{SC} - \overrightarrow{SB}\right) \\ &= \overrightarrow{SA} - \overrightarrow{SB} + \overrightarrow{SC}.\end{aligned}$$

3. Assume that the plane α does intersect the edge SD at some point N, so this can be written as $SD \cap \alpha = N, \ \ SN = \lambda SD, \ \ \overrightarrow{SN} = \lambda \overrightarrow{SD}$.
4. Using the previous vector relation, we obtain that

$$\overrightarrow{SN} = \lambda \left(\overrightarrow{SA} - \overrightarrow{SB} + \overrightarrow{SC}\right). \tag{1.95}$$

5. On the other hand, the same vector can be written as

$$\overrightarrow{SN} = \alpha \overrightarrow{SK} + \beta \overrightarrow{SL} + \gamma \overrightarrow{SM}, \ \ \alpha + \beta + \gamma = 1. \tag{1.96}$$

Using (1.94) and (1.96) we obtain that

$$\overrightarrow{SN} = \frac{2\alpha}{3}\overrightarrow{SA} + \frac{\beta}{3}\overrightarrow{SB} + \frac{3\gamma}{4}\overrightarrow{SC}. \tag{1.97}$$

6. Comparing (1.97) and (1.95) we have
$\lambda = \frac{2\alpha}{3} = \frac{\beta}{3} = \frac{3\gamma}{4}$, from which we can find λ,

$$\begin{aligned}&\alpha = \tfrac{3\lambda}{2}, \ \beta = -3\lambda, \ \gamma = \tfrac{4\lambda}{3}, \\ &\alpha + \beta + \gamma = \tfrac{3\lambda}{2} - 3\lambda + \tfrac{4\lambda}{3} = 1 \\ &\lambda = -6.\end{aligned}$$

7. Because $\lambda = -6 < 0 \Rightarrow \overrightarrow{SN} = -6\overrightarrow{SD}$, which means that point N does lie on the line SD but it does not belong to segment SD, so the cross section by the plane KLM does not intersect the edge SD. Our construction by hand showed the cross-sectional pentagon correctly!

Answer. Plane KLM does not intersect edge SD of the pyramid.

1.7 Polyhedral Angles

A **dihedral angle** is a figure formed by two half-planes with a common straight line bounding them. Half-planes are called faces, and the straight line bounding them is an edge of a dihedral angle (Fig. 1.55). A plane perpendicular to the edge of a dihedral angle intersects its faces along two half-lines. The angle formed by these half-lines is called the **linear angle** of the dihedral angle. The measure of the dihedral angle is taken as the measure of the corresponding linear angle. If we take a point on the

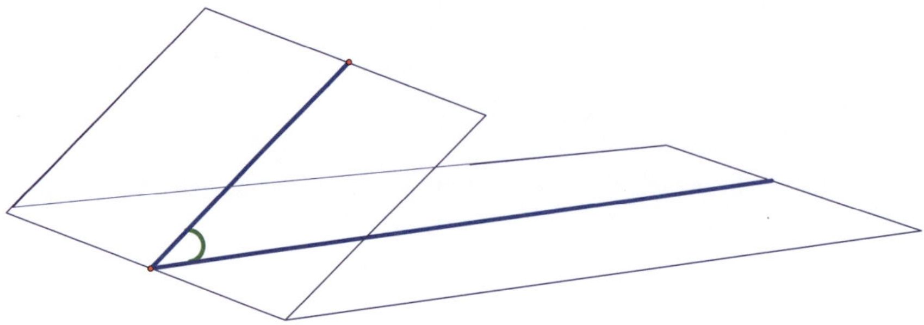

Fig. 1.55 Dihedral angle with its linear angle shown

edge of a dihedral angle (vertex) and drop a perpendicular to this line at each of the two faces, then the resulting angle will be the linear angle of the dihedral angle. All linear angles of a dihedral angle coincide during parallel translation, and therefore are equal. Consequently, the measure of a dihedral angle does not depend either on the choice of the linear angle or on the position of its vertex on the edge. Two dihedral angles are equal if they are superimposed on each other in parallel translation so that their edges coincide. If there is no complete overlap, then the smaller angle is considered to be the dihedral angle that will lie inside the other.

A **trihedral angle** is a part of space bounded by three plane angles with a common vertex and pairwise common sides that do not lie in the same plane. In this case, that part of the space is selected that does not entirely contain any straight line. Let's consider three rays a, b, and c emanating from the same point (S) and not lying in the same plane. A trihedral angle abc is a figure composed of three plane angles ab, bc, and ac Fig. 1.56). These angles are called faces of a trihedral angle and their sides are called edges. The common vertex (S) of plane angles is called the vertex of a trihedral angle. Dihedral angles formed by the faces of a trihedral angle are called dihedral angles of a trihedral angle. Similarly, the concept of a polyhedral angle is defined.

A **polyhedral angle** is a spatial analog of a polygon. Recall that a polygon on a plane is a figure formed by a simple closed broken line and the internal region bounded by it. We can consider a ray in space as an analog of a point on a plane, and a plane angle in space as an analog of a segment on a plane. Then an analog of a simple closed broken line on the plane is a surface formed by a finite set of plane angles $A_1SA_2,\ A_2SA_3,\ \ldots,\ A_{n-1}SA_n,\ A_nSA_1$ with a common vertex S (Fig. 1.57), whose adjacent angles have no common points except the points of the common ray, and nonadjacent angles have no common points except the common vertex.

The figure formed by the specified surface and one of the two parts of the space bounded by it is called a **polyhedral angle**. The common vertex S is called the **vertex of the polyhedral angle**. The rays $SA_1, \ldots, SA_n$ are called the **edges** of the polyhedral angle, and the planes of the angles $A_1SA_2,\ A_2SA_3,\ \ldots,\ A_{n-1}SA_n,\ A_nSA_1$ are called the **faces of the polyhedral angle**. A polyhedral angle is denoted by the

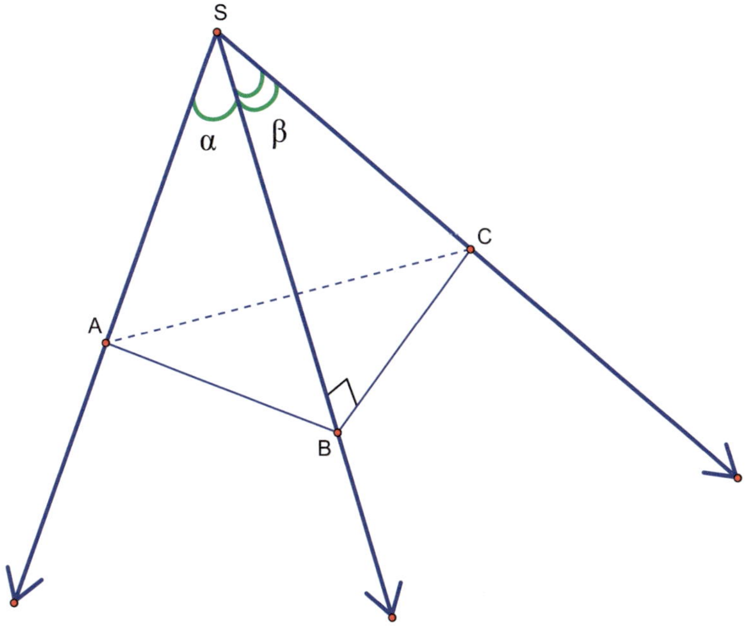

Fig. 1.56 Trihedral angle

Fig. 1.57 Polyhedral angle

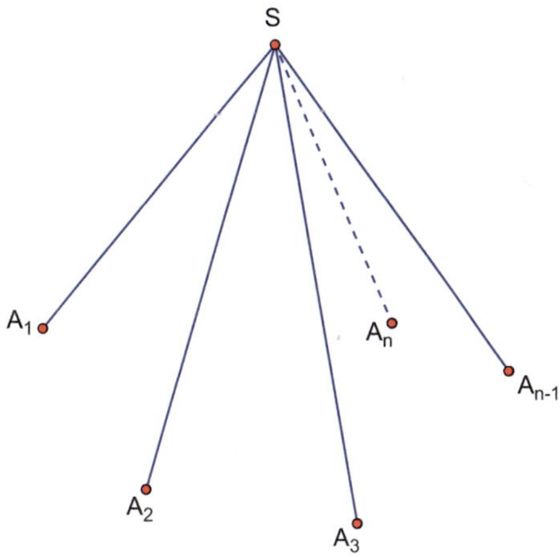

letters $SA_1 \ldots A_n$, indicating the vertex and points on its edges. Depending on the number of faces, polyhedral angles are called trihedral, tetrahedral, pentahedral, etc.

A polyhedral angle is called convex if it lies entirely on one side of the planes of each of its faces. If we intersect all faces of a polyhedral angle by a plane, then in this plane a polygon is formed. If a polyhedral angle is convex, then the polygon is also convex. Consider a polyhedral angle *Sabcmn* is sketched in Fig. 1.58. We can see

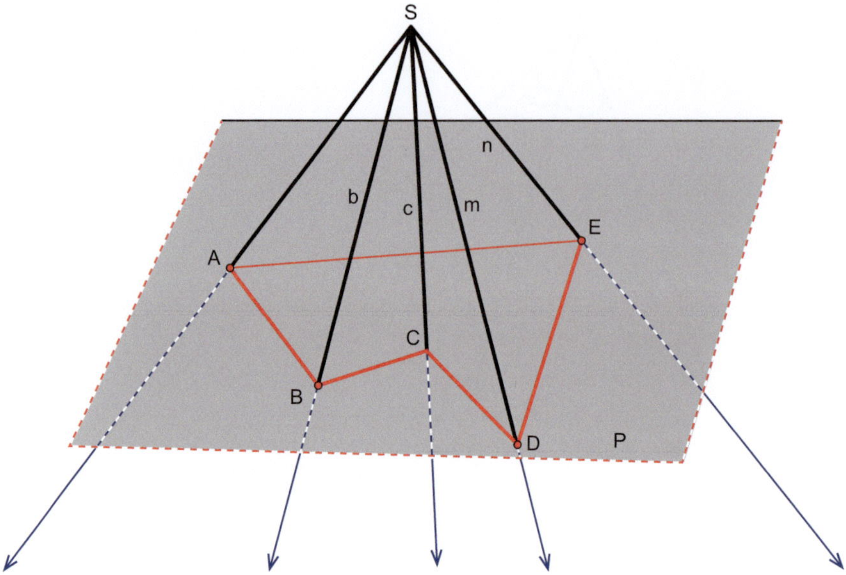

Fig. 1.58 Nonconvex polyhedral angle

that plane p intersects the angle by a nonconvex pentagon $ABCDE$. Therefore, the given polyhedral angle is also nonconvex. A polyhedral angle is called **convex** if it is a convex figure, if together with any two of its points, it also contains the segment connecting them. Thus, trihedral angles are always convex. In this text unless the opposite is specified, we will deal with only convex polyhedral angles.

Problem 53 What is the smallest number of non-intersecting trihedral angles that space can be divided into?

Solution Let O be a point inside the triangular pyramid $ABCD$. Then four trihedral angles $OABC$, $OABD$, $OACD$, and $OBCD$ with the common vertex O satisfy the condition of the problem. It is clear that four is the minimal number of trihedral angles. Space cannot be divided into two and three non-intersecting trihedral angles.

Answer. The minimal number of non-intersecting trihedral angles is four.

Problem 54 Given a tetrahedral angle. How does this angle divide the space?

Solution Consider a tetrahedral angle with vertex O. Let us draw four planes, respectively, parallel to the given four planes. They bound a certain tetrahedron. Let us first prove that these planes divide space into 15 parts. Let point N be an arbitrary point in space located outside the tetrahedron. Then the segment NO intersects either the face of the tetrahedron (4 options), or its edge (6 options), or passes through the vertex of the tetrahedron (4 options). Each option corresponds to exactly one (different from the interior of the tetrahedron) part of the space from those into which the space is divided by the indicated planes. Thus, the total number of the possible parts is $4 + 6 + 4 + 1 = 15$.

However, the number of parts into which the four original planes divide space is one less than what was found (the tetrahedron "turned" into a point, and one part disappeared). At the same time, six parts of space that corresponded to the edges of the tetrahedron turned into tetrahedral angles, and $4 + 4 = 8$ parts that corresponded to the vertices and faces turned into trihedral angles.

Answer. Fourteen parts, eight trihedral and six tetrahedral.

1.7.1 Properties of Polyhedral Angles

Any trihedral angle is completely defined by its plane angles, i.e., α, β, and γ. A trihedral angle is a rigid structure and cannot be deformed. It can be easily explained. We measure equal distances from the vertex of the trihedral angle S and mark points A, B, and C on each edge. Triangle ABC is unique for the given angles of a trihedral angle because the triangle is entirely defined by its three sides. If we take, for example, a tetrahedral angle, adding δ angle to three other angles (α, β and γ), then such a **tetrahedral angle** is easily deformed due to the fact that the quadrilateral is not defined by its sides. So, for example, a rhombus and a square can have the same sides, but a square has an angle between adjacent sides of $90°$, while a rhombus can have any. Next we will look at some properties of polyhedral angles, paying special attention to the properties of trihedral and tetrahedral angles. You'll see how knowing the properties of triangles and quadrilaterals can help you learn similar properties of trihedral and tetrahedral angles.

Property 1 (*Triangle Inequality*) Each side of a triangle is less than the sum of the other two sides.

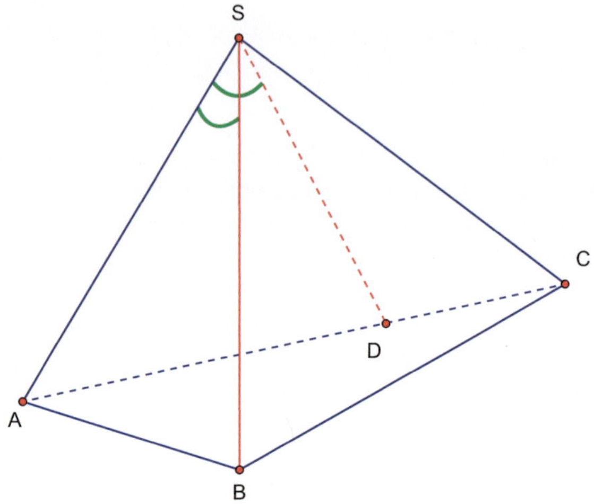

Fig. 1.59 Proof of Theorem 24

A similar property exists for trihedral angles.

Theorem 24 *Each plane angle of a trihedral angle is less than the sum of the other two plane angles.*

Proof Consider the trihedral angle $SABC$ Fig. 1.59. Let the largest of its plane angles be the angle ASC. Then for other two angles, the following inequalities hold:

$$\angle ASB \le \angle ASC < \angle ASC + \angle BSC; \ \angle BSC \le \angle ASC < \angle ASC + \angle ASB.$$

Thus, we still have to prove the inequality $\angle ASC < \angle ASB + \angle BSC$. Let us construct the angle ASD on the face ASC equal to ASB and choose the point B so that $SB = SD$ (Fig. 1.59). Then the triangles ASB and ASD are congruent (by two sides and the angle between them) and, therefore, $AB = AD$.

Applying the triangle inequality to the triangle ABC, we can state:

$$AC < AB + BC.$$

Subtracting from both its parts $AD = AB$, we obtain that

$$AC - AD < AB + BC - AB.$$

Since $AC - AD = DC$, then we obtain the inequality

$$DC < BC.$$

In triangles DSC and BSC, one side is common (SC), $SD = SB$ and $DC < BC$. In this case, a larger angle lies against the larger side and, therefore,

$$\angle DSC < \angle BSC.$$

Adding to both sides of this inequality the angle ASD equal to ASB, as

$$\angle DSC + \angle ASD < \angle ASD + \angle BSC$$
$$\angle ASC < \angle ASD + \angle BSC.$$

Since $\angle ASD = \angle ASB$, we obtain the required inequality

$$\angle ASC < \angle ASB + \angle BSC.$$

The theorem is proven. □

The following theorem can be considered as corollary of Theorem 24.

Theorem 25 *The sum of the plane angles of a trihedral angle is less than* 360°.

Proof Let $SABC$ be a given trihedral angle. We can use Fig. 1.59. Consider a trihedral angle with vertex A, formed by faces ABS, ACS, and angle BAC. By virtue of the proved Theorem 24, the inequality $\angle BAC < \angle BAS + \angle CAS$ takes place. Similarly, for trihedral angles with vertices B and C, the following inequalities hold:

$$\angle ABC < \angle ABS + \angle CBS, \quad \angle ACB < \angle ACS + \angle BCS.$$

Adding these inequalities and considering that the sum of the angles of triangle ABC is 180°, we get

$$180^\circ < \angle BAS + \angle CAS + \angle ABS + \angle CBS + \angle BCS + \angle ACS$$
$$= 180^\circ - \angle ASB + 180^\circ - \angle BSC + 180^\circ - \angle ASC.$$

The second row in the formula above was obtained by regrouping the pairs of angles related to one of the three triangles as follows:

$$\angle BAS + \angle ABS = 180^\circ - \angle ASB \quad (\triangle\ ASB)$$
$$\angle CAS + \angle ACS = 180^\circ - \angle ASC \quad (\triangle\ ASC)$$
$$\angle CBS + \angle BCS = 180^\circ - \angle BSC \quad (\triangle\ BSC).$$

Therefore,

$$\angle ASB + \angle BSC + \angle ASC < 2 \cdot 180^\circ = 360^\circ = 2\pi.$$

The sum of plane angles in a trihedral angle is less than 2π. □

Note. This theorem is very important, for example, for answering the following simple question: Is there a trihedral angle with the plane angles $\alpha = 105^\circ$, $\beta = 145^\circ$, and $\gamma = 135^\circ$? The answer is no, because $\alpha + \beta + \gamma = 105^\circ + 145^\circ + 135^\circ = 385^\circ > 360^\circ$ is impossible for a trihedral angle.

A trihedral angle is a particular case of a convex polyhedral angle. Next, we will state and prove that in any convex polyhedral angle the sum of the plane angles is always less than 360°.

Theorem 26 *The sum of the plane angles of a convex polyhedral angle is less than* 360°.

Proof Consider a polyhedral angle $SA_1A_2A_3...A_{n-1}A_n$ and intersect it by a plane p. The cross section is a convex polygon $A_1A_2A_3...A_{n-1}A_n$. This is shown in Fig. 1.60. Although the angle at the vertex S of the resulting spatial figure is polyhedral, the angles at the vertices $A_1,\ A_2,\ A_3, ...,\ A_{n-1},\ A_n$ are trihedral. The sum of the flat angles, S_n, is

$$S_n = \angle A_1SA_2 + \angle A_2SA_3 + ... + \angle A_{n-1}SA_n + \angle A_nSA_1.$$

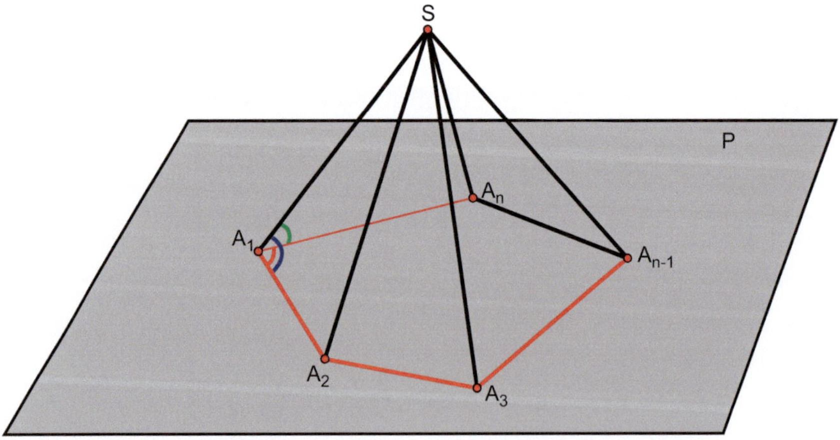

Fig. 1.60 Convex polyhedral angle

Each of the angles can be expressed as the difference of 180° and the sum of two other angles of the corresponding triangle. For example, $\angle A_1SA_2$ can be expressed from ΔA_1SA_2 as $\angle A_1SA_2 = 180° - (\angle SA_1A_2 + \angle A_1A_2S)$. Now, for the sum of n flat angle we can write

$$\begin{aligned}S_n =& n \cdot 180° - (\angle SA_1A_2 + \angle A_1A_2S) - (\angle SA_2A_3 + \angle A_2A_3S) - ...\\&-(\angle SA_{n-1}A_1 + \angle A_{n-1}A_nS) - (\angle SA_nA_1 + \angle A_nA_1S).\end{aligned}$$

After regrouping the angles, we obtain

$$\begin{aligned}S_n =& n \cdot 180° - [(\angle SA_1A_n + \angle SA_1A_2) + (\angle A_1A_2S + \angle SA_2A_3) + ...\\&+ (\angle SA_nA_{n-1} + \angle SA_nA_1).]\end{aligned} \quad (1.98)$$

Considering what we have proven in the Theorem 24, for any trihedral angle, any flat angle is less than the sum of two other flat angles. Hence, for example, for the trihedral angle with vertex at A_1 we can write that

$$\angle SA_1A_n + \angle SA_1A_2 > \angle A_nA_1A_2,$$

where $\angle A_nA_1A_2$ is an interior angle of the polygon $A_1A_2...A_n$. A similar inequality can be written for all quantities inside the parentheses in (1.98), so that the equality will become the inequality. Considering the fact that the sum of all interior angles in an n-gon is equal to $(n-2) \cdot 180°$, then (1.98) is rewritten as

$$S_n < n \cdot 180° - (n-2) \cdot 180° = 360°.$$

Therefore, for any polyhedral convex angle, the sum of the flat angles is less than 360°. □

Problem 55 All plane angles of a convex polyhedral angle are equal to each other and their degree measure is 47°. What is the maximum number of faces that a given polyhedral angle can have?

Solution The sum of all face angles must be less than 360 degrees, then we need to divide 360 by 47 and round the answer to the nearest integer, that is, 7.

Answer. A convex polyhedral with equal plane angles has at most seven faces.

Corollary 2 *The sum of the dihedral angles of a trihedral angle is greater than* 180°.

Proof Let $SABC$ be a trihedral angle. We choose some point P inside it and drop the perpendiculars PA_1, PB_1, PC_1 from it on the faces that are opposite to the vertices A, B, and C, respectively, Fig. 1.61.

To better understand the idea of the proof, let's place the pyramid on its face SBC and consider only two perpendiculars from point P to face SAC (PB_1), and another one from P to face SBC (segment PA_1). This new Fig. 1.62 additionally shows a dihedral angle with edge SC (labeled by one arc) and $\angle A_1PB_1$ (labeled by two arcs), both inside the quadrilateral with two other right angles ($\angle PA_1M = \angle PB_1M = 90°$).

From quadrilateral A_1PB_1M we can state that plane angle A_1PB_1 is complementary to the dihedral angle B_1MA_1. This can be written as $\angle B_1MA_1 = 180° - \angle A_1PB_1$. Now we can state that all plane angles B_1PC_1, A_1PC_1, A_1PB_1 complement the corresponding dihedral angles with edges SA, SB, SC up to π. Therefore, the sum of these dihedral angles (Σ) is equal to $\Sigma = 540° - (\angle B_1PC_1 + \angle A_1PC_1 + \angle A_1PB_1)$. Since $\angle PA_1B_1C_1$ is also a trihedral angle with vertex P, by Theorem 25 the sum of its plane angles (the sum inside parentheses) is less than 360°, which means that the sum of the dihedral angles of the original trihedral angle (Σ) is greater than 180°. □

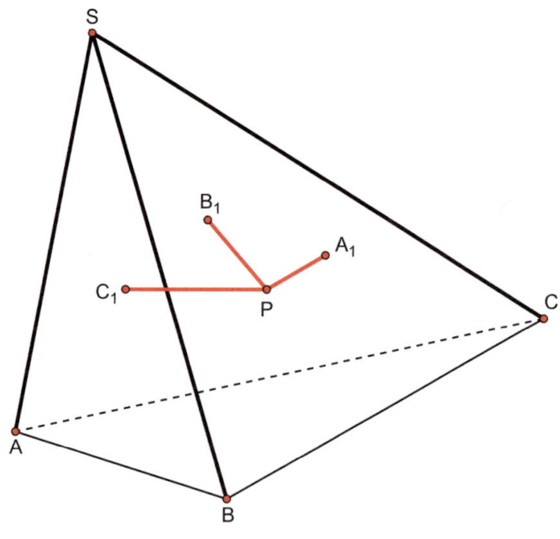

Fig. 1.61 Sketch to prove Corollary 2

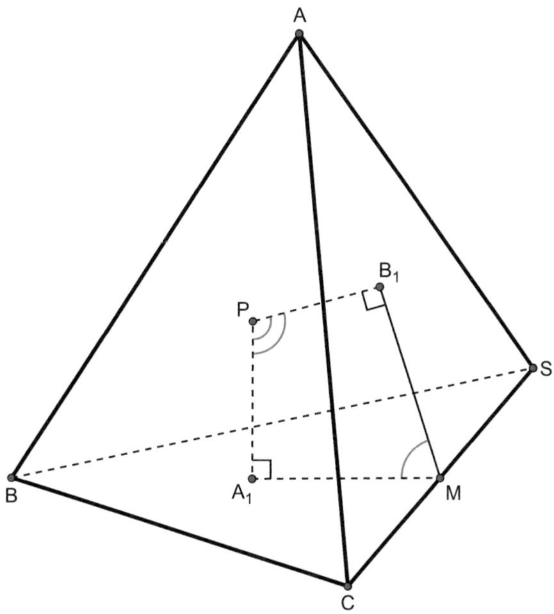

Fig. 1.62 An additional sketch to prove Corollary 2

Problem 56 A trihedral angle has two plane angles whose measures are 80° and 120°, respectively. Which of the following can be the measure of a third face angle?

$$20^\circ,\ 40^\circ,\ 80^\circ,\ 90^\circ,\ 160^\circ,\ 170^\circ.$$

Solution Denote the unknown third angle by α, and also $\beta = 80^\circ$ and $\gamma = 120^\circ$. We know that the sum of all face (plane) angles of a trihedral angle must be less than 360° and that the measurement of each face angle must be less than the sum of two other angles. This can be written as the system (degree notation omitted):

$$\begin{cases} \alpha + \beta + \gamma < 360 \\ \alpha < \beta + \gamma \\ \beta < \alpha + \gamma \\ \gamma < \alpha + \beta \end{cases} \Leftrightarrow \begin{cases} \alpha + 80 + 120 < 360 \\ \alpha < 80 + 120 \\ 80 < \alpha + 120 \\ 120 < \alpha + 80 \end{cases} \Rightarrow \begin{cases} \alpha < 160 \\ \alpha < 200 \\ \alpha > -40 \\ \alpha > 40 \end{cases}.$$

Solving this system, we obtain the double inequality, $40^\circ < \alpha < 160^\circ$. There are only two values of the angles (80° and 90°) that satisfy this inequality.

Answer. The third angle can be either 80° or 90°.

Problem 57 All plane angles of a trihedral angle are equal to each other and their degree measure is expressed as an integer. What is the largest degree measure, α of each of these angles?

Solution Since all face angles have the same measure, $3 \cdot \alpha < 360^\circ$, so $\alpha < 120^\circ$. Hence, then largest integer less than 120° is 119°.

Answer. $\alpha = 119^\circ$.

Problem 58 The two plane angles of a trihedral angle have degree measures of 105° and 75°, respectively. How many different integer values can the degree measure of the third plane angle take?

Solution We can denote the unknown angle by α, then by analogy with one of the previous problem we have to solve several inequalities that will lead us to the following restrictions, $30^\circ < \alpha < 180^\circ$. We can see that there are exactly 149 integer numbers satisfying this inequality (31°, 32°, ..., 179°).

Answer. There are 149 possibilities.

Problem 59 Determine whether a trihedral angle can have face angles with the following measures and give reasons for your answer.

1. 80°, $110^\circ, 140^\circ$;
2. 72°, $128^\circ, 156^\circ$;
3. 45°, $45^\circ, 90^\circ$;
4. 60°, $60^\circ, 60^\circ$;
5. 140°, $170^\circ, 171^\circ$;
6. 105°, $118^\circ, 130^\circ$.

Solution We need to check the validity of the system of inequalities for each case:

$$\begin{cases} \alpha + \beta + \gamma < 360^\circ \\ \alpha < \beta + \gamma \\ \beta < \alpha + \gamma \\ \gamma < \alpha + \beta \end{cases}$$

1. The three angles: $80°, \ 110°, \ 140°$ may be the face angles of a trihedral angle.
2. Since $72° + 128° + 156° = 356° < 360°$ and all other inequalities are valid then $72°, \ 128°, 156°$ can be the face angles of a trihedral angle.
3. Since the last inequality of the system is violated (the third angle is $90° = 45° + 45°$, but must be less than the sum of the other two angles), then $45°, \ 45°, \ 90°$ cannot be planar angles of a trihedral angle.
4. $60°, \ 60°, 60°$ may be three plane angles of a trihedral angle.
5. The sum of the angles is $140° + 170° + 171° = 481° > 360°$. This set is invalid.
6. $105°, \ 118°, \ 130°$ may be three plane angles of a trihedral angle.

Answer. The triples in Cases 1, 2, 4, and 6 may be the face angles of a trihedral angle. Cases 3 and 5 are invalid.

It is time to review more properties of polyhedral angles.

Property 2 The bisectors of a triangle intersect at one point.

Lemma 3 *The bisector planes of dihedral angles in a trihedral angle intersect along one straight line.*

Proof The proof is similar to the plane case. Namely, let $SABC$ be a trihedral angle. The bisector plane of a dihedral angle SA divides the angle into two equal parts and is the locus of all points equidistant from the faces ASC and ASB of the angle. Similarly, the bisector plane of a dihedral angle SB is the locus of the points equidistant from its faces BSA and BSC. The line of their intersection SO will be equidistant from all faces of the trihedral angle and, therefore, the bisector plane of the dihedral angle SC will pass through it. □

Property 3 The perpendicular bisectors to the sides of a triangle intersect at one point.

Lemma 4 *The planes passing through the bisectors of the faces of a trihedral angle and perpendicular to these faces intersect along one straight line.*

The proof is similar to the proof of the previous property.

Property 4 The medians of a triangle intersect at one point.

Lemma 5 *The planes passing through the edges of the trihedral angle and the bisectors of opposite faces intersect along one line.*

Proof Consider the trihedral angle $SABC$, where points A, B, and C, such that $SA = SB = SC$ (Fig. 1.63). Then the bisectors SA_1, SB_1, and SC_1 of the angles BSC, ASC, ASB are the medians of the corresponding triangles. Therefore AA_1, BB_1, CC_1 are medians of triangle ABC. Let O be their intersection point. The line SO is contained in all three considered planes and is so the line of their intersection. □

Property 5 The altitudes of a triangle intersect at one point.

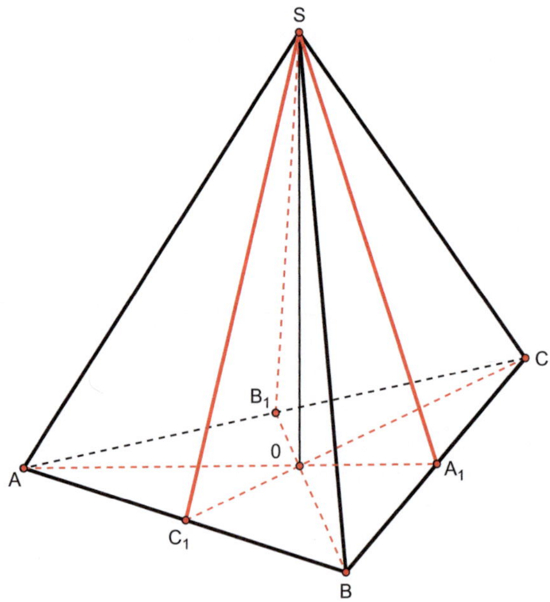

Fig. 1.63 Proof of Lemma 5

Lemma 6 *The planes passing through the edges of a trihedral angle and perpendicular to opposite faces intersect along one straight line.*

Proof Consider a trihedral angle with vertex S and edges $a,\ b,\ c$. Let $a_1,\ b_1,\ c_1$ be the lines of intersection of faces with planes passing through the corresponding edges and perpendicular to these opposite faces (Fig. 1.64). Select a point C on the edge c and drop perpendiculars CA_1 and CB_1 from it to the lines a_1 and b_1. Let A and B be the intersections of lines CA_1 and CB_1 with lines a and b, then SA_1 is the projection of AA_1 onto the face BSC. Since BC is perpendicular to SA_1, it is also perpendicular to AA_1. Similarly, AC is perpendicular to BB_1. So AA_1 and BB_1 are the altitudes of triangle ABC. Let O be their intersection point. The planes passing through the lines a and a_1, b and b_1 are perpendicular to the plane ABC and, therefore, the line of their intersection SO is perpendicular to ABC. So, SO is perpendicular to AB. On the other hand, CO is perpendicular to AB. Therefore, the plane passing through the edge c and SO will be perpendicular to the opposite face (ASB). □

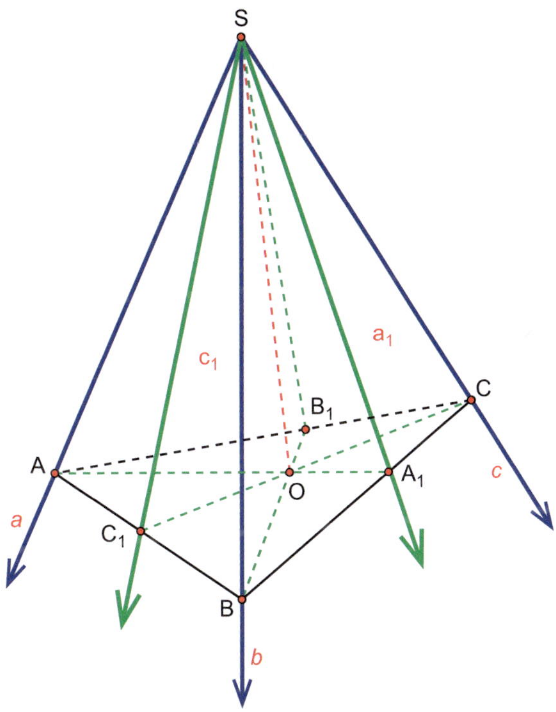

Fig. 1.64 Lemma 6. Planes passing edges ⊥ to the opposite faces intersect at one line

Property 6 (*Law of Sines*) In triangle ABC with sides $a,\ b,\ c$, we have

$$\frac{a}{\sin A} = \frac{b}{\sin B} = \frac{c}{\sin C}.$$

Theorem 27 (The First Law of Sines Theorem for a Trihedral Angle) *Let $\alpha,\ \beta,\ \gamma$ be plane angles of a trihedral angle with $a,\ b,\ c$ being the corresponding opposite dihedral angles, then*

$$\frac{\sin\alpha}{\sin a} = \frac{\sin\beta}{\sin b} = \frac{\sin\gamma}{\sin c}.$$

Proof Let $SABC$ be a trihedral angle. Let us drop perpendicular CC_1 from point C to plane ASB and perpendicular CA_1 to edge SA (Fig. 1.65). Then the angle CA_1C_1 will be the linear angle of the dihedral angle a. $CB_1 \perp SB$. The angle CB_1C_1 is the linear angle of the dihedral angle b. Therefore

$$CC_1 = CA_1 \cdot \sin a = SC \cdot \sin\beta \cdot \sin a.$$

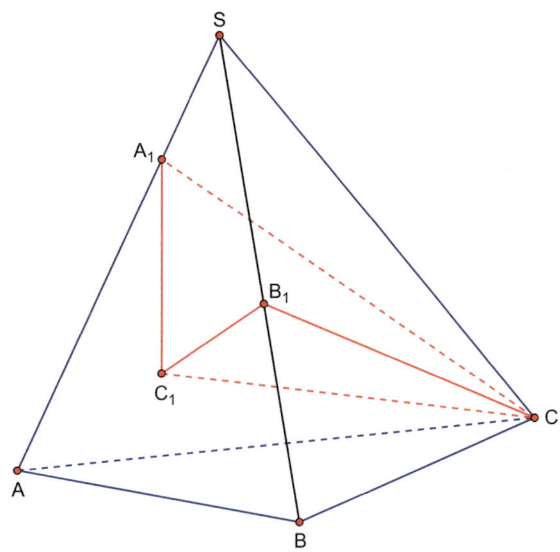

Fig. 1.65 Proof of Theorem 27

Similarly, we can show that

$$CC_1 = CB_1 \cdot \sin b = SC \cdot \sin\alpha \cdot \sin b.$$

So, we have the equality, $\sin\beta \cdot \sin a = \sin\alpha \cdot \sin b$ and then also $\frac{\sin\alpha}{\sin a} = \frac{\sin\beta}{\sin b}$. Similarly, it can be proven that the equality $\frac{\sin\beta}{\sin b} = \frac{\sin\gamma}{\sin c}$ holds. □

Theorem 28 (The First Law of Cosines Theorem for a Trihedral Angle) *Let α, β, γ be the plane angles of a trihedral angle, and A be the dihedral angle opposite to the plane angle α, then*

$$\cos\alpha = \cos\beta \cdot \cos\gamma + \sin\beta \cdot \sin\gamma \cdot \cos A. \tag{1.99}$$

Proof Let us construct the unit vectors SA_1, SB_1, and SC_1 on the sides SA, SB, and SC of the trihedral angle $SABC$. Consider the dihedral angle at the edge SA and construct $MA_1 \perp SA$ and $NA_1 \perp SA$ as it is shown in Fig. 1.66. Here $\angle MA_1N$ is a linear angle of the dihedral angle SA. Let us find its cosine.

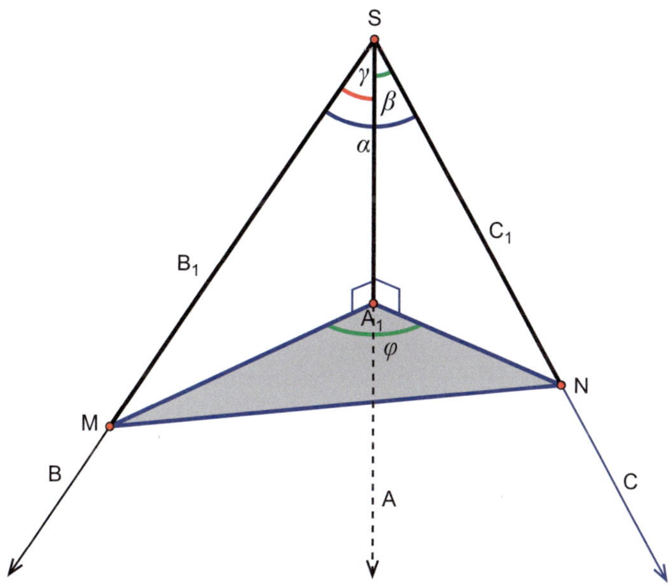

Fig. 1.66 Proof of the First Cosine formula (Theorem 28.)

Using the scalar product property, we can express cosine of the dihedral angle SA as

$$\cos\angle SA = \cos\varphi = \frac{\overrightarrow{A_1M}\cdot\overrightarrow{A_1N}}{\|A_1M\|\cdot\|A_1N\|} = \frac{(A_1M, A_1N)}{\|A_1M\|\cdot\|A_1N\|}. \tag{1.100}$$

From the right triangle SA_1M, we have

$$A_1M = SB_1\cdot\tan\gamma = 1\cdot\tan\gamma = \tan\gamma. \tag{1.101}$$

From the right triangle SA_1N, we have

$$A_1N = SA_1\cdot\tan\beta = \tan\beta. \tag{1.102}$$

Substituting (1.101) and (1.102) into (1.100) we obtain an intermediate formula for the cosine of the dihedral angle SA:

$$\cos\angle SA = \cos\varphi = \frac{(A_1M, A_1N)}{\tan\gamma\cdot\tan\beta}. \tag{1.103}$$

Now we can find the scalar product of the fraction in formula (1.103). Using the rules of vectors, we have

$$\left.\begin{array}{l}\overrightarrow{A_1M} = \overrightarrow{SM} - \overrightarrow{SA_1}\\ \overrightarrow{A_1N} = \overrightarrow{SN} - \overrightarrow{SA_1}\end{array}\right\} \Rightarrow (A_1M, A_1N) = (SM - SA_1, SN - SA_1).$$

Applying the scalar product property again (vector symbol omitted):

$$\begin{aligned}(A_1M, A_1N) &= (SM, SN) - (SM, SA_1) - (SA_1, SN) + (SA_1, SA_1)\\ &= \|SM\|\cdot\|SN\|\cdot\cos\alpha - \|SM\|\cdot\|SA_1\|\cdot\cos\gamma\\ &- \|SA_1\|\cdot\|SN\|\cdot\cos\beta + \|SA_1\|^2.\end{aligned} \tag{1.104}$$

The lengths of SM and SN can be obtained from the right triangles SA_1M and SA_1N, respectively. Each of these triangles has a one unit side, $SA_1 = 1$. Taking advantage of the fact that in a right triangle the tangent of an angle is equal to the ratio of the side opposite the angle to the adjacent side, we obtain

$$SM = \frac{1}{\cos\gamma},\quad SN = \frac{1}{\cos\beta}. \tag{1.105}$$

Substituting (1.105) into (1.104), we get

$$(A_1M, A_1N) = \frac{1}{\cos\gamma}\cdot\frac{1}{\cos\beta}\cdot\cos\alpha - \frac{1}{\cos\gamma}\cdot\cos\gamma - \frac{1}{\cos\beta}\cdot\cos\beta + 1 = \frac{\cos\alpha}{\cos\gamma\cos\beta} - 1$$

which can be written as

$$(A_1M, A_1N) = \frac{\cos\alpha - \cos\gamma\cos\beta}{\cos\gamma\cos\beta}. \tag{1.106}$$

Now we can find the dihedral angle SA by substituting (1.106) into (1.103):

$$\cos\varphi = \frac{\cos\alpha - \cos\gamma\cos\beta}{\cos\gamma\cos\beta\tan\gamma\tan\beta} = \frac{\cos\alpha - \cos\gamma\cos\beta}{\sin\gamma\sin\beta} = \cos\angle SA. \tag{1.107}$$

Formula (1.107) can be solved for $\cos\angle SA = \cos A$ to complete the proof:

$$\cos\alpha = \cos\beta\cdot\cos\gamma + \sin\beta\cdot\sin\gamma\cdot\cos A.$$

□

Remark. The formulas for all three dihedral angles in terms of the plane angles α, β and γ of a trihedral angle are

$$\begin{aligned} \cos A &= \tfrac{\cos\alpha-\cos\gamma\cos\beta}{\sin\gamma\sin\beta} = \cos\angle SA \\ \cos B &= \tfrac{\cos\beta-\cos\alpha\cos\gamma}{\sin\alpha\sin\gamma} = \cos\angle SB \\ \cos C &= \tfrac{\cos\gamma-\cos\alpha\cos\beta}{\sin\alpha\sin\beta} = \cos\angle SC. \end{aligned} \tag{1.108}$$

Theorem 29 (The Second Law of Cosines Theorem for a Trihedral Angle) *Let α, β, γ be its plane angles of a trihedral angle and A, B, C be the dihedral angles composed of planes of angles β and γ, α and γ, and α and β, respectively, then*

$$\begin{aligned} \cos A &= -\cos B\cos C + \sin B\sin C\cos\alpha \\ \cos B &= -\cos A\cos C + \sin A\sin C\cos\beta \\ \cos C &= -\cos A\cos B + \sin A\sin B\cos\gamma. \end{aligned} \tag{1.109}$$

Proof Without loss of generality we will prove only the first relationship in (1.109) and show that $\cos A = -\cos B\cos C + \sin B\sin C\cos\alpha$. Also, we can use the First Cosine Theorem 28 to prove the Second Cosine Theorem 29. Let us rewrite (1.99) as $\cos\alpha = \cos\beta\cos\gamma + \sin\beta\sin\gamma\cos A$. Assume that $SABC$ is a given trihedral angle (Fig. 1.67). Let us mentally drop the perpendiculars from the interior point of the trihedral angle to its faces and obtain a new trihedral angle (dual to the given one). You can also look again at the proof of Corollary 2 and Figs. 1.61 and 1.62.

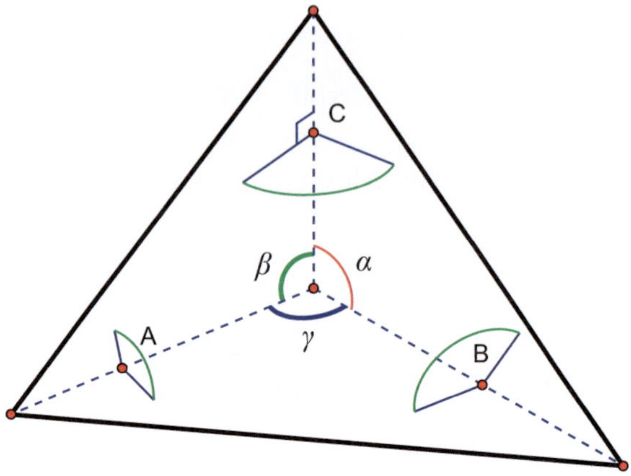

Fig. 1.67 Proof of the Second Cosine Theorem 29

The plane angles of one trihedral angle complement the dihedral angles of another, and the dihedral angles of one angle complement the plane angles of another up to 180°. That is, the plane angles of the dual trihedral angle are, respectively, equal to $180° - A$; $180° - B$; $180° - C$ and the dihedral angles are $180° - \alpha$; $180° - \beta$; $180° - \gamma$. Let's write the First Cosine Theorem 28 for it

$$\cos(\pi - A) = \cos(\pi - \alpha)\sin(\pi - B)\sin(\pi - C) + \cos(\pi - B)\cos(\pi - C)$$

and after simplifications we get

$$\cos A = \cos\alpha \sin B \sin C - \cos B \cos C.$$

□

Property 7 Convex quadrilateral. If a circle can be inscribed in a convex quadrilateral, then the sums of opposite sides are equal.

Lemma 7 *If a sphere can be inscribed in a convex tetrahedral angle, then the sums of opposite plane angles of this tetrahedral angle are equal.*

Proof Assume that we can inscribe a sphere into a tetrahedral angle $SABCD$ shown in the Fig. 1.68. Let points K, L, M, and N be the points at which the sphere is tangent to the face angles ASB, BSC, CSD, , and DSA, respectively. Here points A, B, C, and D will be vertices of the quadrilateral $ABCD$ circumscribed about the largest circle of the sphere. This quadrilateral is sketched in Fig. 1.68 on the right. The sketch is not in scale and it is shown only for better visualization. Recall that

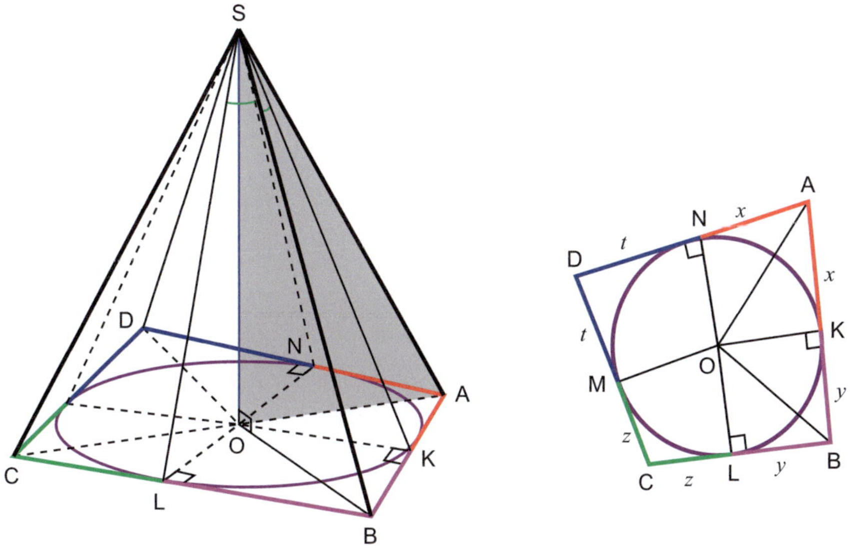

Fig. 1.68 Proof of Lemma 7

for a circumscribed quadrilateral the sum of opposite sides is equal, which is based on the properties of the equal segments of tangent drawn to a circle from the same vertex. Thus, you can see that since $AB = x + y$, $AD = t + x$, $DC = t + z$, and $CB = z + y$, then $AB + DC = x + y + z + t = AD + CB$.

Let us return to a tetrahedral angle. The tangent points N and K are symmetric with respect to plane SOA, tangent points L and M are symmetric with respect to plane SOC, etc. The following is true:

$$\angle ASK = \angle ASN, \quad \angle BSK = \angle BSL, \quad \angle LSC = \angle SCM, \quad \angle MSD = \angle DSN. \tag{1.110}$$

Now we can express the sum of two opposite plane angles of the tetrahedral:

$$\begin{aligned}\angle ASB + \angle CSD &= (\angle ASK + \angle BSK) + (\angle CSM + \angle MSD) \\ &= \angle ASN + \angle BSL + \angle CSL + \angle DSN \\ &= \angle ASN + (\angle BSL + \angle CSL) + \angle DSN \\ &= \angle ASD + \angle BSC,\end{aligned}$$

where we step by step replaced certain angles with equal angles according to the formula (1.110). □

1.7.2 Problems on Polyhedral Angles

Let us solve some problems applying the properties of dihedral, trihedral, and tetrahedral angles.

Problem 60 The opposite edges of a tetrahedron are equal in pairs. Prove that all faces of this tetrahedron are congruent acute triangles.

Proof Draw a tetrahedron $ABCD$ and color the opposite (skew) edges in pairs with different colors, for example, red, blue, and green. You can immediately see that all the triangles are equal (each has a red, blue, and green side). Let us prove these triangles are acute, not obtuse. Consider a tetrahedron $ABCD$ in which $AC = BD = a$, $AD = BC = b$, and $AB = CD = c$. We can see that all triangles are equal by three sides:

$$\triangle ADC = \triangle CBD = \triangle ABD = \triangle ABC.$$

This means that the angles of these triangles lying opposite equal sides are pairwise equal. Let us denote

$$\begin{aligned}\angle ADC = \angle DCB = \angle ABC = \angle DAB = \alpha\\ \angle DCA = \angle CDB = \angle CAB = \angle DBA = \beta\\ \angle DAC = \angle DBC = \angle ADB = \angle ACB = \gamma.\end{aligned}$$

Next, consider plane angles at any vertex because each trihedral angle in this tetrahedron is the same having the plane angles $\alpha,\ \beta$, and γ. Assume that each face of the tetrahedron is an obtuse triangle and let $\alpha > 90°$. Using Theorem 24, each plane angle of a trihedral angle is less than the sum of two other plane angles, then $\alpha < \gamma + \beta$. Then $\gamma + \beta > \alpha > 90°$, and then $\alpha + \beta + \gamma > 90° + 90° = 180°$, which is impossible, because the sum of the angles of any triangle is exactly $180°$. Likewise, $\beta < 90°$ and $\gamma < 90°$. Therefore triangles $\triangle ADC,\ \triangle DBC,\ \triangle ABC$, and $\triangle ADB$ are equal acute triangles. □

Problem 61 Using Theorem 28 prove that any plane angle of a trihedral angle is less than the sum of two other plane angles.

Proof Without loss of generality we can use for our proof only the relationship (1.99):

$$\cos\alpha = \cos\beta \cdot \cos\gamma + \sin\beta \cdot \sin\gamma \cdot \cos A. \tag{1.111}$$

In this formula α, β, and γ are the plane angles and A is the dihedral angle opposite to the plane angle α. It follows from the properties of the cosine function and of a dihedral angle that $-1 < \cos A < 1$. Consider the function $y = f(x) = \cos x$ on the interval $(0, \pi)$. We do not need to consider any other interval because the measure of a plane angle of a trihedral angle cannot be greater than 180°. On this interval, the cosine function is monotonically decreasing taking values $-1 < \cos x < 1$. Hence, if we replace on the right-hand side of (1.111) $\cos A$ by -1, then we have

$$\cos\alpha > \cos\beta \cdot \cos\gamma - \sin\beta \cdot \sin\gamma. \tag{1.112}$$

Next, applying the trigonometric formula of the cosine of the sum of two angles to the right-hand side of (1.112), we can rewrite it as

$$\cos\alpha > \cos(\beta + \gamma). \tag{1.113}$$

Because $y = \cos x$ is decreasing on the interval $x \in (0, \pi)$, then from (1.113) it immediately follows that $\alpha < \beta + \gamma$. □

Problem 62 Give a different proof to Theorem 24 (Triangle inequality for a trihedral angle.) Any plane angle of a trihedral angle is less than the sum of the other two plane angles.

Proof Consider the trihedral angle $DABC$ (Fig. 1.69). Let's denote its plane angles as $\alpha,\ \beta,\ \gamma$. Consider the ray DA' opposite to the ray DA. The plane angles of the trihedral angle $DA'BC$ are equal to $\alpha,\ \pi - \beta,\ \pi - \gamma$. By Theorem 25, we have $\alpha + (\pi - \beta) + (\pi - \gamma) < 2\pi$. Hence $\alpha < \beta + \gamma$ and the statement is proven. □

Problem 63 From two points A and B lying on different faces of a dihedral angle, the perpendiculars (AA_1 and BB_1, respectively) are dropped onto the edge of the angle. Find the length of the segment connecting points A and B, if $AA_1 = a$, $BB_1 = b$, $A_1B_1 = c$ and the dihedral angle is α.

Solution Let us draw a dihedral angle and place points A and B on its faces. Then drop perpendiculars AA_1 and BB_1 as it is required by the problem. Next, we will draw $A_1C \| BB_1$ and $BC \| A_1B_1$ (Fig. 1.70). The quadrilateral A_1B_1BC is a parallelogram, so $A_1C = BB_1 = b$ and $A_1B_1 = BC = c$. The line A_1B_1 is perpendicular to the plane of the triangle AA_1C because it is perpendicular to two lines from this plane

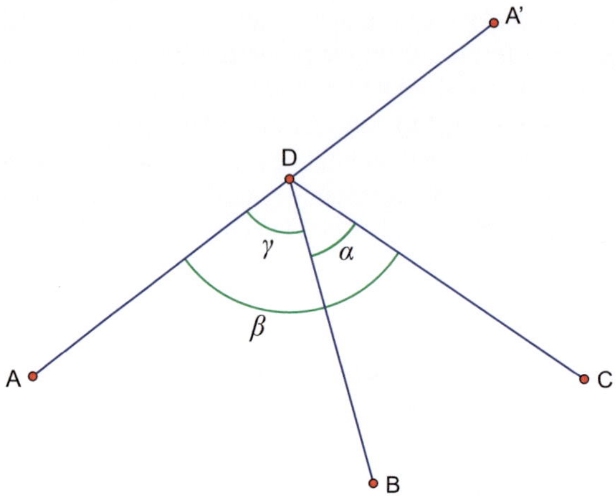

Fig. 1.69 Triangle inequality for a trihedral angle

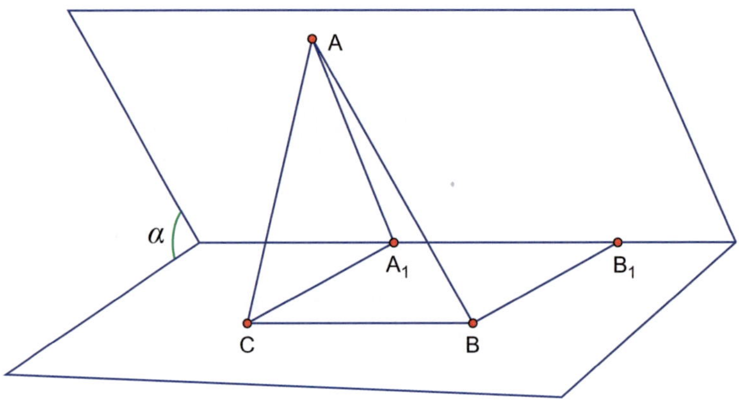

Fig. 1.70 Dihedral angle (Problem 63)

(AA_1 and CA_1). Hence, the line BC is parallel to this line and perpendicular to the plane of the triangle AA_1C. Therefore, triangle ABC is a right triangle with the right angle C (Theorem 20 on three perpendiculars). Applying the cosine theorem to triangle CA_1A, we obtain for the side opposite to angle α,

$$AC^2 = AA_1^2 + A_1C^2 - 2AA_1 \cdot A_1C \cdot \cos(\alpha) = a^2 + b^2 - 2ab\cos(\alpha).$$

Finally, using the Pythagorean theorem for triangle ACB, we find the side AB:

$$AB = \sqrt{AC^2 + BC^2} = \sqrt{a^2 + b^2 + c^2 - 2ab\cos(\alpha)}.$$

Answer. $AB = \sqrt{a^2 + b^2 + c^2 - 2ab\cos(\alpha)}.$

Problem 64 On the edge of a dihedral angle of 120°, a segment of length c is taken and from its ends perpendiculars of length a and b are restored to it in different faces. Determine the length of the line segment connecting the ends of these perpendiculars.

Solution Let us sketch a dihedral angle formed by two half-planes α and β. Consider that $NM = c$, $MB = a$ and $AN = b$. Note that lines AB and MN are not in the same plane and they are skew lines. (See Fig. 1.71). By the polygonal rule for vectors, we have

$$\vec{a} = MB, \ \vec{b} = AN, \ \vec{c} = NM,$$

$$AB = \vec{b} + \vec{c} + \vec{a}.$$

Or using vector symbol it can be written as $\overrightarrow{AB} = \overrightarrow{b} + \overrightarrow{c} + \overrightarrow{a}$. Using property of the scalar product:

$$\begin{aligned} (\vec{u}, \vec{u}) &= \vec{u} \cdot \vec{u} = \|u\|^2 \\ (\vec{u}, \vec{v}) &= \vec{u} \cdot \vec{v} = \|u\| \cdot \|v\| \cos \varphi. \end{aligned} \tag{1.114}$$

Fig. 1.71 Dihedral angle (Problem 64)

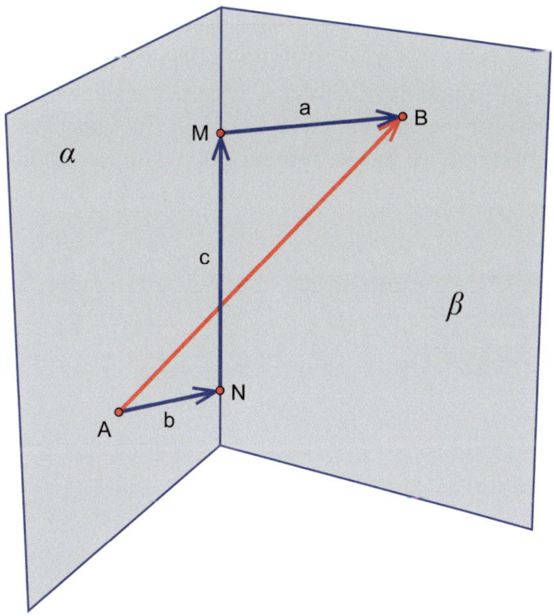

We can find the square of the length of side AB as

$$\begin{aligned}\|AB\|^2 &= \left(\vec{a}+\vec{b}+\vec{c}, \vec{a}+\vec{b}+\vec{c}\right)\\ &= \|a\|^2+\|b\|^2+\|c\|^2+2\,(\vec{a},\vec{c})+2\left(\vec{a},\vec{b}\right)+2\left(\vec{b},\vec{c}\right)\\ &= a^2+b^2+c^2+2\cdot\|a\|\,\|b\|\cos 120^\circ.\end{aligned}$$

Here we consider that $\vec{a}\perp\vec{c}$, and $\vec{b}\perp\vec{c}$. Therefore, the length of the segment equals the magnitude of the corresponding vector, $\|AB\| = \sqrt{a^2+b^2+c^2-ab}$.

Answer. $\|AB\| = \sqrt{a^2+b^2+c^2-ab}$.

Let's solve another problem involving dihedral and trihedral angles.

Problem 65 At the trihedral angle *abc* the dihedral angle at edge c is right, the dihedral angle at the edge b equals φ, and the plane angle bc is γ. For both angles, angles φ and γ are less than 90°C. Find two other plane angles α equal to ab and β equal to ac.

Solution Let angle *abc* be a trihedral angle. Given that dihedral angle c is right, the dihedral angle with edge b is φ, and that the plane angle $\angle bc = \gamma$. We have to find $\alpha = \angle(ab)$ and $\beta = \angle(ac)$. Let us draw a trihedral angle. For this we take a vertex S and sketch three rays a, b, c started at this vertex. Take a point A on the ray a and drop perpendicular AB to edge b. We know that $AC\perp c$ by the condition of the problem. Then by the theorem 20 on three perpendiculars we can state that $CB\perp b$. In our Fig. 1.72, $\alpha = \angle ASB,\ \ \gamma = \angle CSB,\ \ \beta = \angle ASC$.

Fig. 1.72 Plane angles of a trihedral angle

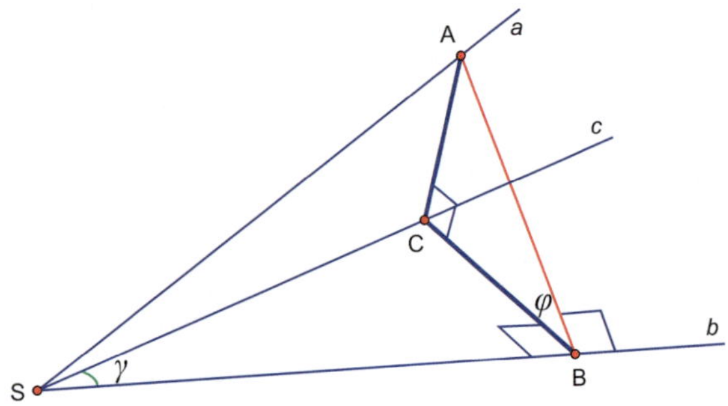

Consider the right triangles SCA, SBC, SAB, and ACB:

$$\tan\alpha = \frac{AB}{SB} = \frac{BC}{\cos\varphi} \div \frac{BC}{\tan\gamma} = \frac{\tan\gamma}{\cos\varphi},$$

$$\tan\beta = \frac{AC}{SA} = BC \cdot \tan\varphi \div \frac{BC}{\sin\gamma} = \tan\varphi \cdot \sin\gamma.$$

These formulas allow us to find the plane angles of a trihedral angle.

Problem 66 Let α, β, and γ be the plane angles of a trihedral angle and let the dihedral angles opposite the plane angles be A, B, and C. Prove the First Law of Sines Theorem 27 using the First Law of Cosines Theorem 28 for a trihedral angle.

Proof In order to prove that

$$\frac{\sin A}{\sin\alpha} = \frac{\sin B}{\sin\beta} = \frac{\sin C}{\sin\gamma}, \tag{1.115}$$

let us use the First Cosine Theorem 28 for a trihedral angle. As we know,

$$\cos A = \frac{\cos\alpha - \cos\beta\cos\gamma}{\sin\beta\sin\gamma}.$$

From here we can express the square of the sine of $\angle A$ as

$$\sin^2 A = 1 - \cos^2 A = 1 - \left(\frac{\cos\alpha - \cos\beta\cos\gamma}{\sin\beta\sin\gamma}\right)^2.$$

Next let us divide both sides of the equality above by $\sin^2\alpha$,

$$\frac{\sin^2 A}{\sin^2\alpha} = \frac{\sin^2\beta\sin^2\gamma - (\cos\alpha - \cos\beta\cos\gamma)^2}{\sin^2\alpha\sin^2\beta\sin^2\gamma}.$$

After simplification, we obtain

$$\frac{\sin^2 A}{\sin^2\alpha} = \frac{1 - \cos^2\alpha - \cos^2\beta - \cos^2\gamma + 2\cos\alpha\cos\beta\cos\gamma}{\sin^2\alpha\sin^2\beta\sin^2\gamma}.$$

The expression on the right-hand side of this equality is symmetric. Hence, it will remain the same under the circular replacement of letters $\alpha \to \beta \to \gamma \to \alpha$ and, respectively, $A \to B \to C \to A$. Therefore we state that

$$\frac{\sin^2 A}{\sin^2 \alpha} = \frac{\sin^2 B}{\sin^2 \beta} = \frac{\sin^2 C}{\sin^2 \gamma}, \tag{1.116}$$

which is equivalent to (1.115), since each angle belongs to the interval $(0, \pi)$. □

Problem 67 Given a trihedral with the plane angles α, β, and γ. Prove that the sum of the cosines of the plane angles is greater than $-\frac{3}{2}$.

Proof Assume that we have a trihedral angle $OABC$ and that $\angle AOB = \alpha$, $\angle BOC = \beta$, and $\angle COA = \gamma$. Then we need to prove that

$$\cos\alpha + \cos\beta + \cos\gamma > -\frac{3}{2}. \tag{1.117}$$

Let us take the unit vector on each side of the trihedral

$$\overrightarrow{e_1} = \overrightarrow{OA},\ \overrightarrow{e_2} = \overrightarrow{OB},\ \overrightarrow{e_3} = \overrightarrow{OC}$$

and consider the sum of these vectors that is not equal to zero, $\overrightarrow{e_1} + \overrightarrow{e_2} + \overrightarrow{e_3} \neq 0$. The scalar product of this vector by itself $(||u||^2)$ is greater than zero. Using the property of the scalar product and replacing then scalar product of the pairs of unit vectors by the cosines of the plane angle and omitting the vector symbol, for example, as for $(\overrightarrow{e_1}, \overrightarrow{e_2}) = (\overrightarrow{e_1} \cdot \overrightarrow{e_2}) = \cos\alpha$ we can simplify the expression:

$$\begin{aligned} &e_1^2 + e_2^2 + e_3^2 + 2 \cdot (e_1 \cdot e_2 + e_2 \cdot e_3 + e_1 \cdot e_3) > 0 \\ &= 3 + 2 \cdot (\cos\alpha + \cos\beta + \cos\gamma) > 0. \end{aligned}$$

The last row is equivalent to $\cos\alpha + \cos\beta + \cos\gamma > -\frac{3}{2}$. □

Problem 68 Let the plane α pass through the base AB of the isosceles triangle ABC, the point C' be the projection of C onto plane α. (It is assumed that the point C does not belong to this plane.) Prove that the angle $AC'B$ is greater than the angle ACB.

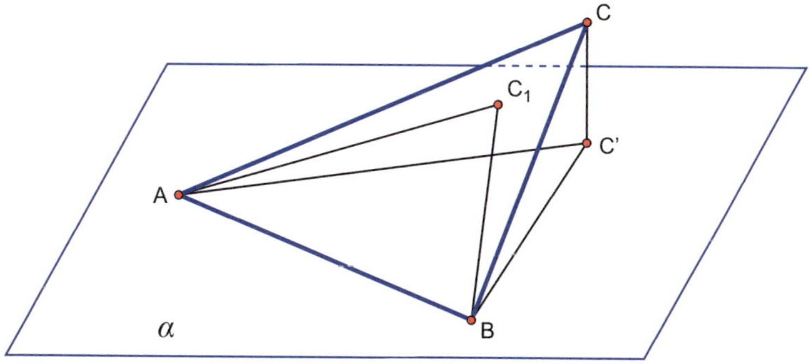

Fig. 1.73 Orthogonal projection of triangle ABC

Proof Consider Figure 1.73. Let AC' and BC' be projections of equal segments AC and BC onto plane α. Hence, $AC' = BC' < AC = BC$, i.e., the triangle ABC' is an isosceles with sides AC' and BC' that are smaller than AC and BC. Therefore, inside the triangle ABC there is a point C_1 such that $AC_1 = BC_1 = AC' = BC'$. Hence, $\angle AC'B = \angle AC_1B$, which is greater than $\angle ACB$. □

Let us see how results of the previous problem can be used to solve the next one.

Problem 69 Is it possible to intersect an arbitrary trihedral angle with a plane so that the cross section contains a regular triangle?

Solution This can always be done if all the flat angles in a trihedral angle are equal—we simply lay out segments of the same length from the vertex of such a trihedral angle and connect the ends of these segments, forming an equilateral triangle. However, we need to find out if this is possible for any given trihedral angle. I offer here two different proofs that such a situation is impossible. If this were possible, then, regardless of the type of trihedral angle, we could always find a plane that would cut the trihedral angle and form a regular triangle in its cross section. This is impossible for arbitrary trihedral angles, if one can give at least one example of a trihedral angle for which this is impossible. Therefore, in both proofs, I consider the same special case of a trihedral angle.

Method 1. Let us take a trihedral angle with two plane angles equal to 90° and one angle $AC'B < 60$. A secant plane can cut off different triangles inside this trihedral angle. Let a secant plane cutting this trihedral angle form an isosceles triangle with the vertex angle, β. Without loss of generality we can use the diagram given in Fig. 1.73, assuming that $\beta = \angle ACB$, $C'ABC$ is the given trihedral angle and then $\angle AC'B$ is the orthogonal projection of $\angle ACB$. Because $\alpha = \angle AC'B < 60°$, then $\beta < \alpha < 60°$. Therefore, the triangle of the cross section cannot be equilateral.

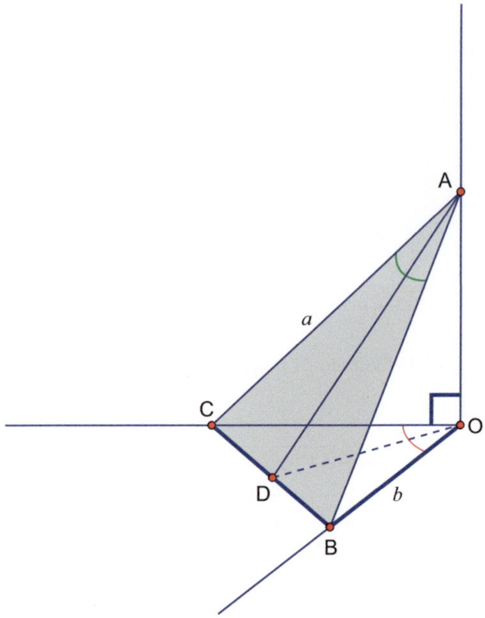

Fig. 1.74 Orthogonal projection of the cross-sectional triangle ABC

Method 2. Let us take a trihedral angle with two plane angles equal to 90° and one angle β less than 60°. Assume that a plane cuts it forming an equilateral triangle, $\triangle ABC$ (Fig. 1.74). Assume that in this triangle, $AB = AC = a$ and then $\triangle COB$ is an isosceles, with side $CO = OB = b$. Denote $AO = c$, $\angle COB = \beta < 60°$, and $\angle CAB = \alpha$. From the right triangle AOC, we obtain that $a^2 = b^2 + c^2$. Applying the Law of Cosines to triangle COB, we have $BC^2 = 2b^2 - 2b^2 \cdot \cos\beta$ which can be rewritten as $BC^2 = 2b^2(1 - \cos\beta)$. On the other hand, CB can be expressed from triangle ABC as $BC^2 = 2a^2(1 - \cos\alpha)$. Because in a right triangle the hypotenuse is always longer than any leg, then $a > b$. Since the left sides are the same, $1 - \cos\alpha < 1 - \cos\beta$ or $\cos\alpha > \cos\beta$, hence $\alpha < \beta < 60°$. Therefore triangle ABC cannot be a regular (equilateral) triangle.

Answer. No, it is impossible.

Problem 70 Let $\alpha = 60°$, $\beta = 45°$, and $\gamma = 30°$ be plane angles of a trihedral angle and $\angle A$, $\angle B$, and $\angle C$, respectively, be opposite dihedral angles. Find the value of the dihedral angle opposite to angle α.

Solution Using the first formula of (1.108), we obtain

$$\cos A = \frac{\cos(60^\circ) - \cos(30^\circ)\cos(45^\circ)}{\sin(30^\circ)\sin(45^\circ)} = \sqrt{2} - \sqrt{3}.$$

Because the sign is negative, then the dihedral angle is obtuse.

Answer. $\angle A = \arccos(\sqrt{2} - \sqrt{3}) \approx 108.5^\circ$.

Problem 71 All plane angles of a trihedral angle are right angles. Find:

1. All its dihedral angles.
2. The distance from the vertex of the trihedral angle O to a point lying inside the trihedral angle and equidistant its faces at a distance a.
3. The distance from the vertex of the trihedral to a point lying inside the trihedral and equidistant from all its edges by a distance $a\sqrt{2}$.
4. The angle, which a ray lying inside a given trihedral angle and making equal angles with the faces.
5. The angle between bisectors of plane angles.

Solution 1. Usually we draw a trihedral angle with a vertex located at the top of the page and draw rays emanating from that one vertex, as if drawing a pyramid. However, for this problem, it would be interesting to draw a trihedral angle differently. Let's look at the condition of the problem again: all plane angles are equal to 90°, so if we denote the trihedral angle $OABC$ with the vertex O, then the ray OA is perpendicular to the ray OB, which means that the linear angle for the dihedral angle at the edge OB is 90 °C. Similarly, you can find dihedral angles at the edges OC and OA, they are also equal to 90 °C!

2. Now we need to find such a point (for example, point M) inside the trihedral angle, which is at the same distance a from all faces of this trihedral angle. Let's draw a situation and figure out what can be done. In Fig. 1.75, we show the trihedral angle $OABC$ with point M inside. What does it mean that point M is equidistant from all three faces of the trihedral angle? A distance from a point to a plane is the distance of the perpendicular dropped from this point to the plane. By the condition of the problem, all distances $MF = ME = MD = a$ are the same. How can we draw these perpendiculars? Obviously, they must be parallel to the corresponding sides of the trihedral angle! So, $MD \| OC$, $ME \| OA$, and $MF \| OB$.
3. If we inspect our sketch in more detail, we can notice that it reminds us a cube with the edge of a. Let us continue and draw such a cube in which OM is the main diagonal (Fig. 1.76). So $MO = a\sqrt{3}$.

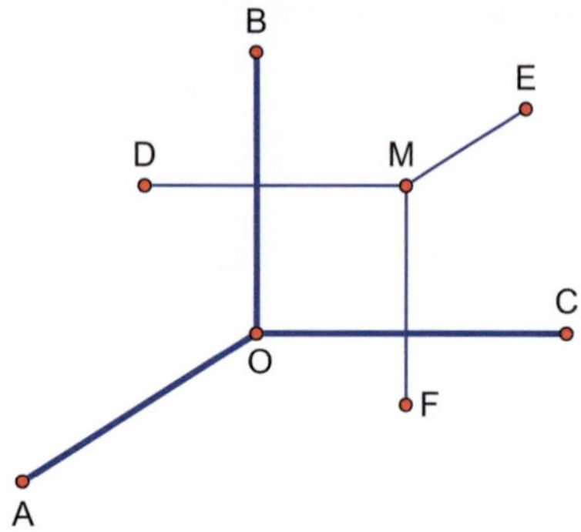

Fig. 1.75 A point inside the given trihedral angle

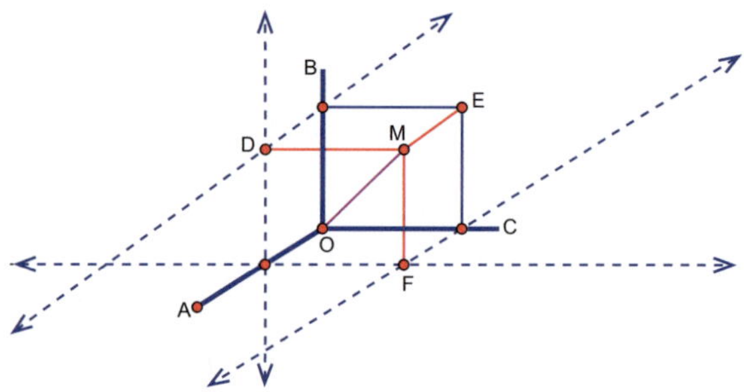

Fig. 1.76 A point M inside the cube with edge of a

4. Now let us find the distance from point M to all edges of the trihedral angle $OABC$. By Theorem 20 on three perpendiculars $BM \perp OB$, $MC \perp OC$, and $MA \perp OA$, then $MB = MC = MA = a\sqrt{2}$. Therefore, point M is equidistant not only from the faces of the trihedral but also from the edges of the trihedral. Additionally, please prove that any point of line OM is also equidistant from all faces and edges of the trihedral angle $SABC$. Angle MOB is the angle between OM and the face $ADBO$ (Fig. 1.77). $\cos\angle MOB = \frac{OB}{OM} = \frac{a}{a\sqrt{3}} = \frac{1}{\sqrt{3}}$.
5. The bisectors of the faces of the trihedral angle are the diagonals of the cube's faces. All these bisectors have the length $a\sqrt{2}$. Three bisectors form an equilateral triangle with the angle at 60°. Therefore, the angle between the bisectors of plane angles is 60°. The segments DO, OF, and OE are the bisectors of the plane angles of trihedral angle $OABS$: $\angle DOF = \angle DOE = \angle EOF = 60°$ (Fig. 1.78).

Problem 72 Prove that for any triangular pyramid, the sum of the plane angles meeting at its apex is less than 360°.

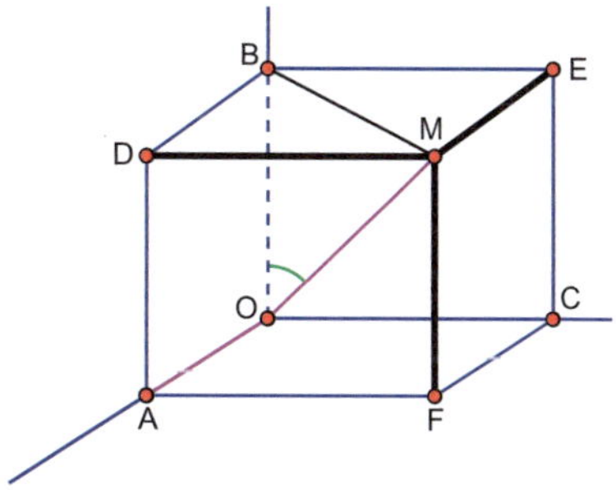

Fig. 1.77 M is equidistant from both the faces and the edges of the trihedral angle

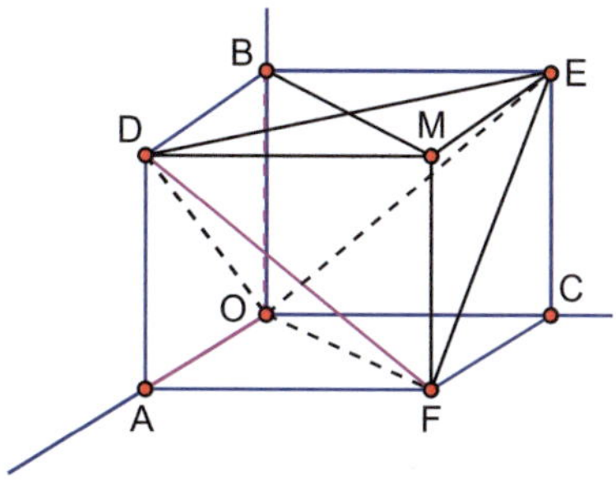

Fig. 1.78 Angles between the face bisectors of the trihedral angle

Proof By proving this we will be able to provide a different proof to Theorem 25 (the sum of plane angles of a trihedral angle is less than 2π).

Consider a triangular pyramid $OABC$ and trihedral angle with vertex O. Let us plot equal segments on its edges: $OA = OB - OC$ (Figure 1.79). Let $\angle BOC = \alpha$, $\angle COA = \beta$, $\angle AOB = \gamma$. Denote the projection of O onto the plane ABC by O'. We have $O'A = O'B = O'C$. Based on the auxiliary statement proven in one of the previous problem, $\angle BO'C > \alpha$, $\angle CO'A > \beta$, and $\angle AO'B > \gamma$. If point O' is inside triangle ABC (see Fig. 1.79, on the left), then

$$\alpha + \beta + \gamma < \angle BO'C + \angle CO'A + \angle AO'B = 2\pi.$$

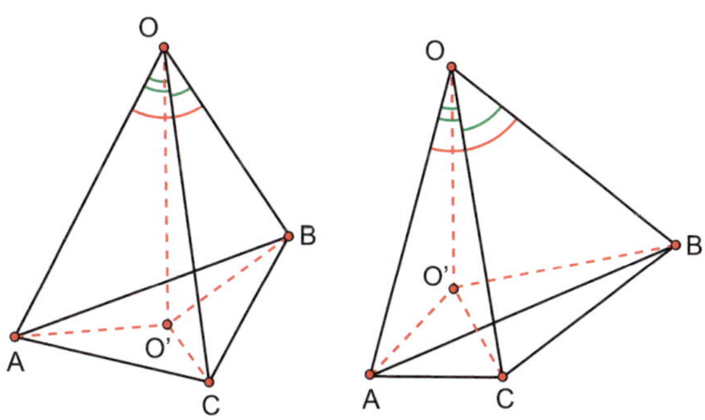

Fig. 1.79 Sum of the plane angles of a trihedral angle

If O' is outside the triangle ABC (for example, as in Fig. 1.79, on the right), then the sum $\angle\, BO'C\, +\, \angle\, CO'A\, +\, \angle\, AO'B$ is equal to twice the largest of these angles (it is $\angle BO'A$ in Fig. 1.79, on the right), so it is less than 2π (since any angle in a triangle is less than π). □

1.8 Homework on this Chapter. Problems to Solve

1. Given two points A and B, lying on opposite sides of the plane α. Find a point M on α such that the sum of the distances from it to these two points is minimal.

   Hint. Sketch the situation. In Problem 41, we considered a similar problem about two points lying on the opposite sides of a given plane.

   Solution. Drop perpendiculars from points A and B to the plane α, $AA_1 \perp \alpha$, and $BB_1 \perp \alpha$. Obviously, $AA_1 \parallel BB_1$. Denote the plane passing through the parallel lines as β. Connect the feet of the perpendiculars (points A_1 and B_1). This segment $A_1B_1 \in a$ lies at the intersection of the given plane and the plane perpendicular to this plane, so passing through two parallel lines of perpendiculars which can be written as $\alpha \cap \beta = a$. Connect points A and B. The point of intersection of line AB and the line of intersection of two planes (A_1B_1) is the required point M. $M = (AB) \cap (A_1B_1)$.

   Figure 1.80 can be helpful.
   Let us prove that the distance $MA + MB$ is minimal. Assume that there is another point $N \in \alpha$. Connect N with A and B. It follows from the triangle ANB that for any location of point N, the sum of two sides of a triangle is always greater than the third side, $AN + NB > AM + MB = AB$.

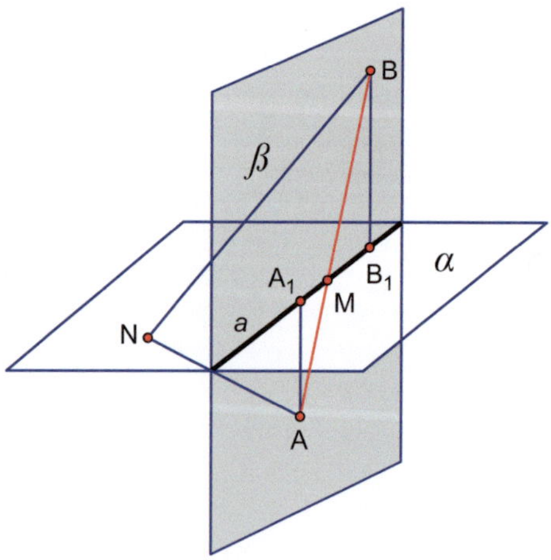

Fig. 1.80 Sum of distances $AM + MB$ is minimal

2. Vertices of the pyramid have the following coordinates: $A(3, 2, -8)$, $B(3, 5, 1)$, $C(0, 4, 1)$, and $D(1, 2, 3)$. Express the vector $d = (2, -3, 4)$ in terms of vectors $\vec{a}$, $\vec{b}$, $\vec{c}$, where

$$\begin{aligned}\vec{a} &= \overrightarrow{AD} - 3\overrightarrow{CD} + \overrightarrow{DB}\\ \vec{b} &= \overrightarrow{AB} - 3\overrightarrow{CB} + 4\overrightarrow{CD}\\ \vec{c} &= -4\overrightarrow{CA} + \overrightarrow{BA}.\end{aligned}$$

Hint. Simplify the vectors,

$$\begin{aligned}\vec{a} &= (AD + DB) + 3CD = (-3,\ 9,\ 3)\\ \vec{b} &= AB + 3BD + CD = (-5, -8,\ 17)\\ \vec{c} &= 4AC + BA = (-12,\ 5,\ 27).\end{aligned}$$

Assume that $\vec{d} = c_1\vec{a} + c_2\vec{b} + c_3\vec{c}$ and solve the system of three equations in three variables that is equivalent to reducing the following augmented matrix and to obtain the corresponding solutions:

$$\begin{pmatrix} -3 & -5 & -12 & 2\\ 9 & -8 & 5 & -3\\ 3 & 17 & 27 & 4\end{pmatrix} \Rightarrow c_1 = \frac{293}{27},\quad c_2 = \frac{77}{9},\quad c_3 = -\frac{58}{9}.$$

Answer. $\vec{d} = \frac{293}{27}\vec{a} + \frac{77}{9}\vec{b} - \frac{58}{9}\vec{c}$.

3. Prove that if all plane angles of a trihedral angle are equal, then the line making equal angles with all its faces makes equal angles with all its edges and that any point of this line is equidistant from its faces and its edges. Explain your constructions.

Hint. Consider a trihedral angle $SABC$ and without loss of generality assume that $SA = SB = SC = 1$. Hence, $SABC$ is a regular triangular pyramid, $\triangle ABC$ is equilateral. Let O be the orthocenter of $\triangle ABC$. Line SO is the requested line described in the condition of the problem. Mark the midpoints of AB, BC, and CA by M, N, K, respectively. CM, AN, and BK are simultaneously medians, bisectors, and heights. Make a sketch, for example, similar to one in Fig. 1.81.

Proof Because $OK = OM = ON$, then $BK \perp AC$, $AN \perp CB$, and $CM \perp AB$. The following triangles are congruent: $\triangle SOM = \triangle SON = \triangle SOK$ and $\triangle SCM = \triangle SAN = \triangle SBK$. Let us prove that any point P, $P \in SO$ is at the same distance from all lateral edges and the lateral faces.

Consider triangle ASN shown in blue (Fig. 1.81). The distance from point P to any of the three edges is equal to PP_1. Because triangle SNA is an isosceles, if H is the midpoint of edge AS, then $NH \perp AS$ so we can connect N and H and draw a parallel line through point P, $PP_1 \| NH$, $PP_1 \perp AS$. Using the same isosceles triangle we can find the distance from point P to the face of the trihedral angle. It

Fig. 1.81 Point equidistant from all faces and edges

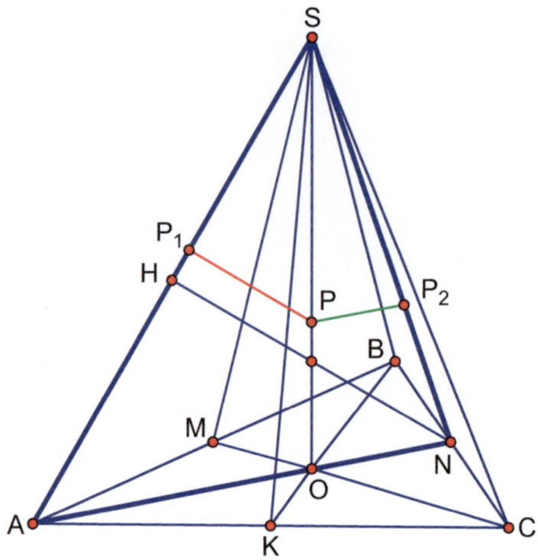

will be the segment PP_2 that represents the distance from point P to the face SBC. Since $AN \perp BC$, then to construct a perpendicular from P to the plane SBC we can simply draw in the plane of triangle SAN the line through P parallel to the height of the base AN that is also one of the sides of triangle ANS, $PP_2 \perp SN$, $PP_2 \| AN$. Repeating the same reasoning for the isosceles triangles SMC and SBK, we obtain two segments for each triangle. Since the triangles are congruent, we can conclude that any point of the line OS is equidistant from the edges and from the faces of the trihedral angle. □

4. In a trihedral angle, two plane angles are equal to 60° and the third angle is right. Find the angle of inclination of the edge opposite the right plane angle to the plane of this angle. Note that the angle between a line and a plane intersecting this line and not perpendicular to it is called the angle between the line and its projection onto the plane.

Solution. Let's draw a trihedral angle with vertex D and take points A, B, C on its sides, such that $AD = BD = CD = 1$. This way we get triangle ABC as shown in Fig. 1.82.

Using the problem statement, we can state that $\triangle ABD = \triangle BDC$ is equilateral with side 1, and $\triangle ADC$ is a right isosceles triangle with right angle ADC and hypotenuse $AC = \sqrt{2}$. Therefore, $\triangle ABC$ is also a right isosceles and equal to $\triangle ADC$. Denote $E \in AC$ as the midpoint of AC. Connect D and E and then E and B. The angle of inclination of the edge DB to the plane ADC is equal to $\varphi = \angle BDE$. The right triangle DEB has identical legs, $DE = EB = \frac{1}{\sqrt{2}}$, so $\varphi = 45°$.

Answer. 45°

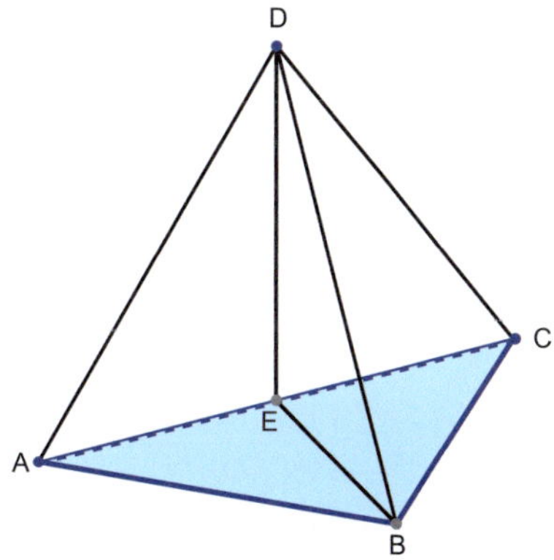

Fig. 1.82 About a trihedral angle

5. A cube's edge is 36. Find the shortest distance between the diagonal of the cube and the diagonal of the base of the cube intersecting it.

   Answer. $6\sqrt{6}$

6. Find the line of intersection of two planes given by the equations $5x + y + z = 0$ and $2x + 3y - 2z + 5 = 0$.

   Answer. $\frac{x-0}{-5} = \frac{y+1}{12} = \frac{z-1}{13}$.

7. Find the equation of the line that passes through points $M = (2, -3, 6)$ and $N = (4, -4, -10)$.

   Answer. $\frac{x-2}{2} = \frac{y+3}{-1} - \frac{z-6}{-16}$.

8. Given a regular hexagonal prism $ABCDEFA_1B_1C_1D_1E_1F_1$ with all sides and edges equal to 1. Find the distance from point A to plane DEF_1.

   Hint. You can solve this problem using the coordinate method and formula (1.79) or by considering triangular pyramid $ADEF_1$. The height dropped from vertex A to the plane of triangle DEF_1 is the requested distance.

   Answer. $d = \frac{2\sqrt{21}}{7}$.

9. Given a cube $ABCDA'B'C'D'$ with edge of 1. Evaluate the angle between line AD' and plane that passes points A', N, and M, where M is the midpoint of edge $C'D'$ and point N that divides edge DD' is the ratio $2 \div 1$, starting from vertex D'.

   Hint. Place the center of a coordinate system at vertex B of the cube so that the direction AB will coincide with the X-axis, BC with the Y-axis, and BB' with the Z-axis. Use the formula for an angle between a line and the plane, (1.86).

   Answer. $\arcsin \frac{5}{\sqrt{58}}$.

10. The sum of two face angles of a tetrahedral angle is 310°. What is the maximum measure of the third face angle?

    Answer. It must be less than 50°. Since the sum of all three angles is at most 360° and the sum of two angles is 310°, it follows that the measure of the third is at most 50°.

11. Determine whether a trihedral angle can have face angles with the following measures and give reasons for your answer:

    a. 140°, 170°, 171°
    b. 105°, 118°, 130°
    c. 130°, 78°, 103°.

    Hint. See Problems 56 or 59.

12. Find the condition for which points $A = (0, 0, a)$, $B = (0, 1, 4)$, $C = (1, 1, 5)$, and $D = (1, 0, b)$ lie in the same plane.

    Hint. Consider vectors $\overrightarrow{AB}$, $\overrightarrow{AC}$, and $\overrightarrow{AD}$ and place them as rows of a 3 by 3 matrix with determinant equal to zero (see Problem 11).
    Answer. $b = a + 1$.

13. Given a tetrahedron $ABCD$. Prove that $(\overrightarrow{AB}, \overrightarrow{CD}) + (\overrightarrow{AC}, \overrightarrow{DB}) = (\overrightarrow{AD}, \overrightarrow{CB})$.

    Hint. See Problem 4.

14. Given cube $ABCDA'B'C'D'$ with side a, points K, M, and N are the midpoints of edges DC, AA', and $B'C'$, respectively. Prove that MNK is the equilateral triangle and that its center coincides with the center of the cube.

    Hint. Introduce a coordinate system with center at point B and the X-axis in the direction of $\overrightarrow{BA}$, the Y-axis in the direction of $\overrightarrow{BC}$, and Z-axis in direction of vector $\overrightarrow{BB'}$. Express the center of mass of the triangle in terms of vectors $\overrightarrow{BK}$, $\overrightarrow{BM}$, and $\overrightarrow{BN}$. Find the coordinates of the center of the cube and show that two points are the same.

Proof We have the following coordinates: $B(0, 0, 0)$, $N(0, \frac{a}{2}, a)$, $K(\frac{a}{2}, a, 0)$, and $M(a, 0, \frac{a}{2})$. If P is the mass center of triangle MNK, then $\overrightarrow{OP} = \frac{1}{3}(\overrightarrow{BM} + \overrightarrow{BN} + \overrightarrow{BK}) = (\frac{a}{2}, \frac{a}{2}, \frac{a}{2})$. If O is the center of the cube, then $O = (\frac{a}{2}, \frac{a}{2}, \frac{a}{2}) = P$. □

15. Given points $A(1, 2, 3)$, $B(2, 1, -1)$, and $C(-1, -2, 1)$. Find the intersection of plane α containing points A, B, and C with the Y-axis.

    Hint. Use the determinant (1.50)

$$\begin{vmatrix} x-1 & y-2\,, & z-3 \\ (2-1) & (1-2) & (-1-3) \\ (-1-1) & (1-2) & (1-3) \end{vmatrix} = 0$$

to find the equation of the plane α, $x - y + z = 2$. Any point on the Y-axis has coordinates $N(0, y, 0)$ so we have to replace $x = z = 0$ in the equation of the plane and find the answer, $y = -2$.

Answer. $N(0, -2, 0)$.

16. Given points $A(-2, 3, -4)$ and $B(0, 1, 2)$, find how line AB intersects plane $3x - 2y - z = -2$.

    Hint. Use formula (1.63) to find the equation of the line using coordinates of two points, $\frac{x-(-2)}{2} = \frac{y-3}{1-3} = \frac{z-(-4)}{2-(-4)}$. Then rewrite it in the form of two intersecting planes by equating, for example, the first and the second terms of this formula with the first and the third. Then solve the system of three linear equations,

$$\begin{cases} x + y = 1, \\ 3x - z = -2, \\ 3x - 2y - z = -2, \end{cases}$$

    obtaining the solution $x = 1$, $y = 0$, and $z = 5$.

    Answer. Line AB intersects the plane at point $M(1, 0, 5)$.

17. The plane angles of a trihedral angle equal α, β, and γ. The edges opposite the angles form the angles a, b, and c with the planes of the faces. Prove the Second Law of Sines theorem for a trihedral angle: $\sin\alpha \cdot \sin a = \sin\beta \cdot \sin b = \sin\gamma \cdot \sin c$.
18. Let $\angle A$, $\angle B$, $\angle C$ be the dihedral angles of a trihedral angle and α be the plane angle opposite to the dihedral angle, A. Prove the Second Law of Cosines Theorem for a trihedral angle: $\cos A = -\cos B \cdot \cos C + \sin B \cdot \sin C \cdot \cos\alpha$.
19. Given points $A(0, 0, 0)$, $B(2, 0, 4)$, $C(-2, 4, 2)$, and $D(0, -2, 6)$, find the equation of the plane equidistant from each of the given points such that A and B are above the plane and C and D are below the plane.
    Hint. The plane must pass through points K, L, and M, the midpoints of segments AC, AD, and BD, respectively. See also Problems 31 and 42.

Solution Let's find the coordinates of the above-mentioned midpoints: $K = (-1, 2, 1)$, $L = (0, -1, 3)$, and $M = (1, -1, 5)$. Using equation of the plane passing through three points (1.50), we have to evaluate the determinant:

$$\begin{vmatrix} (x+1) & (y-2) & (z-1) \\ 1 & -3 & 2 \\ 2 & -3 & 4 \end{vmatrix} = 0.$$

After simplification we obtain the equation of the plane $2x - z + 3 = 0$. Thus, the normal vector to the plane is $\overrightarrow{n} = (2, 0, -1)$ and the plane is parallel to the Y-axis. We can also find the coordinates of point $P = (0, 2, 3)$, the midpoint of segment BC. Points $K,\ P,\ M,\ L$ are vertices of the parallelogram $KPML$. We can additionally check that $\overrightarrow{KL} = (1, -3, 2)$ and $\overrightarrow{PM} = (1, -3, 2)$ are equal. Moreover, using formula (1.79) it is not difficult to check that the distances from all given points to the plane are the same and equal to $d = \frac{3}{\sqrt{5}}$.

Answer. $2x - z + 3 = 0$.

20. Given the points: $A(1, 2, 3)$, $B(0, -1, 3)$, $C(4, -2, 0)$, and $D(1, 1, 1)$, find the volume of pyramid $ABCD$.

    Hint. Find vectors $\overrightarrow{DA}$, $\overrightarrow{DB}$, and $\overrightarrow{DC}$ and evaluate the determinant in which the rows are coordinates of the three vectors:

$$\begin{vmatrix} 0 & 1 & 2 \\ -1 & -2 & 2 \\ -2 & -3 & -1 \end{vmatrix} = -13.$$

    Taking the absolute value and dividing by 6 we obtain that the volume of pyramid $ABCD$ is equal to 13/6.

    Answer. $V = \frac{13}{6}$.

21. Given points $A(2, 0, 0)$, $B(0, 2, 0)$, $C(0, 0, 4)$, and $D(4, 2, -2)$, find the equation of the plane equidistant from each of the given points.

    Solution
    Method 1. Of course, we can solve this problem algebraically by finding the midpoints M, N, and K of segments AB, CB, and CD, respectively. $M(1, 1, 0)$, $N(0, 1, 2)$, and $K(2, 2, 1)$, then using (1.50) we can find the equation of the plane that goes through these three points as

$$\begin{vmatrix} x-1 & y-1 & z \\ 1 & 0 & 1 \\ -1 & 0 & 2 \end{vmatrix} = 0, \Rightarrow y = 1.$$

    $y = 1$ is the equation of the plane that satisfies the condition.

    Method 2. We could get the answer differently, if instead of using determinant and coordinates of the midpoints for three segments, we simply sketched each given point in the coordinate system and then we could get the answer geometrically. The plane shown in gray in Fig. 1.83 is the required plane and its equation is $y = 1$. The distance from each point to this plane is 1 unit.

    Answer. $y = 1$.

Fig. 1.83 Plane equidistant from the four given points

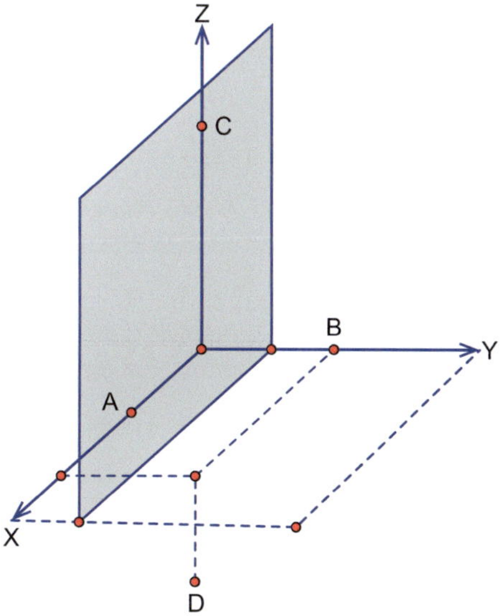

22. Find the equation of a plane that is equidistant from points $A = (-5, 6, -1)$ and $B = (3, 4, 1)$.

    Hint. Any plane that contains the midpoint of the segment AB will be equidistant from the ends of the segment. Let M be the midpoint, then $M = (-1, 5, 0)$ and the equation of the plane passing this point is $ax + by + cz = 5b - a$.

    Answer. For example, plane $4x - y + z + 9 = 0$ is a particular case of such a plane.

23. What is the sum of all flat angles of a triangular pyramid?

    Solution. The surface of a triangular pyramid consists of four triangles. The sum of the angles of each triangle is 180 degrees. Therefore the answer is 720°.

24. Given a cube $ABCDA_1B_1C_1D_1$, find

    (a) the equation of plane $\beta = A_1C_1D$;
    (b) the distance between two parallel planes $\beta = A_1C_1D$ and $\alpha = ACB_1$.

Solution. Consider a cube of unit length. Let us introduce a coordinate system so that its center is at vertex B, and the X-axis is located along the vector BA, the Y-axis along the vector BC, and the Z-axis along the vector BB_1. In this coordinate system, the vertices have the coordinates: $A_1 = (1, 0, 1)$, $C_1 = (0, 1, 1)$, $D = (1, 1, 0)$, $C = (0, 1, 0)$, $B_1 = (0, 0, 1)$, and $A = (1, 0, 0)$. The β plane has the equation $x + y + z = 2$, and the α plane is given by $x + y + z = 1$. Therefore, the distance between the two planes is 1 unit.

25. Find the coordinates of a point symmetric to point $M = (5, 8, -3)$ with respect to the line in space given by points $A = (3, 4, 1)$ and $B = (1, 6, -7)$.

    Hint. See Problem 10.

    Answer. $S = (-1, 2, -3)$

26. Points K, L, and M are non-collinear. Describe and construct projections of these points onto a straight line perpendicular to the plane defined by three given points.

    Hint. Prove that if point P is the intersection of a plane and a line perpendicular to it, then P is the projection of each of the three points onto this line.

27. Given two points $A(1, 1, 1)$ and $B(3, -1, -3)$, find

    (a) the equation of the plane that is equidistant from these points;
    (b) the locus of points in which any point of this set is at the same distance from the ends of the segment AB.

Solution.

(a) Any plane that passes through the midpoint $M(2, 0, -1)$ of segment AB will be equidistant from both points. There are infinity many planes in the form $ac + by + cz = d$ such that $2a - c = d$ and $b \in \Re$.

(b) We need to find the equation of the plane that is perpendicular to AB and passes point through M, the midpoint of AB. This plane is $x - y - 2z = 4$. Obviously, this plane is one of the many planes passing the midpoint of segment AB satisfying condition a). You can check that $a = 1, b = -1, c = -2, d = 4$ and that $2a - c = 2 \cdot 1 - (-2) = 4 = d$.

28. Four planes are drawn through a point in space, no three of which have a common line. Into how many parts do these planes divide space? Name the resulting parts of space.

    Answer. Fourteen parts: eight trihedral and six tetrahedral.

    Solution. A plane passing through a vertex of a trihedral angle and a point inside the trihedral angle divides the trihedral angle into a trihedral and tetrahedral. Consider the planes α and β passing through a point O. They break the space into 4 parts. Let's draw through the point O and the point A, not lying in the planes α and β, the plane γ. The planes α, β, and γ divide the space into eight trihedral angles with a common vertex O. Draw the plane δ through the points O and B which do not lie in any of the planes α, β, and γ. We get four planes, no three of which have a common line. The plane δ passes through the point O and intersects exactly six of the eight trihedral angles formed by the planes α, β, and γ. This means that six more parts of space are added to the already existing eight parts of space. Thus, there will be $8 + 6 = 14$ parts in total, 8 of which are trihedral angles and 6 of which are tetrahedral.

29. Given a tetrahedral angle $SABCD$, whose plane angles are $\angle ASB = 60°$, $\angle CSB = 100°$, $\angle DSC = 85°$, and $\angle ASD = 110°$, is it possible to inscribe a ball into this angle?

    Hint. Check with Lemma 7. Because $60 + 85 \neq 100 + 110$, then a ball cannot be inscribed into this tetrahedral angle. The angle does exist because $80 + 85 + 100 + 110 = 350 < 360$.

    Answer. It is impossible.

30. Within what limits can the flat angle of a trihedral angle change if the other two flat angles are, respectively, equal:

    (a) $70°$ and $100°$;
    (b) $130°$ and $150°$?

   Answer.

   (a) $30° < \alpha < 170°$;
   (b) $20° < \alpha < 80°$.

   Solution.

   (a) Any flat angle of a trihedral angle is less than the sum of the other two flat angles; the sum of the planar angles of a trihedral angle is less than $360°$. Let α be the third plane angle of the trihedral angle. Then $\alpha < 100° + 70°$, $100° < \alpha + 70°$, and $70° < \alpha + 100°$ from which we find that $30° < \alpha < 170°$.
   (b) $\alpha < 130° + 150° = 280°$ and $130° < \alpha + 150°$ additionally, we have to consider the inequalities for two other cases, $150° < \alpha + 130°$ and $\alpha + 130° + 150° < 360°$ so we obtain that $20° < \alpha < 80°$.

31. The opposite edges of a triangular pyramid are equal in pairs. Prove that all faces of this pyramid are congruent acute triangles.

    Hint. Sketch a triangular pyramid (a tetrahedron) and color opposite sides of it by the same color. Since by the condition of the problem, the tetrahedron has three pairs of edges of equal length, then each of its face will have sides colored in three different color. (For example, red, blue, and green.) Hence, each face is a scalene triangle and all four faces are equal.

32. A, B, C, and D are four points in space. It is known that $AD = BD = CD$, $\angle ADB = 90°$, $\angle ADC = 60°$, and $\angle BDC = 120°$. Find the angles of triangle ABC.

    Hint. Prove that $\triangle ADC$ is equilateral and $\triangle CDB$ is an isosceles with base angle of $30°$. Without loss of generality, we can assume that if $AD = BD = CD = 1$, then $AB = \sqrt{2}$ as the hypotenuse of right $\triangle ADB$ with $AC = 1$ and $CB = 2\cos 30° = \sqrt{3}$. Next, $\triangle ABC$ is the right triangle with sides 1, $\sqrt{2}$, and $\sqrt{3}$, hence it is a right triangle with $\angle BAC = 90°$. The other two angles can be found using trigonometry.

Answer. $\arcsin \frac{\sqrt{6}}{3}$, $\arcsin \frac{1}{\sqrt{3}}$, $90°$.

33. All plane angles of a trihedral angle are $60°$. Find the angles formed by the edges of this trihedral angle with the planes of opposite faces.

Answer. $\alpha = \arccos \frac{1}{\sqrt{3}}$.

Solution. Denote the vertex of trihedral angle by D (see Fig. 1.84) and mark points A, B, and C on each side of the trihedral angle such that $DA = DB = DC = 1$. Because the plane angles are all equal to $60°$, then tetrahedral $ABCD$ is regular with each face as an equilateral triangle. Mark the midpoints of AB and DC as M and N, respectively. ΔMDC is an isosceles triangle in which $MD = MC = \frac{\sqrt{3}}{2}$. By the condition of the problem we have to find the angle between edge DC and face ADB, that is, the angle $\alpha = \angle MDC$. This angle can be found from the right triangle MND: $\cos\alpha = ND \div MD = \frac{1}{2} \div \frac{\sqrt{3}}{2} = \frac{1}{\sqrt{3}}$.

34. The vertices of two trihedral angles coincide, while the edges of the first of them are entirely located inside the second. Prove that the sum of the plane angles of the first trihedral angle is less than the sum of the plane angles of the second.

35. Consider the plane p passing through the bisectors of the plane angles AOB and BOC of the trihedral angle $OABC$. Prove that the plane p intersects the plane COA in a straight line perpendicular to the bisector of angle COA.

36. Is it true that as a result of projecting a plane angle onto a plane, an angle is always obtained whose value is not less than the value of the original angle?

Hint. See Problem 68.

37. Four planes are drawn through one point in space, no three of which have a common line. How many parts do these planes divide space into? What are the resulting parts of space?

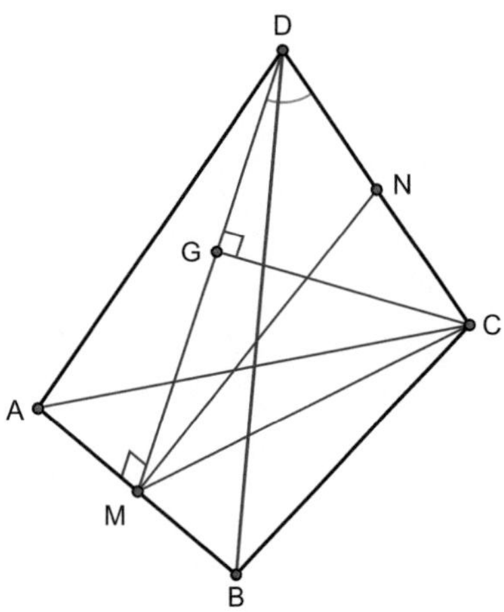

Fig. 1.84 The angle formed by the edges of trihedral angle $ABCD$ with the planes of opposite faces

Solution.

Method 1. A plane passing through a vertex of a trihedral angle and a point inside the trihedral angle divides the trihedral angle into a trihedral and tetrahedral. Consider the planes α and β passing through the given point O. They divide the space into four parts. Let's draw through point O and point A, not lying in the planes α and β, or a new plane γ. The planes α, β, and γ divide the space into eight trihedral angles with a common vertex O (think of coordinate planes centered at O). Draw the plane δ through points O and B, which do not lie in any of the planes α, β, or γ. Now we have constructed four planes, no three of which have a common line. The plane δ passes through point O and intersects exactly six of the eight trihedral angles formed by the planes α and β and γ. This means that six more parts of space are added to the already existing eight parts of space. Thus, there will be 14 parts in total, 8 of which are trihedral angles, and 6 are tetrahedral.

Answer. The space is divided into 14 parts.

Method 2. Consider a standard coordinate system in three-dimensional space—the intersection of three mutually perpendicular planes. We see that these three planes divide space into eight parts. Consider an additional plane that does not go through the origin, for example, the plane shown in Fig. 1.18 in yellow. This is additional to the eight parts because this plane divides three parts into two parts, so there will be additional six parts plus one bounded by the tetrahedron for a total of 15 parts.

38. Vectors $\vec{m}$, $\vec{n}$, and $\vec{p}$ are unit vectors, such that $(\vec{p}, \vec{n}) = (\vec{p}, \vec{m}) = 0$ and the angle between vectors $\vec{n}$ and $\vec{m}$ is $120°$. It is known that $\vec{a} = 4\vec{n} - 2\vec{m} - \vec{p}$, $\vec{b} = 4\vec{n} + 6\vec{m} + \vec{p}$ Is the angle between vectors $\vec{a}$ and $\vec{b}$ obtuse, acute, or right?

    Hint. Use the scalar product of two vectors and see Problem 7.

    Answer. Because the scalar product of $\vec{a}$ and $\vec{b}$ is equal to -5 (negative), then the angle between two vectors is obtuse.

39. Find the equation of the line as the intersection of two planes:

$$\begin{cases} 2x - y + 3z = -4 \\ x + 5y - 3z = 7 \end{cases}.$$

    Hint. See Problem 39 or use formula (1.77).

Solution.

Method 1. Using Gaussian elimination reduce the augmented matrix to

$$\begin{bmatrix} 1 & 0 & \frac{12}{11} & -\frac{13}{11} \\ 0 & 1 & -\frac{9}{11} & \frac{18}{11} \end{bmatrix}$$

and obtain the equation of the lines as

$$\begin{cases} x = -\frac{12}{11}t - \frac{13}{11}; \\ y = \frac{9}{11}t + \frac{18}{11}; \\ z = t, \quad t \in \Re. \end{cases}$$

Method 2.

a. Find the coordinates of one point $M_0 \in l$, where l is the desired line of intersection of two planes that needs to be found. If $y = y_0 = 0$, then the corresponding two other coordinates can be found from the system:

$$\begin{cases} 2x + 3z = -4 \\ x - 3z = 7 \end{cases}.$$

Thus, $M_0 = (1, 0, -2)$, so the position vector of the line is $\overrightarrow{p} = (1, 0, -2)$.

b. The direction vector $\overrightarrow{v}$ of the line $l = \overrightarrow{p} + t\overrightarrow{v}$ must be collinear to the cross product of the normal vectors of the two planes, so we find it as

$$\overrightarrow{v} = \vec{n_1} \times \vec{n_2} = \begin{vmatrix} \overrightarrow{i} & \overrightarrow{j} & \overrightarrow{k} \\ 2 & -1 & 3 \\ 1 & 5 & -3 \end{vmatrix} = -12\overrightarrow{i} + 9\overrightarrow{j} + 11\overrightarrow{k}.$$

We can see that the dot product of the direction vector with each normal vector is zero: $(\overrightarrow{v}, \overrightarrow{n_1}) = (\overrightarrow{v}, \overrightarrow{n_2}) = 0$. Using formula (1.62) we will find the equation of the line.

Answer. The equation of the line is $\frac{x-1}{-12} = \frac{y}{9} = \frac{z+2}{11}$.

40. The plane passes the vertex B_1 of the cube $ABCDA_1B_1C_1D_1$ and the points M and N on the edges C_1C and AD, respectively. It is known that $C_1M \div MC = 2 \div 1$, and N is the midpoint of the edge AD. Find how the cutting plane divides edge DC.

Solution. Sketch the cube $ABCDA_1B_1C_1D_1$ placing points M and N on its edges. Let $B_1M \cap BC = T$, then N and T belong to the base of the cube $ABCD$ and let $NT \cap DC = S$. Consider similar right triangles $\Delta B_1BT \sim \Delta MCT$, then denoting $x = CT$ we have the ratio of the corresponding sides: $\frac{1}{\frac{1}{3}} = \frac{1+x}{x}$

and $x = \frac{1}{2}$. Next, $\triangle CST = \triangle SDN$ by two angles and one side, so S is the midpoint of DC.

Answer. The plane goes through the midpoint of segment DC.

41. The plane α passes the vertex B_1 of cube $ABCDA_1B_1C_1D_1$ and the points M and N on the edges C_1C and AD, respectively. It is known that $C_1M \div MC = 2 \div 1$, and that N is the midpoint of the edge AD. Plane β cuts the cube passing through vertices A, B_1, and C. Describe each plane algebraically and represent the line of intersection of planes α and β by the system of two linear equations.

    Hint. Let us introduce a coordinate system with the origin at point B and the axes X, Y, and Z along the vectors BA, BC, and BB_1, respectively, and find the coordinates of the points belonging to the described planes: ($B_1 = (0, 0, 1)$, $M = (0, 1, 1/3)$, and $N = (1, 1/2, 0)$). The equations of the planes can be found using the formulas (1.49) and (1.50).

    Answer. Line l of the intersection is represented by the following system:

$$\begin{cases} 2x + 2y + 3z = 3 \\ x + y + z = 1 \end{cases}.$$

    Line equation: $(x, y, z) = (-t, t, 1),\ t \in \Re$. This line passes through the point $B_1 = (0, 0, 1)$ ($z = 1$ and is parallel to $y = -x$). The planes will intersect outside the cube along the above line passing through the vertex of the cube B_1.

42. The teacher drew a regular hexagonal pyramid $ABCDEDA_1B1C_1D_1E_1F_1$ in which all edges are equal to one and asked the students whether the lines containing the given segments, BF_1 and A_1C, intersect. Pete connected two points in pairs, intersected the line segments on a two-dimensional drawing, got point M and declared, "Yes, they intersect at M!" Is Pete correct? How can you prove or disprove his claim?

    Hint. You can use the ideas and sketch of the prism given in Problem 49. For each line, we can find the direction vector: $\overrightarrow{BF_1} = (\frac{-3}{2}, \frac{\sqrt{3}}{2}, 1)$, $\overrightarrow{A_1C} = (\frac{3}{2}, \frac{\sqrt{3}}{2}, -1)$. Because the direction vectors of two lines are not collinear, then lines are not parallel. This is also clear from the sketch in Fig. 1.51. Let us check if the four given points are coplanar. For this we can find one more vector with one point on one line and the other—on the second line, $\overrightarrow{A_1F_1} = (-\frac{1}{2}, \frac{\sqrt{3}}{2}, 0)$. Next, we consider the triple scalar product (1.25) of three vectors. Since the quantity is not zero, the lines containing the segments do not intersect. Pete was wrong.

43. Find two planes intersecting on the line, $\frac{x-1}{1} = \frac{y-2}{-1} = \frac{z-3}{5}$.
Solution. Since all three terms in the equation of the line are equal, we can equate, for example, the first and second terms of the given formula, and then the second and third terms. After simplification, we obtain a system of two linear equations:

$$\begin{cases} x + y = 3 \\ 5y + z = 13 \end{cases}.$$

Answer. The planes $x + y - 3 = 0$ and $5y + z - 13 = 0$ intersect by the given line.

Chapter 2
Polyhedra and Their Properties

Knowledge about polyhedra has been used since ancient times by the civilizations of Egypt, Mesopotamia, and Africa. For example, archaeologists have found polyhedral jewelry dating back several thousand years, as well as dodecahedron-shaped dice dating back to 1000 BC. It is interesting that in ancient times the greatest attention was paid to the study of perfect structures, i.e., symmetrical bodies. Ancient people believed in the superpower of such structures. Pythagoras of Samos (about 582–507 BC) created a cosmological doctrine that linked regular polyhedra with the structure of the universe. The Pythagoreans believed that the elements of primitive beings have the form of regular polyhedra, namely, fire is a tetrahedron, earth is a hexahedron, air is an octahedron, and water is an icosahedron. The entire universe, according to the ancients, had the shape of a dodecahedron [33]. The first theory about five regular polyhedra also belongs to the Greek mathematician and geometer Theaetetus of Athens (417 BC). His main discoveries are set forth in *Euclid's Elements* (415–369 BC). They open with a description of the construction of a regular triangle and end with the study of five regular polyhedral solids. In modern times, they are known as the platonic solids. The ancient Greek scientist Plato described in detail the properties of regular polyhedra in his famous book *Dialogs of Timaeus*, written 360 BC [30]. Platonic solids are a three-dimensional analog of plane regular polygons.

There are infinitely many different regular polygons, but only five different regular polyhedra: tetrahedron, hexahedron (cube), octahedron, dodecahedron, and icosahedron. Archimedes of Syracuse (c. 287 BC) left behind many inventions, 13 works (such as *On the Sphere and the Cylinder*, *Measurement of the Circle*, *Regular Heptagon*, and others). The contribution of Archimedes to the theory of polyhedra is a description of 13 semi-regular convex homogeneous polyhedra (Archimedean solids). Johannes Kepler (1571–1639) German astronomer and mathematician was the first to publish a complete list of the 13 Archimedean solids and gave them the names known today as truncated tetrahedron, truncated octahedron, truncated

E. Grigorieva, *Methods of Solving Solid Geometry Problems*,
https://doi.org/10.1007/978-3-031-86969-3_2

icosahedron, truncated cube, truncated dodecahedron, cuboctahedron, icosidodecahedron, rhombicuboctahedron, rhombicosidodecahedron, truncated cuboctahedron, truncated icosidodecahedron, snub cuboctahedron, and snub dodecahedron.

Modern stereometry studies figures in space, called solid bodies. It is clear that a geometric body must be thought of as a part of space occupied by a physical body and bounded by a surface. In general, a polyhedron is a body whose surface consists of a finite number of flat polygons. A polyhedron has a set of elements: an edge, a face, a vertex, a dihedral angle, and a diagonal.

- The polygons that make up a polyhedron are called its **faces**.
- The sides of the faces are called **edges**, and the ends of the edges are called the **vertices** of the polyhedron.
- Faces of a polyhedron meeting at one common edge form a **dihedral angle**.
- A segment connecting two vertices that do not belong to the same face is called the **diagonal** of the polyhedron.

There are many definitions of the concept of a polyhedron found in popular textbooks [1, 6, 9, 13, 18, 42]. The common properties of these definitions can be summarized in our Definition 9. Accordingly, the set of 18 squares in Fig. 2.1 is not a polyhedron, because conditions (1) and (3) are violated by the edge AB (two adjacent polygons lie in the same plane with AB common to four different squares). Moreover, the vertex C does not satisfy condition (5) because the chain of polygons converged at this vertex and having a common edge does not exist (see also [15]).

Definition 9 A **polyhedron** as a geometric body whose surface consists of a finite set of flat polygons that satisfy the following conditions:

1. No two adjacent polygons lie in the same plane.
2. The union of all polygons is a two-dimensional manifold [9, 13].
3. Each side of any polygon in the set is a side of exactly one more polygon from the set.
4. From each polygon in the set to any other, one can go through a chain of polygons in the set, in which every two consecutive polygons are adjacent along a common side.
5. If two polygons have a common vertex, then this vertex is common with at least one other polygon that also meets at this vertex. From one polygon to another, converging at the same vertex, one can go along the chain.

Considering Fig. 2.1 of three cubes, we can sort and label the leftmost cube as number one, the middle cube as number two, and the last cube as number three. So, if we remove the third cube altogether, then the figure from the first two cubes will still not be a polyhedron, because four faces of the structure have a common edge, AB. Moreover, AB will lie on the same plane, which is a violation of the definition of a polyhedron.

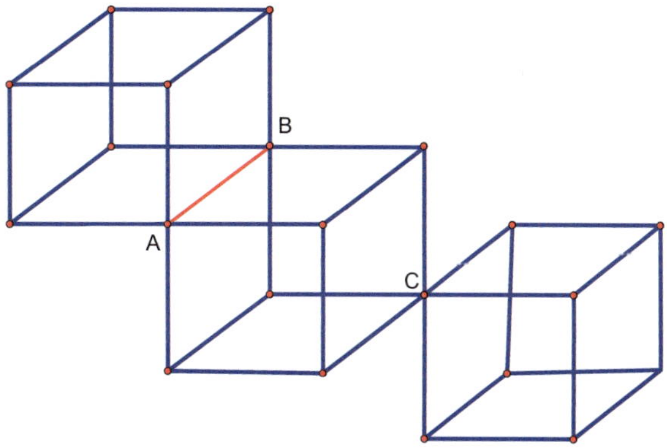

Fig. 2.1 Not a polyhedron

Now take the third cube of Fig. 2.1 and place it immediately below the first, then this third cube and the second cube will have one face in common and the third cube and the first one will have another common face. Let's now delete these two faces. The result is a spatial figure shown in Fig. 2.2. It is a polyhedron. What kind of a polyhedron? Can we be more specific? A polyhedron is called **convex** if it lies entirely on one side of the plane of each of its faces. A polyhedron is called convex if it is a convex figure, i.e., together with any two of its points, it entirely contains the segment connecting them. This directly implies that the faces of a convex polyhedron are convex polygons. Therefore, the polyhedron shown in Fig. 2.2 is nonconvex.

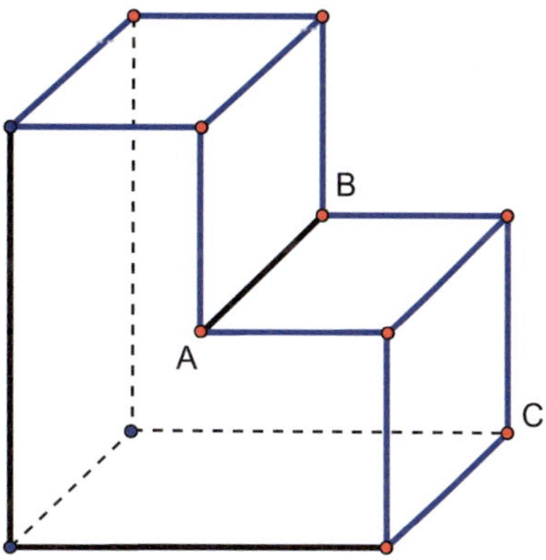

Fig. 2.2 Nonconvex polyhedron

This chapter offers many challenging problems related to the simplest convex polyhedra—prisms and pyramids—which will be the main object of our study. Euler's theorem for a convex polyhedron, Alexandrov's theorem on the development of polyhedra, properties of regular and semi-regular polyhedra, and interesting and complex problems related to them will also be presented. Some natural or synthesized molecules or crystals shaped like regular or semi-regular polyhedra are discussed, with explanations of their structure based on chemistry and stereometry.

2.1 Pyramids

Definition 10 A **pyramid** is a polyhedron with a polygonal base, all other faces of which are triangles meeting at one vertex, called an **apex**.

A pyramid is a polyhedron, where any n-gon lies at the base and the apex is projected to any point in the plane of the base. Perhaps the most famous of pyramid is the Cheops pyramid—the Great Pyramid of Giza—a regular quadrangular pyramid with a height of about 136 m and a base length of 230 m. In the construction of this pyramid, more than two million blocks were used, each of which had a volume of approximately one cubic meter. Cheops was originally about 146 m high, but the lining was removed by the Arabs in the Middle Ages to construct the mosques of Cairo. The second great pyramid is the Khafre Pyramid which is more than 130 m high with the remains of the lining on the top still intact (Fig. 2.3).

Fig. 2.3 The Great Egyptian Pyramids of Giza (Courtesy of Professor Olga Tomashevich)

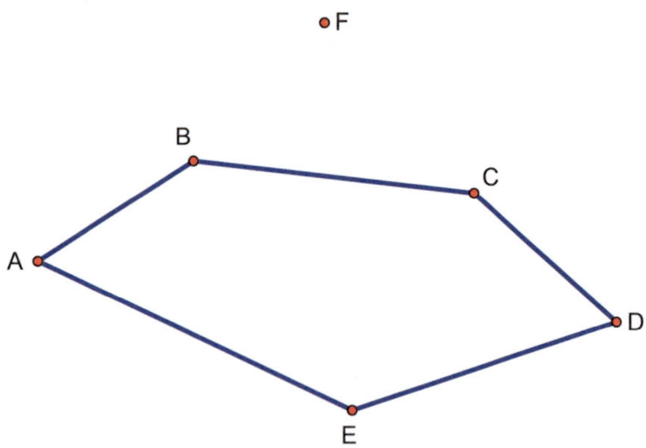

Fig. 2.4 Drawing a pyramid

How can we draw a pyramid? I recommend that we start from drawing its base and showing its apex (vertex) (Fig. 2.4). Next, connect each vertex of the polygon with the vertex of the pyramid. Then decide which side is seen and if it is invisible, then it should be dashed. Suppose that the base is a non-regular pentagon with different sides. Our picture must be as general as possible so when we construct the base we want to place the vertex F in such a position that every edge of $ABCDE$ is clearly visible. By connecting F to each vertex of the polygon when we are sure that the drawing is satisfactory, then we decide which side of the pyramid we see and which is hidden. Hidden sides will be shown as dotted lines (Fig. 2.5).

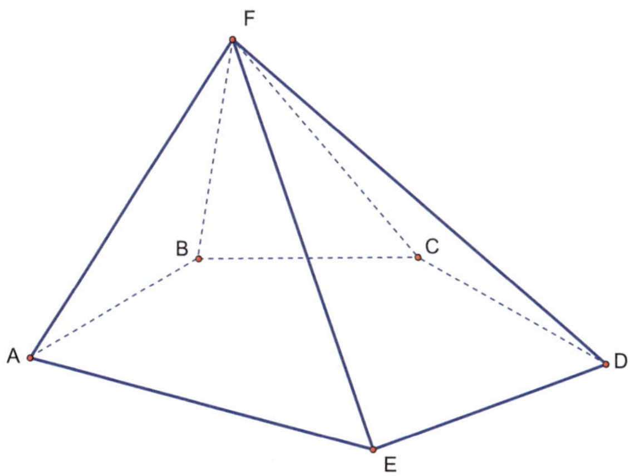

Fig. 2.5 Pentagonal pyramid

The **height of a pyramid** is the segment dropped from its vertex to the base. In Fig. 2.6, the height FG is shown as a red dotted line. Since FG is perpendicular to the plane of the base, it will also be perpendicular to all segments connecting the

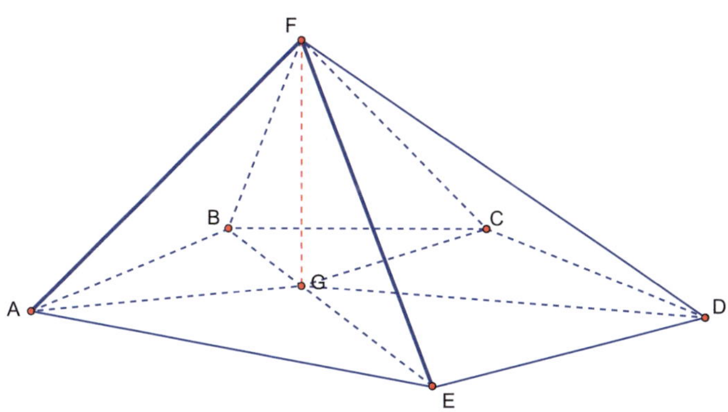

Fig. 2.6 Height of a pyramid

foot of the height with each vertex of the base. So, all triangles FGA, FGB, FGC, FGD, and FGE are right triangles. The height of the side (lateral) face of a regular pyramid drawn from its top is called the **apothem**.

The **total surface area of a pyramid** is equal to the sum of the area of its base and the area of the lateral surface, $S = S_{\text{base}} + S_{\text{lateral}}$. If the base is a triangle, then only in this case does a pyramid has four vertices and each of them can be taken as its apex and it is called a **triangular pyramid**. A **quadrangular pyramid** is a pyramid with a quadrilateral at the base, and so on. A **regular pyramid** is a pyramid whose base is a regular n-gon, and the vertex is projected to the center of the base of this n-gon. The important point is that all other properties of a regular pyramid, such as equality of lateral edges, are a consequence of the information given in the definition of a regular pyramid.

In a regular pyramid,

1. the lateral edges are equal;
2. the lateral faces are equal isosceles triangles;
3. dihedral angles at base edges are equal;
4. dihedral angles at lateral edges are equal;
5. flat angles at the vertex are equal; and
6. the heights of the lateral faces dropped onto the edges of the base are equal.

The area of the lateral surface of a regular pyramid is equal to the product of the semi-perimeter of the base and the apothem. It can be found as follows:

$$S_{\text{lateral}} = L \cdot p, \tag{2.1}$$

where L is the apothem and p is the semi-perimeter of the base.

$$S_{\text{lateral}} = \frac{S_{\text{base}}}{\cos\alpha}, \tag{2.2}$$

where the angle α is the dihedral angle at the edge of the base.

Proof The apex of any regular pyramid is projected onto the center of a regular n-gon of the base. Let us connect the center of the base polygon with each of the vertices of the base. The area of the base is equal to the sum of the areas of the n isosceles triangles obtained by partitioning. The area of such a triangle is equal to $1/2$ of the height dropped from the center of the base to the side of the base, multiplied by the side a of the base, i.e., $\frac{1}{2} \cdot a \cdot h_b$. Now consider a right triangle one leg of which is the height of the pyramid, the other leg is the height of the small isosceles triangle at the base and its hypotenuse as the apothem L of one of the triangles of the lateral surface that shares the same side with the small triangle. The angle between the hypotenuse L and the height of the small triangle of the base h_b is the given dihedral angle α. Then $L = \frac{h_b}{\cos\alpha}$. The lateral surface area of the pyramid is equal to

$$S_{\text{lateral}} = n \cdot \frac{1}{2} \cdot L \cdot a = \frac{na}{2} \cdot L = \frac{\frac{1}{2}na \cdot h_b}{\cos\alpha},$$

where the second expression proves formula (2.1). Since the numerator of the third fraction is the area of the base, then this proves (2.2). □

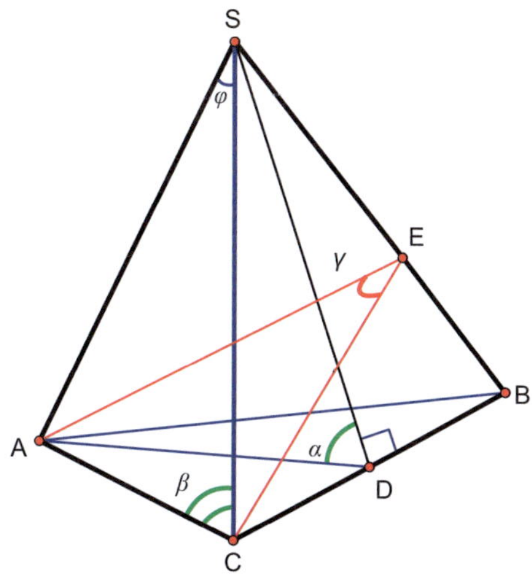

Fig. 2.7 Angles of a regular triangular pyramid

Consider a **regular triangular pyramid** $ABCS$ with apex S (Fig. 2.7). Note that the side edges SA, SB, and SC are equal to each other, but not necessarily equal to the base edge AB. In the case of a regular pyramid, the dihedral angle at the base of the pyramid α is not equal to the angle at the base of the lateral face, β. The dihedral angle between adjacent lateral faces, γ, is not equal to the planar angle at the apex, φ. A **tetrahedron** is a pyramid with all four faces being triangles. Any of the four faces can be taken as the base of the tetrahedron and any vertex is an apex of a tetrahedron. A tetrahedron and a triangular pyramid are exactly the same thing. A **regular tetrahedron** is a regular triangular pyramid whose lateral and base edges are equal.

Remark. In general, in a regular triangular pyramid, the lateral edges are not equal to the edges at the base, so a regular tetrahedron, and a regular triangular pyramid are not the same thing!

Volume of a pyramid is equal to $1/3$ of height, h times the area of the base, B:

$$V = \frac{1}{3}hB. \tag{2.3}$$

Since in a tetrahedron, a height can be dropped from any of the four vertices and the volume formula can be written in four different ways:

$$V = \frac{1}{3}h_1B_1 = \frac{1}{3}h_2B_2 = \frac{1}{3}h_3B_3 = \frac{1}{3}h_4B_4. \tag{2.4}$$

Any part of this formula can be used when we have to find a height dropped to a face with known area and when the volume of pyramid can be evaluated using a different pair of height and face area. The following lemma is true.

Lemma 8 *The areas of the lateral faces of a tetrahedron are inversely proportional to the heights dropped onto them*

$$B_i = \frac{3V}{h_i}, \quad i = 1, 2, 3, 4.$$

A **bi-pyramid** is obtained by taking a polygon in the plane, taking one point above the polygon and one point below it, and connecting both points with the vertices of the polygon.

A **truncated pyramid** is obtained by cutting the top of the pyramid by a plane parallel to the base of the pyramid. When solving problems on truncated pyramids, I recommend to construct the whole pyramid first, then take a point on one of the edges (for example, point A_1 on AF, Fig. 2.8) and draw the top of the truncated pyramid by keeping the bases parallel. I also advise to keep the top of the original pyramid shown by dashed lines. A truncated pyramid is called a **frustum of a pyramid**, which is the part of a pyramid between the two parallel bases.

Note. Construction of a truncated pyramid can be considered as **homothety** with center F that transforms a small pyramid $A_1B_1C_1D_1E_1$ into a larger pyramid and vice versa. Homothety is not rigid motion; it is a case of central similarity, so homothetic segments of corresponding figures are parallel. A homothety with center F maps a point X to point X' such that $\frac{FX}{FX'} = k$ so that the transformation produces similar 3d figures. Obviously, the ratio of the heights or of any corresponding sides (such as $k = \frac{A_1B_1}{AB}$, etc.) of the small and big pyramids is equal to the **coefficient of similitude** of transformation. The ratio of the areas of the bases equals the square of this measure and the ratio of their volumes equals this coefficient cubed.

Consider two similar spatial objects. Assume that a_1 and a_2 are the corresponding sides; B_1 and B_2 are the areas of the selected corresponding two-dimensional (plane) figures, such as faces; and V_1 and V_2 are the volumes of the similar objects. Denote k as the coefficient of the similitude, then

$$k = \frac{a_1}{a_2} = \sqrt{\frac{B_1}{B_2}} = \sqrt[3]{\frac{V_1}{V_2}}. \tag{2.5}$$

The **volume of a truncated pyramid** with bases B_1, B_2, and height h is

$$V = \frac{1}{3}h\left(B_1 + \sqrt{B_1B_2} + B_2\right). \tag{2.6}$$

Fig. 2.8 Truncated pyramid

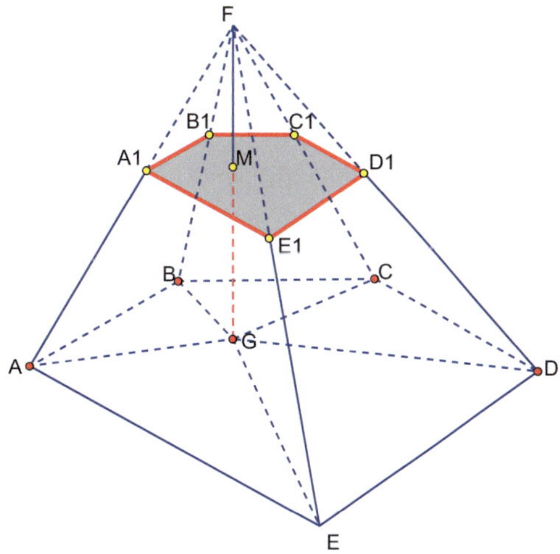

Remark. As you probably know for two positive numbers m and n, the quantity $G = \frac{m+\sqrt{mn}+n}{3}$ is called the **Heronian mean**. So the volume of the truncated pyramid is the product of its height and the Heronian mean of the bases of the frustum.

In 1893, the Russian nobleman Golenishev purchased an Egyptian papyrus scroll that was 18 feet long and 3 inches wide and gave it as gift to a Moscow museum. The ancient Egyptians wrote on papyrus and that one was approximately 4000 year old. The content of the Moscow papyrus had 25 solved problems; one of them illustrated how to evaluate the volume for the frustum of the truncated square pyramid with the measures of the lengths of its sides a and b, and height h. The ancient Egyptian had the following formula, $V = \frac{1}{3}h\left(a^2 + ab + b^2\right)$. Of course, the ancient Egyptians did not have the general formula (2.6) for a truncated pyramid. However, their mathematical abilities were impressive. Many of us still wonder how this could have been obtained 4000 years ago... Now it's time to start solving some problems related to the properties of pyramids.

Problem 73 Find the volume of a truncated pyramid with height h and with B_1 and B_2 representing the areas of the bottom and top bases of the frustum.

Solution Denote the height of the truncated pyramid as h, then $h = FG - FM = H_1 - H_2$, or the difference of the heights of the big and small pyramids (Fig. 2.8). So, the volume of the truncated pyramid is the difference of the volumes of the corresponding pyramids that can be written as

$$V = V_1 - V_2 = \frac{1}{3}H_1B_1 - \frac{1}{3}H_2B_2. \tag{2.7}$$

We also know that

$$H_1 = h + H_2. \tag{2.8}$$

And from the similarity of the pyramids, the ratio of the heights equals the square root of the corresponding bases (2.5), $\frac{H_1}{H_2} = \sqrt{\frac{B_1}{B_2}}$. From which we can get that

$$H_1 = \frac{H_2\sqrt{B_1}}{\sqrt{B_2}}. \tag{2.9}$$

Combining (2.8) and (2.9) we obtain that

$$\begin{aligned} \frac{H_2\sqrt{B_1}}{\sqrt{B_2}} &= h + H_2; \\ H_2\sqrt{B_1} &= h\sqrt{B_2} + H_2\sqrt{B_2}; \\ H_2\left(\sqrt{B_1} - \sqrt{B_2}\right) &= h\sqrt{B_2}. \end{aligned}$$

From this we can express the heights of small and big pyramids in terms of h and the bases:

$$H_2 = \frac{h\sqrt{B_2}}{\sqrt{B_1} - \sqrt{B_2}}, \quad H_1 = \frac{h\sqrt{B_1}}{\sqrt{B_1} - \sqrt{B_2}}. \tag{2.10}$$

Substituting formula (2.10) into (2.7), the volume of the truncated pyramid is

$$V = \frac{1}{3}h\left(\frac{\sqrt{B_1}}{\sqrt{B_1} - \sqrt{B_2}} \cdot B_1 - \frac{\sqrt{B_2}}{\sqrt{B_1} - \sqrt{B_2}} \cdot B_2\right) = \frac{1}{3}h\left(\frac{\left(\sqrt{B_1}\right)^3 - \left(\sqrt{B_2}\right)^3}{\sqrt{B_1} - \sqrt{B_2}}\right).$$

Applying the difference of cubes formula on the top of the fraction, we can simplify this expression to a much nicer form, (2.6), $V = \frac{1}{3}h\left(B_1 + \sqrt{B_1 B_2} + B_2\right)$. The beauty of the proof given here is that we have derived this formula in the general case for any truncated pyramid, using the property of similar figures.

Problem 74 Prove that each condition describes the same pyramid. What pyramid is it?

(a) the lateral edges are the same;
(b) the lateral edges are equally inclined to the plane of the base;
(c) the lateral edges form the same angles with the height of the pyramid;
(d) a circle can be circumscribed around the base of the pyramid, and the height of the pyramid passes through the center of this circle.

Fig. 2.9 Properties of a pyramid

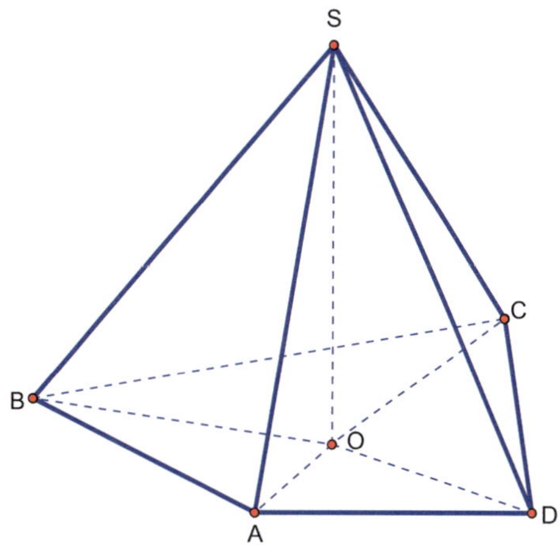

Proof Let the first condition be satisfied, that is, the side edges are equal. Then let us prove that all other conditions are also satisfied. In Fig. 2.9, the segment SO is the height of the pyramid.

If the first condition is met, then $SB = SA = SD = SC$ and $\triangle SBO = \triangle SAO = \triangle SDO = \triangle SCO$ (by the hypotenuse and two legs). From this it follows that $\angle SBO = \angle SAO = \angle SDO = \angle SCO$, which means that the side edges are equally inclined to the base plane. Also, $\angle BSO = \angle ASO = \angle DSO = \angle CSO$, which means that the side edges form the same angles with the height of the pyramid: $OB = OA = OD = OS$. Thus all four statements describe a pyramid, the base of which can be inscribed in a circle (the base of the pyramid is a cyclic polygon) and the height of the pyramid passes through the center of this circle. We should proceed in a similar way, taking each other condition in turn as the initial one.

□

2.1.1 *The Beautiful Tetrahedron*

Consider a regular tetrahedron $ABCD$ with side $a = AB = BC = AC = CD$ (Fig. 2.10), then all its faces will have the same area as its base area $[ABC]$:

$$B = [ABC] = \frac{a^2\sqrt{3}}{4}. \tag{2.11}$$

The total surface area of a regular tetrahedron is four times this value, $S_{ABCD} = a^2\sqrt{3}$.

Let us find its volume. In Fig. 2.10, we dropped the height DO_1 from vertex D onto the base triangle ABC such that $DO_1 = h$. Because $ABCD$ is a regular tetrahedron, then the foot of the perpendicular O_1 is the point of intersections of the medians

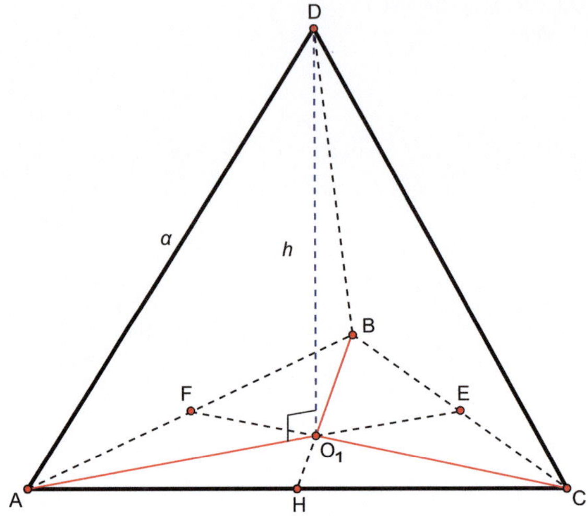

Fig. 2.10 Regular tetrahedron

(that are also the heights of the equilateral triangle ABC) such that $O_1 = AE \cap CF$. Moreover, point O_1 divides each median in the ratio $2 \div 1$, starting from its vertex so $AO_1 \div O_1E = 2 \div 1$. Thus,

$$AO_1 = \frac{2}{3} \cdot AE = \frac{2}{3} \cdot \frac{a\sqrt{3}}{2} = \frac{a\sqrt{3}}{3},$$

and

$$AO_1 = BO_1 = CO_1 = \frac{a\sqrt{3}}{3} = R_1. \tag{2.12}$$

Hence, $r = R_1$ is the radius of the circle circumscribed about the base triangle. Next, the height of the tetrahedron can be found from the right triangle AO_1D as one of its legs:

$$DO_1^2 = AD^2 - AO_1^2;$$
$$h^2 = a^2 - \left(\frac{a}{\sqrt{3}}\right)^2.$$

We obtain that

$$h = \frac{a\sqrt{2}}{\sqrt{3}}. \tag{2.13}$$

Finally, the volume of a regular tetrahedron with side a can be obtained using (2.11) and (2.13) as $V = \frac{1}{3}hB = \frac{1}{3} \cdot \frac{a\sqrt{2}}{\sqrt{3}} \cdot \frac{a^2\sqrt{3}}{4} = \frac{a^3\sqrt{2}}{12}$, so

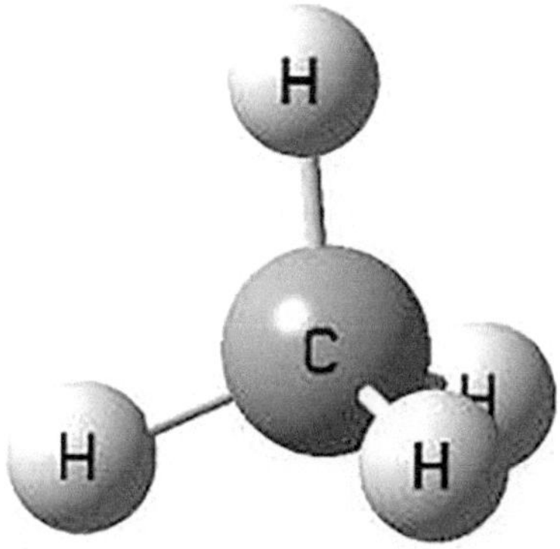

Fig. 2.11 Methane molecule

$$V_{\text{tetrahedron}} = \frac{a^3\sqrt{2}}{12}. \tag{2.14}$$

There are some objects in nature that have the form of a regular tetrahedron, for example, a molecule of methane CH_4 has a carbon atom in the center and hydrogen atoms at the vertices (Fig. 2.11). This molecule is an example of *sp*3 hybridization. External orbits of carbon get transformed into one *s* and three *p* electrons (Figs. 2.12 and 2.13). The concept of hybridization of valence atomic orbitals was proposed by the American chemist **Linus Pauling** in 1931 to answer the question why, if the central atom has different (s, p, d) valence orbitals, the bonds formed by it in polyatomic molecules with the same ligands are equivalent in energy and spatial characteristics? In 1954, the Nobel Committee awarded L. Pauling the Prize in Chemistry "for his research into the nature of the chemical bond and its application to the elucidation of the structure of complex substances."

Ideas about hybridization are central to the valence bond method. Hybridization itself is not a real physical process, but only a convenient model that allows one to explain the electronic structure of molecules, in particular, the hypothetical modifications of atomic orbitals during the formation of a covalent chemical bond, in particular, the alignment of chemical bond lengths and bond angles in the molecule.

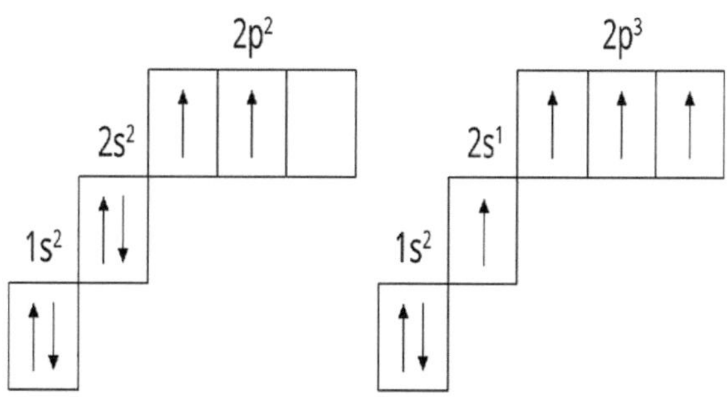

Fig. 2.12 Electron structure of carbon

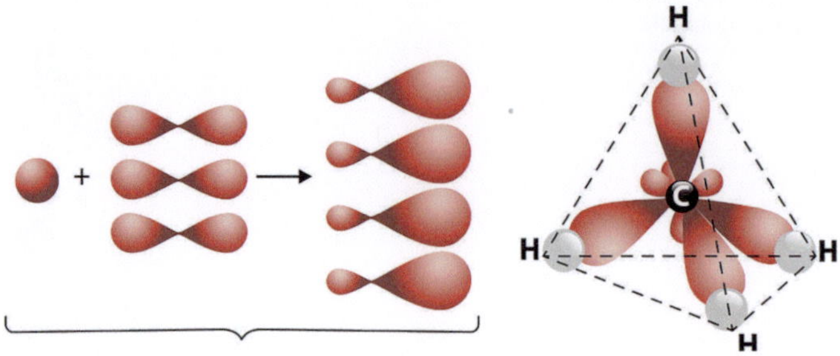

Fig. 2.13 sp3 hybridization

Hybridization is the alignment of electron clouds in shape and energy. The concept of hybridization was successfully applied to the qualitative description of simple molecules, but was later extended to more complex ones. Unlike molecular orbital theory, it is not strictly quantitative; for example, it is not able to predict the photoelectron spectra of even simple molecules such as water. Currently, it is used mainly for methodological purposes and in synthetic organic chemistry.

Obviously, a regular tetrahedron that can model the molecule of methane has nothing inside. How can we determine the center of the tetrahedron, at which the carbon atom is located? We can expect atom of C be on the height of the tetrahedron equidistant from each vertex (atoms of H). The model of molecule CH_4 is shown in Fig. 2.14.

Fig. 2.14 A tetrahedron inscribed into a sphere

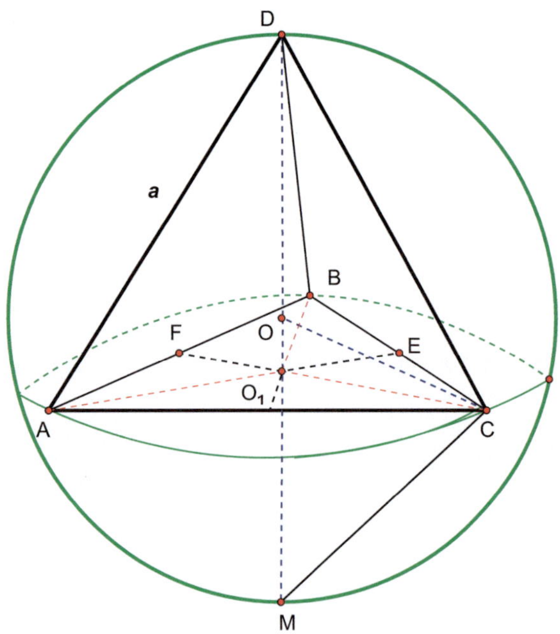

If all vertices of the tetrahedron $ABCD$ are equidistant from some point inside it, then this point is the center of the sphere circumscribed about the tetrahedron. Denote it by O and make a sketch of the situation. In Fig. 2.14, $ABCD$ is inscribed into a sphere of radius $R = OD = OC = OA = OB$. However, only segments $OD = OC$ and shown in blue. Each vertex of the tetrahedron lies on the sphere and the triangle of the base $\triangle ABC$ is inscribed into a circle with radius smaller than the big circle that can be written as $r = O_1A = O_1B = O_1C < R$ (shown in red). In order to find the angle between the center of the sphere and two vertices of the tetrahedron $ABCD$—this angle models the angle formed by the two hydrogen atoms and the central carbon atom—we need to find the radius of the circumscribed sphere. Let us continue the height of the tetrahedron DO_1 down until it intersects the sphere at point M and then let us consider the triangle MCD, one of sides, DM, is the diameter of the sphere. It follows from Thales's theorem that the triangle MCD is a right triangle with the right angle $\angle DCM = 90°$.

Triangle MCD is inscribed into one of the big circles of a sphere as shown in Fig. 2.15. Since $CO_1 \perp DM$ and $DC \perp CM$ we have that $O_1C = r = \frac{a\sqrt{3}}{3}$, $DO_1 = h = \frac{a\sqrt{2}}{\sqrt{3}}$, and $a = DC$. Let us evaluate the radius of the sphere using properties of similar right triangles (for simplicity we marked equal angles by one arc). So $\Delta DCM \sim \Delta DO_1C$ by the three angles and then the ratios of the sides corresponding to the same angles are equal to the coefficient of similitude, $\frac{DC}{DM} = \frac{DO_1}{DC}$. We can express the hypotenuse DM as $DM = \frac{DC^2}{DO_1}$. Substituting the formulas for DC and DO_1, we obtain $2R = DM = \frac{a\sqrt{6}}{2}$ so that the radius of the sphere is equal to $R = \frac{a\sqrt{6}}{4}$. Now we can use the Law of Cosines applied to the isosceles triangle $\triangle DOC$:

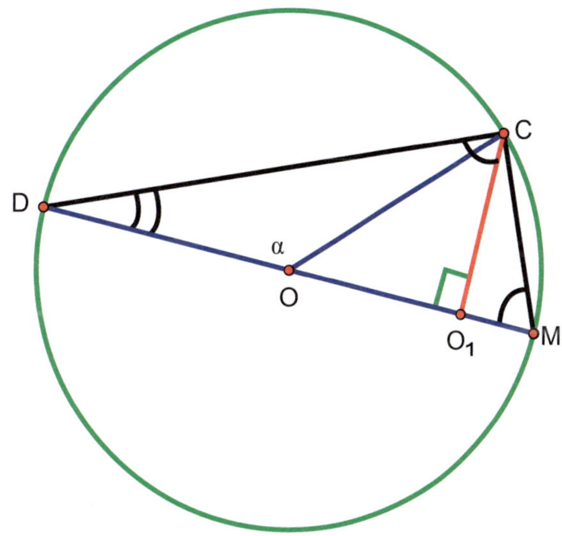

Fig. 2.15 Right triangle *DCM* inscribed into a big circle of a sphere

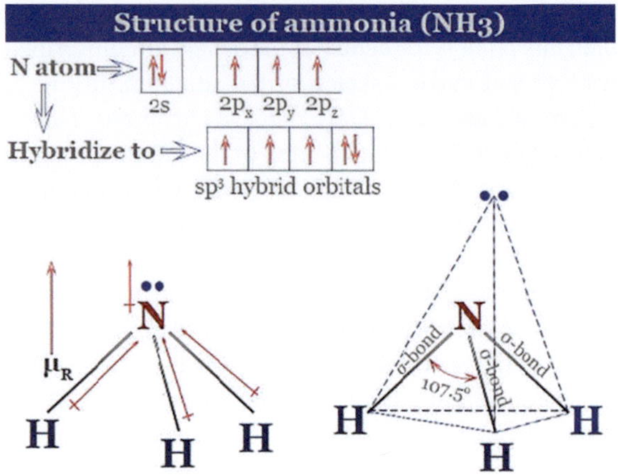

Fig. 2.16 Ammonia molecule

$$DC^2 = 2DO^2 - 2DO^2 \cdot \cos\alpha;$$

$$a^2 = 2 \cdot \left(\frac{a\sqrt{6}}{4}\right)^2 \cdot (1 - \cos\alpha);$$

$$1 = \frac{3}{4} \cdot (1 - \cos\alpha);$$

$$\cos\alpha = -\frac{1}{3};$$

from which we have the angle, $\alpha = \angle DOC = \cos^{-1}\left(-\frac{1}{3}\right) \approx 109^\circ$. This is exactly what is observed in the molecule of methane!

An example of a non-regular tetrahedron is the ammonia molecule NH_3. As a result of *sp*3-hybridization, the atoms form the shape of a regular triangular pyramid shown in Fig. 2.16. Here the nitrogen atom N occupies the vertex position, and the hydrogen atoms H are located at the vertices of the equilateral triangle of the base. The lone electron pair of nitrogen can be considered as the fourth vertex of a regular tetrahedron. However, this unshared electron pair has a stronger effect on the chemical bonds, putting more pressure on them so the bond angle for this so-called "trigonal pyramidal" structure is 107.5°, which is less than 109° for a regular tetrahedral structure. Similar trigonal pyramidal structures are observed in other molecules: PF_3, PBr_3, and SbH_3. On the other hand, the ammonia cation NH_4^+ has a completely regular tetrahedral structure like the methane molecule.

2.1.2 *Some Properties of a Tetrahedron*

The tetrahedron has many unique properties. Let's discuss some of them here before tackling the related problems. **Medians of a tetrahedron** are the segments each of

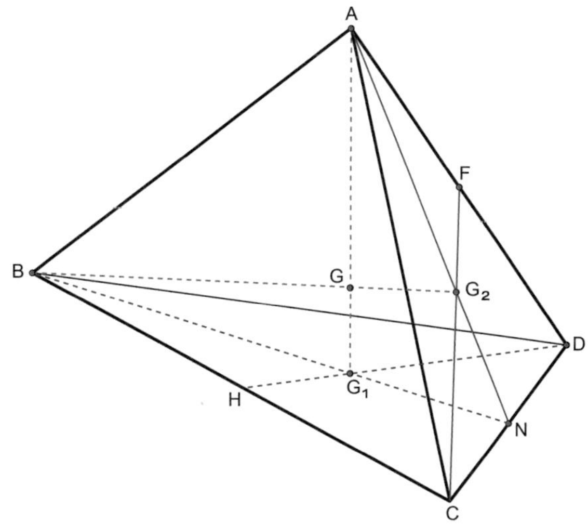

Fig. 2.17 Medians of a tetrahedron

which connects the vertex of a tetrahedron with the centroid of the opposite face. Consider a tetrahedron *ABCD* shown in Fig. 2.17. It has four medians that can be drawn from each vertex to the opposite to this vertex face. For example, it would be AG_1, BG_2, CG_3 and DG_4, where G_1, G_2, G_3, G_4 are the centroids of the faces.

Consider the relationship between vectors AG_1, *AB*, *AD*, and *AC* of Fig. 2.17 and we can show that Theorem 30 is true.

Theorem 30 *Let AG_1 be a median of the tetrahedron ABCD, and $\overrightarrow{AG_1}$ the corresponding vector, then* $\overrightarrow{AG_1} = \frac{1}{3}\left(\overrightarrow{AB} + \overrightarrow{AD} + \overrightarrow{AC}\right)$.

Proof Consider Fig. 2.17 and triangle AG_1B. By addition of vectors we have $\overrightarrow{AG_1} = \overrightarrow{AB} + \overrightarrow{BG_1}$. On the other hand, vector $\overrightarrow{BG_1}$ is collinear with vector *BN* and because G_1 is the centroid of the face *BCD* we can write that $\overrightarrow{BG_1} = \frac{2}{3} \cdot \overrightarrow{BN}$. Next, since *BN* is a median of triangle *BCD*, then $\overrightarrow{BN} = \frac{1}{2} \cdot \left(\overrightarrow{BC} + \overrightarrow{BD}\right)$. Now substitute this into the expression for vector $\overrightarrow{AG_1} = \overrightarrow{AB} + \frac{2}{3} \cdot \left(\frac{1}{2}(\overrightarrow{BC} + \overrightarrow{BD})\right)$. After simplification we obtain that $\overrightarrow{AG_1} = \overrightarrow{AB} + \frac{1}{3}\left(\overrightarrow{BC} + \overrightarrow{BD}\right)$. Using, triangles AG_1D and AG_1C we can express the same vector in three different ways:

$$\overrightarrow{AG_1} = \overrightarrow{AB} + \frac{1}{3}\left(\overrightarrow{BC} + \overrightarrow{BD}\right);$$
$$\overrightarrow{AG_1} = \overrightarrow{AC} + \frac{1}{3}\left(\overrightarrow{CB} + \overrightarrow{CD}\right);$$
$$\overrightarrow{AG_1} = \overrightarrow{AD} + \frac{1}{3}\left(\overrightarrow{DB} + \overrightarrow{DC}\right).$$

By adding the left- and the right-hand sides of the formulas above and using the fact that the sum of opposite vectors is zero vector (for example, $\overrightarrow{BC} + \overrightarrow{CB} = 0$), we obtain that $3 \cdot \overrightarrow{AG_1} = \overrightarrow{AB} + \overrightarrow{AC} + \overrightarrow{AD}$. Dividing by 3 we complete the proof,

$$\overrightarrow{AG_1} = \frac{1}{3} \cdot \left(\overrightarrow{AB} + \overrightarrow{AC} + \overrightarrow{AD}\right).$$

□

Remark. Theorem 30 can be proven differently.

The center of mass of a triangle is its center of gravity. One may ask, "Why are we talking about a flat figure, a triangle, if the tetrahedron is a three-dimensional figure?" Let me explain. Recall the definition of center of masses. Let n equal masses be placed at vector positions: $\overrightarrow{x_1}, \overrightarrow{x_2}, ..., \overrightarrow{x_n}$, then the center of masses is $\frac{\sum \overrightarrow{x_i}}{n}$. If we have only three points (for example, triangle BDC), then the center of their masses (i.e., denoted by point G_1) will be

$$\overrightarrow{AG_1} = \frac{\overrightarrow{x_1} + \overrightarrow{x_2} + \overrightarrow{x_3}}{3}; \text{ where } \overrightarrow{x_1} = \overrightarrow{AB},\ \overrightarrow{x_2} = \overrightarrow{AD},\ \overrightarrow{x_3} = \overrightarrow{AC}.$$

So the center of masses of triangle BDC is

$$\overrightarrow{AG_1} = \frac{1}{3}\left(\overrightarrow{AB} + \overrightarrow{AD} + \overrightarrow{AC}\right). \tag{2.15}$$

Point A is the center of a coordinate system but in reality wherever this point is, the formula will not change! Look at the tetrahedron in Fig. 2.17. Vertex A can be considered as the center of some coordinate system, then all together we can say that vector AG_1 represents the center of masses of triangle of the base BCD and simultaneously it is the median of tetrahedron $ABCD$. Let the centroid of triangle BCD be a point M and N be the midpoint of side CD. Since we know that the medians in a triangle are divided by the centroid (point of their intersection) in the ratio $2 \div 1$, then $\overrightarrow{NM} = \frac{1}{3}\overrightarrow{NB}$ and $\overrightarrow{AN} = \frac{1}{2}(\overrightarrow{AC} + \overrightarrow{AD})$. For the centroid of triangle BCD, we can write $\overrightarrow{AM} = \overrightarrow{AN} + \frac{1}{3}\overrightarrow{NB} = \overrightarrow{AN} + \frac{1}{3}(\overrightarrow{AB} - \overrightarrow{AN})$. After simplification,

$$\overrightarrow{AM} = \frac{1}{3}\left(\overrightarrow{AB} + \overrightarrow{AD} + \overrightarrow{AC}\right). \tag{2.16}$$

Comparing (2.15) and (2.16), we conclude that $\overrightarrow{AG_1} = \overrightarrow{AM}$, and so $M = G_1$.

In principle, the proof of Theorem 30 can proceed from the fact that the center of mass of any triangle is its center of gravity (the intersection point of three medians). Let us see how Theorem 30 can be used in solving Problem 75.

Problem 75 Given a pyramid $SABC$ with the base ABC, the sum of the squares of the sides of the base is equal to a^2 and the sum of the squares of the lateral edges is equal to b^2. Find the distance from the vertex S to the centroid O of the base triangle.

Solution Assume that pyramid trihedral angle $SABC$ has the plane angles: $\angle ASC = \alpha$, $\angle CSB = \beta$, and $\angle BSA = \gamma$ (Fig. 2.18). Since the centroid O is the point of intersection of the medians in triangle ABC, $\overrightarrow{SO} = \frac{1}{3} \cdot (\overrightarrow{SA} + \overrightarrow{SB} + \overrightarrow{SC})$. The length of SO can be evaluated as $SO = \sqrt{(\overrightarrow{SO}, \overrightarrow{SO})}$. Using the properties of the scalar product, let us omit the vector symbol and expand the scalar product as

$$\|SO\|^2 = \frac{1}{9}\left(SA^2 + SC^2 + SB^2 + 2SA \cdot SC\cos\alpha + 2SC \cdot SB\cos\beta + 2SB \cdot SA\cos\gamma\right). \tag{2.17}$$

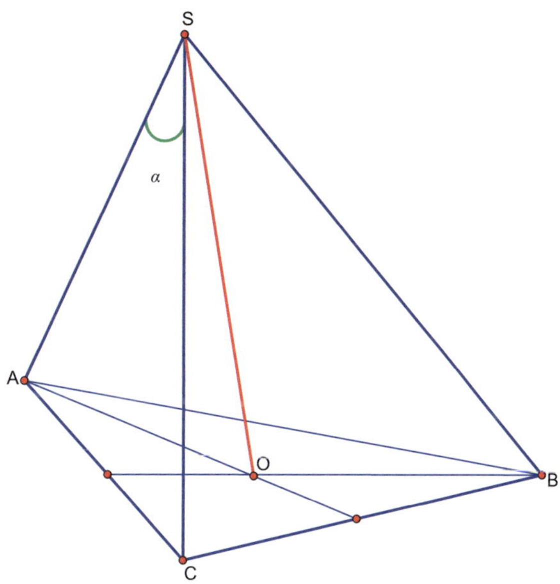

Fig. 2.18 The distance from the vertex to the centroid of the base of a pyramid

Substituting the given information, ($SA^2 + SC^2 + SB^2 = b^2$), we simplify (2.17) as

$$\|SO\|^2 = \frac{1}{9} \cdot \left(b^2 + 2SA \cdot SC \cos\alpha + 2SC \cdot SB \cos\beta + 2SB \cdot SA \cos\gamma\right). \quad (2.18)$$

Since we do not know the cosines of any of the angles, we will find them using the Law of Cosines from the appropriate triangles:

$$\begin{aligned} \triangle\, ASC : \cos\alpha &= \frac{AS^2 + SC^2 - AC^2}{2AS \cdot SC}; \\ \triangle\, ASB : \cos\gamma &= \frac{AS^2 + SB^2 - AB^2}{2AS \cdot SB}; \\ \triangle\, SCB : \cos\beta &= \frac{CS^2 + SB^2 - CB^2}{2SC \cdot SB}. \end{aligned} \quad (2.19)$$

Next, we will substitute (2.19) into (2.18) and simplify

$$\begin{aligned} \|SO\|^2 &= \tfrac{1}{9} \cdot \left(b^2 + \left(AS^2 + SC^2 - AC^2 + CS^2 + SB^2 - CB^2 + AS^2 + SB^2 - AB^2\right)\right); \\ &= \tfrac{1}{9} \cdot \left(b^2 + \left(2AS^2 + 2SC^2 + 2SB^2 - \left(AC^2 + CB^2 + AB^2\right)\right)\right); \\ &= \tfrac{1}{9} \cdot \left(b^2 + \left(2b^2 - a^2\right)\right); \\ &= \tfrac{1}{9} \cdot \left(3b^2 - a^2\right). \end{aligned}$$

Answer. $SO = \frac{1}{3} \cdot \sqrt{3b^2 - a^2}$.

Segments connecting midpoints of opposite edges are called **bi-medians** of the tetrahedron. Opposite edges of a tetrahedron are those pairs of edges that are skew.

Theorem 31 *Bi-medians of a tetrahedron intersect at one point that divides them into two equal parts.*

Proof Let MN, EF, and PQ be the bi-medians of tetrahedron $ABCD$ that correspond to the pairs of edges AB and CD, AC and BD, and BC and AD as shown in Fig. 2.19. Because the segments ME and FN are parallel to BC and half its length, the quadrilateral $EMFN$ is a parallelogram. Point O is the intersection of its diagonals MN and EF and divides them by half. If follows from the fact that $EPFQ$ is a parallelogram that the midpoints of bi-medians EF and PQ coincide at O. □

Note. The figure $NFQEPM$ formed as a result of our proof is an octahedron inscribed in the given tetrahedron.

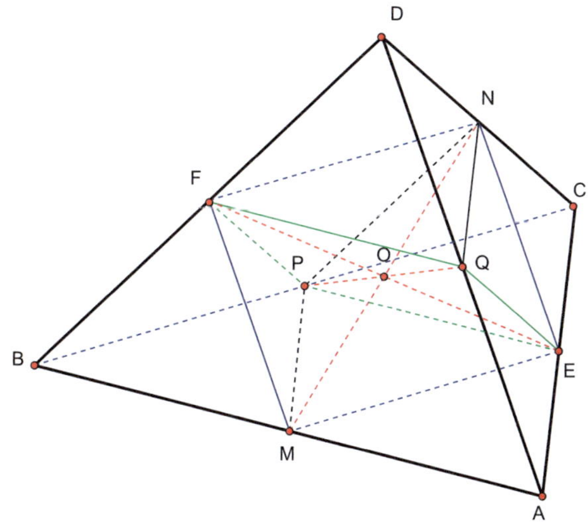

Fig. 2.19 Bi-medians of a tetrahedron (Theorem 31)

Problem 76 Prove that the segment connecting midpoints of the two opposite edges of a regular tetrahedron is the common perpendicular of these edges.

Proof Let us sketch a regular tetrahedron $ABCD$ as shown in Fig. 2.20. Let M and N be the midpoints of the opposite edges AB and DC, respectively. Let us introduce three vectors: $\overrightarrow{a} = \overrightarrow{DA}$, $\overrightarrow{b} = \overrightarrow{DB}$, and $\overrightarrow{c} = \overrightarrow{DC}$. We will omit vector symbol but will operate with vectors. Vector MN can be written as

$$\overrightarrow{MN} = \overrightarrow{DN} - \overrightarrow{DM} = \frac{1}{2}c - \frac{1}{2}(a+b) = \frac{1}{2}(c-a-b),$$

where we used the fact that vector DM connects the vertex of a triangle ADB and the midpoint of the opposite side AB, then it equals half of the sum of vectors DA and DB. In order to prove that segment MN is perpendicular to edge DC, we can demonstrate that the scalar product of vectors MN and DC equals zero:

$$\overrightarrow{DC} \cdot \overrightarrow{MN} = \frac{1}{2}c \cdot (c-a-b) = \frac{1}{2}\left(\|c\|^2 - \|a\|\,\|c\| \cos 60^\circ - \|b\|\,\|c\| \cos 60^\circ\right),$$

where $\|a\| = \|b\| = \|c\| = a$ are the magnitudes of the vectors because all sides are equal. Since the cosine of a 60° angle is 0.5, $\overrightarrow{DC} \cdot \overrightarrow{MN} = 0$ and hence $\overrightarrow{DC} \perp \overrightarrow{MN}$. The orthogonality of edges MN and AB can be proven similarly. □

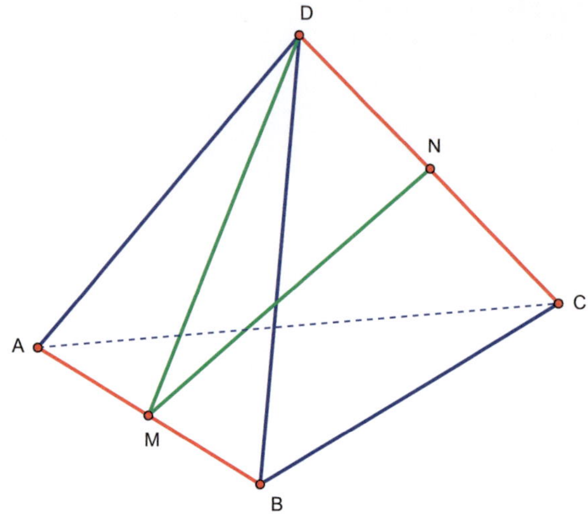

Fig. 2.20 Common perpendicular in a tetrahedron

Theorem 32 *Medians of a tetrahedron intersect at unique point inside the tetrahedron, at its centroid, and are divided by this point in the ratio* $3:1$ *measured from the vertices.*

Proof Consider Fig. 2.21. Here G_1 and G_2 are the centroids of the faces BCD and ACD and belong to the medians BN and AN of these faces, respectively, of tetrahedron $ABCD$. By the property of the centroid of a triangle we can state that $\frac{AG_2}{G_2N} = 2$ and $\frac{BG_1}{G_1N} = 2$. Using Thales's theorem we know that $G_1G_2 \| AB$. According to the property of the trapezoid, the lines AG_1 BG_2 and MN must intersect at one point, say K. Applying Menelaus's theorem (see, for example, [2] or [11]) for triangle AMN and line BG_2, we have

$$\frac{AB}{BM} \cdot \frac{MK}{KN} \cdot \frac{NG_2}{G_2A} = -1,$$

from which we obtain that $-2 \cdot \frac{MK}{KN} \cdot \frac{1}{2} = -1$. Therefore $\frac{MK}{KN} = 1$. This means that point K is the midpoint of segment MN, so it is the centroid of the tetrahedron, point G. Thus, two medians AG_1 and BG_2 intersect at centroid G. Hence, all four medians intersect at this point. Applying Menelaus's theorem for the triangle ABG_1 and line MN we obtain that

$$\frac{AM}{MB} \cdot \frac{BN}{NG_1} \cdot \frac{G_1G}{GA} = -1$$

or $1 \cdot (-3) \cdot \frac{G_1G}{GA} = -1$, so $\frac{G_1G}{GA} = \frac{1}{3}$. This formula can be obtained for any of the four medians of a tetrahedron. □

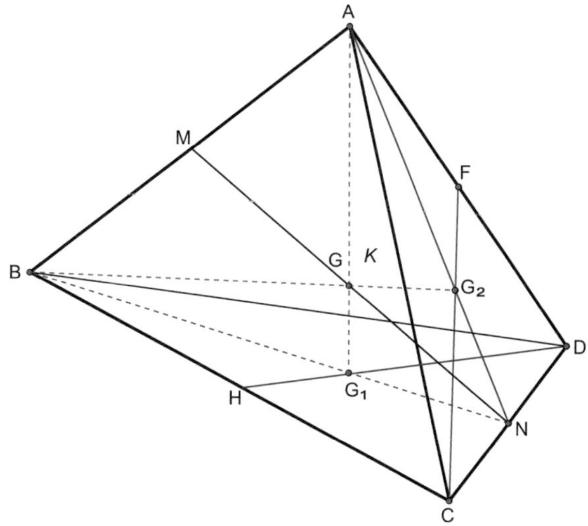

Fig. 2.21 Proof of the medians theorem 32

2.1.3 Problems on Pyramids

Problem 77 Prove that the sum of the distances from any point of the base of a regular pyramid to its side faces is a constant value.

Proof We will prove this for a regular triangular pyramid. The idea of the proof for all other regular pyramids is very similar. Draw a triangular pyramid $MABC$ and place a point K somewhere inside the base ABC. We can evaluate its volume in four different ways since each regular triangular pyramid has four potential apexes. Let H be the height dropped to the face MBC from vertex A, then the volume of the pyramid is

$$V = \frac{1}{3}[MBC] \cdot H.$$

Connect K with all vertices of the pyramid as shown in Fig. 2.22.

We can recognize three smaller pyramids: $KMBC$, $KAMB$, and $KAMC$, the sum of the volumes of which equals the volume of $MABC$. Denote h_1 as the height of pyramid $KMBC$ from vertex K to face MBC and similarly h_2 the height of $KAMB$ and h_3 the height of $KAMC$. Then $V_1 = \frac{1}{3}[MBC] \cdot h_1$, $V_2 = \frac{1}{3}[AMB] \cdot h_2$, and $V_3 = \frac{1}{3}[AMC] \cdot h_3$ so that $V = V_1 + V_2 + V_3 = \frac{1}{3}[MBC] \cdot h_1 + \frac{1}{3}[AMB] \cdot h_2 + \frac{1}{3}[AMC] \cdot h_3$. Since the area of the lateral faces are all equal ($[AMC] = [MBC] = [AMB]$), we have

$$V = \frac{1}{3}[MBC] \cdot (h_1 + h_2 + h_3).$$

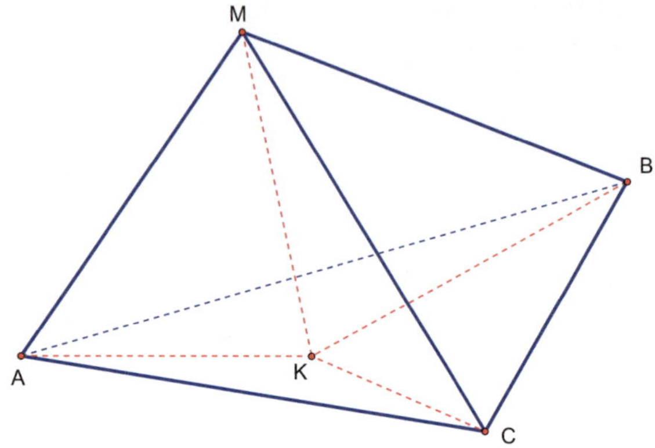

Fig. 2.22 The sum of the distances is the same

Comparing the very first formula and last one, we can see that $H = h_1 + h_2 + h_3$. Therefore, the height of a regular pyramid is equal to the sum of the heights of the small pyramids into which it is divided by connecting any point of its base with each vertex of the pyramid, i.e., $H = \sum h_i$, $i \geq 3$. □

> **Problem 78** The midpoint of the height of a regular tetrahedron is connected by segments to the vertices of the base of the tetrahedron. Find the plane angle of the trihedral angle at the top of the resulting pyramid.

Solution Please refer to the sketch in Fig. 2.23 where we let the sides of the tetrahedron be $AB = BC = AC = AD = DB = DC = a$ and drop a perpendicular DO from vertex D to the base of triangle ABC.

Denote $DO = h$ where O is the foot of the perpendicular and let the midpoint of the height be E so that $EO = \frac{h}{2}$. Connect points A to E and E to C. In the pyramid $EABC$, the angle AEC is the required plane angle that is to be found so denote it by $\angle AEC = \alpha$. Now, consider the triangle of the base—it is an equilateral so $OA = OB = OC = R$ where R is the radius of the circumscribed circle. Segment AO can be found from the right triangle AOG as $AO = \frac{AC}{2\cos 30^\circ} = \frac{a}{\sqrt{3}}$. Let us evaluate the height of the regular tetrahedron in terms of its side. From the right triangle AOE, we have $h = \sqrt{AD^2 - AO^2} = \sqrt{a^2 - \frac{a^2}{3}} = \frac{a\sqrt{6}}{3}$ so $EO = \frac{a\sqrt{6}}{6}$. In triangle AEC we know one side and that it is isosceles, hence we need to evaluate side $AE = EC$. From the right triangle AEO, we obtain $AE = \sqrt{EO^2 + AO^2} = \sqrt{\frac{a^2}{6} + \frac{a^2}{3}}$. Applying the Law of Cosines Theorem to AEC, we have

Fig. 2.23 Finding a plane angle in a tetrahedron

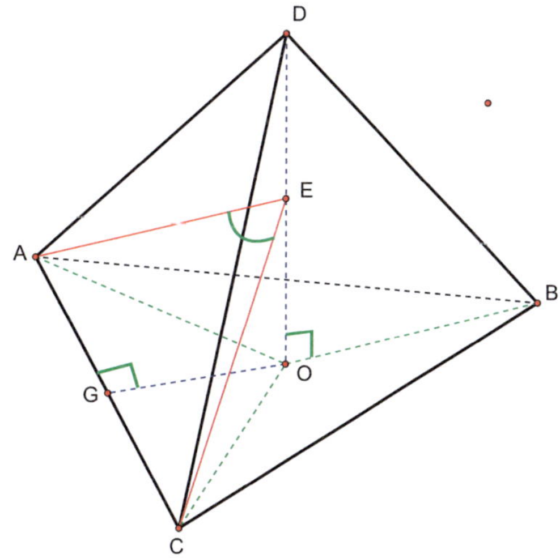

$$AC^2 = AE^2 + EC^2 - 2 \cdot AE \cdot EC \cdot \cos\alpha;$$

$$a^2 = \frac{a^2}{2} + \frac{a^2}{2} - 2 \cdot \frac{a}{\sqrt{2}} \cdot \frac{a}{\sqrt{2}} \cdot \cos\alpha;$$

so $\cos\alpha = 0$, i.e., $\alpha = 90°$. Therefore, the plane angle AEC is the right angle.
Answer. The plane angle of the trihedral angle is 90°.

Let us solve the problem from the nineteenth century that was first published in 1807 in *Sammlung geometrischen Aufgaben* journal [23]. We can call it the Pythagorean theorem for space.

Problem 79 (*Popov*) Prove that in a rectangular tetrahedron, the square of the area of the face opposite the right trihedral angle is equal to the sum of the squares of the areas of the other three faces of the tetrahedron.

Proof A rectangular tetrahedron is a tetrahedron in which one of the trihedral angles has all right plane angles. In our sketch in Figure 2.24 such a trihedral angle is at vertex D. Let us drop the height CE of triangle ABC.

Using our sketch (Fig. 2.24), we can state that $CE \perp AB$, $DE \perp AB$, and $DC \perp AB$. Denote the following segments for simplicity, $a = CB$, $b = AC$, $c = AB$, $n = AD$, $m = DB$, and $l = DC$. The area of face triangle ADB is $[ADB] = \frac{1}{2}nm = \frac{1}{2}c \cdot DE$ from which $DE = \frac{nm}{c}$. From $\triangle CDE$, we have $CE^2 = l^2 + DE^2 = l^2 + \frac{n^2m^2}{c^2} = \frac{l^2c^2+n^2m^2}{c^2}$. Since $c^2 = m^2 + n^2$, we see that $CE^2 = \frac{l^2m^2+l^2n^2+n^2m^2}{c^2}$. The area of the face triangle ABC is evaluated as $[ABC] = \frac{1}{2}AB \cdot CE$ and its square is equal to

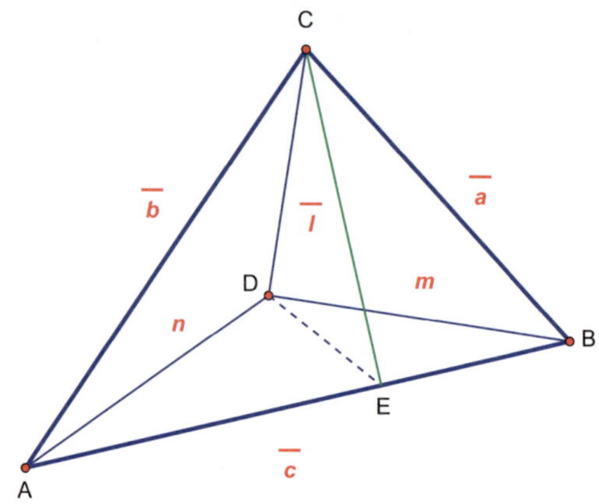

Fig. 2.24 A problem of Samuel Girsch about a rectangular tetrahedron

$[ABC]^2 = \frac{l^2m^2}{4} + \frac{l^2n^2}{4} + \frac{n^2m^2}{4}$. The square of the area of face CDB is $[CDB]^2 = \left(\frac{lm}{2}\right)^2 = \frac{l^2m^2}{4}$, the square of face CDA is $[CDA]^2 = \left(\frac{ln}{2}\right)^2 = \frac{l^2n^2}{4}$, and the square of face ADB is $[ADB]^2 = \left(\frac{nm}{2}\right)^2 = \frac{m^2n^2}{4}$. Adding these three squares of the areas $[ABC]^2 = [ADC]^2 + [CDB]^2 + [ADB]^2$ will complete our proof. □

Problem 80 Find the volume of a pyramid whose base is a triangle with angles α and β if the height of the pyramid h forms an angle φ with each lateral edge.

Solution Let us sketch the situation with all known information (Fig. 2.25). Let $h = DO$, $DO \perp ABC$, and $\angle ADO = \angle BDO = \angle CDO = \varphi$ so that $OA = OB = OC = h\tan\varphi$. Hence point O is the center of the circumscribed circle around triangle $\triangle ABC$ such that $R = OA = OB = OC = h\tan\varphi$ is its radius.

Next, let us consider the triangle of the base, $\triangle ABC$. Using the formulas of the area of this triangle written in four different ways applying the Law of Sines Theorem for each pair of the adjacent sides and the formula for the area of the triangle using its three sides and the radius of the circumscribed circle, we have

$$[ABC] = \frac{1}{2}cb\sin\alpha = \frac{1}{2}ca\sin(180 - \alpha - \beta) = \frac{1}{2}ab\sin\beta = \frac{abc}{4R}.$$

These formulas can be simplified as

$$\frac{1}{2}cb\sin\alpha = \frac{1}{2}ca\sin(\alpha + \beta) = \frac{1}{2}ab\sin\beta = \frac{abc}{4R},$$

and then we can take them by pairs and express all the unknown variables, if we wish. Let us take the first and the last of them, $\frac{1}{2}cb\sin\alpha = \frac{abc}{4R}$ from which we have

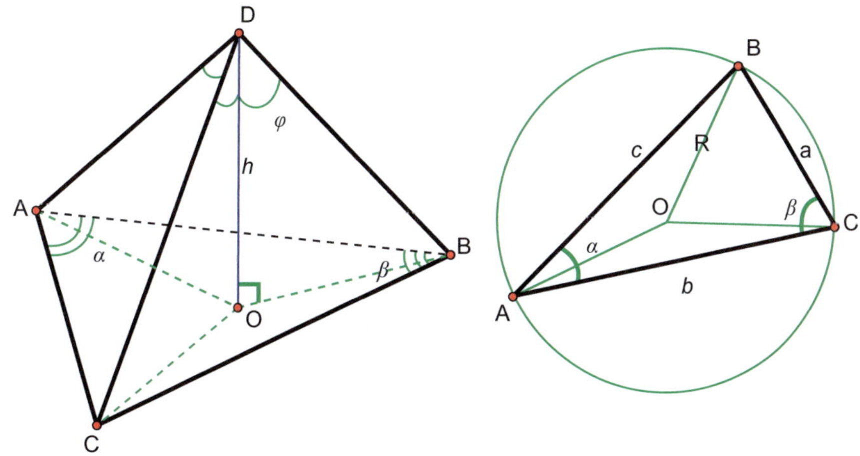

Fig. 2.25 Finding the volume of a pyramid

$$\sin\alpha = \frac{a}{2R} \Rightarrow \boxed{a = 2R\sin\alpha}.$$

The second and the last formulas $\frac{1}{2}ca\sin(\alpha+\beta) = \frac{abc}{4R}$ give

$$\sin(\alpha+\beta) = \frac{b}{2R} \Rightarrow \boxed{b = 2R\sin(\alpha+\beta)}.$$

Next, we will take the third and the last formulas: $\frac{1}{2}ab\sin\beta = \frac{abc}{4R}$. Removing the same factors on both sides, we obtain that

$$\sin\beta = \frac{c}{2R} \Rightarrow \boxed{c = 2R\sin\beta}.$$

Substituting the formula for R in terms of the known information in each of the emphasized formulas, we have

$$\begin{aligned} a &= 2h\tan\varphi\sin(\alpha+\beta); \\ b &= 2h\tan\varphi\sin\alpha. \end{aligned}$$

And substituting these formulas into the equation for the area of the triangle ABC, we have $[ABC] = \frac{1}{2}ab\sin\beta = \frac{1}{2}\cdot 2h\tan\varphi\sin(\alpha+\beta)\cdot 2h\tan\varphi\sin\alpha\sin\beta$, which simplifies to

$$[ABC] = 2h^2\tan^2\varphi\sin\alpha\sin\beta\sin(\alpha+\beta).$$

Finally, the volume of the pyramid is 1/3 the product of its height and the area of the base, which can be written as

$$V = \frac{2}{3}h^3 \tan^3 \varphi \sin \alpha \sin \beta \sin(\alpha + \beta).$$

Answer. The volume of the pyramid is $V = \frac{2}{3}h^3 \tan^3 \varphi \sin \alpha \sin \beta \sin(\alpha + \beta)$.

Problem 81 Given a regular hexagonal pyramid with height h such that all seven faces have an equal area, find the volume and total surface area of the pyramid.

Solution In this problem, we are given only one thing—the height of the regular hexagonal pyramid—so we need to use the condition of the problem to figure out how to find the other parameters and to evaluate the total surface area and the volume of the pyramid. The sketch of the pyramid is shown in Fig. 2.26.

Since $SO \perp ABCDEF$ and the pyramid is regular, then $SO = h$ and O is the center of the regular hexagon $ABCDEF$ such that $AO = BO = CO = DO = EO = FO = a$, where $a = AB = BC = CD = DE = EF = FA$ is the side of the hexagon. Let us evaluate the height of the lateral face—the isosceles triangle ASF by denoting $h_1 = SM$. The height $h_1 = SM = \sqrt{AS^2 - AM^2}$ (from the right triangle ASM). Obviously, point M is the midpoint of AF, so $AM = \frac{AF}{2} = \frac{a}{2}$. On the other hand, side AS can

Fig. 2.26 About a regular hexagonal pyramid

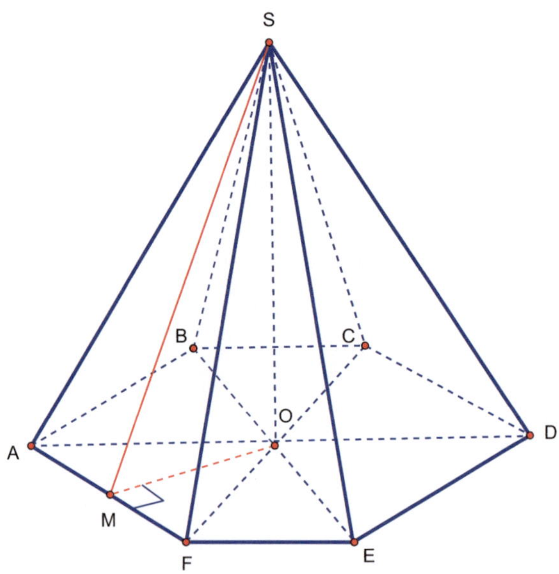

be found as the hypotenuse of the right triangle AOS, so $AS = \sqrt{a^2 + h^2}$. Now we can see that $h_1 = \sqrt{a^2 + h^2 - \frac{a}{4}^2} = \frac{\sqrt{3a^2+4h^2}}{2}$. Let us now use the condition of the problem that the areas of all faces are the same. This means that, for example, the area of triangle ASF is equal to the area of the regular hexagon, $ABCDEF$, that in turn equals the sum of six equal equilateral triangles AOF. The following equality holds:

$$\frac{1}{2}\frac{\sqrt{3a^2+4h^2}}{2} \cdot a = 6 \cdot \frac{a^2\sqrt{3}}{4};$$

$$\sqrt{3a^2+4h^2} = 6a\sqrt{3};$$

$$a^2 + 4h^2 = 36a^2 \cdot 3;$$

$$\boxed{a^2 = \frac{4h^2}{105}}$$

.

The surface area of the pyramid equals the seven times the area of triangle ASF,

$$S = 7 \cdot \frac{6a^2\sqrt{3}}{4} = \frac{21}{2} \cdot \frac{4h^2}{105} \cdot \sqrt{3} = \frac{2h^2\sqrt{3}}{5}.$$

The volume of the pyramid is $1/3$ of its height times base:

$$V = \frac{1}{3} \cdot h \cdot 6 \cdot \frac{a^2\sqrt{3}}{4} = \frac{2h^3\sqrt{3}}{105}.$$

Answer. $V = \frac{2h^3\sqrt{3}}{105}$, $S = \frac{2h^2\sqrt{3}}{5}$.

Problem 82 The height SO of a regular triangular pyramid $SABC$ is $5/7$ of the height SM of the lateral face SAB. Find the angle between the plane of the base of the pyramid and its lateral edge.

Solution Let us sketch the pyramid $SABC$ (Fig. 2.28). The pyramid is regular, then its base is an equilateral triangle, the height is projected to the center of the base, that is, to the intersection point of the medians, heights, and bisectors of the triangle. All the edges of the pyramid are inclined to the plane of the base at the same angle. We know that the angle between a line and a plane is the angle between a line and its projection onto the plane.

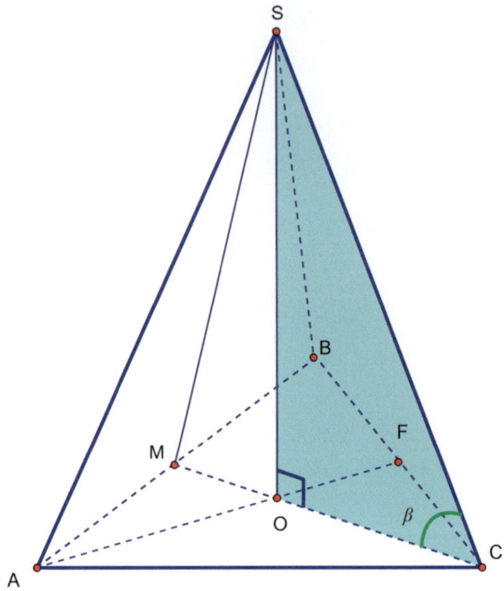

Fig. 2.27 Finding angle β in a regular pyramid

Consider the angle between the edge SC and the plane of the base ABC. The angle SCO is the angle between the edge SC and its projection CO onto the base. Therefore, $\angle SCO$ is the angle between the base of the pyramid and its side edge. Denote it by β. By the condition of the problem, the height SO of a regular triangular pyramid $SABC$ is equal to $5/7$ of the height SM of the lateral face SAB. Let $SM = x$, then $SO = 5x/7$ (see Fig. 2.27). Let's express the length MO in terms of x. Consider right triangle MSO and apply the Pythagorean theorem:$MO = \sqrt{x^2 - \left(\frac{5x}{7}\right)^2} = \frac{2\sqrt{6}x}{7}$. Next, let us find the length of segment OC. Point O is the point of intersection of the medians of triangle ABC, therefore, $OC = 2MO = \frac{4\sqrt{6}x}{7}$. Consider the right triangle OSC:

$$\tan \angle SCO = \tan \beta = \frac{OS}{OC} = \frac{\frac{5x}{7}}{\frac{4\sqrt{6}x}{7}} = \frac{5\sqrt{6}}{24}.$$

Answer. The angle between the plane of the base of the pyramid and its lateral side edge is $\beta = \arctan \frac{5\sqrt{6}}{24}$.

Problem 83 In a regular n-gonal pyramid ($n > 3$), the area of the base is S and each lateral face forms an angle α with the plane of the base. Find the total surface area of the pyramid.

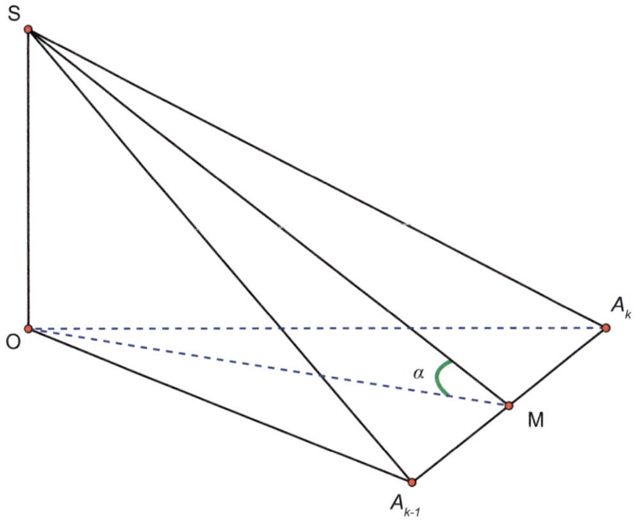

Fig. 2.28 The nth section of a regular n-gonal pyramid

Solution Consider $\frac{1}{n}$ part of the regular pyramid and denote it as $OSA_{k-1}A_k$ (see Fig. 2.28). Like pieces of a pie, we can put together these n small equal right pyramids to obtain the original regular one. Let's find the area of $\frac{1}{n}$ part of the pyramid:

$$[A_kSA_{k-1}] = \frac{1}{2}A_kA_{k-1} \cdot SM.$$

Since $SM = \frac{OM}{\cos\alpha}$, then

$$[A_kSA_{k-1}] = \frac{1}{2}A_kA_{k-1} \cdot \frac{OM}{\cos\alpha}.$$

Adding the areas of all n triangles together, we obtain the lateral surface of the big pyramid and also prove formula (2.2):

$$S_{\text{lateral}} = \frac{n \cdot A_kA_{k-1} \cdot OM}{\cos\alpha \cdot 2} = n \cdot \frac{[A_kOA_{k-1}]}{\cos\alpha} = \frac{S}{\cos\alpha}.$$

The total surface area is

$$S_{\text{full}} = S + \frac{S}{\cos\alpha} = \frac{S\,(1+\cos\alpha)}{\cos\alpha}.$$

Answer. $S_{\text{full}} = \frac{S(1+\cos\alpha)}{\cos\alpha}$.

The following two problems are optimization problems that can be solved using the relationship between arithmetic and geometric means.

Problem 84 All plane angles at vertex D of tetrahedron $ABCD$ are right. Some point K on the face ABC is located at distances of 3, 4, and 5 from the other faces BCD, ACD, and ABD, respectively. Find the minimum value of the volume of this tetrahedron.

Solution Let us solve this problem using analytic geometry. This method works very well since all the plane angles at the tetrahedral angle with vertex D are 90°. We place the tetrahedron vertex at the origin and direct the coordinate axes OX, OY, and OZ along the lateral edges of the tetrahedron DA, DB, and DC, respectively, so point K will have coordinates $K = (3, 4, 5)$. Let the lengths of the side edges of the tetrahedron be $DA = a$, $DB = b$, and $DC = c$ (Fig. 2.29).

Using formula (1.49) the equation of the plane ABC will be written as $\frac{x}{a} + \frac{y}{b} + \frac{z}{c} = 1$. Since this plane passes through the point $K(3, 4, 5)$, then its coordinates satisfy the plane's equation:

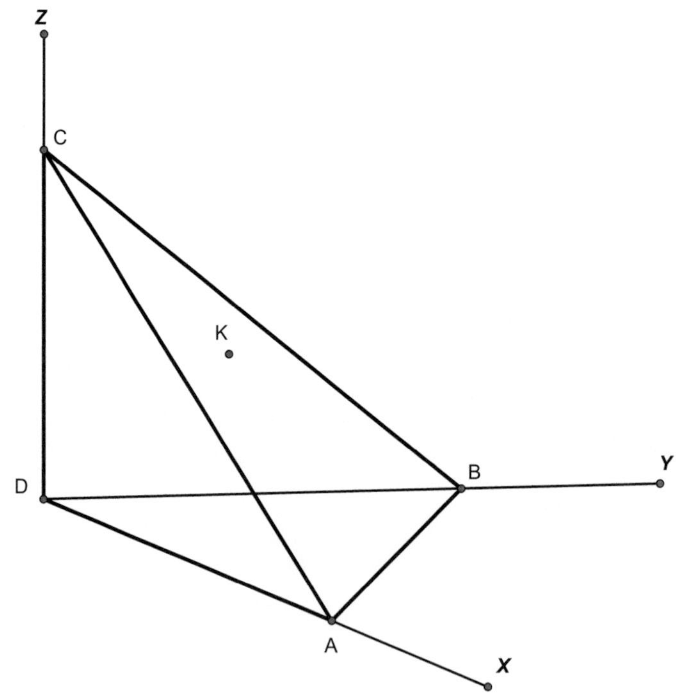

Fig. 2.29 Tetrahedron with all plane angle at the vertex D of 90°

$$\frac{3}{a}+\frac{4}{b}+\frac{5}{c}=1. \tag{2.20}$$

Denote $\frac{3}{a}=\alpha$, $\frac{4}{b}=\beta$, and $\frac{5}{c}=\gamma$, so the volume of this tetrahedron is

$$V=\frac{1}{3}\cdot\frac{1}{2}abc=\frac{10}{\alpha\beta\gamma}. \tag{2.21}$$

We need to minimize the volume (2.21) ($V\to\min$) under the condition (2.20) which may be rewritten as

$$\alpha+\beta+\gamma=1. \tag{2.22}$$

Obviously, the volume of the tetrahedron will be minimal at the maximal value of the denominator, $\alpha\beta\gamma$ in (2.21). Recall from the relationship between arithmetic and geometric means that if the sum of several positive numbers is constant (in our problem it is (2.22)), then their product ($\alpha\cdot\beta\cdot\gamma$) will be greatest when they are equal to each other. In our case, it means that $\alpha=\beta=\gamma=\frac{1}{3}$. Therefore, the minimum value of the volume of the tetrahedron is $V=10\div\left(\frac{1}{3}\right)^3=270$.
Answer. $V_{\min}=270$.

Problem 85 (*Olympiad Lomonosov 2016*) Find the maximum volume of the triangular pyramid for which the opposite edges are equal and the sum of all edges is $36\sqrt{2}$.

Solution Consider the pairs of equal edges in pyramid $ABCD$, $CD=AB, AD=BC$, and $DB=AC$. Let us draw through each edge of the pyramid the plane parallel to the opposite edge. For example, we can draw plane $GCHD$ through AB parallel to its opposite edge CD and vice versa (plane $AEBF$ parallel to CD). The obtained polyhedron is a parallelepiped as shown in Fig. 2.30. Let us show that this parallelepiped is a rectangular box.

Without loss of generality we can consider only two parallel faces. Thus, by our construction $AGCE\|DHBF$. For $AGCE$, $GC\|AE$, and $AG\|CE$, then $AGCE$ is a parallelogram such that $AGCE=FDHB$. Because $AC=DB$ (two diagonals of the parallelogram are equal), then the parallelograms $AGCE$ and $DHBF$ are equal rectangles. In the rectangular box, we denote $AF=a$, $FD=b$, and $FB=c$. The volume of the box is $V=abc$. The volume of the pyramid $ACBD$ can be obtained from the rectangular box by cutting off four equal triangular pyramids with vertices at H, F, G, and C. The volume of one such pyramid $DFAB$ in which $DF\perp ABF$ is $V_{DABF}=\frac{1}{3}\cdot b\cdot\frac{1}{2}ac=\frac{abc}{6}$. Next, the volume of the pyramid $DABC$ is

$$V_{pyr}=V_{DABC}=abc-\frac{4}{6}abc=\frac{abc}{3}. \tag{2.23}$$

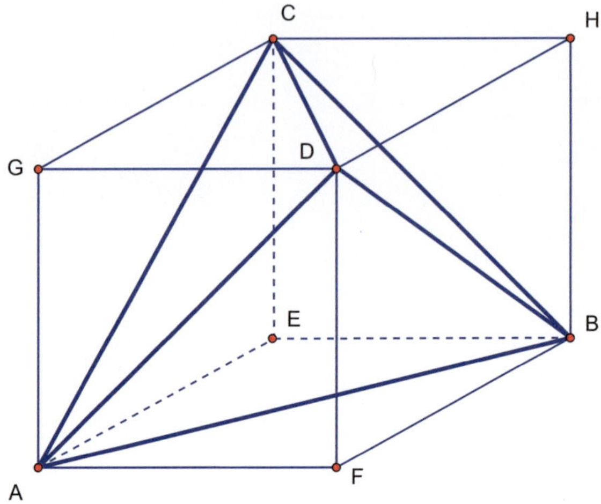

Fig. 2.30 Pyramid of maximum volume when the sum of all edges is given

Let us use the condition for the sum of the edges

$$36\sqrt{2} = 2(\sqrt{a^2+b^2} + \sqrt{c^2+b^2} + \sqrt{c^2+a^2}). \tag{2.24}$$

The relationship between arithmetic and geometric means for positive x and y can be written as $x + y \geq 2\sqrt{xy}$. Applying this to each of the square roots terms inside the parentheses, we can rewrite (2.24) as

$$36\sqrt{2} \geq 2\sqrt{2}(\sqrt{ab} + \sqrt{bc} + \sqrt{ac}). \tag{2.25}$$

Now using the inequality between the arithmetic and geometric means applied to three positive quantities we have $\frac{x+y+z}{3} \geq \sqrt[3]{xyz}$ so that for the expression inside the parentheses of (2.25), we have

$$36\sqrt{2} \geq 2\sqrt{2} \cdot 3\left(\frac{\sqrt{ab} + \sqrt{bc} + \sqrt{ac}}{3}\right) \geq 6\sqrt{2}\sqrt[3]{abc}. \tag{2.26}$$

We can rewrite (2.26) as $6 \geq \sqrt[3]{abc} = \sqrt[3]{3V_{\text{pyr}}}$. If we cube both sides of the inequality $6^3 \geq 3V_{\text{pyr}}$ and divide it by 3, we obtain $V_{\text{pyr}} \leq 72$. The equality occurs if $a = b = c$, i.e., when the pyramid $ACBD$ is a regular tetrahedron.

Answer. The maximum volume is equal to 72.

Problem 86 All faces of a tetrahedron are equal acute scalene triangles with sides a, b, and c. Prove that the volume of the tetrahedron is determined by the formula:

$$V = \frac{\sqrt{2}}{12} \cdot \sqrt{(a^2 + b^2 - c^2)(c^2 + b^2 - a^2)(a^2 + c^2 - b^2)}. \tag{2.27}$$

Proof Let us reuse Fig. 2.30 assuming that we inscribe this tetrahedron in the rectangular box $AEBFGCHD$ with edges $x = AF$, $y = BF$, and $z = DF$. The tetrahedron $ABCD$ satisfies the condition of the problem with sides $AC = DB = a$, $CB = AD = b$, and $AB = CD = c$. As in the previous problem, the volume of tetrahedron $ABCD$ can be found as using formula (2.23), but in our case we have a box with sides x, y, and z, so

$$V = \frac{xyz}{3} = \frac{\sqrt{x^2y^2z^2}}{3}. \tag{2.28}$$

We need to express x, y, and z in terms of the sides of the tetrahedron a, b, and c. Using the Pythagorean theorem we have relationships as a system of three nonlinear equations:

$$\begin{cases} a^2 = y^2 + z^2 \\ b^2 = x^2 + z^2 \\ c^2 = x^2 + y^2. \end{cases}$$

It may look complex at first glance, but we do not need to find x, y, or z. We can simply use their squares! Denote $A = a^2$, $B = b^2$, and $C = c^2$, and also $X = x^2$, $Y = y^2$, and $Z = z^2$, so that it is clear that the underlying system is linear and can be solved by any appropriate linear method. For example, we have the following system, the solution of which can be obtained easily using a simple algebraic or matrix approach:

$$\begin{cases} Y + Z = A \\ X + Z = B \\ X + Y = C. \end{cases}$$

This system has the solution:

$$\begin{aligned} x^2 &= X = \tfrac{B+C-A}{2} = \tfrac{b^2+c^2-a^2}{2} \\ y^2 &= Y = \tfrac{A+C-B}{2} = \tfrac{a^2+c^2-b^2}{2} \\ z^2 &= Z = \tfrac{A+B-C}{2} = \tfrac{a^2+b^2-c^2}{2}. \end{aligned} \tag{2.29}$$

Substituting the values from (2.29) into the formula for the volume of a tetrahedron (2.28), we obtain $V = \frac{1}{3}\sqrt{\frac{(a^2+b^2-c^2)(c^2+b^2-a^2)(a^2+c^2-b^2)}{8}}$ which proves the required statement. □

The next problem shows the advantages of the construction method.

Problem 87 From the point O lying at the base ABC of the triangular pyramid $SABC$, lines OA_1, OB_1, and OC_1 are drawn parallel to the edges of the lateral surface of the pyramid AS, BS, and CS until they intersect with the faces SBC, SCA, and SAB at the points A_1, B_1, and C_1, respectively. Prove that $\frac{OA_1}{SA} + \frac{OB_1}{SB} + \frac{OC_1}{SC} = 1$.

Proof Let's take any point O at the base of the pyramid ABC and draw the necessary segments from it until they intersect with the corresponding faces (shown in red in Fig. 2.31). How can we prove that the sum of the given ratios is equal to one? Let's connect the top of the pyramid S with point O, as well as point O with other vertices A, B, and C. This construction divides the pyramid into three smaller pyramids, the sum of the volumes of which is equal to the volume of the pyramid itself. We write the volumetric decomposition as

$$V = V_1 + V_2 + V_3,$$

where $V_1 = V_{OCSB}$, $V_2 = V_{OASC}$, and $V_3 = V_{OASB}$. Well, the idea is certainly interesting, it would also be interesting to understand why and how we can use this partition

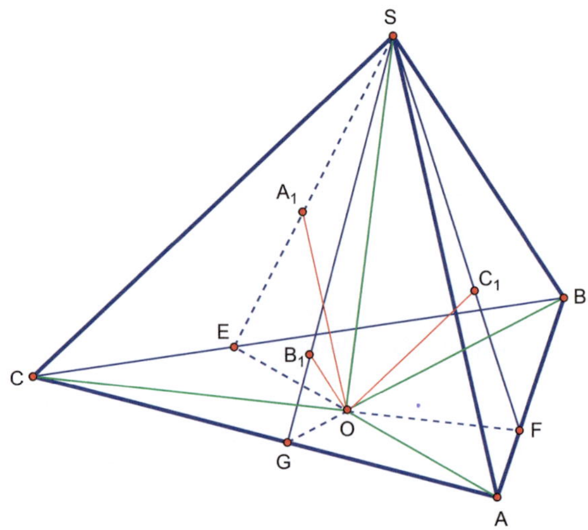

Fig. 2.31 Segments drawn parallel to edges from a point in the base of pyramid

to prove the problem statement. Note that this pyramid is triangular so that any of its faces can serve as a base.

Let's see how the volumes of the *CSAB* pyramid and the *OSAB* pyramid relate to each other. It is obvious that these two pyramids have a common base *SAB*. Since the base is common, the volumes of these two pyramids are related via the corresponding heights drawn from the corresponding vertices to this base. Now let's take another look at Fig. 2.31 carefully. It can be seen that the line *CS* and the line OC_1 are parallel in construction. Denote the heights of the pyramids *OSAB* and *CSAB* dropped from the vertices O and C by $h_3 = ON \perp ASB$ and $H_3 = CM \perp ASB$, respectively. Then $ON||CM$ and then $\Delta ONC_1 \sim \Delta CMS$. Therefore the ratio of the heights of the pyramids will be equal to the ratio of these segments following from the similarity property. Next, the ratio of the volumes of two pyramids can be written as

$$\frac{V_{OASB}}{V_{CASB}} = \frac{V_3}{V} = \frac{h_3[ASB]}{H_3[ASB]} = \frac{h_3}{H_3} = \frac{OC_1}{SC}.$$

That is, it turns out that we do not need to know the height of the large pyramid drawn from the point *A* to the base *SBC*, we simply write the ratio of the volumes of the pyramids *OSBC* and *ASBC* as

$$\frac{V_{OSBC}}{V_{ASBC}} = \frac{V_1}{V} = \frac{h_1[CSB]}{H_3[CSB]} = \frac{h_1}{H_1} = \frac{OA_1}{SA}.$$

Finally, compare pyramids *OSCA* and *BSCA* with common base *ASC*, i.e.:

$$\frac{V_{OSCA}}{V_{BSCA}} = \frac{V_2}{V} = \frac{h_2 \cdot [CSA]}{H_2 \cdot [CSA]} = \frac{h_2}{H_2} = \frac{OB_1}{SB}.$$

Consider only the outer parts of all three formulas and add their left and right sides. A simple manipulation helps to complete the proof

$$\frac{V_1}{V} + \frac{V_2}{V} + \frac{V_3}{V} = \frac{V_1 + V_2 + V_3}{V} = \frac{V}{V} = \boxed{1 = \frac{OA_1}{SA} + \frac{OB_1}{SB} + \frac{OC_1}{SC}}.$$

□

2.2 Prisms

A prism is a polyhedron whose shape most of us can imagine. There are many prisms in our life, from a computer system unit, refrigerator, or washing machine, bedside table and wardrobe to a faceted pencil. Those who have studied physics know that a prism is also an optical element made of a transparent material such as optical glass having flat, polished edges through which light enters and exits.

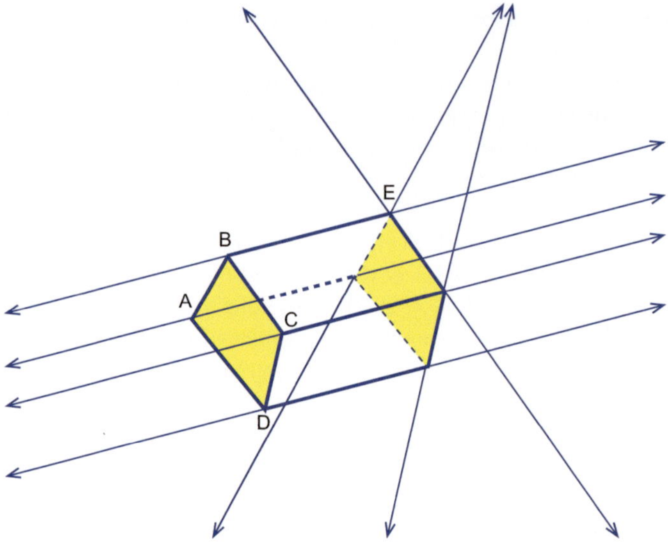

Fig. 2.32 Prism

In order to draw a prism, sketch any flat polygon and translate it by a vector to obtain its image. In Fig. 2.32 vector *BE* translates polygon *ABCD* to the second base of the prism. Connect the corresponding vertices (*B* and *E*, etc.) to obtain the lateral edges of the prism. Translation is one of the four possible transformations—rigid motion by a mapping that preserves shape and size of an object. The other three rigid motions are: reflection, glide reflection, and rotation. Translation moves each point in the space by a fixed vector $\overrightarrow{a}$ and can be written as $T(\overrightarrow{x}) = \overrightarrow{x} + \overrightarrow{a}$. For example, a pentagonal prism $ABCDEA_1B_1C_1D_1E_1$ shown in Fig. 2.33 is obtained by translation of the base-pentagon *ABCDE* by vector $\overrightarrow{BB_1}$ and by connecting the corresponding vertices of the resulting polygons.

When mathematicians describe a geometric figure, we must be clear in our definitions. In my opinion, the following definition, given by Pogorelov [13] based on the idea of translation, describes a prism well.

Definition 11 A **prism** is a polyhedron consisting of two flat polygons lying in different planes, that can be mapped onto each other by translation, and all the segments connecting the corresponding points of these polygons. The polygons are called the bases of the prism, and the segments connecting the corresponding vertices of the polygons are the lateral edges.

Since translation is a rigid motion, so a plane is mapped to a parallel plane (or into itself), so the bases of the prism are equal and lie in parallel planes. Since points

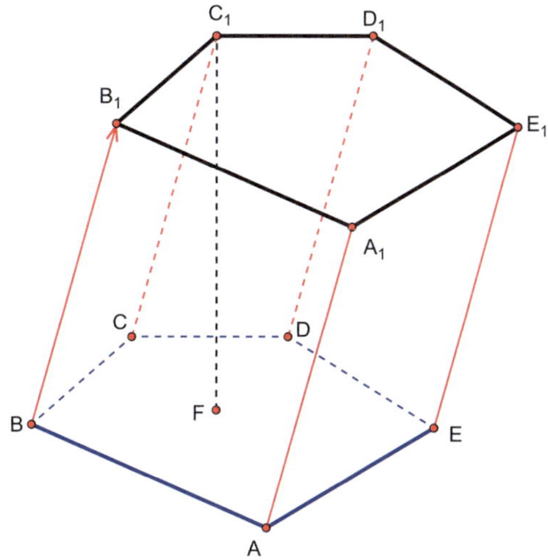

Fig. 2.33 Pentagonal prism

are shifted by translation along parallel or coinciding lines by the same distance, the lateral edges of the prism are also parallel and equal. In order to better appreciate Definition 11, let's compare it to some of those we might run across (Problem 88).

Problem 88 The following definitions of prisms are found in various textbooks. Consider each definition individually and decide whether the definition is correct by our standard. Justify your decision.

1. "*A polyhedron is a prism if and only if it has two congruent faces that are contained in parallel planes and its other faces are parallelograms.*"
2. "*Prism is the solid figure formed by joining two congruent polygonal regions in parallel planes.*"
3. "*A prism is a three-dimensional shape, which consists of two identical polygons, which do not lie in the same plane. For each side of the polygon, there will be a parallelogram.*"
4. "*Prisms are three-dimensional objects or solids, having two congruent polygonal bases that lie in parallel planes.*"
5. "*A prism is a polyhedron with two opposite faces, called bases, that are identical polygonal regions in parallel planes. The other faces are called lateral faces.*"
6. "*A prism is a solid figure formed by taking two congruent figures in parallel planes, with parallel edges, and joining their corresponding vertices. Its faces consist of the two original congruent figures, plus parallelograms joining the corresponding sides.*"

Solution The correctness of each definition is considered in relation to Definition 11.

1. This definition [39] is incorrect. Consider a solid figure called a **rhombic dodecahedron**. All 12 of its faces are equal rhombuses (parallelograms). The planes of the opposite faces are parallel. There are 24 edges and 14 vertices. The edges of the opposite faces are parallel. The polyhedron shown in Fig. 2.34 fits its definition, but obviously it is not a prism. We will learn how to construct a rhombic dodecahedron in the section **Semi-regular and Other Polyhedra**.
   Ironically, even a nonconvex polyhedron shown in Fig. 2.35 would fit this incorrect definition. This solid figure consists of two oblique parallelepipeds symmetric with respect to the plane of their common base and obviously is not a prism. However, this figure has two parallel and congruent "bases" ($ABCD \| EFGH$), all other sides are parallelograms. It has 12 vertices, 20 edges, and 10 faces.)
2. This definition [40] is incorrect. This definition does not say anything about orientations of the congruent polygons. So a polyhedron can have two faces that are identical polygons in parallel planes but rotated with respect to each other. Moreover, this definition says nothing about the lateral faces of the polyhedron. For example, the other faces can be a series of alternately oriented triangles. Consider the polyhedron shown in Fig. 2.36. This solid satisfies the definition but is not a prism. The figure is called an **antiprism** and is obtained by taking a regular polygon in the plane, placing a copy of the regular polygon above the first, but rotated so that each vertex of the upper polygon is above the middle of an edge of the lower polygon, and then joining each upper vertex to the two closest lower vertices.
3. This definition [38] is incorrect since the figure shown in Fig. 2.34 satisfies it, but this polyhedron is obviously not a prism.
4. This definition [37] is incorrect since it describes not only a prism, but also the antiprism shown in Fig. 2.36, which is not a prism.
5. This definition [28] is incorrect by the same token. Definitions 3–5 might be corrected by changing their second part to "A prism is a polyhedron, two of whose faces are equal polygons with correspondingly parallel sides and all other faces are parallelograms intersecting each other along parallel lines."
6. This definition [1] given in the textbook of Hartshorne correctly describes a prism.

When we define a prism, the following definition (not based on the use of the concept of rigid motion) is also appropriate.

Definition 12 A prism is a polyhedron whose bases are equal polygons located in parallel planes with correspondingly parallel sides; the vertices of the polyhedron are the vertices of the bases, and the lateral edges connecting the corresponding vertices of these polygons are equal and parallel.

Fig. 2.34 Rhombic dodecahedron in color

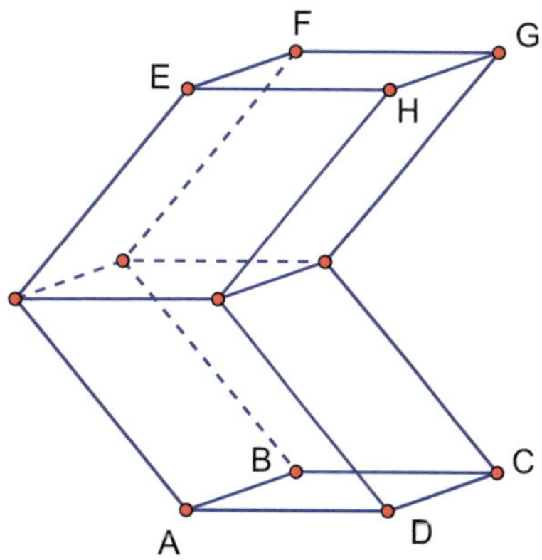

Fig. 2.35 Two oblique parallelepipeds symmetric about a common base. Not a prism

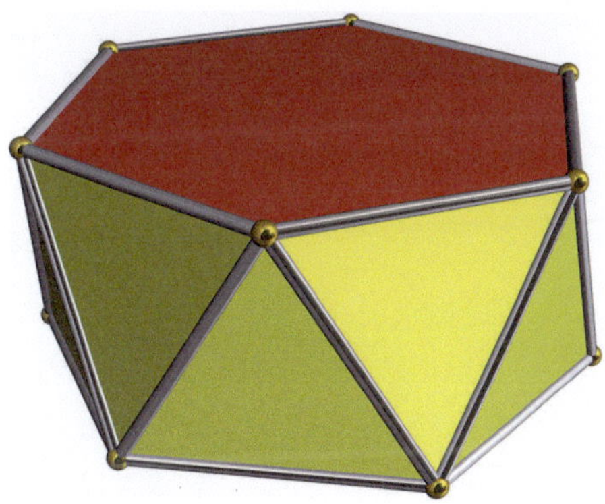

Fig. 2.36 Hexagonal antiprism

See Fig. 2.32 for a prism over a quadrilateral. Can we also say that a prism is obtained by taking a convex polygon on the plane, placing an identical copy of the polygon directly above the first one, and connecting the pairs of corresponding vertices of the two copies of the polygon? Is this right? Please think about it.

The **total surface area of a prism** consists of the bases and the lateral surface. The lateral surface consists of parallelograms; for each of these parallelograms two sides are the corresponding sides of the bases and two others are the adjacent lateral edges. The **height of a prism** is the distance between the planes of its bases. Thus, perpendicular C_1F is the height of the pentagonal prism $ABCDEA_1B_1C_1D_1E_1$ shown in Fig. 2.33. The segment connecting two vertices of the prism that do not belong to the same face is called the **diagonal of a prism**. The prism is called n-gonal if its bases are n-gons. We will consider only prisms whose bases are convex polygons, such prisms are convex polyhedra.

A **right prism** is a prism whose side faces and edges are perpendicular to the bases. A **regular prism** is a right prism whose bases are regular polygons. This definition begins with the straight prism, which means that a regular prism includes all the properties of a straight prism.

Figure 2.37 shows a regular triangular prism. Here triangles ABC and DFE are equilateral triangles. Edges DA, FC, and EB are equal and perpendicular to the base, therefore they are also the height of the prism.

The **volume of a prism** is equal to height of the prism (H) times its base area, B,

$$V = HB. \tag{2.30}$$

The volume also can be found in terms of the orthogonal cross section of a prism $S_\perp$, that is, the cross section by the plane perpendicular to the lateral edge of a prism with its lateral edge of l:

$$V = S_\perp \cdot l. \tag{2.31}$$

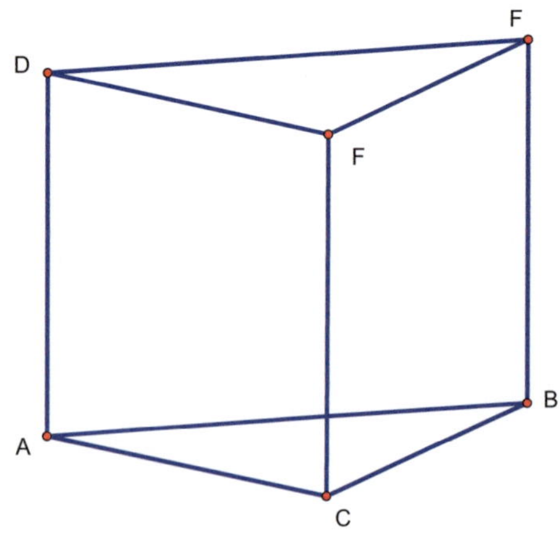

Fig. 2.37 Regular triangular prism

The total surface area of any prism can be found as the sum of the lateral surface area S_{lateral} and twice the area of the base B, i.e.:

$$S_{\text{full}} = S_{\text{lateral}} + 2B.$$

Theorem 33 *The lateral surface of a right prism is equal to the product of the perimeter of the base (P) and the height of the prism (H), that is, the length of the lateral edge:*

$$S_{\text{lateral}} = P \cdot H.$$

A **parallelepiped** is a prism with a parallelogram base. If the parallelogram is a rectangle and the edge is perpendicular to the base, such a parallelepiped becomes a rectangular parallelepiped (rectangular box). A **cube** is a regular quadrilateral prism in which all edges are equal (Fig. 2.38). You can also say that a cube is a particular case of a rectangular parallelepiped when all sides equal. A **diagonal of a parallelepiped** (Fig. 2.39) is a segment connecting two vertices that do not belong to the same face. A **right parallelepiped** is a right prism with parallelogram bases. The diagonals of a right parallelepiped can be evaluated as $d_1^2 = a^2 + b^2 + c^2 + 2ab\cos\alpha$ and $d_2^2 = a^2 + b^2 + c^2 - 2ab\cos\alpha$.

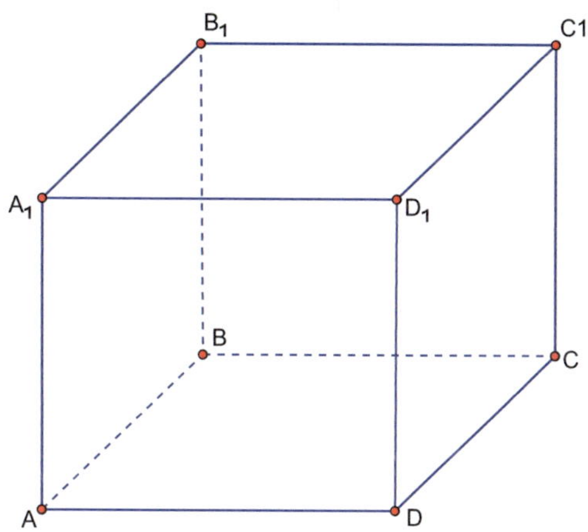

Fig. 2.38 Cube

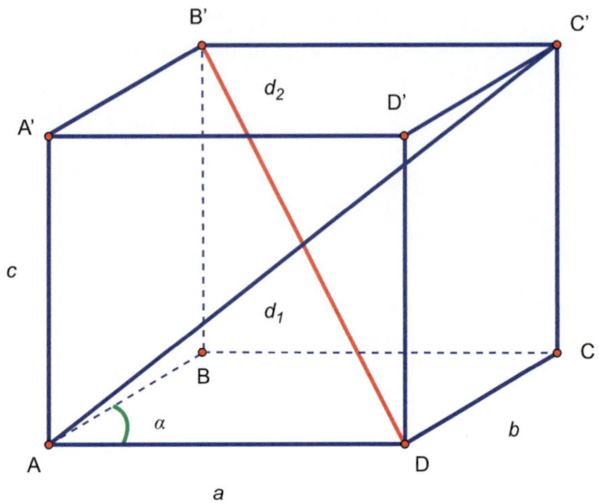

Fig. 2.39 Diagonals of a rectangular parallelepiped

Theorem 34 *The sum of the squares of the diagonals of any parallelepiped equals the sum of the squares of all its edges:*

$$d_1^2 + d_2^2 + d_3^2 + d_4^2 = 4a^2 + 4b^2 + 4c^2. \tag{2.32}$$

Proof Let $AB = a, AD = b, AA_1 = c$. In general, a parallelepiped has four unequal diagonals. Consider the cross- sectional parallelograms AA_1C_1C with diagonals d_1 and d_2 and BB_1D_1D with diagonals d_3 and d_4. Use the fact that the sum of the squares of the diagonals of any parallelogram is equal to the sum of the squares of all its sides (See [2], page 95) and that in a parallelogram the opposite sides are equal. □

The diagonals of a rectangular parallelepiped are equal to each other and can be found as follows:

$$d = \sqrt{a^2 + b^2 + c^2}. \tag{2.33}$$

The volume of a rectangular parallelepiped equals the product of three measurements, $V = abc$ and its full surface area is $S = 2\,(ab + bc + ac)$. A cube or hexahedron is a parallelepiped with all six faces being squares. The volume of a cube is $V = a^3$ and its full surface area is $S = 6a^2$.

2.2.1 Problems on Prisms

There are many interesting problems about prisms. Some of the following problems are related to optimization.

Problem 89 A rectangular parallelepiped is given by three of its measurements a, b, and c. Is it possible to choose such numerical values a, b, and c so that the volume of a rectangular parallelepiped would be numerically greater than $1/3$ of the sum of the volumes of three cubes with edges a, b, and c?

Solution Is it possible to create such a rectangular box with sides a, b, and c, so that the volume of such a box would be greater than the total volume of three cubes with sides a, b, and c?

The volume of a rectangular parallelepiped can be written as $V = abc = \sqrt[3]{a^3b^3c^3}$. By the relationship between the arithmetic mean and geometric mean for three positive numbers a, b, and c, the following is true: $\frac{a^3+b^3+c^3}{3} \geq \sqrt[3]{a^3b^3c^3} = V$. So we have the inequality for the $1/3$ of the sum of the volumes of three cubes with edges a, b, and c, $\frac{1}{3}\left(a^3 + b^3 + c^3\right) \geq V$, which allows us to answer the question in the negative.

Problem 90 Given S the area of the rectangular parallelepiped. Find its maximum volume.

Solution Let a, b, c be the dimensions of the box, S its full surface, and V its volume. Then $S = 2(ab + bc + ac)$ and the volume is $V = abc$. We can express $ab + bc + ac = \frac{S}{2}$ and for their product $ab \cdot bc \cdot ac = a^2b^2c^2 = V^2$. Applying the relationship between arithmetic and geometric means, we have $V^2 = (abc)^2 \leq \left(\frac{ab+bc+ac}{3}\right)^3 = \left(\frac{S}{6}\right)^3$ from which we obtain that $V \leq \sqrt{\left(\frac{S}{6}\right)^3}$. The only case of equality is $a = b = c$. Therefore, a cube has the biggest volume.

Answer. $V_{\max} = \sqrt{\left(\frac{S}{6}\right)^3}$.

Problem 91 (*Kaganov*) A cube-shaped vessel filled with liquid is tilted 30° relative to the base. Obviously, some liquid is dropped out. Find what part of the original volume of the liquid remained in the vessel.

Solution Let the side of the cube be a, then the volume of the cube is equal to a^3. When the cube is tilted, liquid will pour out in the volume of the triangular right prism, $ABCNMK$ (see my sketch in Fig. 2.40). Note that the angle $\measuredangle BCA = 30°$. Next, we have to evaluate the volume of this prism. Let us first find the area of triangle ABC as $[ABC] = \frac{1}{2} \cdot BC \cdot AB = \frac{a \cdot a \cdot \tan 30°}{2} = \frac{a^2\sqrt{3}}{6}$. Next, we will find the volume of the prism $ABCNMK$. Its height equals a, so for the volume we obtain

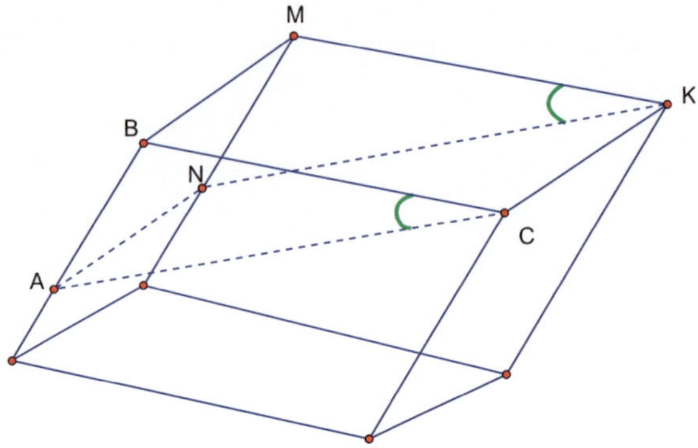

Fig. 2.40 Tilted cube

$$V_{\text{prism}} = \frac{a^3\sqrt{3}}{6}. \tag{2.34}$$

After the water was poured out of the cube in the amount given by (2.34), the remaining volume is

$$V = V_{\text{cube}} - V_{\text{prism}} = a^3 - \frac{a^3\sqrt{3}}{6} = a^3 \cdot \frac{\left(6-\sqrt{3}\right)}{6}.$$

Finally, the ratio of this volume to the volume of the cube can be found as follows:

$$V \div V_{\text{cube}} = a^3 \cdot \frac{\left(6-\sqrt{3}\right)}{6} \div a^3 = \frac{6-\sqrt{3}}{6}.$$

Answer. $\frac{6-\sqrt{3}}{6}$. Approximately 71 % of the water is still in the vessel.

Problem 92 (*Rivkin*) A cube is placed in a parallelepiped with edges a, b, and c so that the cube's vertex C coincides with the parallelepiped's vertex. Find the angle between the diagonals of the cube and the parallelepiped passing through the vertex C.

Solution If one cube is located inside another cube and they have a common vertex (for example, C), then the diagonals of these cubes passing through C lie on the same line, so the angle between diagonals would be zero. Therefore, from all such cubes that can be placed in a parallelepiped, we will choose the maximum. Let $c < a$, $c < b$, then a cube with edge c can be placed into a parallelepiped (box), for example, as it is shown in Fig. 2.41.

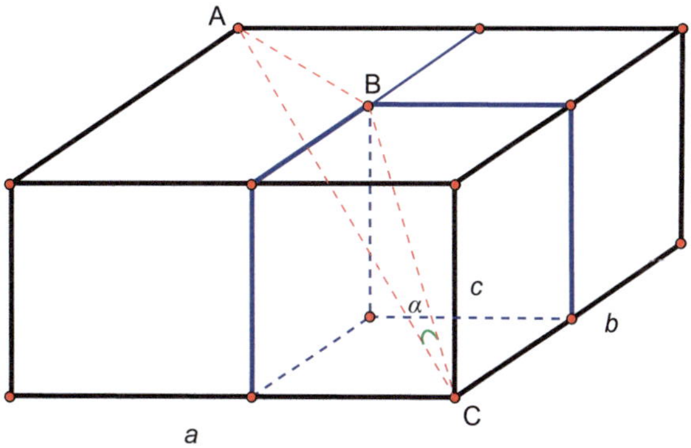

Fig. 2.41 Cube inside parallelepiped

Evaluate the sides of triangle ABC using the Law of Cosines:

$$\begin{aligned} AB^2 &= AC^2 + BC^2 - 2AC \cdot BC \cos\alpha; \\ AC^2 &= a^2 + b^2 + c^2; \\ BC^2 &= 3c^2; \\ AB^2 &= (a-c)^2 + (b-c)^2, \end{aligned}$$

from which we obtain an equation that can be used to determine the cosine of angle α, specifically $\sqrt{3\left(a^2+b^2+c^2\right)}\cos\alpha = a+b+c$. Note that this equation is symmetric in the values of a, b, and c. For any of the values, the cosine of angle α will be always less than one, so $\alpha = \arccos\frac{a+b+c}{\sqrt{3\left(a^2+b^2+c^2\right)}}$.

Answer. $\alpha = \arccos\frac{a+b+c}{\sqrt{3\left(a^2+b^2+c^2\right)}}$.

Problem 93 Find the distance between two non-intersecting diagonals of adjacent faces of a cube whose edge is a.

Solution Let us select diagonals $C'D$ and $B'C$ as they are shown in Fig. 2.42.

Because $B'C \| A'D$ and $C'D \| B'A$, then the planes $A'C'D$ and $AB'C$ are parallel to each other. The distance between diagonals $B'C$ and $C'D$ equals the distance between these planes. These two planes are the bases of two identical tetrahedrons $A'C'DD'$ and $ABCB'$. Both planes $A'C'D$ and $AB'C$ are perpendicular to the cube's diagonal BD'. Then the distance between two planes can be found as the difference between the length of the diagonal BD' and the twice height of the tetrahedron $A'C'DD'$.

Let us find the height of tetrahedron dropped to the base $A'C'D$. Since $A'D' \perp DD'C$, then the edge $A'D'$ is the height of the pyramid $A'C'DD'$ dropped to the base of the right isosceles triangle $DD'C$ with side a. Denote the altitude of the same pyramid

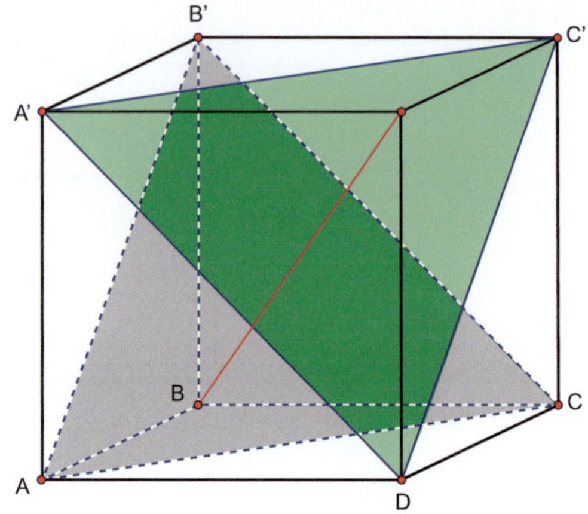

Fig. 2.42 Finding distance between skew diagonals of two adjacent faces of a cube

dropped to the base $A'C'D$ as h. Triangle $A'C'D$ is an equilateral with side of $a\sqrt{2}$. hence its area is equal to $[A'C'D] = \frac{2a^2\sqrt{3}}{4} = \frac{a^2\sqrt{3}}{2}$. For a tetrahedron, the volume can be written in four different ways, so we can state that $B_iH_i = 3V = \text{const}$. In our case, we have $h \cdot \frac{a^2\sqrt{3}}{2} = a \cdot \frac{a^2}{2}$ from which we obtain that $h = \frac{a}{\sqrt{3}}$. Finally, the required distance is $d = a\sqrt{3} - \frac{2a}{\sqrt{3}} = \frac{a}{\sqrt{3}}$. Moreover, we can state that two parallel planes divide the diagonal of the cube into three equal parts!
Answer. $d = \frac{a}{\sqrt{3}}$.

Problem 94 One side of the triangle is equal to a, the difference of the angles adjacent to it is equal to φ. A right prism is built on this triangle, as the base of a prism. A cross section of a prism with area S is obtained by a cutting plane passing through the edge of the prism opposite to side a. This cross section bisects the dihedral angle. Find the volume of the prism.

Solution Let us sketch the situation described in the problem.

In the triangle ABC that is the base of the future right prism, let $AC = a$. Next, in the same triangle, the difference of the angles A and C is equal to φ, $\angle A - \angle C = \varphi$. By drawing parallel lines through the vertices of the bottom base triangle ABC and after making the top base triangle $A'B'C'$, we construct a right triangular prism $ABCA'B'C'$, the volume of which we have to evaluate.

It is known that some cross section drawn through edge BB' divides the dihedral angle into two equal parts. In Fig. 2.43, it is the cross section $DD'BB'$, the area of which equals S. We can see that the volume of prism can be found if we find the areas of triangles ABD and DBC. Let BD be the bisector of angle B in triangle ABC.

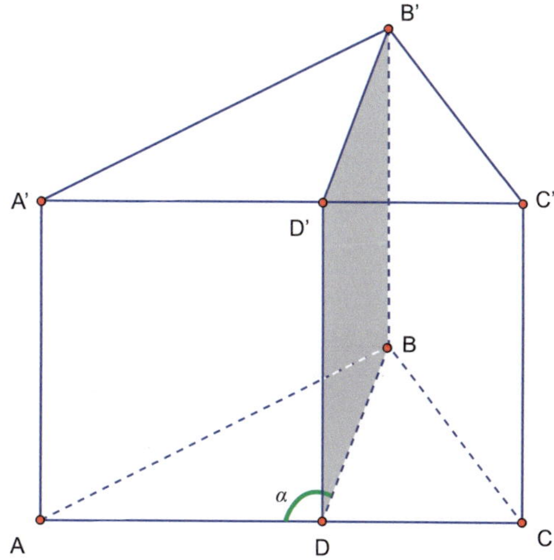

Fig. 2.43 Finding the volume of a given triangular prism

Denote angle ADB by α and evaluate this angle using triangle DBC, and then triangle ABC.

From triangle DBC, we obtain

$$\angle\alpha = \angle\frac{B}{2} + \angle C. \tag{2.35}$$

On the other hand, from triangle ABC, $\angle B = 180° - (\angle A + \angle C)$. Dividing both sides by 2 we obtain the value of the half of angle B:

$$\angle\frac{B}{2} = 90^{\circ} - \angle\frac{A}{2} - \angle\frac{C}{2}. \tag{2.36}$$

Substituting (2.36) into (2.35) we have $\angle\alpha = 90° - \angle\frac{A}{2} - \angle\frac{C}{2} + \angle C = 90° - \frac{A-C}{2}$.

$$\angle\alpha = 90° - \frac{\varphi}{2}. \tag{2.37}$$

The volume of prism $ABCA'B'C'$ equals the product of its edge AA' and the area of the base triangle ABC. The area of triangle ABC equals the sum of areas of triangles ABD and BDC. Using (2.37) we have the following formula:

$$V = AA' \cdot \left(\frac{1}{2}AD \cdot DB \sin\alpha + \frac{1}{2}CD \cdot DB \sin(180^\circ - \alpha)\right);$$
$$= \frac{1}{2}AA' \cdot DB\,(AD + DC)\sin\alpha;$$
$$= \frac{1}{2}\left[AA' \cdot DB\right] \cdot AC \cdot \sin\left(\frac{\pi}{2} - \frac{\varphi}{2}\right).$$

Replacing the product inside brackets by the area of the cross section S and after applying properties of trigonometric functions, we finally obtain the formula for the volume of prism $ABCA'B'C'$, $V_{\text{prism}} = \frac{1}{2}aS \cdot \cos\frac{\varphi}{2}$.
Answer. $V_{\text{prism}} = \frac{1}{2}aS \cdot \cos\frac{\varphi}{2}$.

2.3 Problems on Inscribed Pyramids and Prisms

Problem 95 A pyramid is inscribed into a cube $ABCDA_1B_1C_1D_1$ in such a way that its vertex is the center of the upper face of the cube $A_1B_1C_1D_1$ and the base is the same as the base $ABCD$ of the cube. How does the pyramid divide the volume of the cube?

Solution The center of the top face of the cube, S is the intersection of its two diagonals. The volume of the pyramid is $V_{\text{pyramid}} = \frac{1}{3}hB = \frac{1}{3}a^3$, then it divides the cube as $2:1$.

Note. Do you see in Fig. 2.44 four identical pyramids inside the cube, formed in addition to this pyramid? All five pyramids have a common vertex S. Four identical pyramids have bases AA_1DD_1, AA_1BB_1, BB_1C_1C, and DD_1C_1D, they have equal volumes. These are rectangular pyramids with base a^2 and height $a/2$. The total volume of the four pyramids is $4 \cdot \frac{1}{3} \cdot a^2 \cdot \frac{a}{2} = \frac{2a^3}{3}$. Together with the inscribed regular square pyramid they add up to the total volume of the cube, $V = \frac{2a^3}{3} + \frac{a^3}{3} = a^3$.

Answer. $2 \div 1$.

Lemma 9 *Pyramids with the same base and apex located in a plane parallel to this base have the same volumes.*

Proof A height of a pyramid is the length of the perpendicular dropped from the apex to the plane of the base. If the vertices of two pyramids are located in the plane parallel to the base, then the length of the perpendicular segment equals the distance between two parallel planes, and hence two pyramids have the same height. Therefore their volumes are the same. □

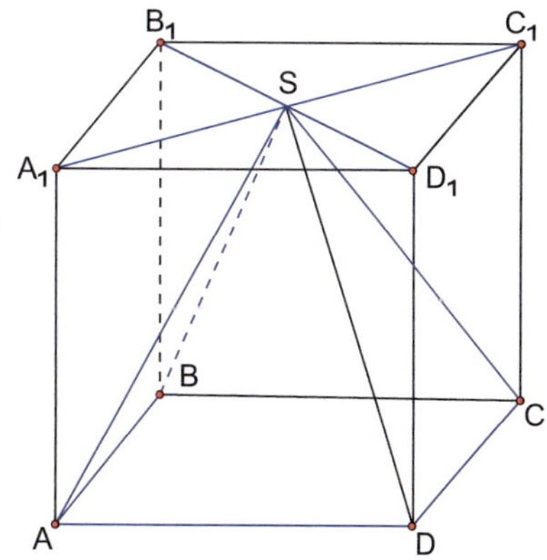

Fig. 2.44 Pyramid inscribed into a cube

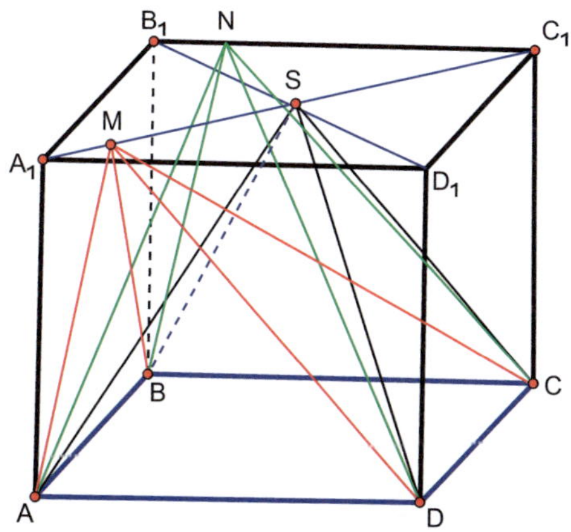

Fig. 2.45 Three pyramids inscribed into a cube

Let us consider the following sketch, on which three different pyramids are shown inscribed into a cube or rectangular parallelepiped (Fig. 2.45).

In Fig. 2.45, pyramids $ABCDM$, $ABCDS$, and $ABCDN$ have different shapes but the same volumes because they share the same base and have the same height.

Lemma 10 *If two pyramids share the same base and their heights connected by the formula* $\frac{h_i}{h_j} = \alpha$*, then their volumes are in the same ratio,* $\frac{V_i}{V_j} = \alpha$.

Proof Assume that two pyramids have base B, then $V_1 = \frac{1}{3} \cdot h_1 \cdot B$, $V_2 = \frac{1}{3} \cdot h_2 \cdot B$ and because $\frac{h_1}{h_2} = \alpha \Rightarrow h_1 = \alpha h_2$, then $V_1 = \frac{1}{3} \cdot \alpha h_2 \cdot B = \alpha V_2$. □

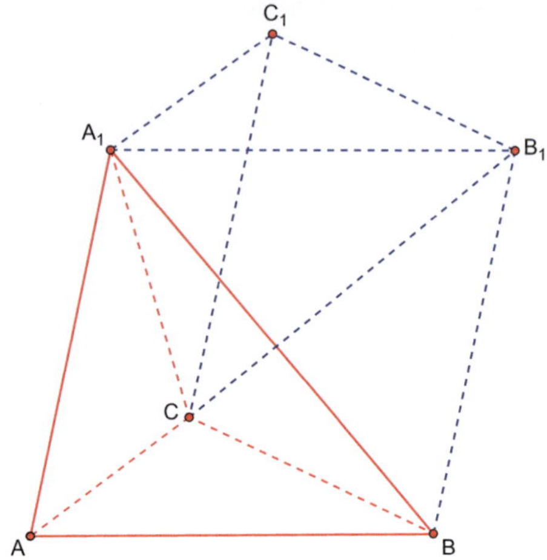

Fig. 2.46 Pyramid built up to a prism

> **Theorem 35** *The volume of a pyramid is equal to one-third of the product of the area of its base and height.*

Proof Without loss of generality, we can prove this statement for a triangular pyramid. Let A_1ABC be a triangular pyramid. Let's build it up to the triangular prism $ABCA_1B_1C_1$ (Fig. 2.46).

Planes passing through points B, C, A_1 and C, B_1, A_1 split this prism into three pyramids A_1ABC, A_1CBB_1, and $A_1CB_1C_1$ with vertices at point A_1. Pyramids A_1CBB_1 and $A_1CB_1C_1$ have equal bases CBB_1 and CB_1C_1, since the diagonal CB_1 splits the parallelogram CBB_1C_1 into two equal triangles. In addition, these pyramids have a common vertex A_1, and their bases lie in the same plane. So these pyramids have a common height. Therefore, these pyramids have equal volumes. Consider now the pyramids A_1ABC and $CA_1B_1C_1$. They have equal bases ABC and $A_1B_1C_1$ and equal heights. Therefore, they have equal volumes. Thus, the volumes of all three pyramids are equal. Given that the volume of the prism is equal to the product of the area of the base B and the height, h, we obtain the formula for the volume of a triangular pyramid as follows $V = \frac{1}{3}Bh$. □

> **Lemma 11** *If two pyramids share the same height and their bases related by the formula* $\frac{B_i}{B_j} = \beta$, *then their volumes are in the same ratio,* $\frac{V_i}{V_j} = \beta$.

The proof of this lemma is similar to Lemma 10. Please do it on your own.

Problem 96 Propose a construction that divides a cube into

(a) two equal parts,
(b) four equal parts (or eight),
(c) six equal parts, and
(d) three equal parts.

By equal parts we understand figures of equal volume, not necessarily identical or congruent figures.

Solution (a) This is the easiest part of the problem. Consider a cube $ABCDA_1B_1C_1D_1$. You can divide it in many ways. For example, we can draw a plane passing through two diagonals of the parallel faces and two connecting edges (plane AA_1C_1C). This way we will obtain two right triangular congruent prisms. We can divide the cube by the plane that goes through the midpoints of opposite edges of two adjacent faces (for example, points M, N, M_1 and N_1. This way we obtain two congruent rectangular parallelepiped (like $ABNMA_1B_1N_1M_1$ and $MNCDM_1N_1C_1D_1$). For all constructions see Fig. 2.47. It is fascinating that if we dissect the cube with any plane passing through the center of the cube, then it will always divide the cube into two figures of equal volume! This fact will be proven in Chap. 3.

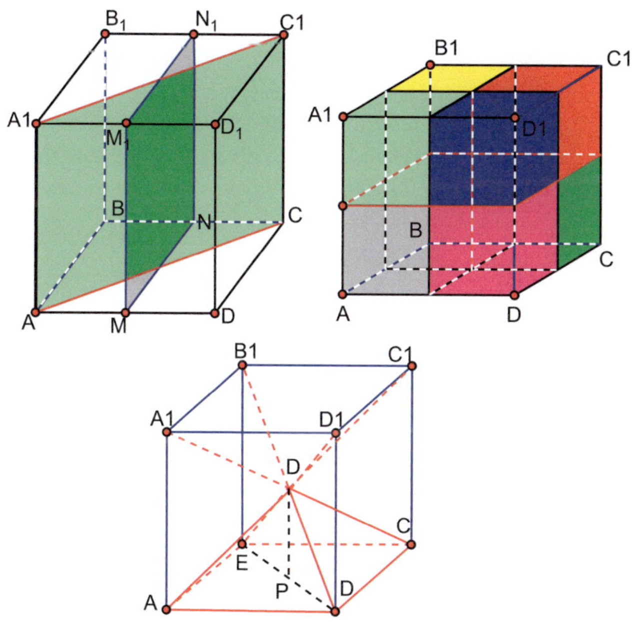

Fig. 2.47 Dissections of a cube

(b) The answer may vary. For example, planes AA_1C_1C and BB_1D_1D divide the cube in Fig. 2.47 into four equal parts, then each part can be divided into two more equal parts, and so on.
(c) Let us connect the center of the cube O with each vertex of the cube. This way we will obtain six regular quadrangular pyramids, each has a square of side a as the base and the height is equal to half of the edge of the cube. Hence, the volume of each pyramid is $V = \frac{1}{3}a^2 \cdot \frac{1}{2} \cdot a = \frac{a^3}{6}$.
(d) If we take two such pyramids that share one triangular face, we obtain the figure the volume of which is equal to $1/3$ of the volume of the cube.

Problem 97 Rectangular parallelepiped $ABCDA_1B_1C_1D_1$ with the volume V is given. $ABCD$ and $A_1B_1C_1D_1$ are the bases, $AA_1 \| BB_1 \| CC_1 \| DD_1$. Point M is taken on segment AB_1 such that $AM/MB_1 = 2/3$. Express the volume of pyramid AA_1MC_1 in terms of V.

Solution Assume that $AA_1 = a$, $A_1B_1 = b$, and $BC = c$, then place point M on side AB_1. Connect points M, A, and A_1 with the vertex C_1. You should have a sketch similar to one in Fig. 2.48.

Let us look at pyramid AA_1MC_1 and consider triangle AA_1M as its base and C_1 as its vertex. Denote its volume as V_1 to be found. Let us now get the use of the lemmas above. Compare this pyramid and pyramid $AA_1B_1C_1$, then volume of which (V_2) can be easily evaluated. Denote its base area (the area of triangle AA_1B_1) as $[AA_1B_1] = B_2$ and its height (side $B_1C_1 = AD$) as h. Obviously that $h = c$ and $B_2 = \frac{1}{2} \cdot a \cdot b = [AA_1B_1]$. $V_2 = \frac{1}{3}c \cdot \frac{1}{2} \cdot a \cdot b = \frac{1}{6} \cdot V$. Pyramid AA_1MC_1 has the same height h as pyramid $AA_1B_1C_1$, but different base located in the same face of the parallelepiped. Let us find how these bases are related to each other. Assume that the area of base $AA_1M = B_1$. Since triangles AA_1M and AA_1B_1 have the same vertex, then the ratio

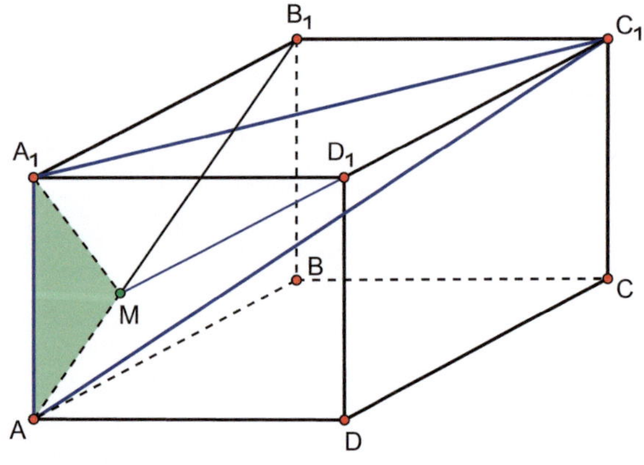

Fig. 2.48 Pyramid inscribed into a parallelepiped

of their areas equals the ratio of the corresponding sides to which the same height is dropped (see [2], p. 72). The following is true: $\frac{B_1}{B_2} = \frac{AM}{AB_1}$. By the condition of the problem $AM/MB_1 = 2/3$, then $AM = \frac{2}{5} \cdot AB_1$. Hence, $\frac{B_1}{B_2} = \frac{2}{5}$, $B_1 = \frac{2}{5} \cdot B_2$. Since two pyramids have the same height, then the ratio of their volumes will be equal the ratio of their bases: $V_1 = \frac{2}{5}V_2 = \frac{2}{5} \cdot \frac{1}{6} \cdot V = \frac{V}{15}$.
Answer. The volume of the pyramid AA_1MC_1 is $\frac{V}{15}$.

Next we will consider a more general case by solving the following problem.

Problem 98 Rectangular parallelepiped $ABCDA_1B_1C_1D_1$ with the volume V is given. $ABCD$ and $A_1B_1C_1D_1$ are the bases, $AA_1 \| BB_1 \| CC_1 \| DD_1$. Express the volume of pyramid AA_1MN in terms of V, if point M is taken on segment AB_1 such that $AM = \alpha MB_1$, $0 < \alpha < 1$ and point N is on segment A_1C_1 and $A_1N = \beta \cdot A_1C_1, 0 < \beta < 1$.

Solution 1. First, let us draw a rectangular parallelepiped and mark corresponding vertices of the pyramid following the condition of the problem. You may get a picture like Fig. 2.49. The volume of pyramid AA_1MN with base $B_1 = AA_1M$ and height h_1 can be written as $V_1 = \frac{1}{3}h_1B_1$.
2. Next, let us find and mark the height of the pyramid. It is segment PN that goes parallel to side B_1C_1 and so perpendicular to AA_1M. How can we express $h_1 = NP$ in terms of the volume of the box, V? We can find how it is related to one of its sides, B_1C_1. Note that this side is the height (h_2 of pyramid $AA_1B_1C_1$, the volume of which can be easily evaluated as $\frac{V}{6}$. On the other hand, the volume of pyramid $AA_1B_1C_1$ can be denoted by V_2 and written as $V_2 = \frac{1}{3}h_2B_2$. Since we know how V_2 is expressed in terms of V, we can find how B_1, B_2 and h_1, h_2 are related.
3. Triangles $C_1B_1A_1$ and NPA_1 are similar by three angles, so the following is true:

$$\begin{aligned} &\frac{PN}{C_1B_1} = \frac{A_1N}{A_1C_1}, \\ &\frac{h_1}{h_2} = \beta \\ &h_1 = \beta \cdot h_2. \end{aligned} \tag{2.38}$$

4. Now, let us find the relation between bases B_1 and B_2 (the areas of the bases of two indicated pyramids). Because triangle AA_1M shares with triangle AA_1B_1 two sides, then they share the same height and then (see the property in [2]) the ratio of the triangles' areas equals the ratio of the sides AM and AB_1 that is given as parameter α. This can be written as

$$\begin{aligned} &\frac{[AA_1M]}{[AA_1B_1]} = \frac{AM}{AB_1} = \alpha, \\ &\frac{B_1}{B_2} = \alpha \\ &B_1 = \alpha \cdot B_2. \end{aligned} \tag{2.39}$$

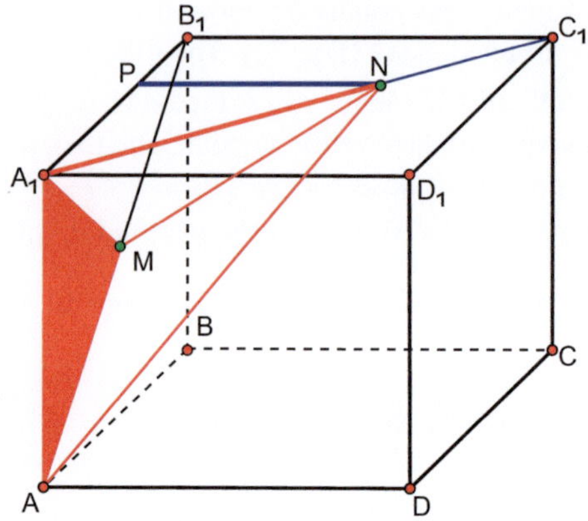

Fig. 2.49 Triangular pyramid inside a parallelepiped

5. After substitution we obtain that $V_1 = \frac{1}{3} \cdot \beta \cdot h_2 \cdot \alpha B_2 = \alpha\beta \cdot [\frac{1}{3} \cdot h_2 \cdot B_2]$. Substituting the expression inside the brackets by the volume V_2 of pyramid AA_1MB_1, we finally have $V_1 = \alpha \cdot \beta \cdot V_2 = \frac{1}{6} \cdot \alpha\beta \cdot V$.

Answer. $V_1 = \frac{1}{6} \cdot \alpha\beta \cdot V$.

Problem 99 A regular tetrahedron is inscribed into a cube with side a as shown in Fig. 2.50. How does this tetrahedron divide the volume of the cube? Find the volume of the tetrahedron.

Solution Let us inspect the regular tetrahedron in Fig. 2.50. Each edge of the inscribed tetrahedron is $a\sqrt{2}$.

We could use plane geometry and find the height of the tetrahedron. However, we can use a different approach. We can see that the volume of the tetrahedron equals the volume of the cube minus the volumes of three identical right triangular pyramids, $ADCD_1$, $AA_1B_1D_1$, and $B_1C_1CD_1$. Each such right pyramid has isosceles right triangle as its base with legs equal a. For example, pyramid $ADCD_1$ has height $DD_1 = a$ and the base area $B = [ADC] = \frac{a^2}{2}$, and the volume of

$$V_{\text{pyr}} = \frac{1}{3} \cdot a \cdot \frac{a^2}{2}.$$

Denote the volume of the tetrahedron by V_{tetr}, then it can be evaluated as follows:

$$V_{\text{tetr}} = a^3 - 3 \cdot \frac{1}{3} \cdot a \cdot \frac{a^2}{2} = \frac{a^3}{2}.$$

Therefore, the inscribed tetrahedron divides the cube into two equal parts.

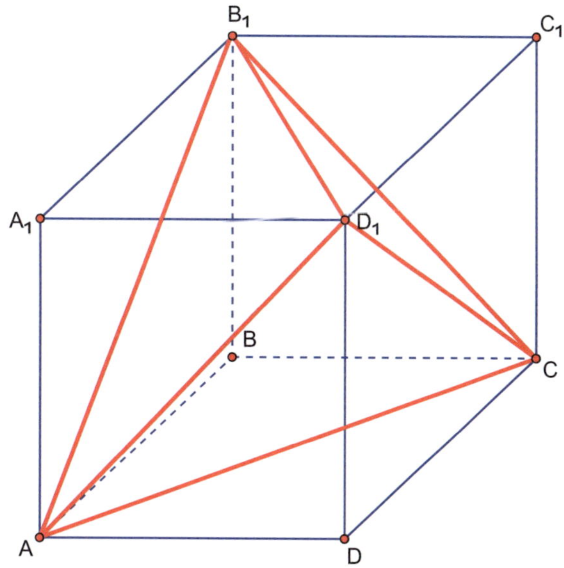

Fig. 2.50 Regular tetrahedron inside a cube

Note. Having solved this problem, we discovered that when placing a tetrahedron inside a cube, the capacity of the cube is halved. If the original cube can hold 1 liter of water, then the cube with the inscribed tetrahedron can only hold 0.5 liter.

Let us solve some application problems.

Problem 100 A solid tetrahedron is sliced off a solid wooden unit cube by a plane passing through two nonadjacent vertices on one face and one vertex on the opposite face not adjacent to either of the first two vertices. The process is repeated for a second cube. The two cubes are then glued together by matching the cut surfaces. Let A be the numerical value of the surface area of the resulting solid, and let V be its volume. What is A/V?

Solution Let us reconstruct one of the identical cubes following by the description of the problem.

We cut the wooden cube by the plane passing through two nonadjacent vertices of the front face (points A and C) and the vertex of the opposite face (B). We can see in the Fig. 2.51 that the portion of each cube removed is a pyramid $ABCD$ with an isosceles right triangular base (DCB) of area $1/2$ and height $AD = 1$, so its volume is $1/6$. Therefore the remaining portion of each cube has volume $5/6$, so the volume of the two glued figures is $V = 10/6 = 5/3$. From the original surface area of six for each cube, three triangles of area $1/2$ each have been removed, so the remaining

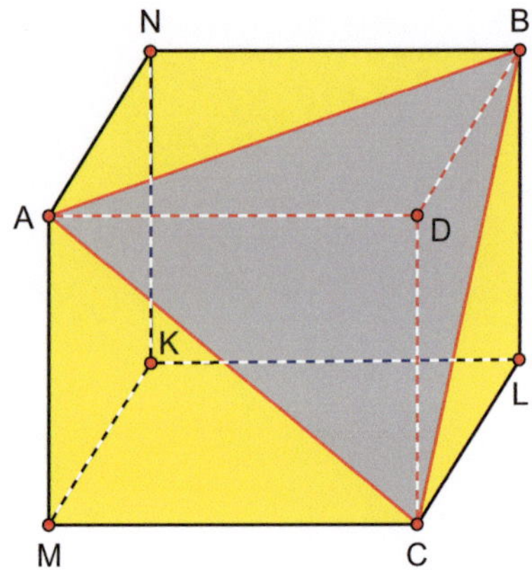

Fig. 2.51 About two cubes glued together

area in each cube is $6 - 1.5 = 4.5$ and the total surface area of the two figures is $A = 9$. Therefore $A/V = 27/5$.

Answer. $A/V = 27/5$.

Problem 101 The faces of a triangular pyramid are numbered 1, 2, 3, and 4. If two of these pyramids are rolled like dice, what is the probability that the sum of the upper numbers will be a multiple of 5?

Solution Let us sketch the possible cases. I showed all four possibilities for one pyramid when it lands on one of its faces. Based on the condition of the problem, each time we can see only three sides of the pyramid as it is shown in Fig. 2.52. The possible sums of the up numbers for one of these pyramids are 9, 8, 7, and 6. Table 2.1 shows that the sum of 15 is the only multiple of 5 and it will occur 4 times, so the probability is 4/16 or 1/4.

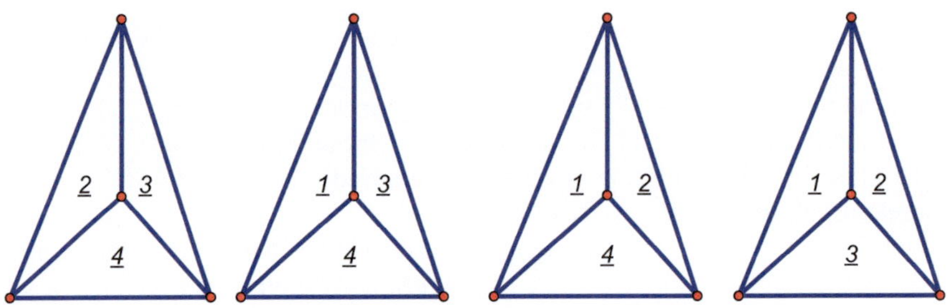

Fig. 2.52 A tetrahedron rolled like a dice

Table 2.1 Rolling dice results

+	9	8	7	6
9	18	17	16	**15**
8	17	16	**15**	14
7	16	**15**	14	13
6	**15**	14	13	12

Answer. $\frac{1}{4}$.

2.4 Convex and Nonconvex Polyhedra

Definition 13 A polyhedron is called **convex** if it lies on one side of a plane containing any of its faces. The faces of a convex polyhedron are flat convex polygons. The sides of the faces are called the edges of the polyhedron, and their vertices are called the vertices of the polyhedron.

Definition 14 A polyhedron is called **nonconvex** if it has a face such that the polyhedron lies on both sides of the plane containing this face.

In 1752, a Swiss mathematician Leonhard Euler formulated a necessary condition for the existence of any convex polyhedron.

Theorem 36 (Euler's Formula) *For any convex polyhedron, the equality*

$$V - E + F = 2 \tag{2.40}$$

holds, where V is the number of vertices, E is the number of edges, and F is the number of faces of the given polyhedron.

Proof There are many proofs of Euler's formula. One was given by Cauchy in 1811. Let us consider a cube. Remove one face of the cube, for example, the back face marked in red in Fig. 2.53.

We remove one face, but keep the edges and vertices. Imagine that the cube is made of some cloth material with wire edges that are the segments with visible vertices. By pulling the edges of the missing face away from each other, we deform

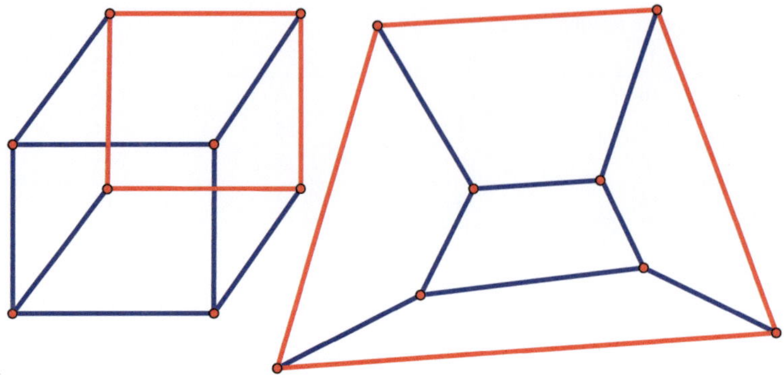

Fig. 2.53 Cube and its planar graph with one face removed

all the rest into a planar graph of points and segments or curves, in such a way that the perimeter of the missing face is placed externally. In the sketch on the right of Fig. 2.53, it is shown in red and looks like the outer border of the graph. So, on the left we have a cube-3D figure and on the right, the planar graph of the deformed figure. After this deformation, the regular faces are generally not regular anymore. Inspect this figure closely. The number of vertices and edges has remained the same, but the number of faces has been reduced by 1.

In general, imagine the surface of a given polyhedron made of an elastic material. Delete (cut out) one of the faces and stretch the remaining surface on a plane. We obtain a planar graph (Fig. 2.53 containing $F' = F - 1$ polygons (which will continue to be called faces), V vertices, and E edges. Therefore, in order to prove Euler's formula (2.40) for a polyhedron, we can prove that for this deformed, planar object:

$$V - E + F' = 1. \tag{2.41}$$

First, let us show that the relation (2.41) does not change if we draw a diagonal in some polygon of the planar graph. Indeed, after drawing such a diagonal, the grid will have V vertices, $E + 1$ edges, and $F' + 1$ faces, and therefore $V - (E + 1) + (F' + 1) = V - E + F'$. Next, if there is a face with more than three sides, draw a diagonal, that is, a curve through the face connecting two vertices that are not yet connected. This adds one edge and one face and does not change the total number of vertices, so it does not change the quantity $V - E + F'$. Continue adding edges in this manner until all of the faces are triangular. In Fig. 2.54, we added several diagonals to our planar object.

Using this property, we draw diagonals in the grid that divide the polygons included in it into triangles (Fig. 2.54), and for the resulting grid we show that the relation (2.41) holds. To do this, we will sequentially remove the outer edges of the grid, reducing the number of triangles in it.

Reapply any of the following two transformations while maintaining the invariant (2.41).

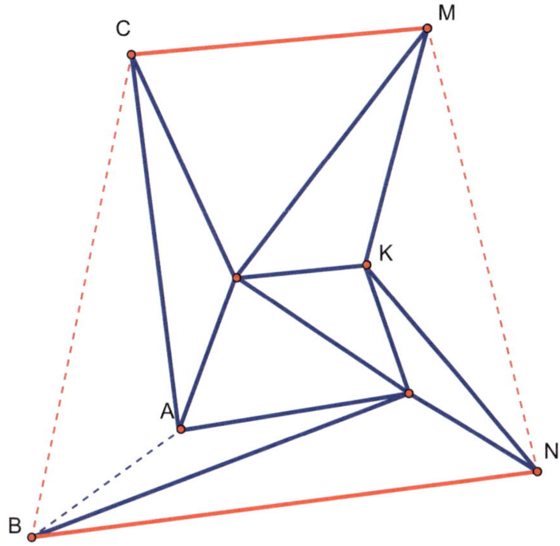

Fig. 2.54 Proof of Euler's formula: cube with two different triangles removed

If this planar graph satisfies the relation (2.41), then the required relation (2.40) will be true for the original polyhedron. The two cases are possible:

1. Delete the triangle with only one edge adjacent to the outside, as shown, for example, for triangle *MKN* in Fig. 2.54. This reduces the number of edges and faces by one each and does not change the number of vertices. Now, we have $E - 1$ edges, $F' - 1$ faces, and V vertices. Consider face *MNK*, by removing segment *MN* we remove face *KMN*, lose one edge (*MN*) but keep all the vertices (M, N, K) so the quantity (2.41) is preserved, $V - (E - 1) + +(F' - 1) = V - E + F' = 1$.
2. Delete the triangle with two edges common to the outer part of the graph, as shown in Fig. 2.54. Consider face *ABC*. Each deletion of a triangle removes a vertex, two edges (*BC* and *AB*), and one face (*ABC*), so that (2.41) again holds $(V - 1) - (E - 2) + (F' - 1) = V - E + F' = 1$.

Thus, removing one triangle by each of the previous two transformations does not change the relation (2.41).

Continuing this process of removing triangles, these transformations eventually reduce the planar graph to a single triangle with $V = 3$, $E = 3$, $F' = 1$, and, consequently, $V - E + F' = 1$. This means that relation (2.41) also holds for the original planar graph. We have shown that $V - E + F' = 1$ for a deformed planar object, thus demonstrating that for the given polyhedron relation (2.40) is valid. This proves the theorem. □

For additional proofs, see [7] by David Eppstein. Multiple proofs, including their flaws and limitations, are used as examples in [8] by Imre Lakatos or another proof is given in [9] by Smirnova.

Table 2.2 Euler characteristic

Polyhedron	Number of vertices, V	Number of edges, E	Number of faces, F	χ
Tetrahedron	4	6	4	2
Cube	8	12	6	2
Octahedron	6	12	8	2
Hexahedron	8	12	6	2
Icosahedron	12	30	20	2
Dodecahedron	20	30	12	2
n-gonal pyramid	n+1	2n	n+1	2
n-gonal prism	2n	3n	n+2	2

Let us introduce the number

$$\chi = V - E + F. \tag{2.42}$$

The number χ is called the **Euler characteristic of the polyhedron**. By Euler's formula for a convex polyhedron this characteristic is equal to 2. The fact that the Euler characteristic is equal to 2 for many polyhedra familiar to us can be seen from Table 2.2.

The integer $\chi - 2$ is called the **genus of a polyhedron.** At the beginning of this section, I stated that Euler's formula (2.46) is a necessary, not a sufficient, condition for a polyhedron to be convex. What does it mean? This means that if χ calculated using the formula (2.42) $\neq 2$, then the polyhedron is nonconvex. However, if $V - E + F = 2$ it does not automatically prove that the given polyhedron is convex. For many nonconvex polyhedra this value is equal to $\chi = 2$.

Let us consider a cube $ABCDA_1B_1C_1D_1$ shown in Fig. 2.55, draw the diagonal of the upper base B_1D_1, and start stretching the cube by moving only in the direction of two edges of the cube. For example, you can take two vertices A_1 and C_1 and move them simultaneously up a certain distance, leaving the other vertices in place, obtaining new vertices denoted by A_3 and C_3, respectively. Thus, we get a spatial figure $ABCDA_3B_1C_3D_1$ that is a nonconvex polyhedron. And if we move the vertices A_1 and C_1 down, for example, by the same distance along the corresponding edges, then we will get a figure $ABCDA_2B_1C_2D_1$, different from the cube, but it will also be convex.

What's interesting: when we performed this procedure, in both cases we added one edge, (B_1D_1), that is, the number of edges became $E = 13$ and the number of faces increased by one face (there are seven of them, $F = 7$). However, if we calculate the number χ, then in both cases, for both a convex and a nonconvex figure, this number will be equal to two. For example, for polyhedron $ABCDA_3B_1C_3D_1$ we have $\chi = V + F - E = 8 + 7 - 13 = 2$.

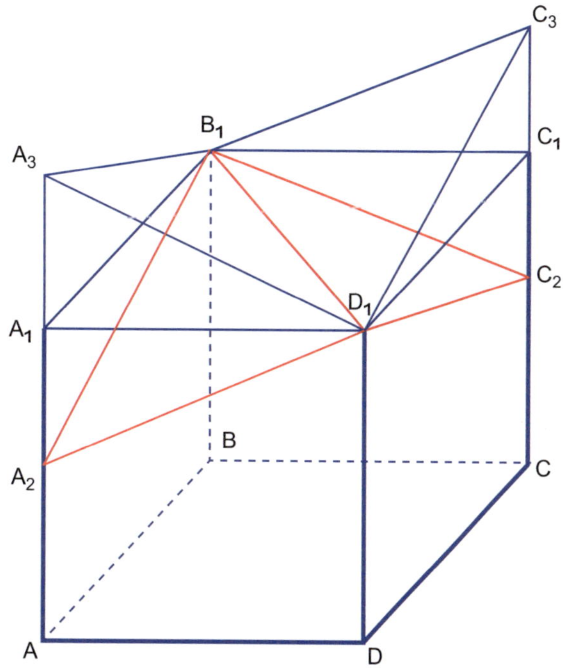

Fig. 2.55 Nonconvex and convex polyhedra obtained from a cube

Lemma 12 *If a convex polyhedron has exactly n faces, then any of its faces is a convex polygon with at most* $(n-1)$ *sides.*

Proof Assume the contradiction and that there exists a polygon with n sides as a face of the given polyhedron. Then this polygon must share its sides with n other polygonal faces. Hence, the given polyhedra would have $n+1$, not n faces. We obtain the contradiction. Our assumption was wrong which proves the lemma. □

Lemma 12 is very important, although almost obvious. For example, any polyhedron with only triangular faces must have at least four faces (for example, a tetrahedron has four triangular faces). On the other hand, if a convex polyhedron has six faces, then only triangles, quadrilaterals, or pentagons can be its faces, not hexagons.

Problem 102 A polyhedron has 12 vertices, 8 faces, and 18 edges. Is this polyhedron convex or nonconvex?

Solution Let us calculate the χ number, for $V = 12$, $F = 8$, and $E = 18$: $\chi = V + F - E = 12 + 8 - 18 = 2$. If χ was not 2, then we could state that the polyhedron is nonconvex. However, in case $\chi = 2$, since Euler's formula is the necessary but not sufficient condition for a polyhedron to be convex, then we cannot state that a polyhedron with $\chi = 2$ is convex. For example, in Fig. 2.2, the polyhedron is sketched that also has 12 vertices, 8 faces, and 18 edges. Its $\chi = 2$, but it is clearly nonconvex.

Problem 103 Prove that in any convex polyhedron there are two faces with the same number of sides.

Proof Suppose that any two faces of some convex polyhedron have a different number of sides. Consider the face F that has the largest number of sides; let it be equal to n. Therefore, the number of sides of any of the remaining faces is strictly less than n. Since the minimum number of sides of a polygon is 3 (for a triangle), then for this polyhedron there will only be $(n - 3)$ different polygons. On the other hand, the face F is adjacent to exactly n other faces of the polyhedron (one on each side). We obtain the contradiction. Therefore, some two faces necessarily have the same number of sides. □

Remark. We can also use the **Dirichlet Principle** or the Pigeonhole principle for proof that states that "*If n items are put into m containers, with n > m, then at least one container must contain more than one item.*"

Consider the face with the largest number of sides. Let us denote this face F and assume that the number of its sides is equal to $n = 12$. Each side of F is adjacent to a face of the polyhedron, for a total of 12 adjacent faces. The number of sides of the remaining faces ranges from 3 to $n - 1 = 11$, for a total of $n - 3 = 12 - 3 = 9$ possibilities (polygons with $3,\ 4,\ 5,\ 6,\ 7,\ 8,\ 9,\ 10$ or 11 sides, no more than one of each type) Since the number of options 9 is less than the required number of 12 adjacent polygons, then according to the Dirichlet principle one of the options will be repeated. Thus, among the faces adjacent to the face F, there are two faces with the same number of sides.

Mathematics Olympiads often present problems involving convex polyhedra. Some of them can be solved using the famous Euler formula (2.40).

Problem 104 Given a convex polyhedron, all faces of which have five, six, or seven sides and all polyhedral angles are trihedral. Prove that the number of pentagonal faces is 12 more than the number of heptagonal faces.

Proof Let x be the number of pentagons, y be the number of hexagons, and z be the number of heptagons in the given polyhedron. Then the polyhedron has the following number of faces:

$$F = x + y + z.$$

If we did not make a polyhedron from these pentagons, hexagons, and heptagons, then they all together would have $5x + 6y + 7z$ vertices and sides. Since these polygons are faces of the polyhedron, then they are connected together. We will take into account the fact that three sides converge at each vertex, because all the angles are trihedral. The following must be true for the total number of the vertices of the polyhedron:

$$V = \frac{5x + 6y + 7z}{3}.$$

The total number of edges can be found similarly, because each side of the polyhedron is common for two polygons:

$$E = \frac{5x + 6y + 7z}{2}.$$

Our F, V, and E must satisfy Euler's formula (2.40), so we obtain

$$6V - 6E + 6F = 12$$

or $10x + 12y + 12z - 15x - 18y - 21 + 6x + 6y + 6z = 12, \implies x - z = 12.$ From the last formula, we obtain that the number of pentagons is 12 more than the number of heptagons. □

Problem 105 Prove that for any convex polyhedron that has F faces, V vertices, and E edges, the following inequalities are true (a) $2E \geq 3F$; (b) $3F \geq E + 6$.

Proof (a) Since we do not have any specific information about a polyhedron, we can assume that our polyhedron consists of V_3 triangles, V_4 quadrilaterals, V_5 pentagons,... and V_m of m-gons. Suppose that we cut them out and places on the table. We can state that the polyhedron has F of total faces:

$$F = V_3 + V_4 + V_5 + \cdots + V_m. \tag{2.43}$$

Right away we can find the number of total vertices in all polygons that is the same as the total number of edges in all polygons. (Thus one triangle face has three edges and three vertices, etc.) This number is

$$3 \cdot V_3 + 4 \cdot V_4 + 5 \cdot V_5 + \cdots + m \cdot V_m.$$

Because two polygons in a convex polyhedron share common edge, then we can evaluate the total number of edges in a polyhedron by dividing the expression above by two. The number of edges (E) in a convex polyhedron is

$$E = \frac{3 \cdot V_3 + 4 \cdot V_4 + 5 \cdot V_5 + \cdots + m \cdot V_m}{2}. \tag{2.44}$$

Unfortunately, we do not know anything about the type of polyhedral angles in our convex polyhedron. However, if a convex polyhedron has all trihedral angles, then there will be exactly three faces meeting at each vertex of the polyhedron, and the total number of vertices of all polygons we would have to divide by three to obtain the number of vertices in the polyhedron. For such hypothetical polyhedron, the number of vertices $\tilde{V}$ would be evaluated as

$$\tilde{V} = \frac{3 \cdot V_3 + 4 \cdot V_4 + 5 \cdot V_5 + \cdots + m \cdot V_m}{3}.$$

Since our polyhedron might have other polyhedral angles, then it is obvious that the actual total number of vertices of a convex polyhedron V would be less than or equal to $\tilde{V}$., $V \leq \tilde{V}$. (We can assume that at some real vertices of the polyhedron more that three faces meet.) This inequality also can be rewritten as follows:

$$3 \cdot V_3 + 4 \cdot V_4 + 5 \cdot V_5 + \cdots + m \cdot V_m \geq 3V. \tag{2.45}$$

According to (2.44), the left side of this inequality is equal to $2E$, then we can rewrite (2.45) as

$$2E \geq 3V. \tag{2.46}$$

This proves the first part of the problem. Note that the equality $2E = 3V$ is valid for any convex polyhedron whose has only trihedral angles at each vertex.
(b) This part is easy to prove by using Euler's formula (2.40) and (2.46). We can solve (2.40) for V as $V = E - F + 2$, and then substitute it in (2.46): $2E \geq 3(E - F + 2)$. After collecting like terms and simplification, we obtain the requested relationship:

$$3F \geq E + 6. \tag{2.47}$$

Since the inequality (2.47) can be rewritten as follows:

$$E \leq 3F - 6, \tag{2.48}$$

then the number of edges of a convex polyhedron is at most three times the number of faces minus six. □

Remark. Let us check the inequality (2.48) for some convex polyhedra from Table 2.2, for example, for a cube and n-gonal prism.

For a cube we have $F = 6$ and $E = 12$. Using (2.48) we check that $12 \leq 3 \cdot 6 - 6$. Yes $12 \leq 12$, which in case of a cube becomes the equality because at each vertex of a cube three faces (squares) meet, and all polyhedral angles are trihedral.

For a prism that has in the base a polygon with n sides, we get from Table 2.2 that $E = 3n$ and $F = n + 2$, let us check that (2.48) is valid, $3n \leq 3(n + 2) - 6$, which is again true, $0 \leq 0$, and becomes an equality because a prism is a convex polyhedron with three faces meeting at each vertex, forming a trihedral angle.

Problem 106 Find all convex polyhedra with six faces.

Solution 1. Because a polyhedron has six faces, $F = 6$, then its faces can be only triangles, quadrilaterals, or pentagons. Let x be the number of triangular faces, y be the number of quadrilaterals, and z be the number of all pentagons. Thus, we have the first equation:

$$x + y + z = 6. \tag{2.49}$$

2. Let us find the number of the edges E of such a polyhedron, remembering that two of the faces have a common side: $\frac{3x+4y+5z}{2} = E$. From this equation, we can conclude that the numerator of the fraction must be divisible by 2 (even). This can be written as follows:

$$3x + 4y + 5z = 2E, \tag{2.50}$$

where E represents the number of edges and can be a natural number.
3. Let us multiply both sides of (2.49) by 3:

$$3x + 3y + 3z = 18. \tag{2.51}$$

Subtracting (2.51) and (2.50), we obtain new constraints for our variables:

$$y + 2z = 2E - 18 = 2p. \tag{2.52}$$

Since the second term is even, the number of the quadrilaterals y must be also even:

$$y = 2\,m. \tag{2.53}$$

4. Now it is time to apply Euler's formula (2.40), in which we will substitute $F = 6$. If $V - E + 6 = 2$, then

$$E = V + 4. \tag{2.54}$$

Here V is the total number of the vertices in our convex polyhedron with six faces. Earlier we proved the inequality (2.46), i.e., $2E \geq 3V$. Substituting in this inequality the value for E from (2.54), we obtain that $2E = 2(V + 4) = 2V + 8 \geq 3V$, which gives us restrictions for the total number of the vertices: $V \leq 8$. Since the minimum number of vertices in a convex polyhedron with six faces is $V = 5$ (for a triangular bi-pyramid), we can get a double inequality for V and then for E using (2.54): $5 \leq V \leq 8$.

$$9 \leq E \leq 12. \tag{2.55}$$

5. Next, using (2.55) we can find the new constraints for the left side of (2.50):

$$18 \leq 3x + 4y + 5z \leq 24. \tag{2.56}$$

6. Putting together (2.49), (2.53), and (2.56) we can find all possible combinations (x, y, z) for the requested polyhedron:

$$\begin{cases} x + y + z = 6 \\ 18 \leq 3x + 4y + 5z \leq 24 \\ y = 2\,m. \end{cases} \tag{2.57}$$

There aren't many options. You can start with the first most important equation. For example, if we assume that the number of triangular faces is maximum and equal to six, then there simply cannot be other faces.

Let's check the second inequality. Yes, everything is fine and it turns out that the answer is the triple $(x, y, z) = (6, 0, 0)$—this is a triangular bi-pyramid.

Next, we consider that the polyhedron has 5 triangular faces ($x = 5$), then we have only one option for another type of polygon, and it must be only one pentagon ($z = 1$). If we check the inequality, then $3 \cdot 5 + 0 \cdot y + 5 \cdot 1 = 20,\ 18 \leq 20 \leq 24$. We obtained a convex polyhedron with $(x, y, z) = (5, 0, 1)$. This is a recognizable shape of a pentagonal pyramid.

Assume that $x = 4$, then y can be either 0 and then $z = 2$ or if $y = 2$, then $z = 0$. For $(4, 0, 2)$ the second inequality is valid, since $3 \cdot 4 + 4 \cdot 0 + 5 \cdot 2 = 22,\ 18 \leq 22 \leq 24$. We obtained a combination of four triangular and two pentagonal faces. However, this combination $(4, 0, 2)$ would not give us a convex polyhedron. Similarly, the following combinations of (x, y, z) satisfy the system of constraints (2.57) but cannot be associated with any known convex polyhedron:

$$(x, y, z) : \{(4, 0, 2), (3, 0, 3), (1, 4, 1)\}. \tag{2.58}$$

Question. Do you know why the combinations of six polygons (2.58) satisfying system (2.57) are not convex polyhedra?

Another triple, $(x, y, z) = (4, 2, 0)$ works since $18 \leq 3 \cdot 4 + 4 \cdot 2 = 20 \leq 24$. This convex polyhedron has four triangles and two quadrilaterals as its faces. It is sketched as convex polyhedron $ABCDEB_1$ in Fig. 2.56.

Next, we can demonstrate that the combination of three triangles, two quadrilaterals, and one pentagon ($(x, y, z) = (3, 2, 1): \ 3 \cdot 3 + 4 \cdot 2 + 5 \cdot 1 = 22$) can make a shape of a convex polyhedron $GHIJKFL$ shown in Fig. 2.56. This polyhedron has $F = 6$, $E = 11$, and $V = 7$. It can be obtained by truncating the quadrilateral pyramid $MGHIL$ with apex L by plane passing through points F, K, J, so that vertex M will disappear. Pyramid $MGHIL$ has five faces ($F = 5$), eight edges ($E = 8$), and five vertices ($V = 5$). The dissecting plane forms a new face FKJ on the object, increasing the number of aces to six ($F' = 6$), the number of vertices by two, and the edges by three, so $V' = 7, \ E' = 11$.

The combination of two triangles, two quadrilaterals, and two pentagons ($(x, y, z) = (2, 2, 2): \ 3 \cdot 2 + 4 \cdot 2 + 5 \cdot 2 = 24$.) will make convex polyhedron like one ($NQRSTOPU$) in Fig. 2.56. This convex polyhedron with 6 faces, 12 edges, and 8 vertices can be obtained by cutting a tetrahedron by two different planes at two vertices of the base triangle. Truncating once at one vertex, we get three additional edges, one additional face, and two additional vertices. Truncating twice, we get the following formulas $F' = F + 2 \cdot 1, \ V' = V + 2 \cdot 2$, and $E' = E + 2 \cdot 3$.

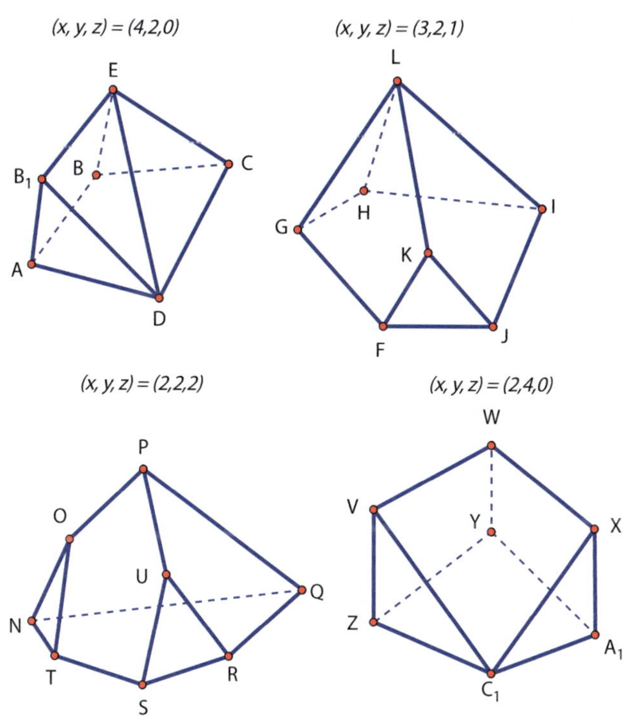

Fig. 2.56 Some less common convex polyhedra with six faces

Since for the tetrahedron $F = 4$, $V = 4$, and $E = 6$, then for our new polyhedra we have $F' = 6$, $V' = 8$, and $E' = 12$. Two examples of convex polyhedra $(x, y, z) = (2, 2, 2)$ can be seen in Figs. 3.91 and 3.92. On the other hand, our polyhedron $(2, 2, 2)$ in Fig. 2.56 can be build up to the tetrahedron from which it was obtained. Let us extend the segment TS and intersect it with line RQ at the point M. At the same time, we can extend the segment PO and intersect it with NQ and TS. All three lines should intersect at one point, say K. Thus, $KPQM$ is a tetrahedron.

A convex polyhedron with two triangular and four quadrilateral faces, for which $(x, y, z) = (2, 4, 0)$ is shown in Fig. 2.56. If $(x, y, z) = (0, 6, 0)$, then we have six quadrilaterals, of which the cube is a special case. More information about construction of new polyhedra using *truncation* can be found in Chap. 3.

Answer. $(x, y, z) : \{(6, 0, 0), (5, 0, 1), (4, 2, 0), (3, 2, 1), (2, 2, 2), (2, 4, 0), (0, 6, 0)\}$.

Problem 107 A polyhedron is called "simple" if each of its vertex is associated with only a trihedral angle. A simple convex polyhedron has 2025 edges. Determine the number of vertices and faces of the polyhedron.

Solution Since three polygons converge at each vertex, then

$$3V = 2E.$$

Note that this relation indicates that for a simple polyhedron, the number of edges must be a multiple of 3. Let us check that $E = 2025$ is a multiple of 3 by adding its digits: $2 + 0 + 2 + 5 = 9 = 3n$. Substituting 2025 for E in the equation above, we can find the number of vertices:

$$V = \frac{2 \cdot 2025}{3} = 1350.$$

Now, using Euler's formula (2.40), we can find the number of faces:

$$1350 - 2025 + F = 2, \quad F = 677.$$

Answer. $E = 2025$, $V = 1350$, $F = 677$. Such simple polyhedron is unique.

2.5 Regular Polyhedra

A convex polyhedron is called **regular** if

- all its faces are equal and regular polygons;
- all its polyhedral angles are equal.

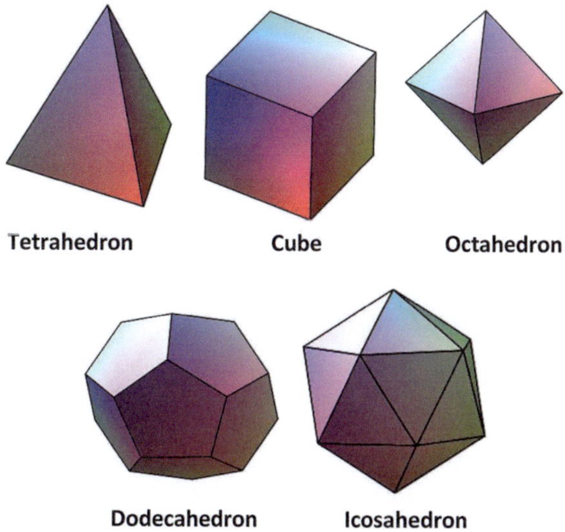

Fig. 2.57 Five regular polyhedra-platonic solids

There are only five types of regular convex polyhedra shown in Fig. 2.57:

- A regular **tetrahedron** has regular triangle faces, three edges converge at each vertex. A tetrahedron is a triangular pyramid with all edges equal.
- A **cube** has all square faces and three edges converge at each vertex. The cube is a rectangular parallelepiped with equal edges.
- An **octahedron** faces regular triangles, but unlike tetrahedra, four edges converge at each vertex.
- A **dodecahedron** faces are 12 regular pentagons, and three edges meet at each vertex.
- An **icosahedron** has 20 equilateral triangles as its faces, but unlike the tetrahedron and octahedrons, five edges converge at each of its vertex.

These regular polyhedra are also known as **Platonic solids**. Platonic solids were studied by the ancient Greeks, who also called these solids cosmic solids and classified them into five types. The five platonic solids found in nature are believed to represent the five elements: earth, air, fire, water, and the universe.

Theorem 37 *There are only five types of regular polyhedra.*

Method 1.

Proof Let m be the number of sides of each face of a regular polyhedron; n be the number of edges of each polyhedral angle. First, consider a regular pentagon as possible face of a regular polyhedra (Fig. 2.58). So, $m = 5$. Assume that we do not know the possible number of n.

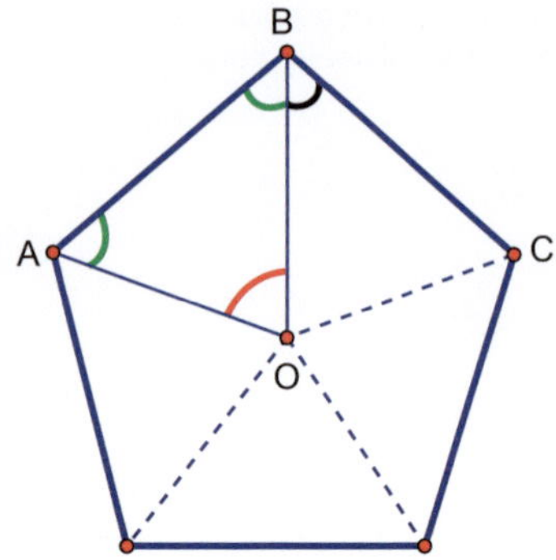

Fig. 2.58 A regular pentagon

For a regular pentagon, the central angle $\angle AOB = \frac{360^\circ}{5} = 72^\circ$. It is marked by red color in Fig. 2.58. If this pentagon is the face of some regular polyhedron, then, for example, at vertex B there must meet n plane angles, such as $\angle ABC$. How can we find the measurement of this angle? From Fig. 2.58, we can see that this angle is $\angle ABC$, that is, double angle of $\angle ABO$ shown by green arc. From the isosceles triangle AOB, we obtain

$$2 \cdot \angle ABO = 180^\circ - \angle AOB = 180^\circ - 72^\circ = 108^\circ.$$

Thus, n of these plane angles must converge at the vertex B. Can we say how many of the plane angles of 108° meet at the vertex of a hypothetical polyhedral? Can there be 3, 4 or maybe 5? The answer comes almost immediately, if we remember that the sum of the plane angles in any polyhedral angle must be less than 360°. Hence, only three such angles can meet at one vertex! Please, check that $3 \cdot 108^\circ = 324^\circ < 360^\circ$.

Using the same sketch (Fig. 2.58), consider a regular polygon with m sides. Then its central angle is $\frac{360^\circ}{m}$ and the plane angle at the polyhedral vertex must be

$$180^\circ - \frac{360^\circ}{m}.$$

Then each plane angle of a polyhedral face can be written as

$$\frac{180^\circ(m-2)}{m}.$$

Assume that at each vertex of a regular polyhedron there n plane angles meet. Because the sum of n plane angles adjacent to one vertex must be less than 360°, then the following must be true:

$$\frac{180^\circ n(m-2)}{m} < 360^\circ.$$

This can be simplified to

$$\frac{n(m-2)}{m} < 2$$

and also after multiplying both sides of the inequality by a positive m it can be written as

$$nm - 2n - 2m < 0.$$

This looks like a good problem in Number Theory. Let us add four to both sides of the inequality above. Then we will obtain the new inequality:

$$nm - 2n - 2m + 4 < 4,$$

the left side of which can be factored by grouping to obtain a symmetric inequality:

$$(n-2)(m-2) < 4. \tag{2.59}$$

Note that the equality is not applicable here. This inequality (2.59) gives the desired solution. Because of its symmetry we understand that if a pair (n, m) satisfies (2.59), then (m, n) also satisfies. Obviously, each of the numbers n and m is greater than or equal to 3. If we consider that both m and n are greater or equal to 4 ($m \geq 4,\ n \geq 4$) then the left side of (2.59) would satisfy the following:

$$(n-2)(m-2) \geq (4-2)(4-2) = 4,$$

which contradicts its right side (2.59). Therefore m and n in a regular polyhedron cannot be simultaneously greater or equal to 4.

Is there any regular polyhedron, such that $n = m$? If it is possible, then

$$(n-2)^2 < 4, \quad n \geq 3.$$

The quadratic inequality above can be factored as difference of squares to obtain

$$n(n-4) < 0,\ n \geq 3.$$

This system has only one solution $n = 3$, so $m = 3$ and this means that if $n = m$, then $n = m = 3$ is the only possible case and this happens only for a regular tetrahedron. If n and m are not equal, assume that one of the numbers m or n is equal to 3. Let's say that it will be m: in the inequality (2.59), the numbers n and m can be interchanged, since it is symmetric with respect to these two numbers. In doing so, we will have

$$n - 2 < 4,\ n > 3, \quad m = 3.$$

Hence,

$$3 < n < 6, \quad m = 3.$$

Therefore, if $m = 3$, then n can only take the values of four and five. The symmetry of inequality (2.59) with respect to the numbers m and n means that each time when m and n are different, we will have a pair of symmetric solutions—all in all we get the following five solutions:

1. $m = n = 3$; tetrahedron that has a triangular face with three edges at each vertex.
2. $m, n = 3,\ 4$; cube $(m = 4,\ n = 3)$ or octahedron $(m = 3,\ n = 4)$
3. $m, n = 3,\ 5$; icosahedron $(m = 3,\ n = 5)$ or dodecahedron $(m = 5,\ n = 3)$.

All these five regular polyhedra are shown in Fig. 2.57. □

Method 2. Let us prove this theorem differently, applying Euler's formula.

Proof Consider again a regular polyhedron with each face having m edges and each vertex having n edges. Indeed, since each edge belongs to exactly two faces, and each face has exactly m edges, then $m \cdot F$ is equal to twice the number of edges E in the polyhedron. So, for the number of faces in terms of the edges, we obtain

$$F = \frac{2E}{m}.$$

Since each edge has exactly two ends, and exactly n edges converge at each vertex, then $n \cdot V = 2E$. This can be also solved for the total number of vertices of a regular polyhedron:

$$V = \frac{2E}{n}.$$

Since a regular polyhedron is a convex polyhedron, then the number of its vertices, edges, and faces must satisfy Euler's formula (2.40):

$$V - E + F = 2.$$

Substituting into this formula our expressions for F and V, we have

$$\frac{2E}{n} - E + \frac{2E}{m} = 2.$$

Solving this formula for E, we obtain

$$E \cdot (2\,m + 2n - nm) = 2nm.$$

Or after simplification:

$$E = \frac{2nm}{2\,m + 2n - nm}. \tag{2.60}$$

We can rewrite this formula a little bit differently, for example, by dividing its numerator and denominator by $2nm$:

$$E = \frac{1}{\frac{1}{n} + \frac{1}{m} - \frac{1}{2}}. \tag{2.61}$$

We know that the total number of the edges in any polyhedron must be a natural number, it follows from formula (2.61) that at least the following must be true:

$$\begin{cases} \frac{1}{n} + \frac{1}{m} - \frac{1}{2} > 0 \\ \frac{1}{n} + \frac{1}{m} - \frac{1}{2} < 1 \end{cases} \tag{2.62}$$

Because both $n \geq 3$ and $m \geq 3$, then $\frac{1}{n} \leq \frac{1}{3}$, $\frac{1}{m} \leq \frac{1}{3}$, the second inequality can be strengthened, so now (2.62) will be rewritten as

$$\frac{1}{2} < \frac{1}{n} + \frac{1}{m} \leq \frac{2}{3}. \tag{2.63}$$

If we assume that $n = m$ and substitute it into (2.63), we obtain

$$\begin{aligned} &\tfrac{1}{2} < \tfrac{2}{n} \leq \tfrac{2}{3} \\ &3 \leq n < 4 \; . \\ &n = m = 3 \end{aligned}$$

We can see that except one case $n = m = 3$, the numbers n and m must be different. Assume that two numbers are different, but one of them $n = 3 \neq m$, substituting it into (2.63), we obtain

$$\frac{1}{2} < \frac{1}{3} + \frac{1}{m} \leq \frac{2}{3}$$

Subtracting $1/3$ from both sides of the double inequality, we have

$$\begin{aligned} &\tfrac{1}{2} - \tfrac{1}{3} < \tfrac{1}{m} \leq \tfrac{2}{3} - \tfrac{1}{3} \\ &\tfrac{1}{6} < \tfrac{1}{m} \leq \tfrac{1}{3} \end{aligned} .$$

From the last inequality, we can obtain that

$$n = 3, \quad 3 \leq m < 6. \tag{2.64}$$

Because of symmetry of (2.60), (2.64) gives all five possible pairs of (n, m),

$$(n, m) : \; \{(3, 3), \;\; (3, 4), \;\; (4, 3), \;\; (3, 5), \;\; (5, 3)\}.$$

□

In accordance with the theorem, we obtain the following regular polyhedra (Platonic solids): tetrahedron, cube (hexahedron), octahedron, dodecahedron, and icosahedron (Table 2.3).

Table 2.3 Regular polyhedra

Regular polyhedron	Number of vertices	Number of edges	Number of faces
Tetrahedron	4	6	4
Octahedron	6	12	8
Hexahedron	8	12	6
Icosahedron	12	30	20
Dodecahedron	20	30	12

Problem 108 For each of the five regular polyhedra find a) the number of edges; b) the number of the vertices.

Solution (a) How can we find the number of the edges for a given regular polyhedron?

Multiply the number of faces bounded the body by the number of the sides of this face and since two faces share common side, divide the result by 2.

For example, icosahedron has 20 triangular faces, each face has 3 sides, then $20 \cdot 3 = 60$ divided by 2 we obtain 30 edges of the icosahedron.

Consider dodecahedron. It has 12 pentagonal faces, so the number of the edges is equal to $\frac{12 \cdot 5}{2} = 30$ hence both, icosahedron and dodecahedron, have 30 edges.

(b) How can we find the number of vertices in a regular polyhedron?

Multiply the number of faces bounded the body by the number of the plane angles of each face and divide by the number of the plane angles converging at one vertex (the number of plane angles of the polyhedral angle).

For example, in the dodecahedron we have 12 pentagonal faces, then the number of the plane angles of each face is 5. There are three pentagons meeting at each vertex, therefore the angle is trihedral. The number of the vertices is

$$\frac{12 \cdot 5}{3} = 20.$$

On the other hand, an icosahedron has 20 triangular faces, and there are five triangles meeting at each vertex (polyhedral angle has five plane angles), hence the number of the vertices in an icosahedron is

$$\frac{20 \cdot 3}{5} = 12.$$

You can check our answers with Table 2.3. This method was described in *Euclid's Book 5* [33].

The following theorem we present without proof [15].

Theorem 38 *Two regular polyhedra, with the property that one of the faces of the first polyhedron (and, consequently, each of its faces) is equal to one of the faces of the second, and one of the polyhedral angles of the first is equal to one of the polyhedral angles of the second, are equal to each other.*

In 1839, Cauchy formulated the following theorem: "*Two convex polyhedra with corresponding congruent and similarly situated faces have equal corresponding dihedral angles.*" In addition, if one of the faces of the first polyhedron coincides with the face of the second, and if both polyhedra are located on the same side of this common face (or, otherwise, if one of the polyhedral angles of the first coincides with the polyhedral angle of the second), then the polyhedra are congruent.

2.5.1 Construction and Properties of Regular Polyhedra

In one of the ancient books that have come down to our time, it is said that the octahedron and icosahedron were discovered by a student of Plato (Theaetetus of Athens, 400 BC), and the cube, tetrahedron and dodecahedron were known to the Pythagoreans long before him [32]. Many historians of mathematics doubted the validity of these words, the fact that the octahedron was discovered later than the dodecahedron caused particular concern, because the Egyptian pyramids were built in ancient times, and by mentally connecting two such pyramids it is easy to get an octahedron... However, more careful research makes us believe the words of the ancient book. These words can hardly be interpreted otherwise than in this way Theaetetus of Athens singled out the class of regular polyhedra, that is, he defined them with some degree of rigor, thereby revealing their common property and proving that there are only five different regular polyhedral. "Theaetetus was a friend of Socrates and Plato and was the central character in Plato's eponymous Socratic dialogue" [31].

The cube, tetrahedron, and dodecahedron attracted the attention of ancient geometers, even before Theaetetus, but only as interesting geometric objects, and not as regular polyhedra. Interest in the cube, tetrahedron, and dodecahedron is confirmed by ancient Greek terminology: these polyhedra had special names; it is not surprising that geometers have always been interested in the cube and tetrahedron, and the dodecahedron should have intrigued the Pythagoreans because their symbol was a regular five-pointed star, naturally fitting into the faces of the dodecahedron. It is possible that ancient people could find in nature something resembling the shape of dodecahedron. Thus, pyrite crystals (chemical formula FeS_2) are found close in shape to a dodecahedron. Iron pyrite was named in ancient times from the Greek word for "fire" because sparks were observed when it hit metal or stone. At the same time, the geometric shape of the dodecahedron is incomparably more attractive and mysterious than the shape of the octahedron.

It is easy to make a regular tetrahedron, cube, or dodecahedron at home using paper or other material. By joining three regular triangles, three squares, or three regular pentagons and continuing this construction, we will eventually arrive at a regular tetrahedron, cube, or dodecahedron. In this case, a rigid structure is obtained at each stage. The rigid structure means that it does not deform if placed, for example, on a plane.

Consider a trihedral angle with vertex S and with the plane angles α, β and γ. Such a trihedral angle is unique. Assume that we go one unit down from its vertex along each side of the trihedral angle. Let $SA = SB = SC = 1$. Triangle ABC is defined by its three given sides. For example, if $AB = a$, $BC = b$, and $AC = c$, then such triangle $\triangle ABC$ is unique. No other trihedral angle with the same measurements of the plane angles exist! Therefore, when three angles meet at a vertex, then such a construction is rigid and cannot be deformed.

What do we have for a cube?

$n = 3$ and $m = 4$, which means that three angles (right angles) coming from three squares (a regular polygon with four sides, $m = 4$) meet at one vertex.

What do we have for a tetrahedron?

Again three angles ($n = 3$), each of 60° meet at each vertex, because of connecting three equilateral triangles (regular three-sided polygon, $m = 3$).

What do we have for a dodecahedron?

In a dodecahedron again three angles meet at one vertex ($n = 3$), each angle of 108° comes from a regular pentagon ($m = 5$).

In cases of a cube, tetrahedron and dodecahedron, by connecting three regular polygons at one vertex, we create a rigid figure, that could wait for our next step in modeling without deformation.

In order to construct an octahedron, we have to make a polyhedral angle with four angles ($n = 4$) of 60° coming from four equilateral triangles ($m = 3$). We know that such polyhedral angle is not rigid structure, because a quadrilateral made of the sides opposite to the corresponding angles is not unique. A quadrilateral is not uniquely defined by its sides. Thus, rhombus and a square with the same side a may have very different shape that depends on the angle between sides.

Finally, in an icosahedron, $n = 5$ and $m = 3$, so at each vertex of a polyhedral angle, five equilateral triangles meet. Such polyhedral angle can be deformed, the construction is not rigid...

Maybe this is why the Greeks first constructed cube, tetrahedron, and dodecahedron, and only later they found octahedron and icosahedron...

To construct a tetrahedron, first, we simply glue three equilateral triangles. Translating this into the language of the earlier introduced n and m, we associate our actions with the pair $(n, m) = (3, 3)$.

To construct a cube, we glue together three squares, obtaining $(n, m) = (3, 4)$.

To construct a dodecahedron, we glue together three regular pentagons, create the pair $(n, m) = (3, 5)$.

Construction of regular octahedron and icosahedron cause some difficulties. Since for the octahedron and icosahedron, four and five triangles have to be fastened, respectively, that is, the initial construction is not rigid. It will eventually deform.

Thus, when we glue four triangles at one vertex, we obtain the pair $(n, m) = (4, 3)$ for the octahedron. By gluing together five equilateral triangles we get a starting piece for the icosahedron, associated with the pair $(n, m) = (5, 3)$.

Nowadays, in principle, it is not at all hard to build a figure of any shape, and the fact that it was very difficult for the ancient Greeks, for example, to build a regular polyhedron from some rods and somehow manage to create a stable structure, now it is not at all difficult—there are a lot of materials available. You can even do this with a 3D printer. Here's an example of what I found in a hotel lobby in Dallas. The icosahedron shown in Fig. 2.59 is cast from a metal that could not be made in the ancient world.

In general, for a known spatial figure, you can get a so-called unfolding (flat net) if you cut this figure along the edges and with the help of which you can then create other similar figures. However, we will not deal with this in this section, but will discuss it in a separate section of this chapter, where we will show how the unfolding method helps solve many complex stereometry problems.

It is known that the ancient Greeks were interested in the possibility of inscribing a known regular polyhedron into another polyhedron and thus studying the properties of this or that polyhedron. The ability to construct a polyhedron inside another polyhedron allows not only to develop spatial imagination but also often makes it possible to solve various problems that are sometimes difficult to solve in another way.

Let us construct a **regular tetrahedron** inside a cube. Take four vertices of a cube for the vertices of tetrahedron. For example, we can take A, B_1, C and D_1 as it is shown in Fig. 2.50. Then six edges connecting vertices are the diagonals of the cube's

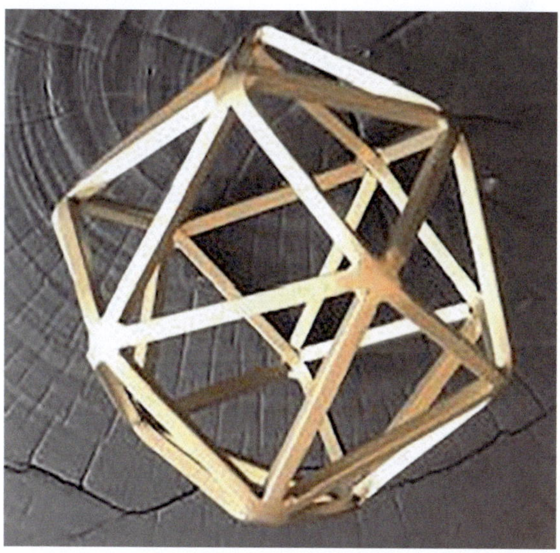

Fig. 2.59 Icosahedron on a table in one hotel's lobby

faces and equal. All faces of the figure are equilateral equal triangles. All polyhedral angles are congruent so as dihedral angles in each of them.

A **regular octahedron** is a polyhedra that has 8 congruent faces that are congruent equilateral triangles, 12 congruent edges, and 6 vertices; an edge is a line segment formed by the intersection of two adjacent faces; a vertex for a regular octahedron is a point where 4 edges meet. You can imagine it by gluing bases of two identical regular quadrangular pyramids.

If we take a regular tetrahedron and connect the midpoints of its edges, we will construct a regular octahedron. Earlier in this chapter we obtained such a octahedron while proving Theorem 31 about bi-medians of a regular tetrahedron. Please look at the solid shown by dashed segments in Fig. 2.19. Obviously, if the side of the tetrahedron was a, then the side b of the inscribed octahedron will be $b = \frac{a}{2}$.

We can also make a regular octahedron by inscribing it into a cube. Find the centers of all six faces of the cube and connect them as it is shown in Fig. 2.60. Mark the centers of the top (L) and centers of two other faces of the cube (N and M). Assume that the cube and octahedron are made of glass so all the edges can be visible from any point. We obtain an octahedron inside the cube (Fig. 2.60).

Problem 109 Prove that in a regular octahedron, opposite faces are parallel.

Proof Let us use the octahedron inscribed into a cube that is shown in Fig. 2.60. Without loss of generality we can only prove that $PNQ \| LTM$. Because $\triangle PNQ \in \triangle A_1DB$

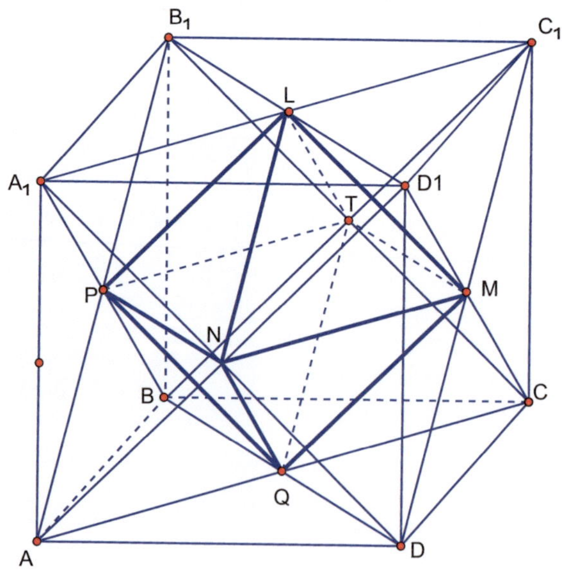

Fig. 2.60 Regular octahedron inscribed into a cube

and $\triangle LTM \in \triangle B_1D_1C$, then we need to show that in the cube the corresponding planes are parallel, i.e.:

$$A_1DB \| B_1D_1C.$$

We know that if two intersecting lines of one plane are, respectively, parallel to two intersecting lines of another plane, then these planes are parallel. It is not difficult to find the pairs of the corresponding intersecting lines in two planes. The following is valid:

$$\begin{Bmatrix} B_1C \| A_1D \\ A_1B \| D_1C \\ BD \| B_1D_1 \end{Bmatrix} \Rightarrow PNQ \| LTM\,.$$

Therefore, in a regular octahedron, opposite faces are parallel. □

It is interesting how we constructed a regular octahedron by using a cube. In geometry, every polyhedron can be associated with a second (dual) structure, where the vertices of one correspond to the faces of the other, and the edges between pairs of vertices of one correspond to the edges between pairs of faces of the other. Thus, if we find the centers of all faces of the octahedron (for example, using Fig. 2.60) and connect them, we will obtain a cube! A cube and an octahedron are **dual polyhedra**. Therefore, the center of the faces of the cube are the vertices of the octahedron and the center of the faces of the octahedron are the vertices of the cube (see Fig. 2.61). Such two polyhedra are called **mutually dual.** It can be shown that a dodecahedron and an icosahedron are also mutually dual polyhedra.

Consider now the **icosahedron** and see how we can make it. We can look at this polyhedron from the point of view of construction and recognize its most remarkable elements (see Fig. 2.62).

Fig. 2.61 Mutually dual octahedron and cube

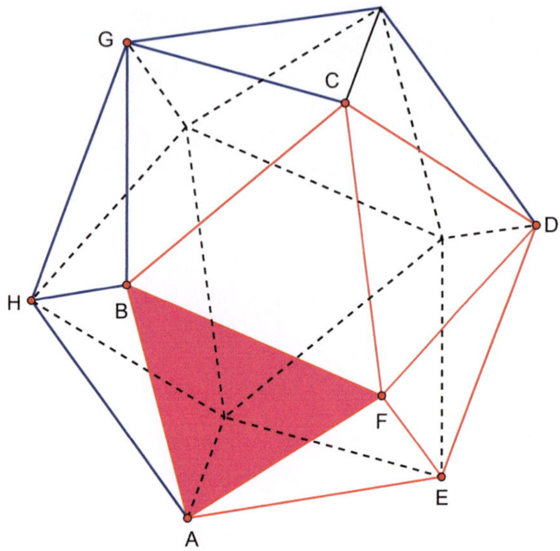

Fig. 2.62 A view of a regular pentagonal pyramid inside an icosahedron

Do you see the pyramid with red edges in Fig. 2.62? This pentagonal pyramid *ABCDEF* can give us an idea of possible construction!

You can start by constructing a regular pentagon (see, for example, my book [2]) *ABCDE* with side 1. Find its center *O* (the center of the circle circumscribed around it) and draw a perpendicular from this point. Using a compass centered on point *A* and radius 1, mark point *F* on this perpendicular (point *F* is the intersection of a line passing through point *O* and perpendicular to the plane of pentagon *ABCDE*, and a circle with center *A* and radius 1). Connect point *F* to each vertex of pentagon *ABCDE*. The following is true:

$$AF = BF = CF = DF = EF = 1.$$

We have constructed the lateral surface of a regular pentagonal pyramid with all edges of 1. Moreover, all lateral faces of this pyramid are equilateral triangles and the dihedral angles between two adjusted triangles are equal.

Next, let us construct the congruent pyramid $A'F'C'GHB'$. Label apex of the this pyramid as B' and then glue the triangle $A'B'F'$ onto triangle ABF, such that points C and C' would coincide. This way we can attach two pyramids together (see Fig. 2.63) and obtain a spatial figure made of eight equilateral triangles, such that its all dihedral angles are equal.

We can repeat the procedure one more time and obtain a 3D figure of ten equilateral triangles. Next, we can make another figure, just like this one, congruent to it with also ten equal equilateral triangles. These two figures must fit together very well since they have equal edges and dihedral angles, we can glue them together to obtain a regular icosahedron.

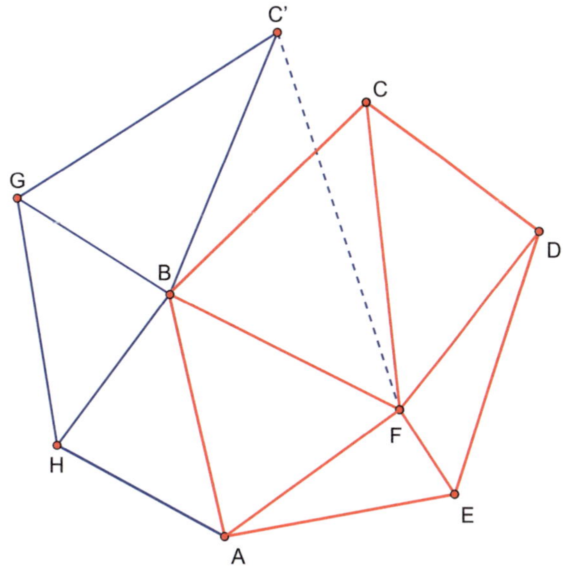

Fig. 2.63 Construction of two pentagonal pyramids of an icosahedron

Let us solve the following problems.

Problem 110 (*Archimedes' problem*) Inscribe a regular octahedron into a regular tetrahedron.

Solution This solution was given by Archimedes. Let us sketch a tetrahedron $ABCD$ (see our sketch in Fig. 2.64). We divide the edges of the tetrahedron in half at the points $E,\ F,\ G,\ H,\ N,\ L$ and connect these points with each other by twelve (12) straight segments, which are obviously equal to each other.

There are eight equilateral triangles—the faces of the future octahedron. Therefore, $EGFHNL$ is the octahedron inscribed in the tetrahedron $ABCD$, because its vertices lie at the midpoints of the edges of the tetrahedron.

Problem 111 A spatial figure is obtained by connecting the centers of the adjacent faces of an octahedron. Prove that the obtained figure is a cube. Find its side if the side of the octahedron is equal to a.

Proof In order to prove this statement we can consider only the top part of the octahedron—the regular quadrangular pyramid with all edges equal to a (see all constructions in Fig. 2.65).

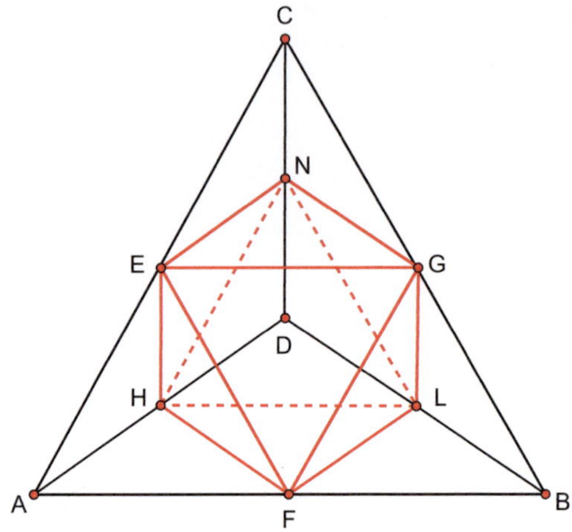

Fig. 2.64 An octahedron inscribed into a tetrahedron

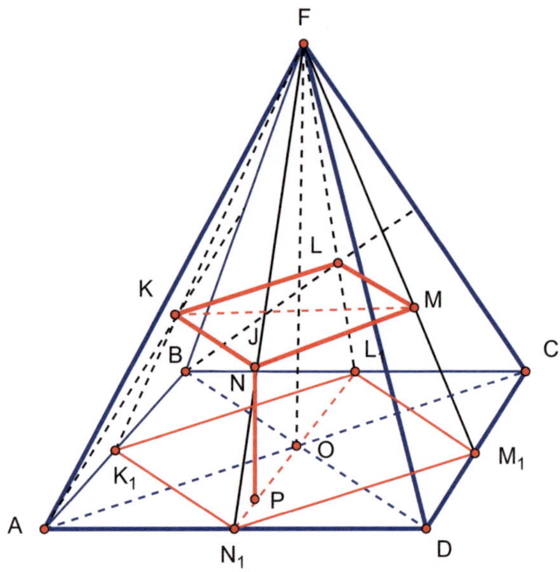

Fig. 2.65 Cube inscribed in an octahedron

Here *ABCDF* is half of the octahedron, its upper part. Let M, N, K, L be the centers of faces *DFC*, *DFA*, *AFB*, and *BFC*, respectively. As we know, the center of a triangle is the point of intersection of its medians, so it is divided in ratio $2 \div 1$ starting from a vertex of a triangle. Connecting these points, we obtain a parallelogram *MNKL*. Let us prove that this parallelogram is a square.

If M and N are the centers of the adjacent triangles, then

$$NM \in \triangle N_1FM_1,$$

where N_1 and M_1 are the midpoints of edges *AD* and *DC*, respectively.

Since the medians of a triangle are divided into ratio $2 \div 1$, then the following is true:

$$NM = \frac{2}{3} N_1 M_1, \; KN = \frac{2}{3} K_1 N_1$$

and

$$KL = \frac{2}{3} K_1 L_1, \; LM = \frac{2}{3} L_1 M_1.$$

Here $M_1 N_1 K_1 L_1$ is the square with side $\frac{a\sqrt{2}}{2}$. Then the side of the square $MNKL$, b can be evaluated as follows: $MN = b = \frac{2}{3} \cdot \frac{a\sqrt{2}}{2}$ or

$$b = \frac{a\sqrt{2}}{3}.$$

Further, the lower face of the cube inscribed in this way can be obtained by reflecting the square $MNKL$ with respect to the plane $ABCD$. This transformation is a rigid motion and transforms the figure into its congruent figure.

Assume that N_2 is the image of point N under this reflection. We will not show point N_2 in the sketch. However, it is clear that if we drop a perpendicular from point N onto plane $ABCD$, and denote the foot of the perpendicular by P, then $NN_2 = 2NP$. Consider triangle $N_1 F L_1$. In this isosceles triangle,

$$N_1 F = F L_1 = \frac{a\sqrt{3}}{2}, \; N_1 L_1 = a,$$

then its height $h = FO$ can be found using Pythagorean theorem:

$$h = \sqrt{\frac{3a^2}{4} - \frac{a^2}{4}} = \frac{a\sqrt{2}}{2}.$$

Point N divides side $N_1 F$ again in ratio $2 \div 1$. Then using Thales theorem we can state that

$$NP = \frac{1}{3} FO = \frac{a\sqrt{2}}{6}.$$

Now we can evaluate the edge NN_2 as twice distance NP:

$$NN_2 = \frac{a\sqrt{2}}{3}.$$

Because all 12 edges of the inscribed solid are equal to $b = \frac{a\sqrt{2}}{3}$ and, respectively, perpendicular and because there are six equal square faces, it proves that the inscribed figure is a cube $MNKLM_2 N_2 K_2 L_2$. □

The mutually dual cube and octahedron constructed using Maple 17 can be seen in Fig. 2.61.

Problem 112 Which polyhedron is the dual with a tetrahedron?

Solution If we connect all the centers of the faces we will find out that a tetrahedron is a self-dual polyhedron.

Answer. A Tetrahedron.

Problem 113 Find the volume of the regular octahedron with side a.

Solution The volume of an octahedron shown in Fig. 1.19 is equal to the double volume of one of the pyramids.

$$V = \frac{2}{3} h a^2 = \frac{1}{3} H \cdot a^2,$$

where h and a are, respectively, the height of one pyramid and side of the square base, respectively. Then under this condition, the height of the pyramid FO will be evaluated from Pythagorean theorem as

$$h = \sqrt{a^2 - \frac{a^2}{2}} = \frac{a}{\sqrt{2}}.$$

Denote by H the octahedron height, then $H = 2h$, and

$$H_{\text{octahedron}} = a\sqrt{2}. \tag{2.65}$$

The **volume of the regular octahedron** with side a is

$$V_{\text{octahedron}} = \frac{1}{3} a\sqrt{2} \cdot a^2 = \frac{1}{3} a^3 \sqrt{2}. \tag{2.66}$$

Answer. $V_{\text{octahedron}} = \frac{1}{3} a^3 \sqrt{2}$.

Remark. Obviously that the volume of the octahedron inscribed into a cube will be also described by this formula, but the length of the edge a and height of the octahedron (H) will depend on the edge of the cube b connected by the formula:

$$a = \frac{b}{\sqrt{2}}, \quad H = 2b.$$

Problem 114 Propose a construction that divides space into 14 parts.

Solution Solution to this problem can vary. Since we are learning about regular polyhedra, let us consider an octahedron $ABCDFH$, for example, drawn in Fig. 1.19, where $ABCD$ is the square with center O, $O = AC \cap BD$. Point O forms with the edges of the octahedron eight trihedral angles.

Denote three planes by $\alpha = AFC$, $\beta = ABCD$, and $\gamma = BFD$. All three planes intersect at point O and divide the space into eight parts:

$$\alpha \cap \beta \cap \gamma = O.$$

Please note that all faces of the octahedron are equal triangles, moreover, there are four pairs of triangles lying in parallel planes. Consider $ABH \| DFC$ or $ADF \| BCH$. We can draw plane δ through point O, such that $\delta \| ADF$. The boundary of this plane $GLMK$ inside the octahedron is shown in green. Such a plane will intersect with three other planes α, β, γ at point O, but will not intersect planes of triangles ADF and BCH. So, δ will divide each trihedral angle into two, except two trihedral angles $ADFO$ and $BCHO$. Finally, we will obtain $8 + 6 = 14$ parts.

Problem 115 Find dihedral angles of a cube, regular tetrahedron, and octahedron.

Solution The dihedral angles of a cube are equal to 90°.

Consider a regular tetrahedron $ABCD$ with edge 1. From vertices A and D, we drop perpendiculars AE and DE to edge BC (Fig. 2.66).

To find the dihedral angle DEA, we use the Law of Cosines Theorem applied to the triangle ADE, in which $AD = 1$, $AE = DE = \frac{\sqrt{3}}{2}$. We have the equality

$$1^2 = \left(\frac{\sqrt{3}}{2}\right)^2 + \left(\frac{\sqrt{3}}{2}\right)^2 - 2\left(\frac{\sqrt{3}}{2}\right)\left(\frac{\sqrt{3}}{2}\right)\cos\varphi.$$

Hence, $\cos\varphi = \frac{1}{3}$, $\varphi \approx 70.5°$.

Next, we evaluate cosine of the dihedral angle of the octahedron with edge 1 referring to the second sketch in Fig. 2.66. To do this, from vertices E and F we will drop perpendiculars EG and FG onto the edge BC. We note that $EG = FG = \frac{\sqrt{3}}{2}$ as heights of equilateral triangles. Quadrilateral $AECF$ is a square with side 1 and therefore its diagonal is equal to $EF = \sqrt{2}$. By the Law of Cosines Theorem, we have

Fig. 2.66 Tetrahedron and octahedron built from tetrahedron

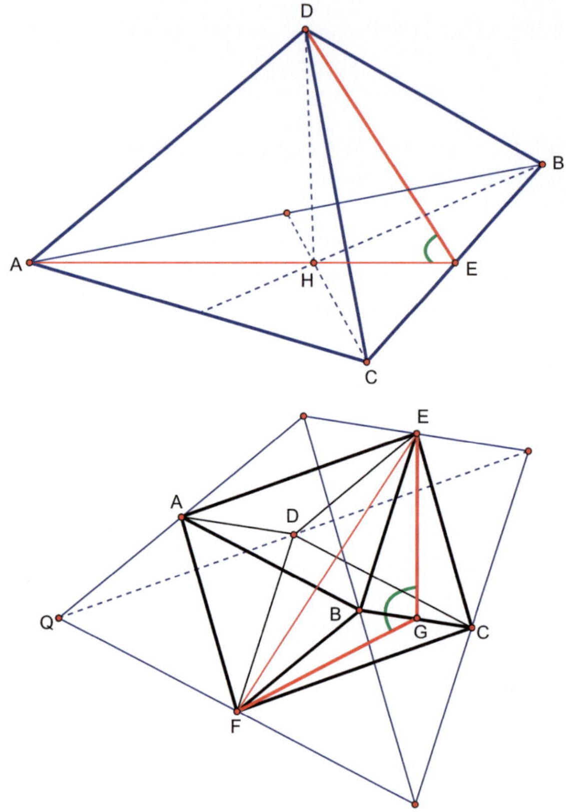

$$\left(\sqrt{2}\right)^2 = \left(\frac{\sqrt{3}}{2}\right)^2 + \left(\frac{\sqrt{3}}{2}\right)^2 - 2\left(\frac{\sqrt{3}}{2}\right)\left(\frac{\sqrt{3}}{2}\right)\cos\psi.$$

From which we obtain

$$\cos\psi = -\frac{1}{3}, \quad \psi \approx 109.5^\circ.$$

Note that the dihedral angles of a tetrahedron and an octahedron add up to 180° ($\varphi + \psi = 180^\circ$). This can be deduced without calculating the dihedral angles but using the fact that the midpoints of the edges of a regular tetrahedron are the vertices of the octahedron.

Answer. $\varphi \approx 70.5^\circ$ and $\psi \approx 109.5^\circ$.

The following problems can help us to evaluate surface areas and volumes of all other regular polyhedra.

Problem 116 Let α be the dihedral angle of an icosahedron. Evaluate cosine and sine of angle α.

Solution Let us evaluate cosine of the dihedral angle of the icosahedron with edge 1. To do this, from vertices A and C we will drop perpendiculars AG and CG onto the edge BD. $AG = CG = \frac{\sqrt{3}}{2}$ as heights of equilateral triangles. AC is the diagonal of the regular pentagon $CDAEF$ with side 1 (see Fig. 2.67).

Let us evaluate the diagonals using, for example, regular pentagon $FKBNC$ sketched in Fig. 2.68. A regular pentagon has five equal diagonals. Let us find the diagonal BC. Connect vertices K and N. We can state that $KN \parallel FC$. Let M be the point of intersection of diagonal BC and segment KN:

$$KN \cap BC = M.$$

Next, because the inscribed angles supported by equal arcs are equal (we can mark three angles in our sketch with the same color), then the following is true:

$$\angle BKM = \angle BNK = \angle MBN$$

Denote the side of the pentagon as

$$a = FC = BN = KB = NC = KF$$

and the diagonal by $d = BC = KN$.

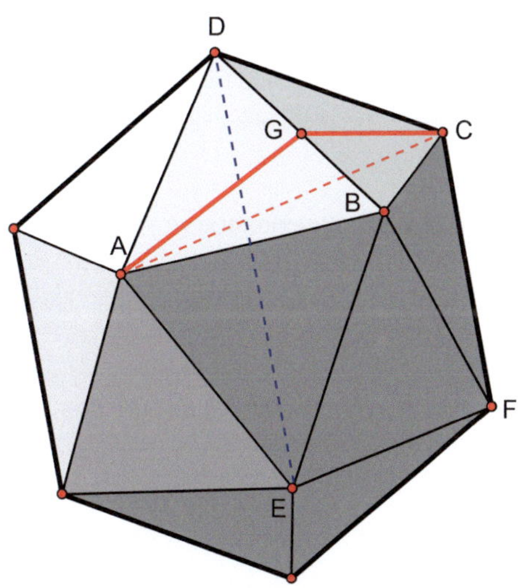

Fig. 2.67 Finding dihedral angle in an icosahedron

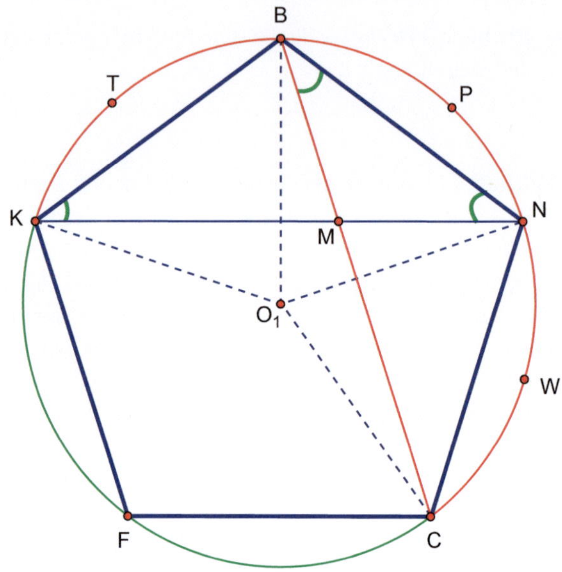

Fig. 2.68 Finding the diagonals of regular pentagon *FKBNC*.

Since also $KF \parallel MC$, then quadrilateral *FKMC* is a parallelogram and hence

$$a = KM\,.$$

Additionally, triangles *BMN* and *KBN* are similar triangles by two angles ($\triangle BKN \sim \triangle MBN$). The following is valid:

$$\frac{BN}{KN} = \frac{MN}{BN}$$
$$BN^2 = KN \cdot MN\,.$$

This can be rewritten as

$$a^2 = d\,(d - a)\,.$$

This is a quadratic equation in variable d:

$$d^2 - ad - a^2 = 0$$
$$d = \frac{a \pm a\sqrt{5}}{2}.$$

By taking only positive value, we finally obtain the formula for the diagonal of a regular pentagon with side a:

$$d = \frac{a\left(1+\sqrt{5}\right)}{2}. \tag{2.67}$$

If the side of a pentagon is equal to 1, then from (2.67) its diagonal BC is equal to the diagonal AC of the pentagonal face of the icosahedron shown in Fig. 2.67.

$$AC = \frac{1+\sqrt{5}}{2}.$$

Applying the Law of Cosines to $\triangle AGC$ in Fig. 2.67, we have

$$\left(\frac{1+\sqrt{5}}{2}\right)^2 = \left(\frac{\sqrt{3}}{2}\right)^2 + \left(\frac{\sqrt{3}}{2}\right)^2 - 2\left(\frac{\sqrt{3}}{2}\right)\left(\frac{\sqrt{3}}{2}\right)\cos\alpha.$$

The left and right sides can be simplified by multiplying by four:

$$1 + 2\sqrt{5} + 5 = 3 + 3 - 2 \cdot 3\cos\alpha$$
$$\cos\alpha = -\tfrac{\sqrt{5}}{3}, \quad \alpha \approx 138.19°.$$

Answer. $\cos\alpha = -\frac{\sqrt{5}}{3}$, $\sin\alpha = 2/3$.

Problem 117 Let θ be the dihedral angle of a dodecahedron. Prove that

$$\cos\theta = -\frac{\sqrt{5}}{5}, \ \tan\theta = -2.$$

Proof For the unit dodecahedron sketched in Fig. 2.69, we can drop perpendiculars AG and CG onto edge BF.

Here $AB = AF$, where AB is the diagonal of a regular pentagon with side 1, so using formula (2.67), we can state that

$$AB = \frac{1+\sqrt{5}}{2} = AF$$

and $GF = \frac{1}{2}$. From the right triangle AGF, we have

$$AG = CG = \sqrt{AF^2 - FG^2} = \frac{\sqrt{2\sqrt{5}+5}}{2}.$$

On the other hand, AC is the diagonal of a regular pentagon $ABCDE$ with edge $AB = \frac{1+\sqrt{5}}{2}$ so we can evaluate its length using again formula (2.67) as

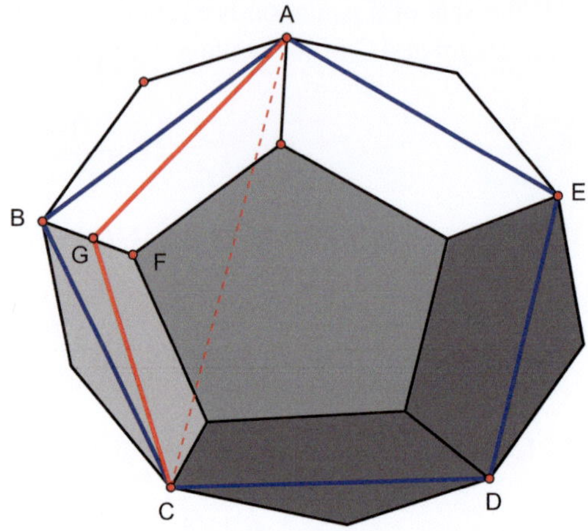

Fig. 2.69 Finding dihedral angle in a dodecahedron

$$AC = \frac{1+\sqrt{5}}{2} \cdot \frac{1+\sqrt{5}}{2} = \frac{3+\sqrt{5}}{2}.$$

Applying the Law of Cosines theorem to $\triangle AGC$ we have

$$AC^2 = 2AG^2 - 2AG^2 \cos\theta$$

$$\left(\frac{3+\sqrt{5}}{2}\right)^2 = 2\left(\frac{\sqrt{2\sqrt{5}+5}}{2}\right)^2 - 2\left(\frac{\sqrt{2\sqrt{5}+5}}{2}\right)^2 \cos\theta.$$

Multiplying both sides by four and after simplification, we obtain

$$14 + 6\sqrt{5} = 2\left(2\sqrt{5}+5\right) - 2\left(2\sqrt{5}+5\right)\cos\theta.$$

We can divide both sides by two and after collecting like terms to obtain

$$\cos\theta = -\frac{2+\sqrt{5}}{5+2\sqrt{5}}.$$

We can use a calculator right away and find that $\theta \approx 116.6^\circ$. However, we can try to simplify the expression for cosine of the dihedral angle by rationalizing the denominator:

$$\begin{aligned}\cos\theta &= -\frac{2+\sqrt{5}}{5+2\sqrt{5}} = -\frac{\left(2+\sqrt{5}\right)\left(5-2\sqrt{5}\right)}{\left(5+2\sqrt{5}\right)\left(5-2\sqrt{5}\right)}\\ &= -\frac{10-4\sqrt{5}+5\sqrt{5}-10}{25-20} = -\frac{\sqrt{5}}{5}.\end{aligned}$$

Thus, the cosine of the dihedral angle of the dodecahedron is equal to

$$\cos\theta = -\frac{\sqrt{5}}{5}.$$

The corresponding sine of the same angle is

$$\sin\theta = \frac{2}{\sqrt{5}},$$

then the tangent of the dihedral angle is the ratio of $\sin\theta$ and $\cos\theta$:

$$\tan\theta = -2.$$

□

Problem 118 Inscribe a regular dodecahedron in the regular icosahedron with edge of a. Find the edge of the inscribed dodecahedron.

Solution As we know, a regular icosahedron has 12 vertices, and we can mentally imagine 12 pentagonal pyramids. Let's take one of these pyramids and find in the face of each of the five triangles converging to one vertex the center of the circle circumscribed about this triangle. Now let's connect all five such centers to make a regular pentagon. If this procedure is repeated with all 12 pentagonal pyramids, then in this way we will get a new polyhedron (dodecahedron) that will be inscribed in our icosahedron. This construction was known to the Ancient Greeks.

If we remember how we built the icosahedron, we can imagine 12 regular pentagonal pyramids inside one icosahedron. Let's consider one of these pyramids with the apex F and base $ABCDE$, similar to the one shown in Fig. 2.62. Five equilateral triangles converge at vertex F. Let us denote the circumcenters of each such triangle as G, H, K, L, and M. On the left in the Fig. 2.70 we seem to be looking at one such pyramid from the top, and on the right we show our ideas for construction using such a pyramid in 3D. By connecting these five points, we get a regular pentagon $GHKLM$. The plane of the pentagon $GHKLM$ is parallel to the plane of the base of the pyramid. The pentagons $GHKLM$ and $G_1H_1K_1L_1M_1$ can be obtained from each another by homothety centered at the vertex F.

Let us find the side of this regular pentagon, b. Continue segment FH and FK until they intersect with the corresponding sides at points H_1 and K_1, respectively. Because triangles BFA and BCF are equilateral and equal, with side of the icosahedron a, then their circumcenters lie at the intersection of the medians (bisectors and height that are the same for an equilateral triangle), then points H_1 and K_1 are the midpoints of sides AB and BC, respectively, and $KH \| K_1H_1$ (see Fig. 2.70 on the right). Then $\triangle FKH \sim \triangle FK_1H_1$ and following is true:

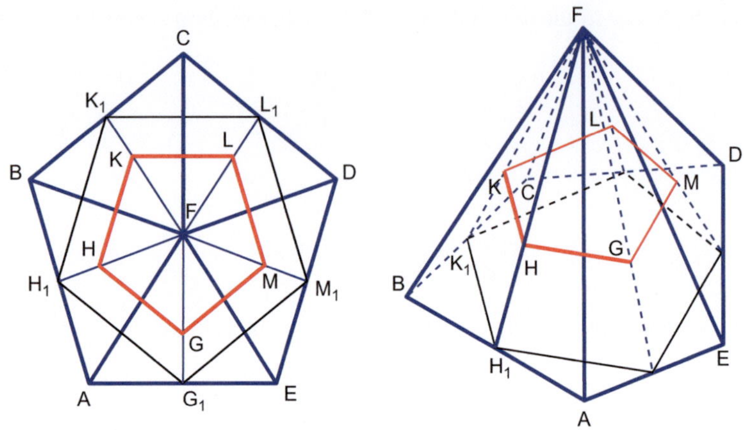

Fig. 2.70 Inscribing dodecahedron in icosahedron

$$\frac{KH}{K_1H_1} = \frac{FH}{FH_1}.$$

Denote the side of the pentagon $GHKLM$ by b, then

$$b = KH = \frac{2}{3} \cdot K_1H_1.$$

Let us mentally connect points A and C, then in the formed triangle ABC, AC is the diagonal of the regular pentagon $ABCDE$ with side a and H_1K_1 is midline segment of $\triangle ABC$. Then

$$H_1K_1 = \frac{1}{2} \cdot AC,$$

AC is the diagonal of the regular pentagon that we evaluated by formula (2.67):

$$KH = \frac{2}{3} \cdot \frac{1}{2} \cdot a \cdot \frac{1+\sqrt{5}}{2}.$$

After simplification the side of the pentagon $GHKLM$ is

$$b = \frac{a(1+\sqrt{5})}{6}.$$

If we do the same with 11 other pyramids of the icosahedron as we did with this pyramid, we will get 11 more regular pentagons with equal sides, which, together with this pentagon, will bound the dodecahedron.

Next, let us illustrate our construction using Fig. 2.62. We can see that, for example, pyramids $FABCDE$ and $BHGCFA$ have two common faces, triangles BCF and

Table 2.4 Surface area and volume of regular polyhedra with edge a

Regular polyhedron	Surface area	Volume
Tetrahedron	$a^2\sqrt{3}$	$a^3\sqrt{2}/12$
Octahedron	$2a^2\sqrt{3}$	$a^3\sqrt{2}/3$
Hexahedron	$6a^2$	a^3
Icosahedron	$5a^2\sqrt{3}$	$\frac{5}{12}a^3(3+\sqrt{5})$
Dodecahedron	$3a^2\sqrt{5(5+2\sqrt{5})}$	$\frac{a^3(15+7\sqrt{5})}{12}$

BAF, so our construction will produce two identical regular pentagons with one common side formed by connecting circumcenters of these two common faces. Twelve pyramids will produce 12 faces, the boundaries of the regular dodecahedron. Because its vertices lie on the face of the icosahedron, it is inscribed in the given icosahedron.

Answer. $b = \frac{a(1+\sqrt{5})}{6}$.

The surface areas and the volumes of regular polyhedra with edge a can be summarized in Table 2.4.

In 1619 Kepler defined **stellation** for polygons and polyhedra as the process of extending edges or faces until they meet to form a new polygon or polyhedron. He stellated the regular dodecahedron to obtain two regular star polyhedra, the small stellated dodecahedron (shown in Fig. 2.71) and great stellated dodecahedron.

Kepler also stellated the regular octahedron to obtain the stella octangula, a regular compound of two tetrahedra (see one in Fig. 2.72 made using Maple 17).

Fig. 2.71 First stellation of dodecahedron obtained by Kepler

Fig. 2.72 First stellation of octahedron

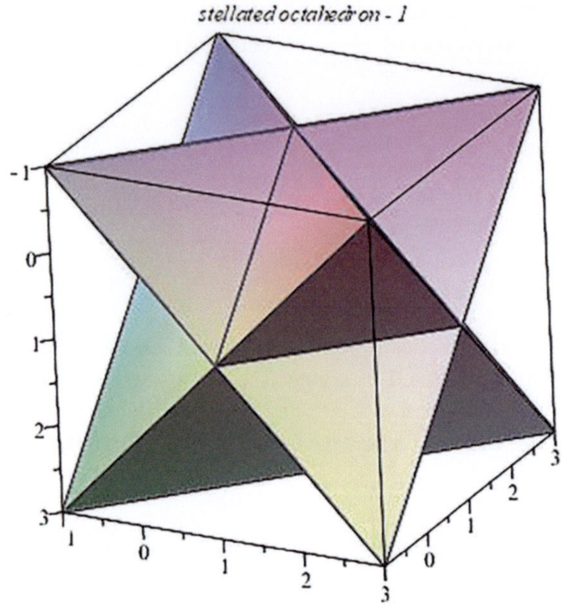

The following problem is not about stellated polyhedron. However, the shape of the new geometric solid somewhere reminds us the one stellated.

Problem 119 A polyhedron is made by attaching pyramids to each face of a regular polyhedron. Each pyramid should have the base congruent to the face of the polyhedron. Assume that a new shape is made out of cube with side a, and that each pyramid is a regular pyramid with all edges equal to a. Find the volume of the obtained spatial figure.

Solution Since a cube has six faces, then the new figure will have the volume of the cube plus six volumes of the pyramids. Each pyramid is like a half of the regular octahedron. The volume of one pyramid (V_1) is equal to

$$V_1 = \frac{1}{3}a^2 \cdot h, \quad h = \frac{a\sqrt{2}}{2}.$$

After simplification we obtain for the volume of six such pyramids:

$$V = 6V_1 = a^3\sqrt{2}.$$

The volume of the new figure V_2 is

$$V_2 = a^3(1 + \sqrt{2}).$$

Answer. $V_2 = a^3(1 + \sqrt{2})$.

2.5.2 *Regular Polyhedra in Chemistry and in Nature*

Regular polyhedra are the most spectacular solid figures, and nature takes advantage of this. This is confirmed by the shape of some crystals. Take at least salt, without which we cannot live just because all of us use it in cooking. It is known that it is highly soluble in water and serves as a conductor of electric current. Thus, salt crystals (NaCl) have the shape of a cube.

In the production of aluminum, aluminum-potassium sulfate (alum) is used, the single crystal of which has the shape of a regular octahedron.

Obtaining sulfuric acid, iron, or special grades of cement is not complete without sulfurous pyrites FeS_2. The crystals of this chemical are shaped like a dodecahedron.

Antimony sodium sulfate $Na_5SbO_4(SO_4)$—a substance synthesized by scientists is used in various chemical reactions. The crystal of antimony sodium sulfate has the shape of a tetrahedron.

In general, diamond crystals found in nature are shaped like an octahedron.

When atoms interact, molecules and crystals can be formed through chemical bonds (covalent, ionic or metallic). We also note that during the formation of a crystal, the type of connection between particles determines what the crystal lattice will be. By influencing the type of crystal lattice, the chemical bond also determines the physical properties of the substance: hardness, volatility, or its melting point.

Stereometry plays an important role in studying the structure of molecules and crystal lattices. Knowledge of stereometry helps chemists both explain the properties of already known molecules or crystals and predict the appearance of new compounds or, for example, look for the possibility of synthesizing them.

2.5.2.1 Molecules in the Shape of Regular Polyhedra

A molecule is already an elementary cell of a substance that has the properties of this substance.

Molecules can be formed by means of covalent or ionic bonds between atoms. Covalent bonds are formed between nonmetal atoms in simple substances (H_2, F_2, O_2), nonmetal oxides (H_2O, CO_2), acids (HF, HCl, H_2SO_4, HNO_3).

An ionic bond is an interaction between atoms in a molecule of a substance, the result of which is the formation and mutual attraction of oppositely charged ions due to Coulomb force. Below we describe some of the molecules having the shape of regular polyhedra.

Tetrahedron. As noted earlier, the CH_4 methane molecule has a tetrahedral structure, such that the atoms of hydrogen are at the vertices and the carbon is at the center of the gravity of the tetrahedron (see Fig. 2.11). However, there are other molecules in the shape of a regular tetrahedron, such as the white phosphorus molecule P_4 that exists in nature. Each vertex is connected to three others, the phosphorus atoms in P_4 are trivalent (see Fig. 2.73).

This molecule has a fairly high energy and easily interacts with other molecules. At the beginning of the last century, it was used in match heads because it ignites easily when rubbed against any surface [27]. White phosphorus is a soft, pale-yellow, very toxic substance with an unpleasant garlicky odor; it has the unique ability to glow in the dark due to slow oxidation by atmospheric oxygen. Since white phosphorus is insoluble in water, in laboratories, it is stored under a layer of water. The melting point of white phosphorus is 44 °C. When white phosphorus is heated above 500 °C without air, one phosphorus–phosphorus bond is broken into tetrahedral molecules which in turn are connected to each other in long chains or rings. These tetrahedral chains are very strong, therefore red phosphorus is less active than white phosphorus, does not glow in the dark and is absolutely harmless. With increasing pressure, white phosphorus transforms into crystalline black phosphorus, the most stable modification of phosphorus. Black phosphorus conducts current and is a semiconductor.

As we know, the valency of carbon is 4 and therefore the question arises: is there a C_4H_4 molecule in which the carbon atoms are located at the vertices of a regular tetrahedron (so they would take the position of the green balls in Fig. 2.73) but each carbon atom has another bond with a hydrogen atom? As far as I know, such molecule does not exist in nature and it has not yet been possible to synthesize a tetrahedrane molecule. The complexity of its synthesis may be due to the fact that the angle between the carbon atoms in C_4H_4 should be only 60°, which is significantly less than the optimal 109.5°. This creates a lot of tension in the molecule and it becomes extremely unstable.

Cube. Hydrocarbons in the shape of a cube are stable. For example, C_8H_8 has such a shape and is called cubane (see Fig. 2.74).

Cubane was obtained in 1964 by a 13-step synthesis. This substance is interesting because it has the highest density among all hydrocarbons. In addition, due to the fact

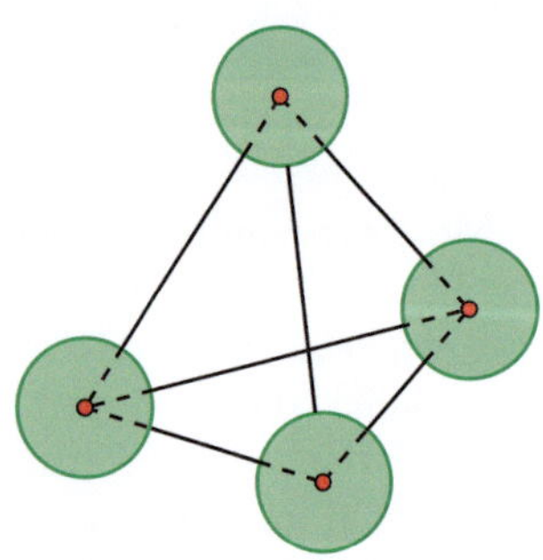

Fig. 2.73 Tetrahedral structure of white phosphorus molecule, P_4

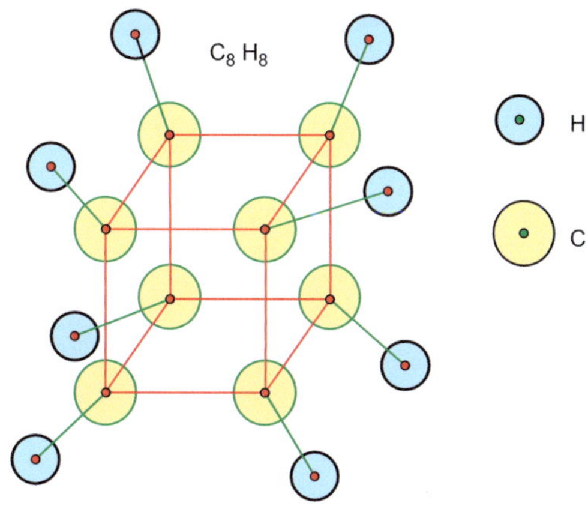

Fig. 2.74 Hexahedral structure of C_8H_8, cubane molecule

that the angle of 90° between the C–C bonds is significantly different from the angle of 109.5°, characteristic of ordinary saturated hydrocarbons, the cubane molecule has a fairly high energy. Coupled with its high density, this makes cubane a promising substance for storing and releasing energy.

Octahedron. Let us imagine a molecule C_6 with an octahedral shape, so each vertex is occupied by the atom of carbon. If we connect each atom with its neighbors, we can see that the valence of each carbon atom is 4. However, nobody ever reported the existence of such molecule. We can see that geometry of a hypothetical molecule plays an important role is prediction and synthesis of new compound but there are other factors that must be considered. For example, the octahedron can be divided into two regular square pyramids having common base, and the angle between two adjacent C–C bonds would be 90° for the square and only 60° for the two carbon atoms above and below the plane of the square. Obviously such molecule is not stable and impossible to synthesize.

In stable molecules having an octahedral shape, the atoms at the vertices of the octahedron are not connected to each other but form chemical bonds with the atom located in the center. This is how, for example, a molecule of sulfur hexafluoride SF_6 is structured (see Fig. 2.75). It can be explained by sp^3d^2 hybridization. Due to the high strength of S-F bonds, fluoride sulfur is very inert: it is a gaseous substance, without color and odor, difficult to enter into chemical reactions. It is used to create an inert atmosphere in electrical engineering.

Icosahedron. Organic molecules in the shape of an icosahedron do not exist, since in this polyhedron each vertex has five neighbors, and carbon cannot be pentavalent. However, a negative ion of this form is known as dodecaborate $B_{12}H_{12}^{2-}$ [26]. The 12 boron atoms are at the vertices of an icosahedron and each B boron atom is connected with a hydrogen atom. Borohydride molecules exhibit a very unusual type of bond—a three-center, two-electron bond. This means that two electrons provide a bond not between two, as usual, but between three atoms, in particular, B-H-B and B-B-B. Such connections are also called "bridge" or "banana." The latter name is because

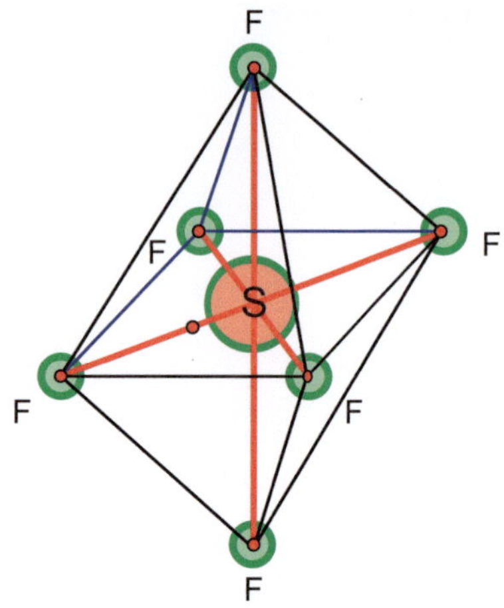

Fig. 2.75 Octahedral structure of SF_6 (sulfur hexafluoride)

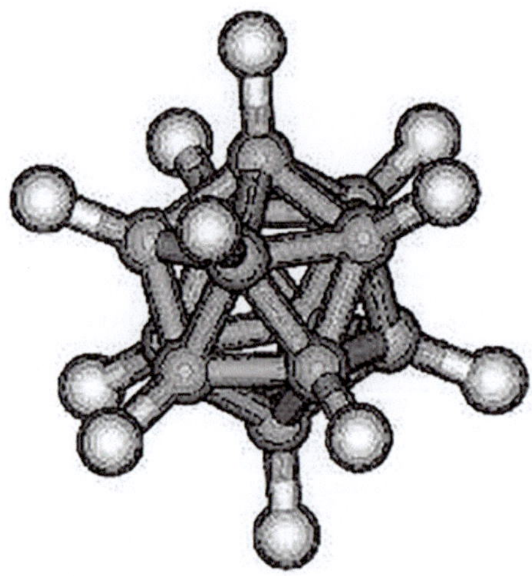

Fig. 2.76 Icosahedral structure of dodecaborate, $B_{12}H_{12}^{2-}$

the distribution of electron density between the three bonded atoms is shaped like a banana (see [27] and Fig. 2.76).

Dodecahedron. Finally, the most complex of regular polyhedra—the dodecahedron—is realized among hydrocarbons. The corresponding compound, dodecahedrane, $C_{20}H_{20}$, was synthesized in 1982. The synthesis route included at least 29 stages! There are **two dodecahedrons** in the $C_{20}H_{20}$ molecule, one is formed by interconnected carbon atoms, and the other, external one, is formed by unbounded hydrogen atoms (see Fig. 2.77).

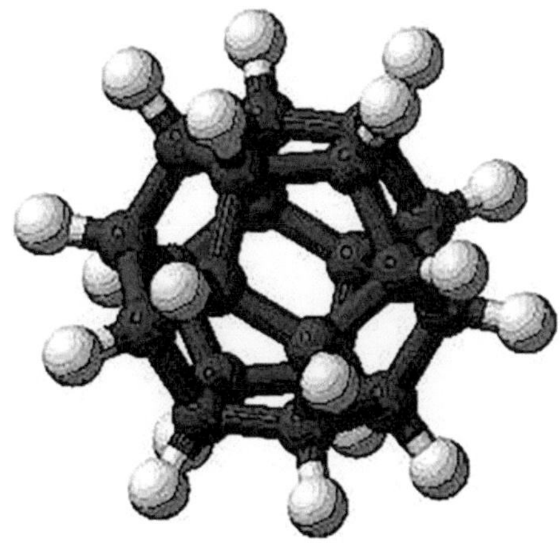

Fig. 2.77 Dodecahedral structure of dodecahedrane, $C_{20}H_{20}$

2.5.2.2 Crystals in the Shape of Regular Polyhedra

Crystal structure is the arrangement of particles (atoms, molecules, ions) in a crystal. Being individual for each substance, the crystal structure refers to the basic physical and chemical properties of this substance. The crystal structure is characterized by the correct (regular) arrangement of particles in strictly defined places in the crystal. By mentally connecting these points with lines, a spatial framework with three-dimensional periodicity is obtained, which is called a crystal lattice. The points at which particles are located are called lattice nodes. These nodes can contain ions, atoms, or molecules. These particles oscillate at lattice nodes around their equilibrium position. If the crystal is heated, the amplitude (range) of vibrations increases and the crystal body expands. Each particle located in a crystal is closely adjacent to only a certain number of neighboring particles. This number of nearest neighboring particles is called **the coordination number**.

Crystals with metallic bonds. A metallic bond is formed between metal atoms in simple substances (Na, K, Cu) and metal alloys. Let's get acquainted with how atoms of metal elements interact with each other. Metals usually do not exist as isolated atoms, but in the form of a piece, ingot, or metal product. What holds metal atoms in a single volume? The atoms of most metals at the outer level contain a small number of electrons 1, 2, or 3. These electrons are easily detached, and the atoms turn into positive ions. The detached electrons move from one ion to another, binding them into a single whole. Connecting with ions, these electrons temporarily form atoms, then break off again and combine with another ion, etc. Consequently, in the volume of the metal, atoms are continuously converted into ions and vice versa. The bond in metals between ions through shared electrons is called metallic. Figure 2.78 schematically shows the structure of a sodium metal fragment.

Each vertex of the cube is occupied by a Na atom, as is its center. Each sodium ion is surrounded by eight nearest neighbors—this is especially clearly visible in

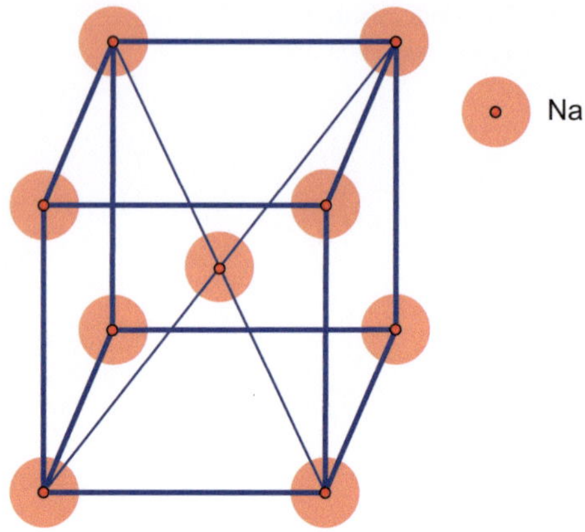

Fig. 2.78 Sodium (Na) metallic crystal

the figure by the atom that is placed in the center of the cube. But in a metal, all sodium atoms are equivalent and have the same environment. Thus, the coordination number of sodium in a metal crystal is 8. Substances with metallic bonds have metallic crystal lattices. At the sites of such lattices there are atoms and ions (either atoms or ions, into which metal atoms easily transform, giving up their outer electrons "for common use"). A metallic bond has some similarities with a covalent bond, since it is based on the sharing of external electrons. However, with a covalent bond, the outer unpaired electrons of only two neighboring atoms are shared, while with a metallic bond, all atoms take part in the sharing of these electrons. That is why crystals with a covalent bond are brittle, but with a metal bond, as a rule, they are ductile, electrically conductive and have a metallic luster. Metallic bonding is characteristic of both pure metals and mixtures of various metals—alloys in solid and liquid states.

Ionic crystal lattices. Ionic crystal lattices are those whose nodes contain ions. They are formed by substances with ionic bonds, which can bind both simple ions such as Na^+ or Cl^-. Consequently, salts and some oxides and hydroxides of metals have ionic crystal lattices. For example, a sodium chloride crystal consists of alternating positive Na^+ and negative Cl^- ions, forming a cube-shaped lattice. The bonds between ions in such a crystal are very stable (Fig. 2.79). Therefore, substances with an ionic lattice are characterized by relatively high hardness and strength, they are refractory and non-volatile (they do not vaporize rapidly).

Typically, an ionic bond appears between metal and nonmetal atoms in compounds such as CsCl, NaF, $CaCl_2$, BaO, NaCl, MgF_2, RbI, and others. An ionic bond is formed between metal and nonmetal atoms in salts (NaCl, KF), metal oxides (K_2O, BaO), and metal hydroxides (NaOH, KOH).

The coordination number of cesium and chlorine ions in a CsCl crystal is exactly the same as in metallic sodium crystal. A CsCl crystal is constructed, as it were, from two different types of cubes nested inside each other. For example, you could draw a cube with a cesium ion at each vertex, and then have a chlorine ion in the center (as shown in Fig. 2.80). If you imagine eight chlorine ions and mentally connect them with lines, you will get a cube with chlorine ions at the vertices, in the center of which there will be a cesium ion.

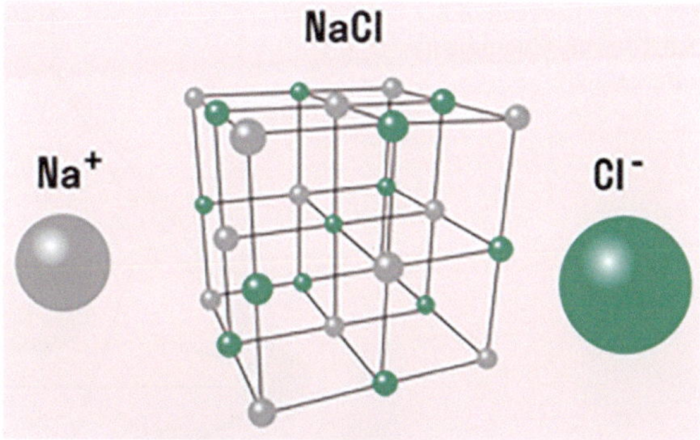

Fig. 2.79 NaCl crystal lattice

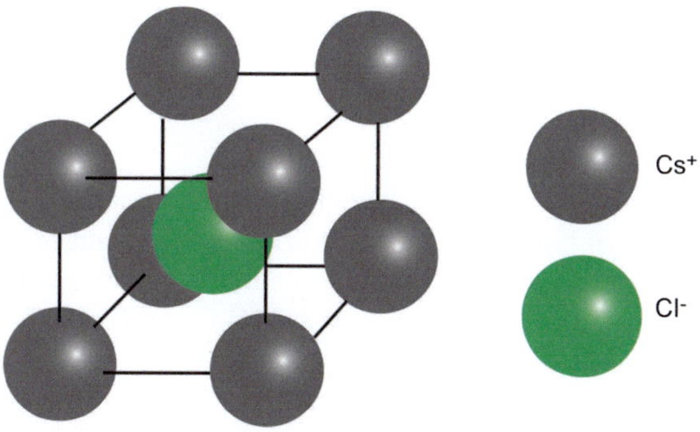

Fig. 2.80 The structure of the cesium chloride crystal CsCl. The coordination number is 8

CsCl and NaCl lattices do not have identical crystal structure because the Na^+ ion is much smaller than Cs^+ ion, which is almost as large as Cl^- ion by size. For a NaCl crystal, where the coordination number is six for both the cation and the anion, the unit cell will look different (see Fig. 2.81). Each unit cell has its own method of close packing. The NaCl unit cell corresponds to a face-centered cubic packing, and the CsCl crystal cell discussed above corresponds to a body-centered cubic packing. The names of the packages speak for themselves: the face-centered one has a particle in the center of each face of the cube and the body-centered one has a particle in the center of the cube.

Atomic Crystal Lattices and Allotropic Modifications

Atomic lattices are called crystal lattices whose nodes contain individual atoms. In such lattices, the atoms are connected to each other by very strong covalent bonds.

Allotropic modifications are simple substances formed by atoms of the same element. Allotropic modifications of one element can differ either in the composition of the molecules, i.e., contain a different number of atoms in the molecule, such as oxygen O_2 and ozone O_3, or in different crystal structures.

An example of substances with this type of crystal lattices is a diamond, one of the allotropic modifications of carbon. Most substances with an atomic crystal lattice have very high melting points (for example, for diamond it is above 3500 °C), they are strong and hard, and practically insoluble.

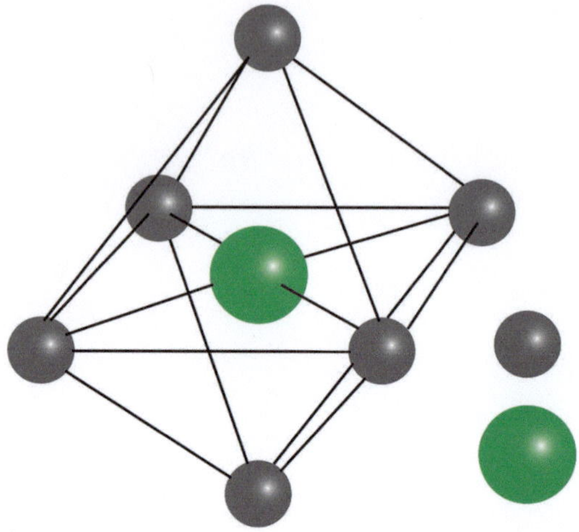

Fig. 2.81 Unit cell of a NaCl crystal. Coordination number is six

Diamond crystal (a beautiful natural crystal in the shape of an octahedron) is shown in Fig. 2.82c. The carbon atoms are arranged tetrahedrally and are in state of *sp*3 hybridization. Each carbon atom is at the center of a tetrahedron and attached to four other carbon atoms with C-C-C bond angle of 109.5° as shown in Fig. 2.82. This is very similar to methane molecule that because of *sp*3 hybridization keeps the structure of a regular tetrahedron.

In an idealized model, a solid diamond crystal can be represented as one huge molecule of this substance. In general, diamond crystals are shaped like a cube, rhombic dodecahedron, tetrahedron, and octahedron.

Due to the different structure of crystal lattices, allotropic modifications of carbon such as diamond and graphite differ significantly in properties. Diamond has an atomic crystal lattice in which each atom is connected to four other carbon atoms, which are located at the corners of the tetrahedron. In diamond, all the valence electrons of carbon atoms are in a state of sp^3 hybridization and participate in the formation of covalent bonds, so diamond does not conduct electric current. Diamond is transparent, colorless, with a high refractive index, the hardest of all known substances. Diamonds are used to make drills, grinding and glass cutting tools. Cut diamonds are used to create jewelry. The mass of diamonds is measured in carats: one carat is equal to 200 mg.

Graphite, like diamond, has an atomic crystal lattice, but its structure differs from the structure of the diamond lattice. Graphite has a layered structure. Carbon atoms in graphite are in a state of sp^2 hybridization. Inside the layer, each carbon atom, due to three hybrid electrons, forms three covalent bonds with neighboring carbon atoms. Graphite is trigonal planar with a bond angle of 120°. The fourth, non-hybrid electron of each carbon atom takes part in the formation of bonds between individual carbon layers. The bonds between individual layers are much weaker than the bonds within a layer. Non-hybrid electrons are mobile and form a single electron cloud, as in the crystal lattice of metals. It is this structure of the graphite crystal lattice that explains

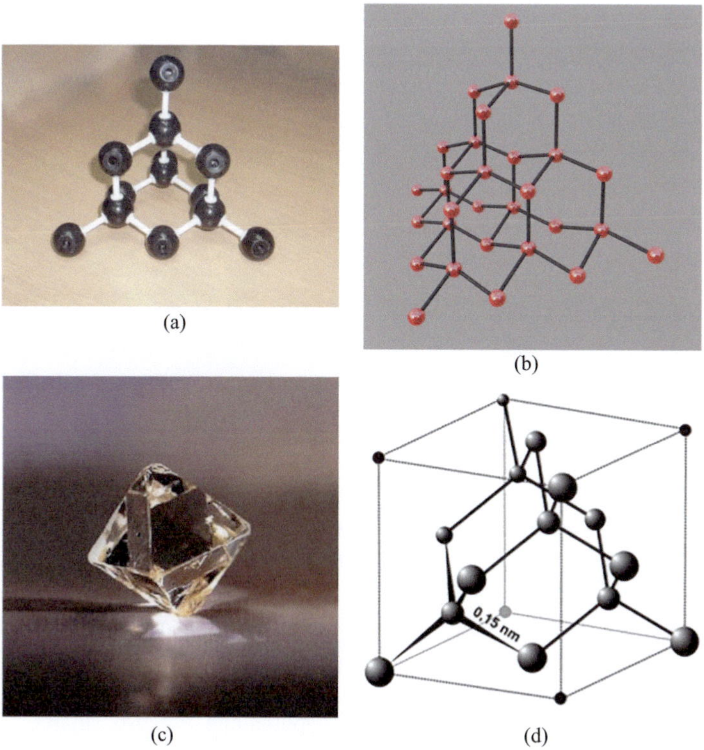

Fig. 2.82 **a** Diamond crystal tetrahedral structure, **b** crystal lattice of diamond, **c** natural diamond crystal in the shape of octahedron, **d** structure of the diamond

its characteristic properties: metallic luster, electrical and thermal conductivities, opacity, and the ability to easily separate into flakes. The properties of graphite determine its use for the manufacture of electrodes, solid lubricants, and pencil leads.

Aluminum atoms can combine into ordered stable clusters mimicking other elements of the periodic table. So, experience has shown that an ordered group of 13 aluminum atoms have an icosahedral structure (Fig. 2.83) and acts in chemical reactions as a single iodine atom, that is, it behaves like iodine, and 14 atomic aluminum cluster seems to turn into beryllium. In chemical terms, 7 aluminum atoms in a cluster give the cluster properties of Ge and it is not yet known about clusters of 23 or 27 aluminum atoms.

The chemical element boron has more than ten allotropic modifications. The boron atoms in them are united in B_{12} groups, which have the shape of an **icosahedron.** So, the icosahedron conveys the shape of the boron crystals. At one time, boron was used to create first-generation semiconductors.

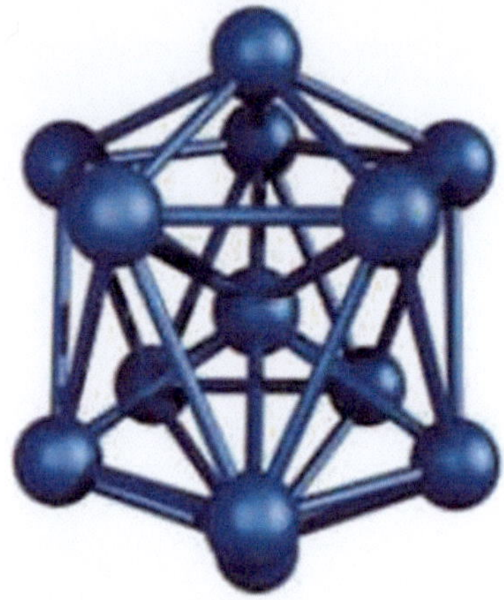

Fig. 2.83 The icosahedral structure of Al_{13} cluster

Molecular Crystal Lattices. Molecular are crystal lattices in which molecules are located at the nodes. Chemical bonds in these molecules can be both polar (HCl, H_2O) and non-polar (O_2, N_2). Despite the fact that the atoms inside the molecules are connected by very strong covalent bonds, weak intermolecular forces of attraction act between the molecules themselves. Therefore, substances with molecular crystal lattices have low hardness, low melting points, and are volatile. Most solid organic compounds have molecular crystal lattices (naphthalene, glucose, sugar).

Regular polyhedra are also found in living nature. For example, the skeleton of the single-celled organism Circogonia icosahedra is shaped like an icosahedron. Most feodaria live in the deep sea and serve as prey for coral fish. But the simplest animal tries to defend itself: 12 hollow spines emerge from the 12 vertices of the skeleton. The ends of the needles have barbs that make the needle even more effective in protection. What caused this natural symmetrization of feodaria? The point, apparently, is that of all polyhedra with the same number of faces, it is the icosahedron that has the largest volume with the smallest surface area. This property helps the marine organism overcome the pressure of the deep sea column.

Interestingly, the icosahedron has been the focus of biologists' attention in debates about the shape of some viruses. For example, the herpes virus has an icosahedral shell. The virus cannot be perfectly round, as previously thought. In order to determine its shape, they took different polyhedra and directed light at them at the same angles as the flow of atoms toward the virus. It turned out that only one polyhedron gives exactly the same shadow—the icosahedron. So, thanks to regular polyhedra, not only the amazing properties of geometric shapes are revealed, but also ways to comprehend natural harmony.

2.6 Semi-Regular and Other Polyhedra

A convex polyhedron is called **semi-regular** if its faces are regular polygons, possibly with a different number of sides, and at each vertex the same number of faces converge [13]. Semi-regular polyhedra—in general, these are various convex polyhedra, which, while not being regular, have some of their features, for example: all faces are equal, or all faces are regular polygons, or there are certain spatial symmetries. The definition can vary and include different types of polyhedra. Semi-regular polyhedra include regular n-gonal prisms, all edges of which are equal and so-called antiprisms with equal edges.

In addition to these two infinite series of polyhedra, there are 14 more semi-regular polyhedra, 13 of which were first discovered and described by Archimedes. **Archimedean solids** are convex polyhedra with three properties:

1. all faces are regular polygons of two or more types (if all faces are regular polygons of the same type, this is a regular polyhedron, or platonic solid);
2. for any pair of vertices, there is a symmetry of the polyhedron (that is, a movement that takes the polyhedron into itself), taking one vertex into another;
3. all polyhedral vertex angles are congruent.

The first construction of semi-regular polyhedra is also attributed to Archimedes, although the relevant works have been lost. The simplest Archimedean solids can be obtained from regular polyhedra using the operation of truncation, which consists of cutting the corners of a polyhedron by planes. If you cut off the corners of a tetrahedron with planes, each of which cuts off a third of its edges coming from one vertex, then we get a truncated tetrahedron. In a similar way, you can get a truncated cube if its corners are cut off by planes, each of which cuts off a third of its edges coming from one vertex.

It would be interesting to try to solve the following problem and determine whether a particular geometric figure is an Archimedean solid by checking whether its description and shape correspond to the definition of an Archimedean solid.

Problem 120 Suppose we have a cube with side a. Consider two cases: 1. Let's add to this cube the top in the form of a regular quadrangular pyramid with edge a, and also add the same pyramid symmetrically, but to the lower base of the cube. We obtained a new geometric figure. Is it an Archimedean polyhedron?
2. Now let's continue gluing the regular quadrangular pyramids and now, as it were, glue them to each of the side faces of the cube. It turns out an interesting symmetrical figure. Is it an Archimedean polyhedron? In both cases evaluate the volumes of the polyhedra.

Solution **Case 1**. The obtained solid is convex, it is formed by only two types of the polygons (squares and triangles) and it is symmetric with respect to the line passing through the apexes of the top and bottom pyramids. In fact, if we, for example, rotate the entire figure around this axis, in a multiple of 90°, we will obtain the same shape, the same figure. However, in order to be Archimedean solid, all the polyhedral angles in the figure must be congruent. This condition is violated and hence the figure is convex and symmetric but not Archimedean solid.
Case 2. If we construct such a solid, it will be a very symmetric and kind of pretty because of the symmetry. However, the obtained figure will not be convex. Therefore, it cannot be Archimedean solid.

In order to find the volume of solids in both cases, we can "disassemble" them. Then the first solid will consist of two regular quadrangular pyramids and one cube, and the second one will consist of six of the same equal pyramids and one cube. However, if you mentally glue the bases of two such pyramids together, you get a regular octahedron with side a shown in Fig. 1.19. Its volume is evaluated in (2.66). If we "disassemble" the second polyhedron, and then "glue" the bases of each two of the six pyramids in pairs, we get three regular octahedra and a cube. Denote the volume of the first polyhedron by V_1, and the volume of the second by V_2. For their volumes, we get

$$V_1 = a^3 + \frac{a^3\sqrt{2}}{3} = \frac{a^3(3+\sqrt{2})}{3}$$

and

$$V_2 = a^3 + 3 \cdot \frac{a^3\sqrt{2}}{3} = a^3(\sqrt{2}+1).$$

Answer. Both polyhedra are not Archimedean solids. $V_1 = \frac{a^3(3+\sqrt{2})}{3}$ and $V_2 = a^3(\sqrt{2}+1)$.

Figure 2.84 created in Maple 17 shows a **cuboctahedron**, a semi-regular polyhedron, whose faces consist of different types of regular polygons—six squares (like a cube) and eight triangles (like an octahedron). In Chap. 4, you will be asked to find the radius of a sphere circumscribed about a unit cuboctahedron.

Of course, in Maple we can construct different spatial figures. However, we need to understand properties of these solids. We want to figure out how to make such a figure or, for example, how to start making it. If you inspect Fig. 2.84 more closely, you may notice that it is sketched inside a cube with side 2. Think of what would happen if we start reducing the side of the cube, until the moment when all the vertices of the cuboctahedron lie on the edges of the cube and take the midpoint position on each edge? This gives an idea that any cuboctahedron can be constructed by first, creating a cube (of course, we can have a solid wooden cube), then by marking midpoints of all edges and drawing segments through the adjacent pairs of such points, and second, by cutting off from each vertex of the cube equal regular

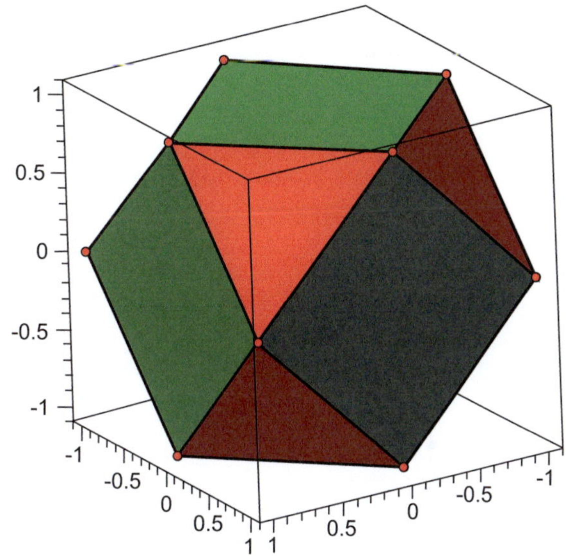

Fig. 2.84 Semi-regular polyhedron, cuboctahedron

pyramid (one of such possible pyramids can be seen in our sketch in red). This way of "construction" of the cuboctahedron will help us to find in Chap. 4 the radius of the circumscribed sphere.

Problem 121 Find the surface area and the volume of the cuboctahedron with side a.

Solution In order to find the total surface area we need to add the areas of six squares and eight areas of equilateral triangles.

$$A = 6a^2 + 8\frac{a^2\sqrt{3}}{4} = 2a^2(3 + \sqrt{3}).$$

Now we will find the volume of cuboctahedron by using the idea of its construction described above and Fig. 2.84. If the cuboctahedron has side a, then a is also the side of the base triangles of eight pyramids that we are cutting off. Let us find the side of the cube, b. Applying Pythagorean theorem to the isosceles right triangle with two legs of $\frac{b}{2}$ and hypotenuse a, we obtain that

$$a^2 = 2\left(\frac{b}{2}\right)^2.$$

Hence, the side of the cube is equal to

$$b = a\sqrt{2}.$$

Next, since each of the eight removed pyramids with the equilateral triangle base can be considered as a pyramid with the vertex at the same vertex of the cube, then the volume of such a pyramid (V_p) we can find as follows:

$$V_p = \frac{1}{3} \cdot \frac{b}{2} \cdot \frac{1}{2}\left(\frac{b}{2}\right)^2 = \frac{b^3}{48} = \frac{a^3\sqrt{2}}{24}.$$

Finally, the volume of the cuboctahedron with side a can be evaluated as

$$V = (a\sqrt{2})^3 - 8 \cdot \frac{a^3\sqrt{2}}{24}$$

$$V = \frac{11a^3\sqrt{2}}{6}. \tag{2.68}$$

Answer. $A = 2a^2(3 + \sqrt{3})$, $V = \frac{11a^3\sqrt{2}}{6}$.

Problem 122 Construct a rhombic dodecahedron, a polyhedron with 12 faces, 14 vertices, and 24 edges.

Solution Let us take two cubes with side a. Using the first cube we will make six regular square pyramids. If you recall we find the center of the cube and then connect this point with each vertex of the cube. Point of intersection of the main diagonals of the cube divide this cube into six congruent pyramids. Each such pyramid has square base and height equal to $a/2$. Next, we will glue one pyramid to each face of the second cube. Rhombic dodecahedron is constructed (see our sketch showing the steps of construction in Fig. 2.85). The whole rhombic dodecahedron created using Maple 17 is shown in Fig. 2.34. It has three rhombi that meet at eight vertices and four rhombi that meet at six vertices.

There are several interesting questions arise. For example, let us prove that the adjacent faces of two pyramids sharing an edge of the inner cube lie in the same plane. It is convenient to introduce a coordinate system so that its center is at point B and the vector BA shows the direction of the X-axis, the vector BC shows the direction of the Y-axis, and the vector BB_1—the Z-axis. Also, for simplicity and better clarity, we will place only two pyramids on the inner cube. In our sketch (Fig. 2.85), the pyramids $A_1B_1C_1D_1E$ and DD_1C_1CF have a common edge D_1S_1.

So, we need to prove that faces ED_1C_1 and D_1C_1F lie in the same plane.

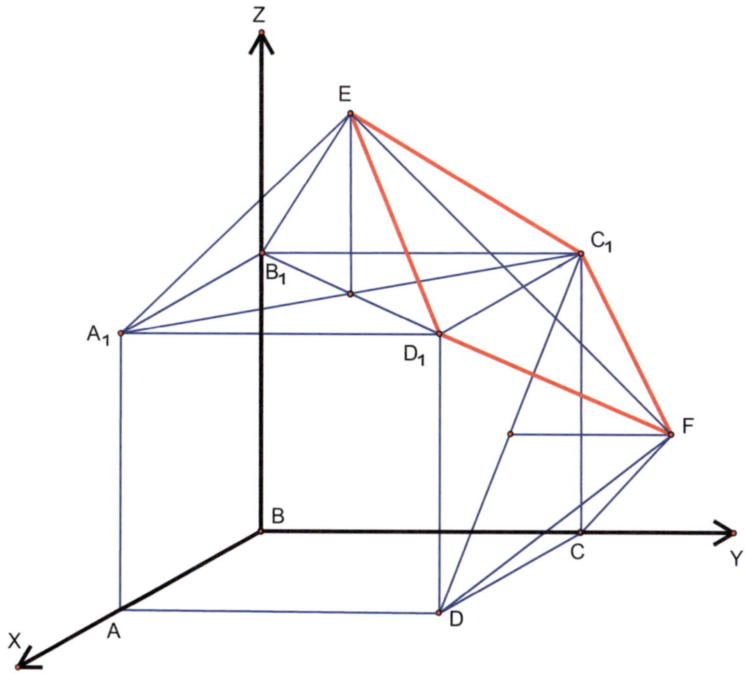

Fig. 2.85 Construction of rhombic dodecahedron

Let us find coordinates of all points.

$$E\left(\frac{1}{2},\frac{1}{2},\frac{3}{2}\right),\quad D_1(1,1,1),\quad C_1(0,1,1),\quad F\left(\frac{1}{2},\frac{3}{2},\frac{1}{2}\right).$$

Now, we can find three vectors that start at point E:

$$\begin{aligned}\overrightarrow{EC_1} &= (-\tfrac{1}{2},\tfrac{1}{2},-\tfrac{1}{2}),\\ \overrightarrow{ED_1} &= (\tfrac{1}{2},\tfrac{1}{2},-\tfrac{1}{2}),\\ \overrightarrow{EF} &= (0,1,-1)\end{aligned}\quad .$$

If four points are coplanar, then the following determinant is zero:

$$\begin{vmatrix} 0 & 1 & -1 \\ \frac{1}{2} & \frac{1}{2} & -\frac{1}{2} \\ -\frac{1}{2} & \frac{1}{2} & -\frac{1}{2} \end{vmatrix} = 0.$$

This determinant is equal to zero, because if we add the last two rows, we will obtain the first row, therefore the rows are dependent, and the determinant is zero. Also, you can simply evaluate this determinant to obtain the same answer (0). Therefore, the two triangular faces (isosceles triangles) that share the same edge lie in the same plane, then together these two triangles form a rhombus D_1EC_1F. The statement is proven.

Imagine that we glued all six pyramids to the corresponding faces of the cube. Now let's look at just these six pyramid's apexes (E, F, etc.). If you connect these points, you get a regular octahedron. Therefore, our rhombic dodecahedron can be considered as a figure having eight vertices from a cube and six from an octahedron.

Remark. Rhombic dodecahedron and cubooctahedron are duel polyhedra.

Problem 123 (*27 th Tournament Lomonosov 2004*) Is there a polyhedron whose faces are isosceles right triangles?

Solution Yes, such polyhedron exists. We will show how one of such polyhedron can be constructed.

1. Cut off a corner from the cube. We choose the cutting plane so that it intersects all three edges of the cube emerging from the cut-off vertex, at the same distance from this vertex. The pyramid-tetrahedron cut in this way will have four faces, three of them are isosceles right triangles of the former part of the faces of the cube, and one more triangle is obviously equilateral, but this is not important for us.
   For example, one of such a tetrahedron sliced of a cube is shown in Fig. 2.86, However, we could cut off from each of the three sides of the cube segments shorter that the cube's edge.
2. Let's make another exactly the same tetrahedron.
3. Glue these two equal tetrahedra along the two faces that are not right triangles; the resulting polyhedron will have only isosceles right triangles as its outer faces, as required. It has six faces ($F = 6$), nine edges ($E = 9$), and five vertices ($V = 5$).

Problem 124 Imagine a bi-pyramid, a polyhedron made up of two equal tetrahedra by combining (gluing) any of their faces. Will this be a regular polyhedron? Why?

Solution A polyhedron composed of two equal regular tetrahedra is not a regular polyhedron because it does not satisfy the definition of a regular polyhedron: although all its faces are equal equilateral triangles, three edges converge at the top and bottom vertices, and four edges converge at each of the other three vertices.

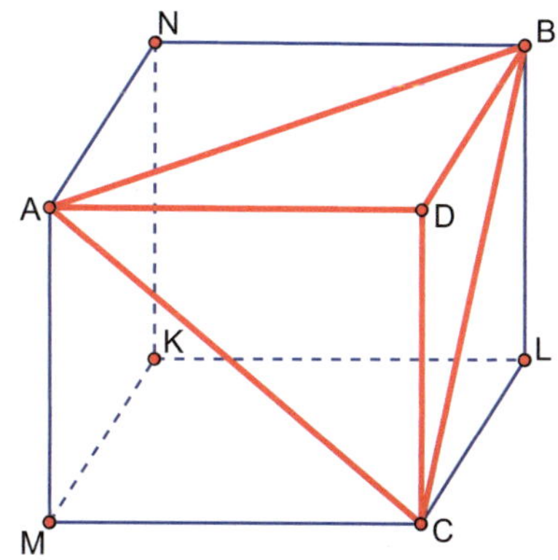

Fig. 2.86 A tetrahedron sliced off a solid wooden cube

Problem 125 Prove that in an octahedron, opposite edges are parallel.

Proof Consider Fig. 2.60 and place the cube into coordinate system centered at point $B(0, 0, 0)$, so the X-axis goes as BA, the Y-axis as BC, and Z-axis as vector BB_1. Let us prove that the opposite sides of the octahedron $LPNMTQ$ are parallel. Without loss of generality, we can only prove that $PQ \parallel LM$.

In this coordinate system, let us find the coordinates of vectors $\overrightarrow{PQ}$ and $\overrightarrow{LM}$. Points $P,\ L,\ Q$ and M have the following coordinates:

$$P = \left(\frac{1}{2}, 0, \frac{1}{2}\right), \quad L = \left(\frac{1}{2}, \frac{1}{2}, 1\right), \quad Q = \left(\frac{1}{2}, \frac{1}{2}, 0\right), \quad M = \left(\frac{1}{2}, 1, \frac{1}{2}\right).$$

Now we can find the vectors:

$$\overrightarrow{PQ} = \left(0, \frac{1}{2}, -\frac{1}{2}\right), \quad \overrightarrow{LM} = \left(0, \frac{1}{2}, -\frac{1}{2}\right) \quad \Rightarrow \overrightarrow{PQ} = \overrightarrow{LM}.$$

Therefore $PQ \parallel LM$. □

Problem 126 Is the spatial cross (a figure composed of a family of seven equal cubes) a convex polyhedron? How many squares bounds its surface and how many vertices and edges does it have?

Solution Imagine this body by mentally constructing it: Prepare seven equal cubes; start with the central cube, then glue one cube to each of its six faces. The central cube will no longer be visible, and six other cubes will stick out. The figure has 30 faces (five from each connected cube), so $F = 30$, 32 vertices (we count the vertices as four times six (24) and then add eight vertices from the hidden central cube), so $V = 32$ and 60 edges ($8 \cdot 6 = 48$ edges from six sticking out cubes that other cubes do not share and 12 edges from the center cube), so total number of sides is $E = 60$. Using Euler's formula that is a necessary, not sufficient condition, we cannot state that such a figure is a convex polyhedron

$$\chi = V - E + F = 32 - 60 + 30 = 2.$$

However, such a figure is a nonconvex polyhedron, since by connecting two points lying on adjacent faces of the cross, which originally belonged to any two adjacent cubes glued to the central cube, we obtain a segment that does not belong to the cross.

Problem 127 Compare the volumes of the regular tetrahedron and octahedron both inscribed into a cube with side c.

Solution Using formula (2.14) we can calculate the volume of the inscribed tetrahedron (V_1) into a cube of side c (see Fig. 2.50). Substituting there $a = c\sqrt{2}$:

$$V_1 = \frac{c^3}{3}.$$

Next, the edge of the octahedron in terms of the edge of a cube c is $\frac{c}{\sqrt{2}}$ (we can use Fig. 2.60). Substituting this into formula (2.66) we obtain that

$$V_2 = \frac{c^3}{6}.$$

Answer. The volume of the regular tetrahedron is twice bigger than the volume of the regular octahedron inscribed into a cube with same edge length.

Problem 128 Is there a convex polyhedron all nine of whose faces are quadrilaterals?

Solution Suppose that all nine faces are quadrilaterals, then $F = 9$, the number of edges is $E = \frac{4 \cdot 9}{2} = 18$. Using Euler's formula, we can get the number of vertices $V = 11$. Since $V = 11 < \frac{36}{3} = 12$, not all vertices of the polyhedron are associated

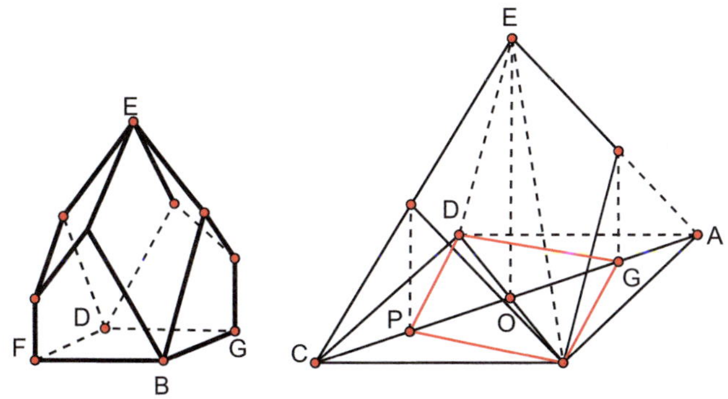

Fig. 2.87 Convex nonahedron

with trihedral angles, and some angles may be tetrahedral. For example, we may have 3 tetrahedral and 8 trihedral angles ($\frac{3\cdot4+8\cdot3}{2} = 18 = E$)

Let's construct such a polyhedron. (See Figure 2.87). Consider a quadrangular pyramid *ABCDE* whose base is the rhombus *ABCD* and whose vertex *E* is projected into the center of the rhombus, *O*.

Consider in the plane of the base a square *FBGD* with a diagonal *BD* (*BD* is the smaller diagonal of the rhombus) and a cube whose lower base is this square. Take the intersection of the cube with pyramid and the part of the pyramid lying above the cube. As a result, we obtain the desired polyhedron. This convex nonahedron has $F = 9$, eleven vertices $V = 11$, and eighteen edges $E = 18$.

Alternatively, we can take 6 congruent kites and 3 congruent squares, so that the longer side of the kite is the side of the square. Take three kites and connect three of them with their short sides, forming a trihedral rigid structure. Repeat for the other three kites. In our polyhedron, these two trihedral angles will be the top and bottom of the polyhedron, and the three squares should be placed in the middle. Place the squares by attaching them to the structure of the top and bottom kites, so that two sides of each square are connected to the longer side of the adjacent kites. Assume that each kite has acute angles $\alpha > \beta$ and two equal obtuse angles γ. Each of the three tetrahedral angles has two angles of 90° and two acute angles with measurement β. In order to exist, the following must hold: $180° + 2\beta < 360°$, so $\beta < 90°$. Note that the six trihedral angles are formed by one 90° coming from the square and two obtuse angles γ, so $\gamma < 135°$. Also, the other two trihedral angles are formed by three equal angles α. Also, for each kite, $\alpha + \beta + 2\gamma = 360°$.

Answer. Yes, such a convex nonahedron (polyhedron with nine quadrangular faces) exists and it is not unique.

Fig. 2.88 Fullerene C_{60}

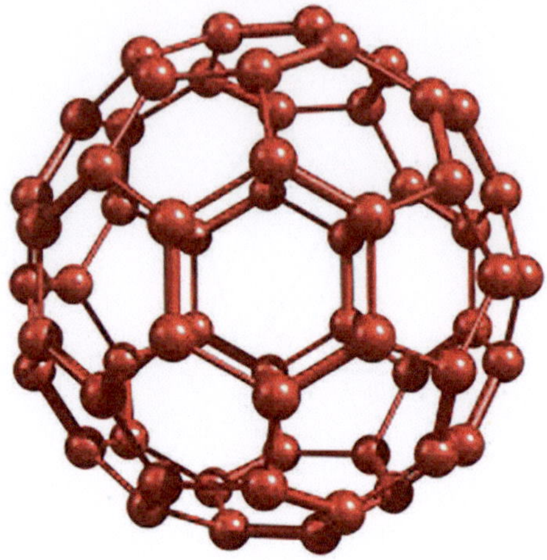

2.6.1 Semi-Regular Polyhedra in Nature and Created by Man

It is interesting that semi-regular polyhedra appear in nature. For example, crystals of garnet have a shape of rhombic dodecahedrons.

The truncated icosahedron is one of the most common semi-regular polyhedra. A classic soccer ball has a shape close to a truncated icosahedron (as you recall its pentagons and hexagons, usually are colored black and white, respectively).

In chemistry, fullerene is a molecular compound that is a convex closed polyhedron consisting of carbon atoms, such that each carbon is bonded to three other carbon atoms. In 1991, fullerene was named Molecule of the Year. Fullerenes owe their name to the engineer and architect Richard Fuller, whose geodetic structures were built according to this principle. The very first and simplest fullerene is the C_{60} molecule. Thus, C_{60} molecule is a truncated icosahedron, in which 60 carbon atoms correspond to 60 vertices of a truncated icosahedron. This polyhedron has 12 regular pentagons and 20 regular hexagons (see Fig. 2.88). It has an icosahedral type of symmetry. At each vertex, two hexagons and one pentagon meet. Each of the pentagons is surrounded on all sides by hexagons.

The existence of such a convex polyhedron constructed from V vertices forming only pentagonal and hexagonal faces can be validated according to Euler's formula for polyhedra, which states the equality

$$V - E + F = 2.$$

How many edges are there in a C_{60} fullerene molecule? According to the number of vertices, edges, and faces, the necessary condition for C_{60} is the presence of exactly 12 pentagonal faces and 20 hexagonal faces.

Let us show that if fullerene C_{2n} exists, then it contains $V = 2n$ vertices (atoms of carbon), $E = 3n$ edges (bonds) and $F = n + 2$ pentagonal and hexagonal faces, of which $n - 10$ hexagonal faces and exactly 12 pentagonal faces. The number of hexagonal faces determines the size of the fullerene molecule.

Problem 129 Show that for any fullerene C_{2n} there are exactly 12 regular pentagonal faces and $(n - 10)$ regular hexagonal faces.

Solution Let x be the number of regular pentagons and y the number of regular hexagons in a fullerene. Denote by F the number of the faces, by V the number of the vertices, and by E the number of the edges in a general fullerene, C_{2n}, a convex polyhedron with $V = 2n$ vertices. Then the total number of faces in the molecule is

$$F = x + y. \tag{2.69}$$

Since each pentagon has five vertices and sides and each hexagon has six vertices and sides, and because each side is shared by two polygons, the following is true for the number of edges of the polyhedron:

$$E = \frac{5x + 6y}{2}. \tag{2.70}$$

Next, because in a fullerene we have $V = 2n$ vertices, and each carbon atom is connected to three other carbon atoms, then each vertex is associated with a trihedral angle, then the total number of vertices is equal to

$$V = \frac{5x + 6y}{3} = 2n,$$

from which we obtain that

$$5x + 6y = 6n. \tag{2.71}$$

Substituting (2.71) into (2.70) we obtain that the total number of the edges can be rewritten as

$$E = 3n.$$

Substituting in Euler's formula the expressions for E and V, we have the following

$$2n - 3n + F = 2$$

that using (2.69) can be simplified as follows:

$$F = x + y = n + 2. \tag{2.72}$$

To find x and y, let us solve the system of two linear equations (2.72) and (2.71) in two variables:

$$\begin{cases} x + y = n + 2 \\ 5x + 6y = 6n \end{cases}.$$

If we multiply the first equation of the system by 6

$$\begin{cases} 6x + 6y = 6n + 12 \\ 5x + 6y = 6n \end{cases}$$

and then subtract the second equation from it, we get that

$$x = 12, \quad y = n - 10.$$

Therefore in any fullerene the number of regular pentagons must be equal to $x = 12$ and only 12, and then the number of regular hexagons is $y = n - 10$. Hence, the original fullerene with formula C_{60} has 12 pentagonal faces and 20 hexagonal faces (since $n = 30$, then $y = 30 - 10 = 20$.).

Answer. $x = 12$ and $y = n - 10$.

Fullerenes are the most powerful antioxidants known today. On average, they exceed the effect of all previously known antioxidants by $100 - -1000$ times. It is assumed that fullerene C_{60}, dissolved in olive oil, can be incorporated into the bilayer lipid membranes of cells and mitochondria and act as a reusable antioxidant. Fullerenes can also be used in pharmacology to create new antiallergic and anticancer drugs and can be used as radioprotective agents [27, 28, 43].

Various fullerene derivatives have shown themselves to be effective agents in the treatment of human immunodeficiency virus: the protein responsible for the penetration of the virus into blood cells—HIV-1 protease—has a spherical cavity with a diameter of 10 Å, the shape of which remains constant with all mutations. This size almost coincides with the diameter of a fullerene molecule [12, 34]. A fullerene derivative has been synthesized that is soluble in water. It blocks the active center of HIV protease, without which the formation of a new viral particle is impossible.

2.7 Unfolding of Polyhedra

I remember how in geometry class the teacher brought models of various polyhedra, and then asked how best (along which what edges) to cut such a polyhedron in order to get its plane net, the unfolding. What is interesting is that if we take an ordinary cube, then it can be cut in several ways, and different ways arose in the minds of very young schoolchildren. However, it is also interesting that, it was much easier, having a model of a polyhedron, to figure out how best to cut it to get its flat configuration

Fig. 2.89 Unfolding 1 of a cube

than, for example, not knowing a priori which polyhedron was taken, to understand from its unfolding what kind of polyhedron it is.

Let us look at the unfoldings of a well-known convex polyhedron such as a cube. Here we will omit drawing any tabs.

(a) Some students cut it into six separate and congruent squares. In order to fold it up back as a cube, we need to connect or glue the corresponding sides.

(b) Some students showed a different unfolding, for example, one that looks like a cross (Fig. 2.89). This unfolding is technically more preferable because one can obtain most of the shape by bending the plane figure and using less of glue. However, these unfoldings of a cube are not unique. For example, in our unfolding (net), we do not even need to have all plane polygons as squares. We can remove the very last square in Fig. 2.89 and instead of it put a right triangle on the top of each other outer square as it is shown in Fig. 2.90. In this sketch, point A is not a vertex of a cube. When we fold up our construction, we will get this point as the center of the top face of the cube, for example, as it is presented in Fig. 2.90.

We can easily prove that point A in Fig. 2.90 is not a vertex of any polyhedron because the sum of the plane angles at any vertex of a convex polyhedral angle must be less than $360°$ or 2π. Since the angle BCA is $90°$, then the sum of all angles at vertex A would be $360°$, which is not possible. As we all know for any cube, the sum of the plane angles at any vertex is exactly $270°$ (because we add up three right angles). There are more (11 total) possible nets for a cube.

A **good unfolding** or development of a polyhedron is the union of plane polygons to a rule (manual) stating how one should glue these polygons together along their sides and vertices to obtain the desired polyhedron.

The following conditions must be satisfied:

- each side of the polygon is attached to at most one side of another polygon;
- two glued polygons's sides must have the same number;
- one can reach any polygon by starting from any other polygon and moving along a sequence of glued polygons.

Fig. 2.90 Unfolding 2 of a cube

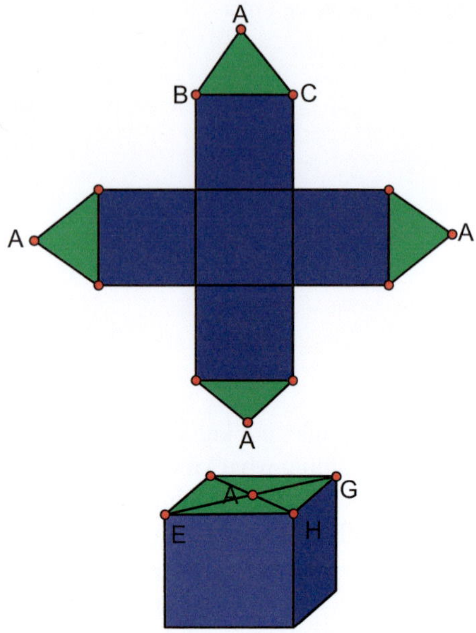

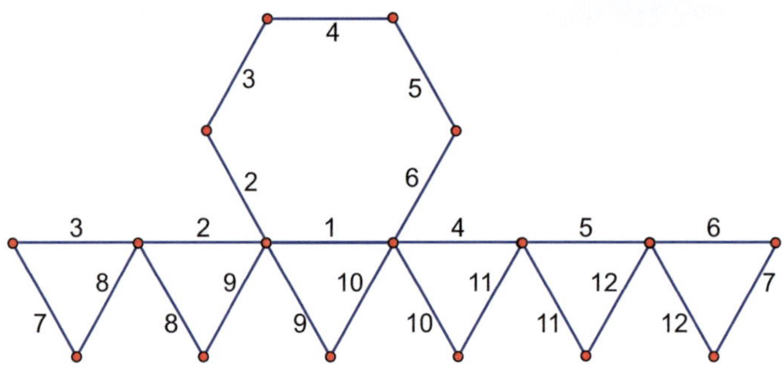

Fig. 2.91 Development of hexagonal pyramid

We can make a unique hexagonal pyramid using the unfolding shown in Fig. 2.91 because the sides that have to be glued are marked with the same numbers. In fact, we can start from connecting any matching sides and still will get the same pyramid.

On the other hand, if we are given just an unfolding of a polyhedra without directions on how to connect the polygons, then such a task can be very hard and even unsolvable sometimes. However, if the development is correct and if the sides of the polygons in it are numbered in the order they must be glued, then we can always get the corresponding polyhedron.

Let us think of how we construct a polyhedron using its net. If any vertex is flattened, the sum of the angles of the polygons jointed at that point must be less than 360°. Otherwise, it cannot be folded to make a polyhedron. Some polyhedra are not difficult to create by making a net in the plane. Let us sketch a net for a regular octahedron. We can start from one face, an equilateral triangle, and then arrange all eight of them in the pattern shown in Fig. 2.92. In this figure, four equilateral triangle faces form an angle of 240°.

Next, we can start from sketching a regular pentagon and using rotation, translation, and reflection, create a net (with tabs) for a dodecahedron. In Fig. 2.93, three pentagonal faces form at each vertex an angle of $3 \cdot 108° = 324°$.

We can make an unfolding for an icosahedron as well. For example, in Fig. 2.94, we demonstrate such a net that after putting the polygon pieces together, we get an icosahedron. At each vertex, five equilateral triangles of different colors (green, brown, yellow, orange, and blue) meet, the sum of the angles of the triangles jointed at each vertex is $5 \cdot 60° = 300° < 360°$.

Let us put three of the Platonic solid's nets in one sketch (see Fig. 2.95). We can see that some of the diagrams are a little bit different from the unfoldings shown in Figs. 2.92, 2.93 and 2.94. This demonstrates that such operation is not unique and that the same polyhedron can have different nets. However, each correct net will correspond to only one unique convex polyhedron.

Given the unfolding of a polyhedron we can find the total surface area of the polyhedron. For example, the net for the icosahedron photographed by me in one of

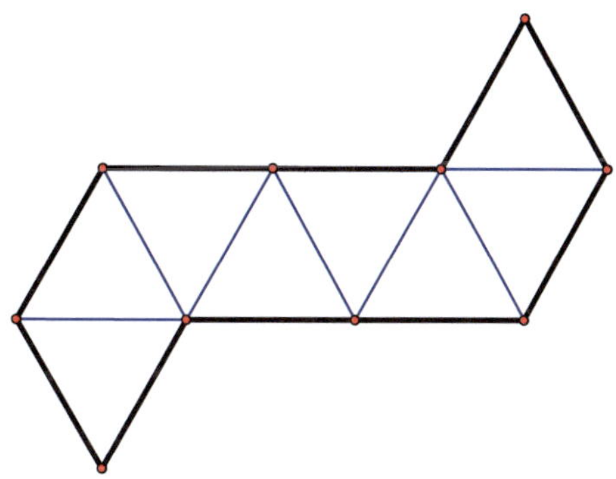

Fig. 2.92 Unfolding of an octahedron

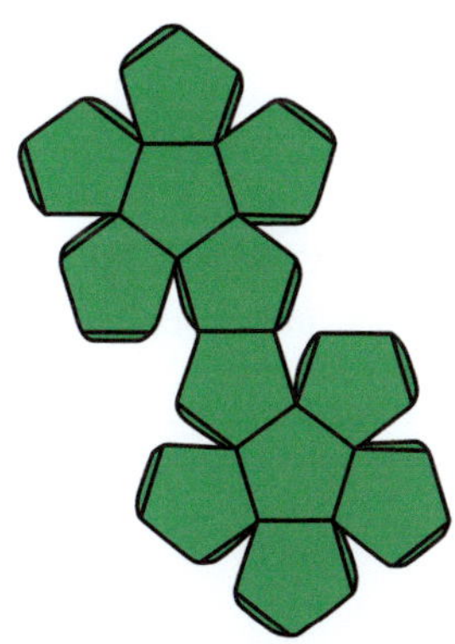

Fig. 2.93 Unfolding of a dodecahedron

Fig. 2.94 Unfolding of an icosahedron

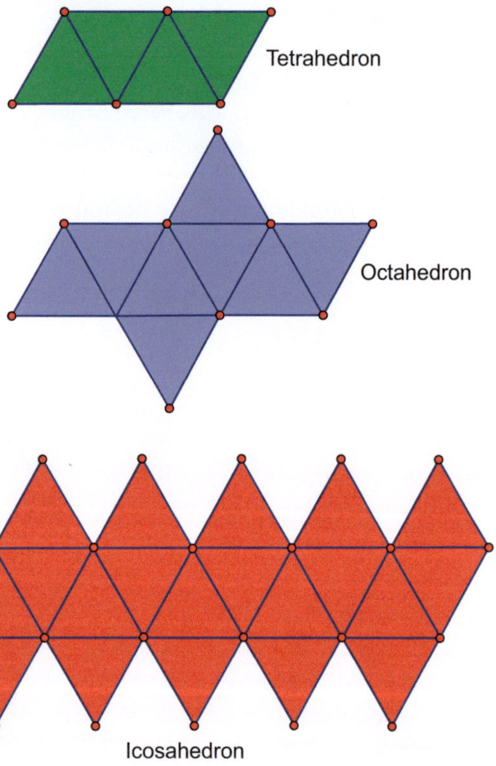

Fig. 2.95 Unfolding of tetrahedron, octahedron, and icosahedron

the hotel's lobby (Fig. 2.59) consists of 20 equilateral triangles with side a, then the surface area of icosahedron equals 20 times $\frac{a^2\sqrt{3}}{4}$ (the area of one such a triangle)

$$A_{\text{icosahedron}} = 5a^2\sqrt{3}. \tag{2.73}$$

This result matches with Table 2.4.

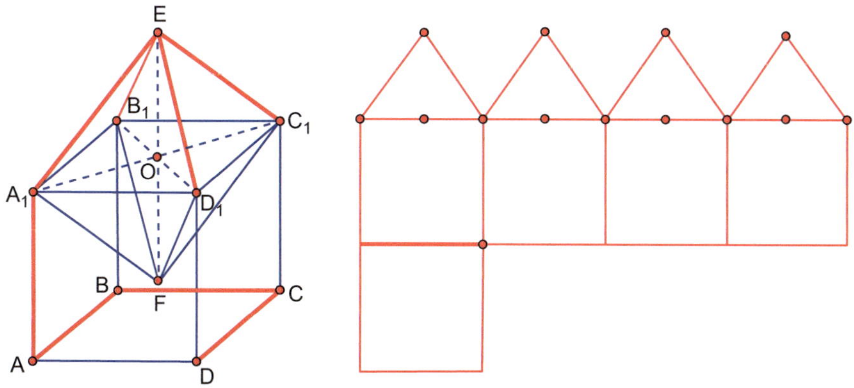

Fig. 2.96 The same net for two different polyhedra

Remark. When we know a polyhedron, then to visualize its different unfolding is not that difficult. In fact, one can simply take scissors and cut the figure by its edges and get the unfolding. Obviously doing such exercise mentally adds to our spatial imagination. It is more difficult to reconstruct the actual polyhedron by looking at its unfolding, especially when no instruction is given.

Additionally, two different polyhedra may have the same nets.

Example 1. Let's imagine that we have the unfolding shown on the right in Fig. 2.96. If we start folding it somehow, then in any case we will get a cube without the top face. Next, the structure of the other parts of the cube can be different: we either have to lift the pieces of the triangular faces up, forming a convex polyhedron, or we can lower the pieces of the triangles down and then we get a completely different figure—a nonconvex polyhedron. Well, interestingly, both of these figures have the same net!

Let us take a cube $ABCDA_1B_1C_1D_1$ and denote the center of the top face $\alpha = A_1B_1C_1D_1$ by O. Next, from point O we can erect the perpendicular to the face α and mark points E and F on this perpendicular such that $OE = OF$. Connecting both points E and F with four other vertices of the top square $A_1B_1C_1D_1$ we make two different polyhedra shown on the left in Fig. 2.96, one ($ABCDA_1B_1C_1D_1E$) is convex and the other one ($ABCDA_1B_1C_1D_1F$) is not convex. However, these two different polyhedra have the same net.

Example 2. Each of us understands that unfolding consisted of four equilateral triangles will produce a regular tetrahedron. Do we recognize right away that using the unfolding presented on the left in Fig. 2.97 we also can fold up a tetrahedron? Think about it if no dashed lines between points B, C, and D are shown.

An unfolding (net) is a plane pattern than can be used to create a model of a three-dimensional figure. Thus, a net of a cube does not have to be made out of only squares.

There are usually three typical problems about a given unfolding:

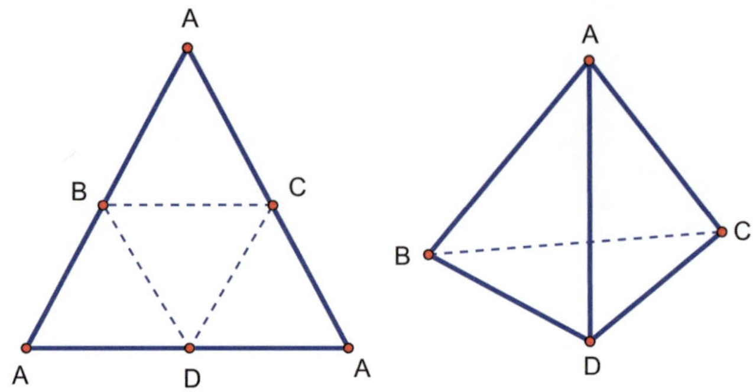

Fig. 2.97 Unfolding of a tetrahedron

1. Which of the given nets could be folded up to make a specific polyhedron (cube, tetrahedron, octahedron, dodecahedron, etc.)?
2. Sketch another net that can form a cube (tetrahedron, octahedron, dodecahedron, etc.).
3. Given a net of an unknown three-dimensional figure. Recognize this figure.

In general, the last problem is more difficult.

Most of you have constructed models of polyhedra from paper cut-outs of joined polygons by gluing solid figures by the tabs. How do we construct such a net? A net for a polyhedron is an unfolding produced by cutting the solid about some of its edges and flattering it into a polygon in the plane. Nets enable to see each face of the polyhedron. For example, in Figs. 2.94 and 2.93, you can see the nets for the icosahedron and dodecahedron, respectively. Obviously, knowing what we are cutting, we can easily do a reverse operation and go back from the plane net of the figure to the figure itself in 3D. However, there is not a one -to-one correspondence between a polyhedron and its net or unfolding.

Theorem 39 (Cauchy) *Two convex polyhedra with pairwise congruent faces arranged in the same order are equal. Two convex polyhedra with the same corresponding faces placed (connected to each other) in the same order are congruent.*

For nonconvex polyhedra, Cauchy's theorem ceases to be true. Since the middle of the last century, there has been the question whether there exists a nonrigid polyhedron composed of rigid faces? Only in 1977 American mathematician R. Connelly built an example of such polyhedron.

Now if we look again at Fig. 2.96, we can see that two different solids obtained from the same net were obtained from the same polygons connected to each other in the same order. But they are unequal to each other because only one of two is convex. There is no other convex polyhedron besides this one with the same net.

The following two theorems will be also given without proof. The first was stated and proved by Russian academician Pogorelov (1949).

Theorem 40 (Pogorelov) *Two convex polyhedra with the same unfolding are congruent.*

The second was formulated and proved by Russian academician Alexandrov. It summarized the work of other mathematicians, for example, the work of Pogorelov.

Theorem 41 (Alexandrov's Theorem, [17]) *Given unfolding of any convex polyhedron that fulfills*

1. *Euler's characteristic,* $\chi = 2$ *and*
2. *the sum of all angles meeting at one point is less than* 2π*, then there exists a unique convex polyhedron that associates with this unfolding.*

Professor Alexandrov not only proved the congruence of two convex polyhedra with the same development, but also listed two conditions that must be fulfilled. This theorem is stronger than the Cauchy theorem. Indeed, if we are given all the faces of a polyhedron, as well as the rule for gluing them along the sides, then, of course, the unfolding is also given. Moreover, according to such a special type of unfolding, by virtue of Cauchy's theorem, a convex polyhedron can be uniquely restored. At the same time, according to the unfolding, which is described in Alexandrov's theorem, nothing can be said about the faces and edges of the future polyhedron, but, nevertheless, a polyhedron is obtained from it unambiguously.

Assume that we have an unfolding without instruction. How to make a polyhedron, what vertices to connect? If we know that we do have an unfolding of a convex polyhedron satisfying conditions 1. and 2. of Theorem 41, then this theorem states that we can make only one unique convex polyhedron. However, the theorem does not tell what exactly would be the shape of the polyhedron. I believe that the problem of finding the shape of a polyhedron given only its unfolding remains an unsolved open problem for an unfolding that have more than four vertices with strictly positive curvature.

The problem of determining a polyhedron by an unfolding, if it has more than four real vertices (whose sum of converging angles is less than 2π), is very difficult. By Alexandrov's unfolding theorem we know that a convex polyhedron exists. According to the Cauchy–Alexandrov uniqueness theorem, it is unique. The question arises: what is it? The vertices of the polyhedron are easy to determine on the unfolding. Each edge of the polyhedron corresponds to the shortest one that connects some vertices of the unfolding. But not all shortest links connecting vertices are edges. Determining from a given unfolding which of the shortest links are the edges, and

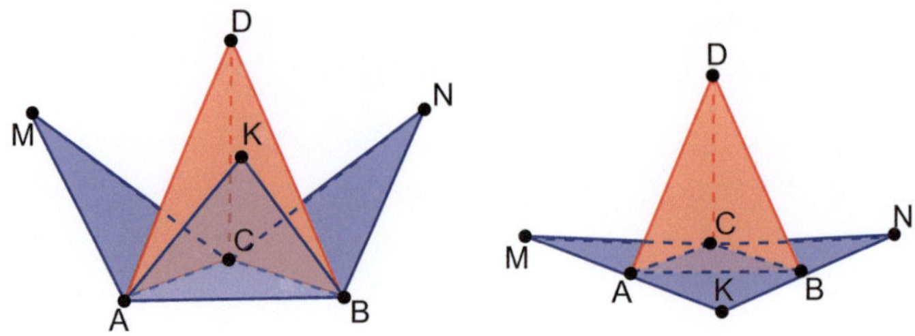

Fig. 2.98 Unfolding of triangular pyramid *ABCD*

consequently to determine all the faces of the polyhedron, is a very difficult, not yet solved problem (see [15]).

Unfortunately, in this book, we cannot provide proofs of these famous theorems, since this would be beyond the scope of the book. However, we can show you how unfolding spatial figures can help us solve some challenging stereometry problems.

2.7.1 Solving Problems Using Unfolding of Polyhedra

Any unfolding of a three-dimensional polyhedron is a sketch on a plane. Sometimes we are given an unfolding, and we need to find or measure something in a real polyhedron. Mental unfolding and folding of a geometric figure develops our spatial imagination and often helps to find an easier way to tackle a problem. Often in solid geometry we want to unfold the known polyhedron, so we can solve the plane geometry problem instead of solid geometry problem. For example, if we know the unfolding of a polyhedron, we can find its surface area by adding the areas of all polygons, the faces. Some complex stereometry problems can be solved using a technique of unfolding. By replacing a three-dimensional sketch by a two-dimensional figure, we can in many situations, find some important or even hidden features of the solid and solve a problem in an easier way.

Problem 130 Any triangular pyramid can be unfolded. (a) Suppose we have unfolded a triangular pyramid as shown in Fig. 2.98, and all four faces, unfolded on the plane, are combined into the shape of one large triangle. For which pyramid will the unfolding consist of only one triangle? (b) Can any triangle be folded into an isohedral triangular pyramid (a triangular pyramid with all equal faces)? Explain.

Solution (a) If an unfolding of a tetrahedron *ABCD* with vertex *D* onto plane *ABC* looks like one triangle, it means that the sum of the angles at each vertex on the

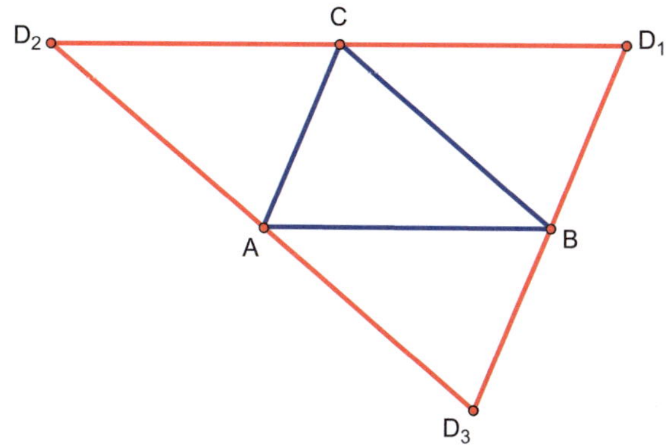

Fig. 2.99 The sum of angles at vertices $A,\ B,\ C$ is 180°

plane unfolding is 180° (Fig. 2.99). If we take any acute triangle and fold it up by the midline segments, we can obtain a pyramid with equal faces-**isohedral** tetrahedron. A particular case for such pyramid is a regular tetrahedron with all faces as equilateral triangles shown in Fig. 2.97.

(b) We can guarantee construction of an isohedral tetrahedron from any acute triangle. Each face of such tetrahedron will be a triangle similar to the net with the coefficient of similitude of $k = \frac{1}{2}$. However, if the original triangle is right or obtuse, then the violation of Theorem 24 may occur, because one of the conditions that must be valid is that each plane angle of the trihedral angle at each vertex must be less than the sum of two other plane angles. This theorem is always valid for any triangle, all angles of which are acute.

Let the net-triangle have angles α, β, and γ. If we can make a pyramid out of this triangle, then each face of this pyramid will be a similar triangle with the same angles and then at the vertex of such a pyramid the trihedral angle will be formed by the plane angles α, β, and γ. The following must be valid for each angle:

$$\begin{cases} \alpha = 180^\circ - (\beta + \gamma), \\ \alpha < \beta + \gamma, \end{cases} \tag{2.74}$$

which can be rewritten as

$$180^\circ - (\beta + \gamma) < \beta + \gamma \Longrightarrow \beta + \gamma > 90^\circ.$$

Hence, if a triangle with angles α, β, and γ can be used as a net to make a triangular pyramid by connecting the midline segments and folding them up, then the sum of each pair of its angles must be greater than 90° and the corresponding third angle must be less than 90°. For example, if $\alpha + \beta > 90^\circ$, then $\gamma < 90^\circ$. Therefore neither right triangle nor obtuse triangle can be fold up by midline segments to make a triangular pyramid with equal faces.

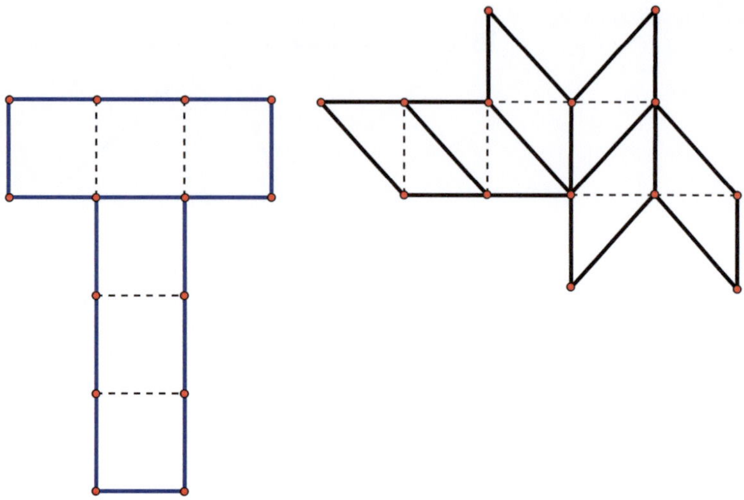

Fig. 2.100 Two unfoldings of a cube

Problem 131 Six identical parallelograms of area 1 pasted over a cube with an edge of length one. Is it possible to say that all parallelograms are squares? Is it possible to say that they are all rectangles? Provide an explanation.

Solution Examples of unfoldings of such cubes are shown in Fig. 2.100: bold lines show the boundaries of parallelograms and dotted lines indicate the edges of the cube. Therefore, not all parallelograms must be squares!

The dotted lines show the lines along which the net should be bent. On both unfoldings, small squares are visible, which will become the faces of the cube.

Problem 132 Let h be the height of an isohedral tetrahedron, h_1 and h_2 be the segments into which one of the heights of a face is divided by the point of intersection of the heights of this face. Prove that $h = 2\sqrt{h_1 h_2}$.

Proof Let $ABCD$ be a given tetrahedron, $DH = h$ its height, DA_1, DB_1, DC_1 the heights of the faces dropped from vertex D to sides BC, CA, and AB. Let's construct the unfolding of this tetrahedron. To do this, take a triangle of the base ($\triangle ABC$) and through each vertex draw lines parallel to the opposite sides. At the intersection, we get a triangle that is the unfolding of this tetrahedron. Here AB, BC, CB are the midline segments of $\triangle D_1D_2D_3$ (please see Fig. 2.101).

The same unfolding can be obtained if we mentally cut the surface of the tetrahedron along the edges $DA,\ DC,\ DB$ and unfold the lateral faces onto the base plane ABC. Obviously, H is the point of intersection of the altitudes of triangle $\triangle D_1D_2D_3$. Let M be the point of intersection of the altitudes of triangle ABC, where $AN,\ BQ$, and PC are the altitudes of this triangle.
$AN \cap CP = M$. $AM = h_1,\ MN = h_2$. Because two triangles are similar

$$\triangle D_1D_2D_3 \sim \triangle ABC$$

with the coefficient of similitude of $k = 2$, and $D_1G \parallel AN$, then

$$D_1H = 2AM = 2h_1.$$

Moreover, since all faces of the tetrahedron are equal, then $\triangle ABC \cong \triangle CDB$ and so their corresponding height are also equal that can be written as

$$DA_1 = D_1A_1 = AN = h_1 + h_2. \tag{2.75}$$

Consider our tetrahedron on the left of Fig. 2.101. From the right triangle DHA_1, using Pythagorean theorem, we obtain

$$DH^2 = DA_1^2 - HA_1^2. \tag{2.76}$$

Using the unfolding again, we can find segment HA_1 as follows $HA_1 = DH - DA_1$, or

$$HA_1 = 2h_1 - (h_1 + h_2) = h_1 - h_2. \tag{2.77}$$

Substituting (2.75) and (2.77) into (2.76), we obtain

$$h^2 = (h_1 + h_2)^2 - (h_1 - h_2)^2 = 4h_1h_2,$$

which can be rewritten as

$$h = 2\sqrt{h_1h_2}.$$

□

Problem 133 An unfolding of a polyhedron looks like one rectangle. Show that any rectangle can be fold up to a tetrahedron. Describe your construction steps and find the dimensions and the volume of the obtained tetrahedron. Assume that the rectangle has dimensions a by b.

Solution I want to explain the general idea of such construction. A rectangle is glued into a cylinder, the horizontal edges of which are then glued in two mutually

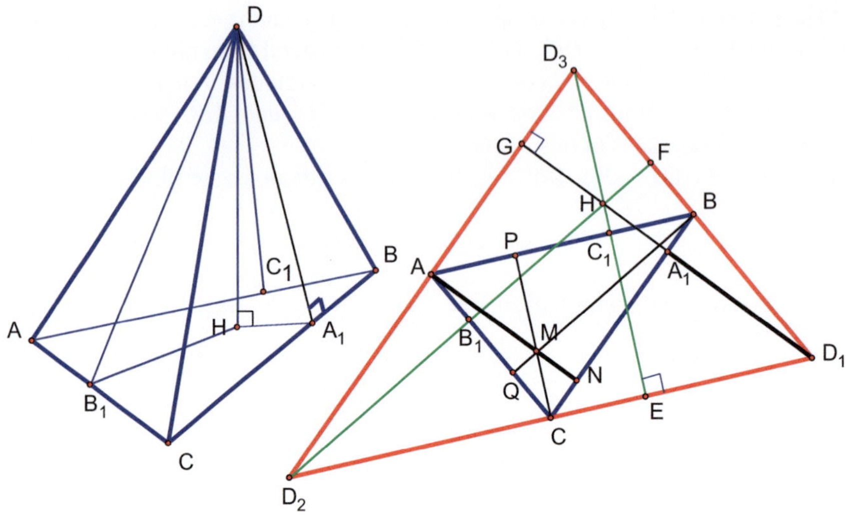

Fig. 2.101 Unfolding of an isohedral tetrahedron

perpendicular planes. Assume that we have a rectangle shown on the left of Fig. 2.102. Two sides that we are going to connect (you do not have to use a glue, you can simply staple two sides to each other) shown in red color and have the same letters. We understand that this way we will obtain a cylindrical surface like shown on the right of the same sketch. Next, connect points by diameter of the green circumference, so two points A will coincide but points C and D become the ends of the obtained segment. Next, staple or glue diameter of the bottom circle, in perpendicular to CD direction, so points E and B will become the ends of the obtained segment.

Fold your figure to make a tetrahedron. You will make something like one shown in Fig. 2.103 on the left. Having fixed two perpendicular sides of the future tetrahedron with a stapler, it is easy to see the other two sides of this tetrahedron being designated. If you saved the vertex markings, which I left as a hint on the initial rectangle (for convenience), then you will see that these edges will just pass between vertices E, C and E, D.

If we unfold our tetrahedron by removing staples and flattering the figure back to its original rectangle (this is why staples are better choice when you try to understand the solution of this problem), you will notice that the actual edges of the tetrahedron will have the trace on the rectangular unfolding. Connect them (draw these sides) using, for example, blue color ink. You can notice how ED, EC, EB, and CD will become the edges of the constructed tetrahedron, $EBCD$, point A will stay as midpoint of side CD (see Fig. 2.103).

Let us calculate the volume and the edges of this tetrahedron. We have the following: $AB = a$ and $AA = b$, A is the midpoint of the bigger side, $AC = CA = AD = DA = \frac{1}{4} \cdot b$. Denote the side of the tetrahedron by c. Then using Pythagorean theorem, from triangle BAC we obtain that

$$c = BC = CE = ED = DC = \sqrt{a^2 + \frac{1}{16} \cdot b^2}.$$

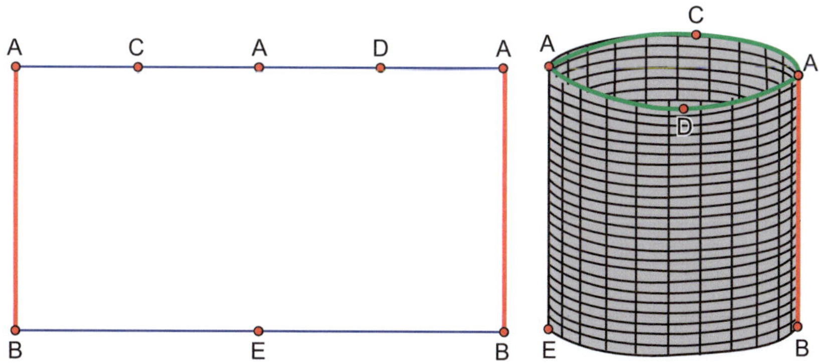

Fig. 2.102 From a rectangle to a cylinder

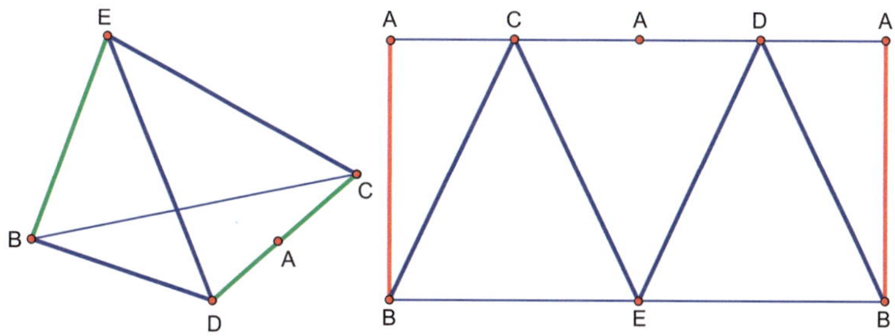

Fig. 2.103 Tetrahedron *EBCD*

The volume of the tetrahedron *EBCD* can be found using formula (2.14):

$$V = \frac{c^3\sqrt{2}}{12}, \quad c = \frac{\sqrt{16a^2 + b^2}}{4}.$$

The unfolding of such a tetrahedron is a rectangle, the sides of which are divided into smaller segments—the edges of the development and are identified in pairs in Fig. 2.102. This unfolding satisfies both conditions of Alexandrov's theorem. You don't even need to check this, because this is the unfolding of a convex polyhedron.

Remark. I remember when I was a child, milk was sold in Tetra-Pak packages. Therefore, milk was poured into packages that did not have the shape of a rectangular box, as we are used to have now, but were poured into something shaped like a regular tetrahedron. Perhaps such tetrahedral containers were easy to make from a rectangular sample.

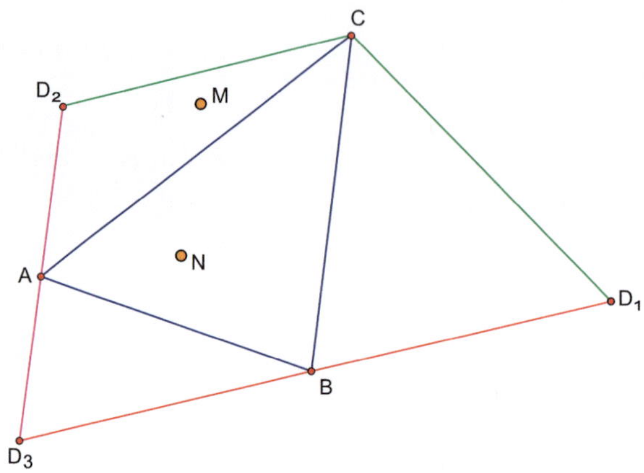

Fig. 2.104 Unfolding of a triangular pyramid with two points on it

Problem 134 Using the unfolding of a triangular pyramid $ABCD$ shown in Fig. 2.104, find the distance between points M and N. Explain what needs to be measured (given) in order to complete the task.

Solution Let us think of a real pyramid that could have such an unfolding and place points M and N on the corresponding faces. It has only one vertex D that look like three different points on the unfolding. We need to find the distance MN. Connect these two points and drop perpendiculars MM_1 and NN_1 to side AC as it is demonstrated in Fig. 2.105.

From the left sketch, we can see that applying Pythagorean theorem to the right triangle (ΔMM_1N) we can evaluate the distance between points M and N:

$$MN^2 = MM_1^2 + M_1N^2.$$

Next, using the right triangle M_1N_1N we can evaluate M_1N^2.

$$M_1N^2 = M_1N_1^2 + NN_1^2.$$

Finally, the distance MN is equal to

$$MN = \sqrt{MM_1^2 + M_1N_1^2 + NN_1^2}.$$

Therefore, we need to know the measurements for three segments MM_1, M_1N_1, and NN_1 shown in blue on the unfolding on the right of Fig. 2.105. Moreover, we can find the distance between two points using only the unfolding of the pyramid.

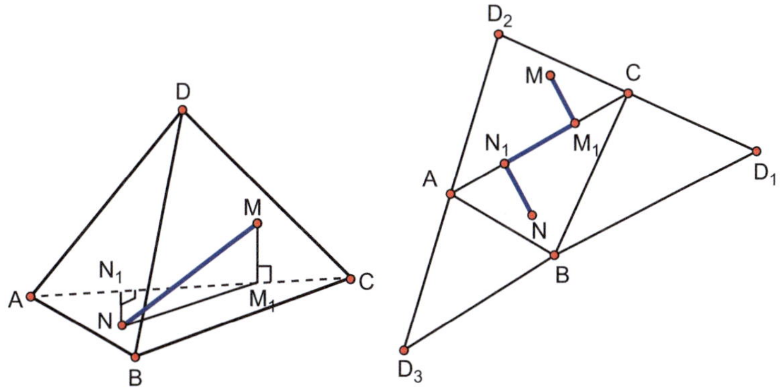

Fig. 2.105 Finding the distance between two given points using the unfolding of a triangular pyramid

Problem 135 Prove that the sum of the plane angles of a trihedral angle is less than 2π.

Proof Let us use in our proof an unfolding of a well-known convex polyhedron—a pyramid. Sketch a trihedral angle with vertex O. Take a segment of unit length on each edge of the trihedral angle, such as $OA = OB = OC = 1$. Let the plane angles of the trihedral be $\angle AOB = \alpha,\ \angle BOC = \beta,\ \angle AOC = \gamma$ (see Fig. 2.106).

Drop perpendicular from the vertex O to the plane ABC and obtain point H. Because the lateral edges of pyramid $ABCO$ are equal, then the foot of the perpendicular H will be the circumcenter of triangle ABC. It follows from the congruence of triangles, $\Delta AHO \backsimeq \Delta BHO \backsimeq \Delta CHO$ and $R = AH = BH = CH$. Hence, H is the intersection of the perpendicular bisectors of triangle ABC.

Next, consider the unfolding of the pyramid $OABC$ shown in Fig. 2.106. Here $HA = HB = HC$ are the projections of the corresponding slants. Obviously, that $HA < OA,\ \ HB < OB,\ \ HC < OC$. Points $O_1,\ O_2,\ O_3$ all represent one vertex of the original trihedral angle, O. Since, $O_2A = O_2C,\ \ HA = HC$, then O_2H is the line of symmetry of segment AC and of triangle AO_2C.

$$\triangle\ O_2AH \cong \triangle\ O_2CH.$$

In $\triangle\ O_2AH$, we see that $HA < O_2A$, then since the larger angles lie opposite larger sides, we can state that

$$\angle AO_2H < \angle O_2HA,$$

which means for the double angle:

$$\angle AO_2C < \angle AHC. \tag{2.78}$$

By analogy using two other triangles, we obtain

$$\begin{array}{l} \angle AO_3B < \angle AHB \\ \angle BO_1C < \angle BHC \end{array}. \tag{2.79}$$

By adding the left sides and the right sides of the inequalities (2.78) and (2.79), we obtain that

$$\begin{array}{l} \angle AOC + \angle AOB + \angle BOC < \angle AHC + \angle AHB + \angle BHC = 360° \\ \alpha + \beta + \gamma < 2\pi \end{array}$$

and compete our proof. □

Problem 136 Given a regular dodecahedron with side of a. Prove that its surface area is equal to $3a^2\sqrt{5(5+2\sqrt{5})}$.

Proof As we know a regular dodecahedron's surface contains 12 regular pentagons. Hence, we need to evaluate the area of one regular pentagon with side a, multiply the answer by 12, and then simplify. Let us look at the regular pentagon using, for example, a sketch in Fig. 2.58. The area of one regular pentagon is equal to the five

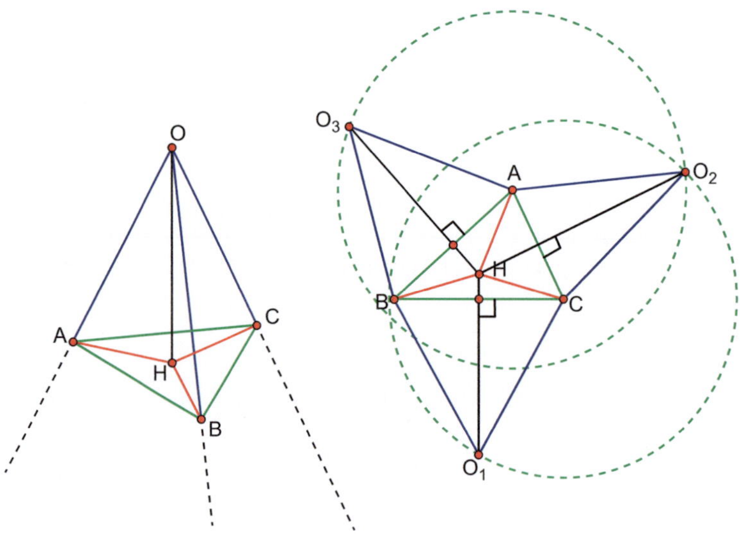

Fig. 2.106 Unfolding of a tetrahedron for proof of Theorem 26

areas of the isosceles triangles AOB with base a and vertex angle of $72°$. Assume that $AD \perp AB$, then AD is the height of $\triangle AOB$, the area of which can be found as

$$[AOB] = \frac{1}{2} AB \cdot AD = \frac{1}{4} \cdot \frac{a^2}{\tan 36°}.$$

What is $\tan 36°$? In my book [2], on p. 212, I showed how to evaluate $\sin 18°$ using the fact that

$$\sin 54° = \cos 36°,$$

and then by solving trigonometric equation $\sin 3x = \cos 2x$, where $x = 18°$, to find that

$$\sin 18° = \frac{\sqrt{5}-1}{4}.$$

Because

$$\cos 2\alpha = 1 - 2\sin^2\alpha, \quad \sin^2\beta = 1 - \cos^2\beta,$$

we obtain the following. Below we will omit degree symbol for simplicity:

$$\cos 36° = 1 - 2 \cdot \left(\frac{\sqrt{5}-1}{4}\right)^2 = \frac{1+\sqrt{5}}{4} \tag{2.80}$$

and

$$\sin^2 36° = 1 - \left(\frac{\sqrt{5}+1}{4}\right)^2 = \frac{10-2\sqrt{5}}{16}.$$

Taking the square root, we will obtain

$$\sin 36° = \frac{\sqrt{10-2\sqrt{5}}}{4}. \tag{2.81}$$

Using the fact that $\tan\alpha = \frac{\sin\alpha}{\cos\alpha}$, dividing the right-hand sides of (2.81) and (2.80), we get that

$$\tan 36° = \frac{\sqrt{10-2\sqrt{5}}}{1+\sqrt{5}}.$$

Next, the area of one triangle like $\triangle AOB$ is equal to

$$[AOB] = \frac{a^2(\sqrt{5}+1)}{4 \cdot \left(\sqrt{10-2\sqrt{5}}\right)}.$$

Then the area of the pentagon is five such areas, and the total surface area of the icosahedron is 60 of such areas, which can be written as

$$A = \frac{15a^2(\sqrt{5}+1)}{\sqrt{10-2\sqrt{5}}}. \tag{2.82}$$

However, (2.82) is different from the formula we are asked to prove. Did we get this formula correctly? Let's do some manipulation of the answer by simply rationalizing the denominator of (2.82):

$$\frac{15a^2\left(\sqrt{5}+1\right)\sqrt{10+2\sqrt{5}}}{\sqrt{10-2\sqrt{5}}\cdot\sqrt{10+2\sqrt{5}}} = \frac{3a^2\left(5+\sqrt{5}\right)\sqrt{10+2\sqrt{5}}}{4}.$$

Above we applied the difference of squares formula in the denominator, reduced the common factors, and inserted square root of 5 inside the first radical quantity.

Let us now place $5+\sqrt{5}$ inside the common radical. We need to use its square:

$$\left(5+\sqrt{5}\right)^2 = 25+10\sqrt{5}+5 = 30+10\sqrt{5}.$$

Then the area can be rewritten as follows:

$$\frac{3a^2\sqrt{\left(30+10\sqrt{5}\right)\left(10+2\sqrt{5}\right)}}{4} = \frac{3a^2\sqrt{20\left(3+\sqrt{5}\right)\left(5+\sqrt{5}\right)}}{4}$$
$$= \frac{3a^2\sqrt{20\left(20+8\sqrt{5}\right)}}{4} = \frac{3a^2}{4}\sqrt{80\left(5+2\sqrt{5}\right)}.$$

Because $80 = 16\cdot 5$ we finally obtain the required formula:

$$A_{12} = 3a^2\sqrt{5\left(5+2\sqrt{5}\right)}.$$

□

Problem 137 Pyramid $DABC$ has the lateral edges of the same length of a. The plane angles of the trihedral angle at vertex D are α, β, and γ. Consider all planes that pass the vertex A of the pyramid and cut other two lateral edges at points M and K. Find the minimum value of the perimeter of the obtained cross section of the pyramid by such planes.

Solution Even if you never learned methods of construction of a cross section by a cutting plane, from the condition of the problem it is clear that any cross section is a

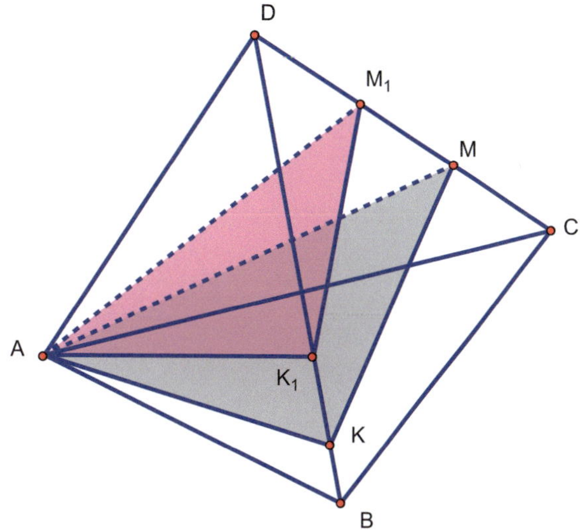

Fig. 2.107 Cross sections of a triangular pyramid

triangle. Two such possible triangles $\triangle AMK$ and $\triangle AM_1K_1$ are shown in Fig. 2.107. Perimeter of a triangle is the sum of the three sides of the triangle and will depend on the location of points M and K on the lateral edges. How can we find its minimal value? Assume that we know the location of points K and M and let's show on each face the sides of the cross-sectional triangle, AK, KM, and MA.

Let us cut the lateral surface of the pyramid by edge AD and unfold the surface. We can obtain something that we sketched in Fig. 2.108. The unfolding of lateral surface of the pyramid consists of three triangles. DA and DA' represent the same lateral edge after we cut the pyramid by this edge and flatten the figure. The sides of the cross section will look like zigzag or broken line. Perimeter of the triangle of the cross section is the sum of the segments of this broken line. The minimal value of the sum will be if all three segments lie on the line AA'. Let us find the length of AA'. Triangle ADA' is an isosceles triangle with side $a = AD = A'D$ and the vertex angle $\angle ADA' = \alpha + \beta + \gamma$ the minimal perimeter of the cross section AKM is equal to the base AA' of triangle ADA' and can be easily found using trigonometry, $P_{\min} = AA' = 2a \cdot \sin\frac{\alpha+\beta+\gamma}{2}$. Obviously, the minimal perimeter will exist only if the sum of the angles at the vertex of trihedral angle of the pyramid is less than 180° that can be written as $\alpha + \beta + \gamma < 180°$.

Answer. $P_{\min} = 2a \cdot \sin\frac{\alpha+\beta+\gamma}{2}$, if $\alpha + \beta + \gamma < 180°$.

Problem 138 Prove that all faces of the tetrahedron $ABCD$ are equal if and only if one of the following conditions is satisfied:

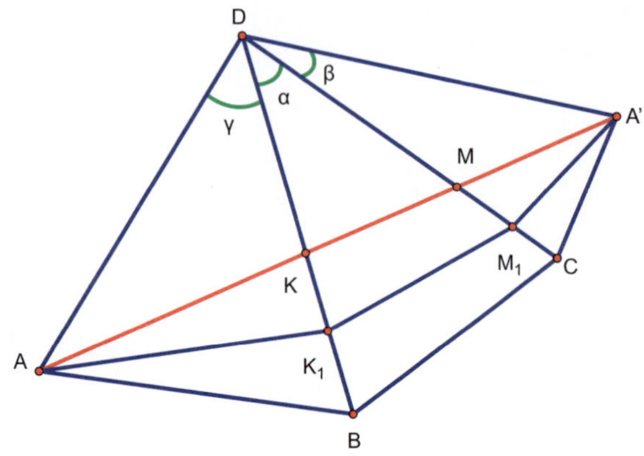

Fig. 2.108 Minimal perimeter of a triangle

1. the sums of plane angles at any three vertices of the tetrahedron are equal to 180°;
2. the sums of plane angles at any two vertices are equal to 180°, and, in addition, any two opposite edges are equal;
3. the sum of plane angles at any vertex is equal to 180°, and, in addition, the tetrahedron has two pairs of equal opposite edges.

Proof Consider the unfolding of a tetrahedron with the described properties.

1. Let the sums of plane angles at vertices A, B,, and C be 180° (Fig. 2.99). Since $D_2C = CD_1$, $D_1B = BD_3$, $D_3A = AD_2$, then the unfolding of the tetrahedron on the plane ABC is a triangle, $D_1D_2D_3$, and the points A, B, and C are the midpoints of its sides. Therefore, all faces of the tetrahedron are equal to the base triangle ABC. On the contrary, if the tetrahedron is isohedral (all faces are equal), then during unfolding, any of its two adjacent faces form a parallelogram. Consequently, the unfolding of the tetrahedron is a triangle, i.e., the sums of plane angles at the vertices of the tetrahedron are 180°. Please note that the tetrahedron $ABCD$ is not regular, and its equal faces can be the same scalene triangles. Obviously, this property will work for any regular tetrahedron as well.
2. Let the sums of plane angles at vertices A and B be 180° and a pair of opposite edges be equal. Consider the unfolding of such a tetrahedron onto the face plane ABC. Can you find opposite edges in a tetrahedron? For example, in Fig. 2.109), we marked in red two opposite edges AB and CD. Also, we can say that edge AD is opposite to BC and edge AC is opposite to DB.

   Let edges AB and CD be equal. By looking at possible unfolding of the tetrahedron $ABCD$, we understand that the edge CD on it is represented twice as segment $CD_1 = CD = AB$ and as segment $CD_2 = CD = AB$. Then $D_1C + D_2C = 2AB = D_1D_2$, so point C is the midpoint of segment D_1D_2. Hence, the sketch shown in Fig. 2.110 is wrong, and the sum of the angles at point C must be 180°.

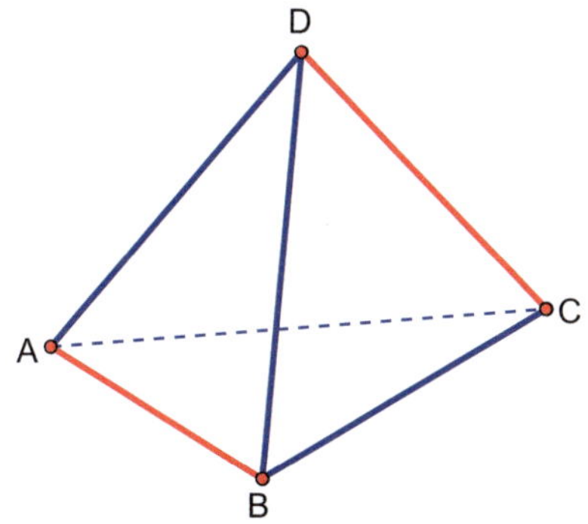

Fig. 2.109 A pair of opposite edges is equal

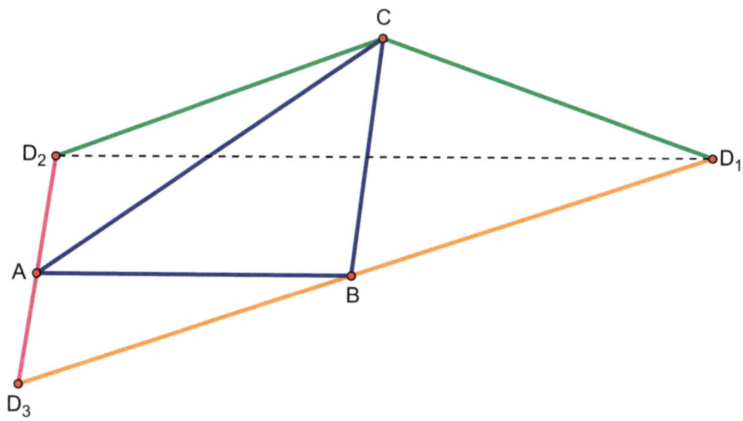

Fig. 2.110 The sum of angles at vertices A, B is $180°$, a pair of opposite edges is equal

3. Without loss of generality assume that the sum of the plane angles at the vertex A is $180°$, $AB = CD$ and $AD == BC$. Let us consider the unfolding of the tetrahedron onto the plane ABC shown in Fig. 2.111. Here we assume that the sum of the angles at two other vertices is not $180°$.

   Let us start from considering the quadrilateral AD_2CB. Since $CD = AB \Rightarrow CD_2 = AB$. Because $AD = BC \Rightarrow AD_2 = BC$. We can see that the opposite sides of quadrilateral AD_2CB are equal, so it is a parallelogram. Therefore, the segments CB and AD_3 are parallel and equal, and hence the quadrilateral $ACBD_3$ is a parallelogram. Therefore, the unfolding of the tetrahedron is a triangle, and A, B, and C are the midpoints of its sides. □

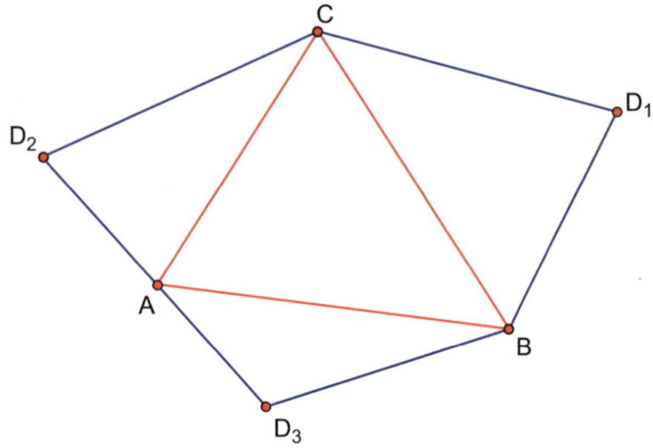

Fig. 2.111 The sum of the angles at one vertex equals 180°

Problem 139 Can the net of a triangular pyramid be a square with side a? If so, find the lateral surface area of the pyramid.

Solution Note that we cannot make a regular quadrangular pyramid from the square net shown on the left of Fig. 2.112, because the sum of the plane angles at its tetrahedral apex would be exactly 360°, which is impossible. Assume that a square with side a is the net of a triangular pyramid. Let us mark the midpoints of any two adjacent sides of the square and then connect these points with each other and with the opposite vertex of the square. We will make an isosceles triangle inside the square as shown in red below. By folding this net up, we will obtain a pyramid with one face as an equilateral triangle with side $a/2$, two faces are right triangles with legs $a/2$ and a, and one isosceles triangle with base of $\frac{a\sqrt{2}}{2}$ and other two sides of length $\frac{a\sqrt{5}}{2}$ (see Fig. 2.112 on the right). The lateral area of such triangular pyramid is equal to the sum of the areas of two equal right triangles and an isosceles right triangle (the hypotenuses are marked in red).

$$A = \frac{1}{2} \cdot \frac{a}{2} \cdot \frac{a}{2} + a \cdot \frac{a}{2} = \frac{5a^2}{8}.$$

Answer. $A = \frac{5a^2}{8}$.

Problem 140 Prove that if any two trihedral angles in a tetrahedron are equal, then all faces of this tetrahedron are equal.

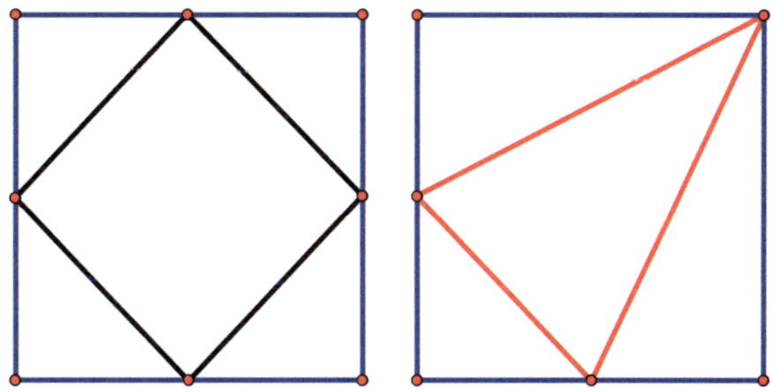

Fig. 2.112 Making a pyramid from a square net

Proof Consider the unfolding of this tetrahedron onto the plane of face ABC sketched in Fig. 2.113. It is triangle ABC, to which triangles $ABD_3,\ BCD_1,\ CAD_2$ are attached. It follows from the condition that the following four quantities are equal: the sum of the triples of angles at the vertices $A,\ B$, and C of the unfolding and the sum of the angles at the vertices $D_1,\ D_2, D_3$. Hence, each of these sums is equal to 180°, since the sum of all 12 considered angles is the sum of the angles of the four triangles. As a result, we get that the unfolding is a triangle in which three middle lines are drawn (we used here the fact that $AD_2 = AD_3,\ BD_1 = BD_3,\ CD_1 = CD_2$). The middle lines divide the triangle into four equal triangles, so the faces of the tetrahedron are equal.

On one hand, the sum of the angles in each triangle of the unfolding is equal to 180°. There are four such triangles $\triangle AD_2C,\ \triangle CD_1B,\ \triangle BD_3A$, and $\triangle ABC$. This can be written as

$$\left.\begin{array}{r}\angle 3 + \angle 4 + \angle 5 = 180^{\circ}\\ \angle 2 + \angle 6 + \angle 10 = 180^{\circ}\\ \angle 7 + \angle 8 + \angle 9 = 180^{\circ}\\ \angle 1 + \angle 11 + \angle 12 = 180^{\circ}.\end{array}\right\}$$

If we add the left and right sides of the equalities above, we obtain

$$(\angle 1 + \angle 2 + \angle 3) + \angle 4 + (\angle 5 + \angle 6 + \angle 7) + \angle 8 + (\angle 9 + \angle 10 + \angle 11) + \angle 12 = 4 \cdot 180^{\circ}.$$

The sums inside the parentheses are all equal, including the sum of the plane angles at vertex D. If we denote $\angle 4 + \angle 8 + \angle 12 = S$, then from the previous equation we have that $4S = 4 \cdot 180°$ and so $S = 180°$. Therefore, the sum of each triple is the same and it is 180°. This means that $A,\ B,\ C$ are the midpoints of the edges $D_3D_2,\ D_1D_3,\ D_1D_2$, respectively, so the unfolding of the tetrahedron is $\triangle\ D_1D_2D_3$ with $\triangle\ ABC$ inside. Obviously,

$$\triangle\ D_1D_2D_3 \sim \triangle\ ABC.$$

Therefore $\triangle ABC = \triangle ACD = \triangle CBD = \triangle ABD.$ □

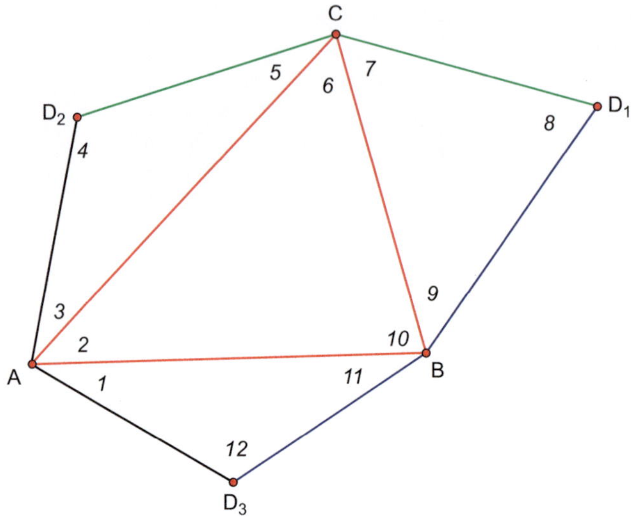

Fig. 2.113 Any two trihedral angles of a tetrahedron are equal

Problem 141 Let S_A, S_B, S_C, and S_D be the sums of the plane angles of the tetrahedron $ABCD$ at vertices A, B, C, and D. Prove that if $S_A = S_B$ and $S_C = S_D$, then $\triangle ABC = \triangle BAD$ and $\triangle ACD = \triangle BDC$.

Proof Since the sum of the angles of each face of a tetrahedron shown in Fig. 2.114 is 180°, then the following is true:

$$\begin{aligned}&\angle 1 + \angle 2 + \angle 3 + \angle 4 + \angle 5 + \angle 6 + \angle 7 + \angle 8 + \angle 9 + \angle 10 + \angle 11 + \angle 12\\&= (\angle 1 + \angle 2 + \angle 3) + (\angle 5 + \angle 6 + \angle 7) + [\angle 9 + \angle 10 + \angle 11] + [\angle 4 + \angle 8 + \angle 12]\\&= S_A + S_B + S_C + S_D = 4 \cdot 180^\circ\end{aligned}$$

.

Since $S_A = S_B$ and $S_C = S_D$, then

$$2S_A + 2S_C = 4 \cdot 180^\circ$$
$$\boxed{S_A + S_C = 360^\circ}.$$

If all trihedral angles equal, then S_A would be equal to $S_C = 180^\circ$. However, the trihedral angles at vertices A and C are not equal.

Let, for definiteness, assume that $S_A < S_C,\ S_C > 180^\circ$. Our sketch shows a similar situation. We can see that the angles at vertices A and B are equal and less than 180°, while the angle at C is greater than 180°. Connect the vertices $D_1 \to D_2 \to D_3 \to D_1$ and obtain triangle $\triangle\, D_1D_2D_3$ (Fig. 2.115).

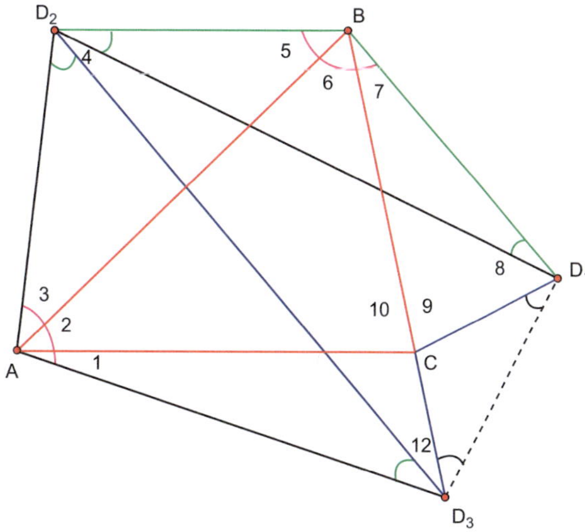

Fig. 2.114 Two trihedral angles of a tetrahedron are equal

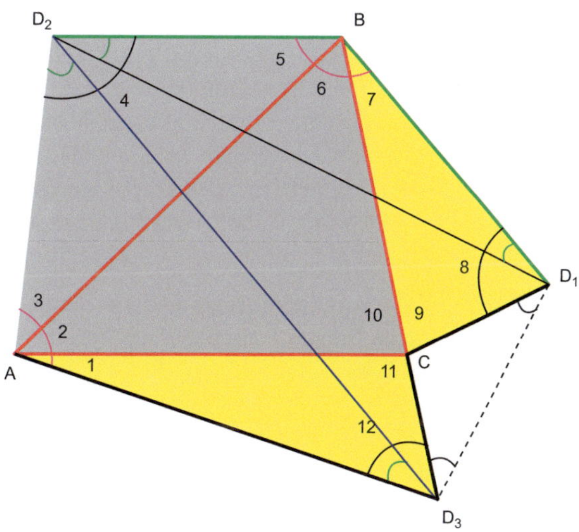

Fig. 2.115 Pairs of equal triangles are shown

Triangles $\triangle\ D_2BD_1$ and $\triangle\ D_2AD_3$ are similar by three angles, because they are both isosceles triangles with the same vertex angle ($S_A = S_B$ or $\angle 1 + \angle 2 + \angle 3 = \angle 5 + \angle 6 + \angle 7$). Let us mark the same base angles of both triangles with black arcs and write the ratio of the similar sides:

$$\frac{D_2B}{AD_2} = \frac{BD_1}{AD_3} = \frac{D_1D_2}{D_2D_3} = \frac{DB}{AD} = 1. \tag{2.83}$$

Next, since $\angle AD_3C\ = \angle D_2D_3D_1$ and $\frac{AD_3}{D_3C}\ =\ \frac{D_2D_3}{D_3D_1}\ ,\ \Rightarrow\ \ \triangle\ ACD_3 \sim\triangle\ D_2D_1D_3$ and the similarity coefficient is equal to the ratio of the side to the base in an isosceles triangle with angle S_A at vertex. Therefore, $AC = D_2B$. Similarly, $CB = AD_2$. Therefore, $\triangle ABC = \triangle BAD_2 = \triangle BAD$. It can be proved similarly with the use of (2.83) that $\triangle ACD = \triangle BDC$. These pairs of equal triangles are shown in Fig. 2.115 in gray and yellow color, respectively. □

2.8 Homework on Chapter 2. Problems to Solve

1. Paul cut a cardboard polyhedron along the edges and sent the resulting set of faces by made to Brian. Brian glued together a convex polyhedron from all these faces. Can it happen so that Paul's and Brian's polyhedra are not the same?

   Hint. Recall information and examples from section devoted to unfolding of polyhedra and important theorems there.

   Solution. The answer can vary.
   Case 1. Assume that Paul's polyhedron was convex. Let us consider a triangular pyramid whose base is an equilateral triangle, the dihedral angles at the base are acute, and the lateral edges are different. From two copies of such a pyramid, gluing them together on a common base, one can make a bi-pyramid in different ways. They all have the same set of faces, but are not equal to each other. If Paul had numbered all the edges and written their numbers next to the edges on each face, then Brian would have glued exactly the same convex polyhedron. It follows from Cauchy theorem: if the faces of one convex polyhedron are, respectively, equal to the faces of another convex polyhedron, then these polyhedra are equal.

   Case 2. If there was no information given about convexity of Paul's polyhedron, then Brian could make a nonconvex polyhedron from the faces similar to one given in Fig. 2.96.
2. It is given that an unfolding of some pyramid serves as an acute-angled triangle, in which three midlines of the triangle are drawn. Prove that there exists a rectangular parallelepiped whose four nonadjacent vertices are the vertices of this pyramid.

   Solution. Let an acute-angled triangle $D_1D_2D_3$ with midline segments AB, AC, and BC be an unfolding of the triangular pyramid $ABCD$ (vertices D_1, D_2, and D_3 are glued to one point D) (you can use Fig. 2.99). If an edge corresponds to half of the side of the triangle $D_1D_2D_3$, then the edge intersecting with it corresponds to the midline parallel to this side, and vice versa. Therefore, the opposite edges of the pyramid are equal.
   We construct a parallelepiped whose four nonadjacent vertices are the vertices of a triangular pyramid. To do this, we draw a plane through each edge of the pyramid parallel to the edge skewed to it. We get three pairs of parallel planes, at the intersection of which a parallelepiped is formed. Since the pairs of opposite edges of the original pyramid are equal, each face of the parallelepiped is a

parallelogram with equal diagonals, that is, a rectangle. A parallelepiped whose faces are all rectangles is rectangular, as required (See Fig. 2.50).

3. Let $ABCDA_1B_1C_1D_1$ be a rectangular box in which $ABCD$ and $A_1B_1C_1D_1$ are squares with side b. Point K is on side AD, such that $\angle KCD = \alpha$, $\alpha < \pi/4$. Find the area of triangle KCD_1 if full surface of the box equals $2a^2$.

   Answer. $S = \frac{b}{2\cos\alpha} \cdot \sqrt{b^2\sin^2(\alpha) + \left(\frac{a^2-b^2}{2b}\right)^2}$.

4. In a regular quadrangular pyramid $ABCDS$ with the base side of $96/17$ and the apothem of 36, a cross-sectional $AMNP$ is drawn. Point M divides the side edge in relation to $SM \div MB = 2 \div 1$. Point N lies on the side edge SC. P divides the side of the base in relation to $DP \div PC = 1 \div 2$. Find the length of the segment MN.

   Answer. $MN = 5$.

5. The base of pyramid $TABCS$ is a rectangle $ABCD$, the diagonal of which equals $\sqrt{5}$. The height of the pyramid (TO) equals $TO = 3$ and passes through the point of intersection of the diagonals of the base, and the distance between the lateral edge TB and the diagonal of the base AC equals $6/7$. Find the volume of the pyramid.

   Answer. $V = 2$

6. In the triangular pyramid $ABCD$ the edges AC and BD are perpendicular to each other. $AB = BD = AD = a$. The midpoint of the edge AC is equidistant from the planes ABD and BCD; the angle between edge AC and face CBD is equal to $\arcsin\frac{1}{\sqrt{3}}$. Find the length of the edge CD, the angle CAD, the angle between the edge BD and the face ACD.

   Answer. a, $45°$, $45°$.

7. A rectangular parallelepiped $ABCDA_1B_1C_1D_1$ has a rectangle $ABCD$ as its base with the sides $AB = 2$ and $BC = 3$. The side edges AA_1, BB_1, CC_1, $andDD_1$ are equal to four. Points P and Q are the midpoints of the edges A_1B_1 and CC_1, respectively. Point M is taken on the line B_1C_1 as such that the distances from P and Q to the line AM are the same. Find the length of the segment B_1M.

   Hint. You can introduce a coordinate system similar to Problem 47 and apply formula (1.89)

   Answer. $B_1M = 2$.

8. The base of a regular triangular pyramid $SABC$ is an equilateral triangle ABC with side of 2. In the plane of the base of the pyramid, a straight line is drawn that is tangent to the circle inscribed into the triangle ABC and intersects the edges AB and AC at points M and N. It is known that the volume of the pyramid $SAMN$ is equal to $\sqrt{66}/90$, and the length of the segment $MN = 7/10$. Determine the length of the lateral edge of the pyramid $SABC$.

   Answer. $\sqrt{34}/3$.

9. In the cube $ABCDA'B'C'D'$ with edge of 1, M is the midpoint of the edge BC. Points P on line $A'C'$ and Q on line $B'M$ are chosen so that line PQ runs parallel to plane $AA'B'B$. Determine the minimal possible length of segment PQ.

   Answer. $\frac{2}{\sqrt{5}}$.
10. At the base of the pyramid lies an isosceles trapezoid, the smaller base of which is $2/5$ of the larger one. All lateral faces of the pyramid form an angle of $60°$ with the base. Find the angle of inclination of the smaller lateral edge of the pyramid to the plane of its base.

    Answer. $\arctan\sqrt{\frac{15}{7}}$.
11. In a regular triangular pyramid, the ratio of the base area to the full surface area of the pyramid is $\sqrt{2}-1$. Find the sine of the angle between the lateral edge of the pyramid and the side of its base.

    Answer. $\sqrt{0.4} = \frac{\sqrt{10}}{5}$.
12. The base of the pyramid is a rectangle of area S, the angle between the diagonals of the base is α, and the height of the pyramid is equal to the perimeter of the base. Find the volume of the pyramid.

    Answer. $V = \frac{2S}{3} \cdot \sqrt{\frac{2S}{\sin\alpha} + 2S}$.
13. Find the volume of a regular triangular prism, if the side of its base is equal to a and the lateral surface is equal to the sum of the areas of the bases.

    Answer. $\frac{a^3}{8}$.
14. The center of the upper base of a regular quadrangular prism and the midpoints of the sides of the lower base serve as the vertices of a pyramid inscribed in the prism whose volume is V. Find the volume of the prism.

    Answer. $6V$.
15. The base of a right prism is an isosceles trapezoid whose bases are equal to a and b ($a > b$), and whose acute angle is equal to α The plane passing through the large base of the upper trapezoid and the smaller base of the lower trapezoid makes an angle equal to β with the plane of the lower base. Find the volume of the prism.

    Answer. $\frac{1}{8}(a-b)^2(a+b)\tan^2\alpha\tan\beta$.
16. In a regular pentagonal pyramid, the angle of inclination of the lateral edge to the plane of the base is equal to α. Find the angle of inclination of the lateral face to the plane of the base.

    Answer. $\arctan\frac{\tan\alpha}{\cos 36°} = \arctan(\frac{4\tan\alpha}{1+\sqrt{5}})$ See (2.80).
17. The base of the pyramid $SABC$ is a right triangle ABC, where C is the vertex of the right angle. All lateral faces of the pyramid are inclined to its base at the same angle equal to $\arcsin(5/13)$. Find the area of the lateral surface of the pyramid, if SO is the height of the pyramid, $AO = 1$, $BO = 3\sqrt{2}$.

    Answer. $91/25$.

18. Given a regular triangular pyramid $SABC$, S is its vertex, $AB = 1, AS = 2$, BM is the median of triangle ABC. AD is a bisector of triangle SAB. Find the length of segment DM.

    Answer. $\frac{\sqrt{31}}{6}$.

19. The base of a pyramid is a triangle with sides of 7, 8, and 9. The lateral edges of the pyramid are inclined to the plane of the base at the angle of 60°. Find the height of the pyramid.

    Answer. $H = \frac{21\sqrt{15}}{10}$.

20. In a regular triangular prism $ABCA'B'C'$ ($AA' \| BB' \| CC'$), the angle between lines AC' and AB' equals α, $AA' = 2$. Find AB.

    Answer. $\frac{4\sin(\alpha/2)}{\sqrt{3-4\sin^2(\alpha/2)}}$.

21. In a regular triangular pyramid, the sum of the lengths of all its six edges is L and the base area is S. Find the area of the lateral surface of the pyramid.

    Answer. $A = \frac{\sqrt{S}}{\sqrt[4]{3}}\sqrt{L^2 - \frac{12L\sqrt{S}}{\sqrt[4]{3}} + 9S\sqrt{3}}$, where $L > 2\sqrt{S}\sqrt[4]{3}(\sqrt{3}+1)$.

22. Triangle ABC is the base of the pyramid $SABC$. In this triangle, $BC = 1,\ CA = 13$, and the height $CE = \sqrt{105}$. Each lateral edge of the pyramid forms an angle α with the base ABC. Determine the area of the base and the volume of the pyramid.
    Hint. The foot of the height of the pyramid is the center of the circle circumscribed about ΔABC.

    Answer. $6\sqrt{105},\ 143\tan\alpha;\ 2\sqrt{105},\ \frac{143}{3}\tan\frac{\alpha}{3}$.

23. Prove that if the sum of the plane angles at the apex of the pyramid is greater than 180°, then each lateral edge of the pyramid is less than the half-perimeter of the base.

    Proof Without loss of generality, we can consider the pyramid $SA_1A_2...A_n$, cut it along one edge SA_1 and unfold its lateral surface, obtaining a polygon $A_1A_2...A_n$ with apex S inside, such that $A_1S = A_1'S$. (It is implied that the points A_1 and A_1' coincide if we connect back the cut edges.) Mark the angle A_1SA_1', for example, in red. Continue A_1S to obtain point B, the intersection point of line A_1S and the boundary of the polygon. This point B divides the boundary of the polygon into two zigzag lines, $A_1A_2...B$ (you can mark it in blue) and $B...A_nA_1'$ (i.e. marked in yellow). Let the length of the blue zigzag line be a, and the length of the yellow line be b. Obviously, $a + b$ is the length of the boundary of the base polygon and the perimeter of the base, $a + b = P$. Moreover, $A_1B = A_1S + SB < a$ and $A_1'S < SB + b$. Since $A_1S = A_1'S$, the following is true

$$A_1S < a - SB$$
$$A_1S < SB + b$$

    Adding the left and right sides of the inequalities completes the proof.

$$A_1S < \frac{a+b}{2} = \frac{P}{2}$$

□

24. In pyramid $ABCD$, the sum of the angles at vertices A and B is equal. Also, the sum of the angles at vertices C and D is also equal. The sum of the areas of the faces ABD and ACD equals S. Find the surface area of the pyramid.

    Hint. Construct its unfolding and see Problem 141 with similar condition.

    Answer. $2S$.
25. Prove that for any convex polyhedron that has F faces, V vertices, and E edges, the following inequalities are true:

    (a) $2F \geq 4 + V$;
    (b) $2V \geq F + 4$.

Proof

(a) Let us use the true inequality (2.46) that was proven and explained in the book, $2E \geq 3V$. Next, multiplying by 2 the left and right sides of Euler's formula (2.40), we obtain

$$2V - 2E + 2F = 4. \tag{2.84}$$

Adding this with the inequality (2.46), we have $2V + 2F \geq 4 + 3V$ that gives us the requested inequality, $2F \geq V + 4$.

(b) Consider (2.43) and multiply this by 3:

$$3F = 3(V_3 + V_4 + V_5 + \cdots + V_m).$$

Compare the right-hand side of this with the right-hand side of (2.44) multiplied by 2:

$$2E = 3V_3 + 4V_4 + 5V_5 + \cdots + mV_m.$$

So

$$2E \geq 3F. \tag{2.85}$$

If we add again (2.85) and (2.84) we obtain that $2V \geq 4 + F$.

□

26. An edge of an octahedron is equal to one unit. Determine the distance between its opposite vertices lying on the axis of the octahedron.

    Hint. Can use formula (2.65).
27. Construct a tetrahedron inscribed in a given tetrahedron.

28. From each vertex of a tetrahedron with an edge of two units, a tetrahedron with an edge of one unit is cut off. What polyhedron will remain? Find its volume.

    Hint. Sketch the situation and you will see how to evaluate the volume of the truncated pyramid. You can use formula (2.7)from this book.
29. How many and which regular polygons can converge (meet) at the vertices of a regular polyhedron?

    Hint. Use the fact that the sum of plane angles at each vertex of a convex polyhedron is less than 360°.
30. What is the edge of the largest tetrahedron that can be placed in a cube with the unit edge?
31. Find the volume of a rectangular parallelepiped, diagonal of which is 14 inch, perimeter of the base is 20 inch, and perimeter of the smallest lateral face is 32 inch.

Solution Consider parallelepiped $ABCDA'B'C'D'$, with $ABCD \| A'B'C'D'$. Denote $D'B = 14, AB = a, BC = b, B'B = c$ and assume that $a > b$. Since

$$D'B^2 = AB^2 + BC^2 + B'B^2 \quad \Rightarrow a^2 + b^2 + c^2 = 196.$$

We also can write the formula for the corresponding perimeters:

$$2(a + b) = 20, \quad 2(b + c) = 32.$$

Next, we need to solve the system of nonlinear equations in three variables a, b, and c:

$$\begin{cases} a^2 + b^2 + c^2 = 196 \\ a + b = 10 \\ b + c = 16 \end{cases}. \tag{2.86}$$

From the first and second equations of the system, we can express a and c in terms of b and substitute into the third equation, $3b^2 - 52b + 160 = 0$. We also will look only for such values of b that are less than 5. (It follows from the second equation of the system and the assumption that $a > b$.) Hence, $b = 4$, $a = 6$, and $c = 12$. The volume of the parallelepiped is $V = abc = 288 \ \text{units}^3$.

Answer. $V = 288 \ \text{units}^3$.

32. A solid tetrahedron is sliced off a solid wooden unit cube by a plane passing through two nonadjacent vertices on one face and one vertex on the opposite face vertices. The resulting solid is then placed on a table with the cut surface face down. What is height of this object?

    Solution. Let us choose two nonadjacent vertices that lie on one face marked as A and C and the third vertex B on the opposite face as it is shown in Fig. 2.86. It is easy to construct the cross section. The sliced surface is bounded by an equilateral triangle ABC with side length $\sqrt{2}$. Mark the fourth vertex of this tetrahedron as D.

In order to find the distance from D to the plane of the triangle ABC, we can equate the volume of pyramid $ABCD$ found in two different ways. First as one with base equilateral triangle ABC and unknown height, h, that will be the actual distance to the plane ABC. And the second, as the volume of the right pyramid $ADCB$ with height $AD = 1$ and the base the right isosceles triangle DCB with legs of 1. We obtain the following relationship:

$$\frac{1}{3}h \cdot \frac{\sqrt{2}\sqrt{3}}{4} = \frac{1}{3} \cdot 1 \cdot \frac{1}{2} \cdot 1 \cdot 1$$
$$h = \frac{\sqrt{6}}{3}.$$

Therefore the height of the resulting solid is $\frac{\sqrt{6}}{3}$, the center of pyramid therefore lies at distance $\frac{\sqrt{6}}{3}$ from the plane ABC.

Answer. $h = \frac{\sqrt{6}}{3}$.

33. Can the faces of a regular polyhedron be regular hexagons?

Solution. Regular polygons with more than five sides cannot be the faces of a regular polyhedra because the sum of the plane angles at each vertex must be less than 360°.

34. Is there a quadrangular pyramid whose two opposite lateral faces are perpendicular to the plane of the base?

Answer. Yes it exists. An example of such a pyramid is shown in Fig. 2.116. It can be constructed using a cube as follows. A triangular pyramid $A'ABC$ is taken, in which the lateral edge $A'A$ is perpendicular to the plane of the base. Its side faces $A'AC$ and $A'AB$ are perpendicular to the base (like the planes passing through the perpendicular AA' to the base). Let us now take arbitrary points M and N on the sides AC and AB of the base, respectively.

Pyramid $A'MNBC$ satisfies the condition of the problem, because faces CMA' and BNA' are orthogonal.

35. The planes of the lateral faces of a triangular pyramid form equal angles with the plane of the base. Prove that the projection of the vertex onto the plane of the base is the center of the inscribed and circumscribed circle of the base.

Proof Let α be the angle between the planes of the side faces and the plane of the base and H be the height of the pyramid, then the distance from the projection of the vertex to the plane of the base to any line containing an edge of the base is equal to $H \cdot \cot\alpha$. Note also that if the dihedral angles at the edges of the base are equal and not just the angles between the planes, then the projection of the vertex is the center of the inscribed circle. □

36. Prove that in any triangular pyramid there is a vertex at which all plane angles are acute.

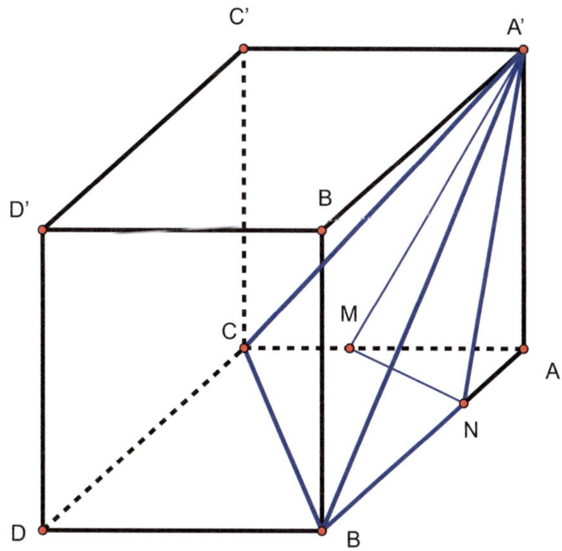

Fig. 2.116 Two opposite lateral faces $CA'M$ and $BA'N$ of pyramid $A'BCMN$ are orthogonal to the base

Proof Suppose that a certain pyramid has an obtuse plane angle at each vertex. Since the sum of two plane angles of a trihedral angle is greater than the third plane angle, then the sum of three plane angles at each vertex of such a pyramid is greater than $180°$. This means that the sum of all 12 plane angles of the pyramid is greater than $4 \cdot 180° = 720°$, which is impossible, because this sum consists of the sums of the angles of four triangles—the faces of the pyramid. Therefore it must be equal to $720°$. □

37. Give an example of a nonconvex polyhedron that has six faces.

Solution In this chapter, in Problem 106, we have found all six convex polyhedra with six faces. If we assume that polyhedron can be nonconvex, then Fig. 2.117 shows an example of such a polyhedron.

It is obtained as follows: on the edge DC of the tetrahedron $ABCD$ (made, for example, of wood), a "notch" is made of two triangular faces—GEH and GFE. It has 8 vertices, 6 faces, and 12 edges. In this polyhedron, two faces are hexagons with two common edges, DF and HC. In a convex polyhedron, this situation is impossible: two faces of a convex polyhedron can have at most one common edge.

38. Prove that the foot of the height of the isohedral tetrahedron and the point of intersection of the altitudes of the face on which this height is dropped are symmetrical with respect to the center of the circle circumscribed about this face.

Proof Let M be the point of intersection of the altitudes of ΔABC, the base of the tetrahedron $DABC$. (We can use Figure 2.101). Let now K be the center of gravity

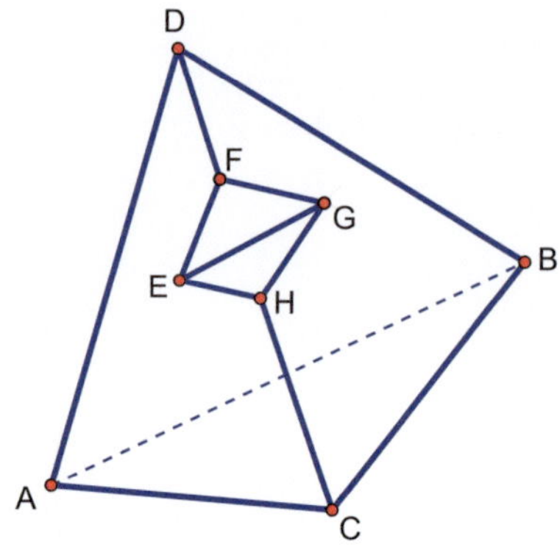

Fig. 2.117 Nonconvex polyhedron with six faces

of triangle ABC (also the center of gravity of $\Delta D_1D_2D_3$), and O be the center of the circle circumscribed around it. It is known that M, K, and O lie on the same line (Euler's straight line), and K is between M and O, $KM = 2KO$. On the other hand, triangle $\Delta D_1D_2D_3$ is similar to ΔABC with center at K and coefficient $k = 2$, which means $KH = 2MK$. It follows from this that $OH = MO$. □

39. Prove that if the sums of plane angles at the three vertices of a triangular pyramid are equal to 180°, then all the faces of this pyramid are equal triangles (i.e., a tetrahedron is isohedral).

Proof Let in the triangular pyramid $ABCD$ the sums of three plane angles at each of the vertices A, B, and C be equal to 180°. Let us consider the unfolding of the pyramid $D_1AD_2CD_3B$ onto the plane of the base triangle ABC, and the points D_1, D_2, D_3 be the vertices of triangles with bases AB, AC, BC, respectively. Since the sum of the three plane angles at each of the vertices of the tetrahedron is equal to 180°, point A lies on the segment D_1D_2, point B lies on the segment D_1D_3, and point C lies on the segment D_2D_3, and A, B, C are the midpoints of these segments. Therefore AB, BC, and AC are the midline segments of the triangle $D_1D_2D_3$. This means that the triangles $\Delta D_1AB = \Delta AD_2C = \Delta BCD_3 = \Delta CBA$. Therefore, the triangles faces of the tetrahedron are also equal $\Delta\ ADB = \Delta\ ADC = \Delta\ BCD = \Delta\ CBA$, i.e., tetrahedron is isohedral and all its faces are equal acute triangles. □

40. In a pyramid $ABCD$ the sum of angles at vertices A and B is equal. Also, the sum of the angles at vertices C and D is also equal. The sum of the areas of the faces ABD and ACD equals S. Find the surface area of the pyramid.
    Hint. Construct unfolding. See Problem 141.
    Answer. $S/2$.
41. Given a flat convex polygon, Mike claims that he can make a pyramid out of any convex polygon by finding the midpoints of adjacent sides of the polygon and then folding it so that the midline segments become the edges of the base of the pyramid. Is his idea correct? Explain. What folds must be made to glue a pyramid together?

Answer. Mike's idea is wrong. Let's assume it's possible. In any convex polygon with n vertices, the sum of the interior angles is $(n-2)\cdot 180°$. After folding, the apex of the pyramid will be the apex of a polyhedral angle consisting of all the plane angles of the polygon. Since any polyhedral angle must be less than $360°$, the following must be true $(n-2)\cdot 180° < 360° \to n < 4$. Therefore Mike's idea would work only for a triangle.

42. Given a triangular pyramid $ABCD$ at the base of which lies an arbitrary scalene triangle ABC with sides $a,\ b$, and c; all lateral edges are of the same length l, and the volume of the pyramid is equal to V. Find the height of the pyramid dropped from vertex D.

Solution Let $h = DH$ be the height of the pyramid dropped from vertex D. Since the lateral edges are the same, then the right triangles AHD, BHD, and CHD are congruent, then $AH = BH = CH = R$, where R is the radius of the circle circumscribed around triangle ABC. The height can be expresses as

$$h = \frac{3V}{B},$$

where B is the area of the base ABC. This area can be written in terms of the circumradius R as follows:

$$B = \frac{abc}{4R},\quad R^2 = l^2 - h^2.$$

Let us square both sides of the formula for h and then substitute these formulas above:

$$h^2 = \frac{9V^2}{B^2} = \frac{144V^2(l^2 - h^2)}{(abc)^2}.$$

This equation can be easily solved for the square of h:

$$(abc)^2h^2 = 144V^2l^2 - 144V^2h^2,$$

from which we factor out h^2,

$$h^2((abc)^2 + 144V^2) = 144V^2l^2,$$

after taking the square root, we obtain h as

$$h = \frac{12Vl}{\sqrt{144V^2 + (abc)^2}}.$$

Answer. $h = \frac{12Vl}{\sqrt{144V^2+(abc)^2}}$.

43. How to cut a convex polyhedral prism into several trihedral prisms?

Answer. The answer may vary. Consider a prism with an n-sides polygon as its base and take a point inside one of the bases of the prism. Connect this point to each vertex, obtaining the corresponding segments and dividing the polygon into n triangles. Using translation parallel to the lateral edge, similarly divide the second base. Connect the corresponding points. In this way, we divide the prism into n smaller prisms.

44. Give an example of a convex polyhedron with nine faces.

    Answer. The answer may vary. For example, an 8-gonal pyramid has 9 faces (eight triangles and one polygon with eight sides) You can also make a regular triangular prism and glue two equal triangular pyramids with the base equal to the prism's base to the top and bottom of the prism. In this case we will again have nine faces (3 rectangles and 6 isosceles triangles)

45. How to cut a convex polyhedron into several pyramids?

    Answer. The answer may vary. For example, you can take any point S inside the polyhedron and connect it to each of its vertices. In this case, any convex polyhedron with F faces will be divided into F pyramids, each with apex S. On the other hand, we can use Fig. 2.50. In this case, we do not take any point inside the cube, but divide the cube into five pyramids.

46. Can a regular triangle be the net of a pyramid? Determine the type of this pyramid and find the area of its lateral surface if the side of the triangle is equal to a.

    Hint. Bend the triangle along the midsegments and align its corresponding vertices and sides.

    Answer. In this case we obtain a regular tetrahedron with edge of $a/2$. The lateral surface is $\frac{3\sqrt{3}a^2}{16}$.

47. Can an obtuse or right triangle be the net of a pyramid? Why is it possible or why not?
    Answer. Each plane angle at the vertex of the pyramid must be less than the sum of the other two. This is not true for a right or obtuse triangle and works only for an acute triangle.

48. Assume that you have a square with side a. Mark the midpoints of each side and connect them obtaining another square. Can this sketch be a net of a regular square pyramid with the small square as the base of it? ? Explain.
    Answer. If we fold up the four triangles, the sum of the four angles will be 360°. Therefore, this construction will not allow us to make a pyramid.

49. Prove that at most one vertex of a tetrahedron has the property that the sum of any two plane angles at this vertex is greater than 180°.

Proof Assume that not one, but two vertices (A and C) have the specified property. Let us consider the unfolding onto the plane of two faces adjacent to the edge AC. We get quadrilateral $ABCD$, which has $\angle DAB > 180°$ and $\angle BCD > 180°$, which means that the sum of all angles of this quadrilateral is greater than 360°, which is impossible. A contradiction is obtained. □

50. Construct a net of a pyramid, with the base as a regular triangle, and the side edge of which is perpendicular to the plane of the base.
Solution. The side of the equilateral triangle (base) and the height of the pyramid are chosen arbitrarily. This net can be constructed using drawing tools, such as compass and ruler. To check whether the net is constructed correctly, you can cut it out and fold the pyramid. One of the possible options is given in Fig. 2.118. Here $FD \perp CE$, $BF = CF = FE$, $AB = BC = AC = CD = DE$.

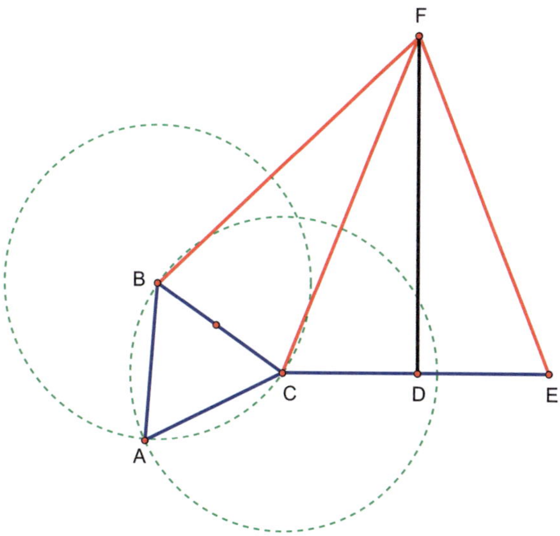

Fig. 2.118 Making a net for a pyramid

Chapter 3
Cross Sections of Polyhedra

Geometry originated in ancient times for solving purely practical problems. Over time, when many geometric facts accumulated, people had a need for generalizations, understanding the dependence of some elements on others, establishing logical connections and proofs, and geometric science was gradually created.

Around the sixth–fifth centuries BC, a new stage in the development of geometry began in Ancient Greece, which is explained by the high level of social, political and cultural life in the Greek states. By experimenting with different shapes and dividing solids into different parts, they established the basic properties of volumes:

- Equal bodies have equal volumes.
- If a body is composed of several bodies in such a way that the internal regions of these bodies do not have common points, then the volume of the body is the sum of the volumes of its constituent bodies.

In the modern world, we are simply overloaded with various knowledge and theories that were created by our predecessors. However, even though almost everything in stereometry has already been obtained before us, sometimes, especially when solving math Olympiad problems, it is necessary to return to the "roots", try to divide a figure into parts or imagine how a plane passing through certain points of the figure will dissect this figure and what will be inside. Such experiments not only help to better understand how, for example, a straight line and a plane, two planes or two lines intersect in space. This helps to see the possible applications of stereometry.

My father was an engineer who developed new diamond cutters for precision machining of materials. When introducing a new instrument, he drew different views of the spatial figure, mentally dissecting it with different planes, so that one could look inside the instrument and understand its unique complex structure. What do architects do when they present their new architectural masterpieces to an esteemed city council committee? They draw cuts! They mentally dissect the building and show its "insides." Why might we need a cross section in our simple life? We encounter them much more often than we think. Let's take a potato and avocado, identical in

E. Grigorieva, *Methods of Solving Solid Geometry Problems*,
https://doi.org/10.1007/978-3-031-86969-3_3

shape. How to show that these two vegetables have different structures inside? We make a cut, and thereby we see that the avocado has a round pit inside, while the potatoes do not. When making a cut, we form a cross section of an object with a plane. Let's look at a glass of water. The upper level of water can be taken as a plane cutting through the body of the glass. By tilting the glass and changing the position of the water, you can see the different cross sections that may appear in it.

We need cross sections both to understand the complex structure of a body inside and to estimate the volumes of complex objects that are homogeneous from the inside by dividing them with a cutting plane into simpler parts of known volumes. How can we construct the cross section? It's easy to cut a piece off a loaf of bread, especially if your job doesn't require precision. In solid geometry, we need precision. Even if we assume that a polyhedron-shaped body is made of a material that is easy to cut, we cannot always cut the entire figure with a knife and know what is in the cut. Depending on through which points of the polyhedron the cut is made, inside we will get polygons of different shapes. Suppose that the cut must pass through three specific points of the polyhedron. Without theoretical knowledge, it will be impossible to correctly cut a figure, even if you have a knife. Think about it: even such a simple figure as a cube can be cut so that the resulting cross section is a triangle, quadrilateral, pentagon or hexagon, and their type will depend on the location of the plane with respect to the cube.

It is known that any three-dimensional figure drawn on a plane looks flat, and if on a flat sketch some segments do not look parallel, then it is easy to make a mistake and intersect them. As we learned in Chap. 1, in a tetrahedron, three pairs of opposite edges lie on skewed lines. Suppose you connect two points lying on opposite edges of a tetrahedron. Now you take two more points, each on a different pair of opposite skew edges, and connect them. In a flat drawing, two lines appear to intersect. However, will the lines containing these segments actually intersect? By applying the methods of stereometry to a planar sketch of a tetrahedron, one can learn that depending on the location of the ends of these segments and whether the four points are coplanar or not, two lines may or may not intersect.

In this chapter, you will learn how to construct cross sections without the aid of electronic devices or computer programs, and how to apply the construction method in various areas—definitely skills needed to succeed in math competitions. You will practice solving some challenging problems of finding cross-sectional areas, determining the angle between a cross-sectional plane and another plane, and between a line in a cross-sectional plane and another line (such as an edge of a polyhedron). Methods for determining the cross-sectional area using its projection onto another plane and methods for finding the ratio of the volumes of spatial figures into which a cutting plane divides a polyhedron, as well as methods for finding how a cutting plane divides an edge of a polyhedron will also be disclosed. This chapter concludes with some interesting applications of stereometry for constructing shadows from various light sources, for constructing new polyhedra from well-known ones, and for solving problems in plane geometry.

3.1 Introduction to Cross Sections

I am sure that most of you had to solve a problem about a plane cutting a polyhedron or about finding a cross section which the secant plane leaves inside a polyhedron. A given secant plane usually goes through three given points of a polyhedron's faces, sides or through a point and a line parallel to some side of a polyhedra, or perpendicular to a side, or simply passing a point and a given edge, etc. Sometimes we have to not only construct the cross section, but also to evaluate its area or find out in what ratio the given second plane divides the volume of a polyhedron. However, for most of such problems we first must construct the correct cross section using the rules of Euclidean geometry.

Circa 2008 a local university was looking for an assistant professor with a background in math education. One of the three finalists was invited for an interview and had to present a lesson showing how he would teach his future students. The presenter selected a talk about geometry and started by saying that "most students usually do not have three-dimensional imagination and need help in solving solid geometry problems". One professor asked him to demonstrate his teaching skills by finding a cross section of a cube by a plane that goes through the midpoints (M, N, K) of the three skew edges $(AA', B'C'$, and $CD)$ of the cube, respectively, as it is shown in Fig. 3.1 on the left. After short thinking, the candidate simply connected three points and obtained the triangle MNK which he pronounced to be the cross section of the cube. (Please, see his wrong construction in Fig. 3.1 on the right.)

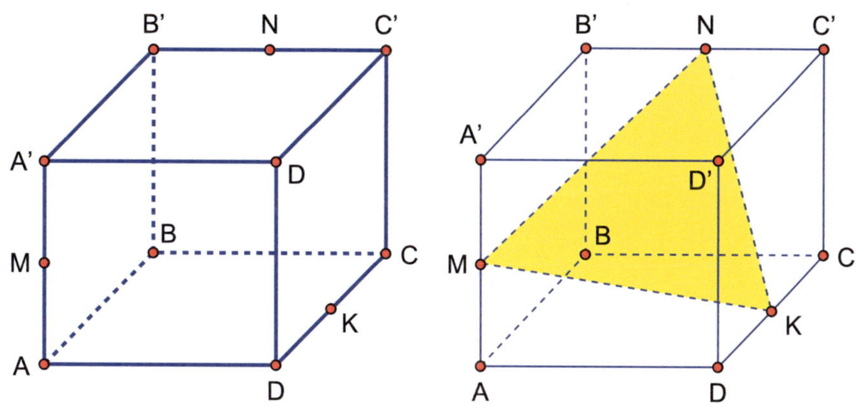

Fig. 3.1 Wrong cross section of a cube

Many people have difficulty constructing cross sections by hand. However, in this case, the candidate contradicted his own statement, since he clearly did not have the deep knowledge of Euclidean geometry to teach others or develop their spatial imagination. Of course, he didn't get the job.

Let us consider the concept of a cross section and its main elements.

What is a cross section? A cross section is a plane figure obtained by mental dissection of a solid body by a plane.

Definition 15 A **cross section** of a polyhedron is a part of the secant plane bounded by the faces of a polyhedron, passing through the given elements.
A plane is called a **secant** or **cutting plane** of a polyhedron if there are points of this polyhedron on both sides of the plane.

A cross section of a polyhedron is a polygon, each side of which lies on a face of the polyhedron. Thus, a cube has six faces, so from common sense we can conclude that its cross section can be a hexagon, a pentagon, a quadrilateral or a triangle (in case when the secant plane cuts only three faces of a cube).

By connecting the three given points, the job candidate drew triangle MNK belonging to the cross section. Points M, N and K are three nonadjacent vertices of that cross section, but the triangle MNK itself is not the cross- section. Soon we will learn how to construct cross sections of polyhedra using different methods. We will solve the aforementioned problem and will find out that its correct cross section is a regular hexagon which passes through the center of the cube and divides the volume of the cube into two equal parts. We will learn how to construct a cross section according to specified conditions, how to recognize a possible cross section dependent on the type of polyhedra and the way the cutting plane intersects it.

To construct a cross section of a polyhedron by a plane means to indicate the points of intersection of the cutting plane with the edges of the polyhedron and connect these points with segments belonging to the face of the polyhedron. The points of intersection of the secant plane with the edges of the polyhedron will be the vertices, and the segments belonging to the faces will be the sides of the polygon. In general, we will use the fact that if any two planes have two distinct common points, then they intersect by the line passing through these two points. Sometimes we do not have two distinct points right away, so we need to help ourselves by an additional construction and by using some well-developed methods and theorems. How can we find the line of intersection of a polyhedron by a cutting plane? This problem can be reduced to the repeated finding of (a) the line of intersection of two planes (faces of a polyhedron and the cutting plane) and (b) the point of intersection of the edge of a polyhedron with the cutting plane.

General algorithm for constructing cross sections of polyhedra:

- determine the faces with which the cutting plane has two common points, and draw straight lines through these points;
- determine the faces with which the cutting plane has one common point, construct a second common point and draw a straight line through them;

- determine the faces with which the cutting plane intersects but has no common points, construct two common points, and draw a line through them;
- select the line segments along which the cutting plane intersects the edges of the polyhedron, and shade the resulting polygon.

3.1.1 *Cross Sections of a Cube*

Assume that a plane cuts a cube. Because a cube has six faces, we can obtain one of the following convex polygons as the resulting cross section.

- a triangle (if three adjacent faces of the cube are intersected);
- a parallelogram (if two pairs of opposite faces are intersected—this includes a rhombus, square or rectangle);
- a trapezoid (if a pair of opposite faces is intersected by a cutting plane);
- a pentagon (if the plane cuts all five faces of a cube);
- a hexagon (if the plane cuts all faces of the cube).

We will not consider the cases when the plane has one common point with the cube or when it passes through only one of its edges because such cases do not fit the definition of the cross section. Assume that the cutting plane α passes vertices B' and D of a cube $ABCDA'B'C'D'$ and also passes the points M and N located on the edges of AA' and CC', respectively. It is not difficult to imagine all these possible cross sections inside a cube. For example, we can prove that if M and N are the midpoints of AA' and CC', respectively, then the quadrilateral $MB'ND$ shown in Fig. 3.2 on the left is a rhombus.

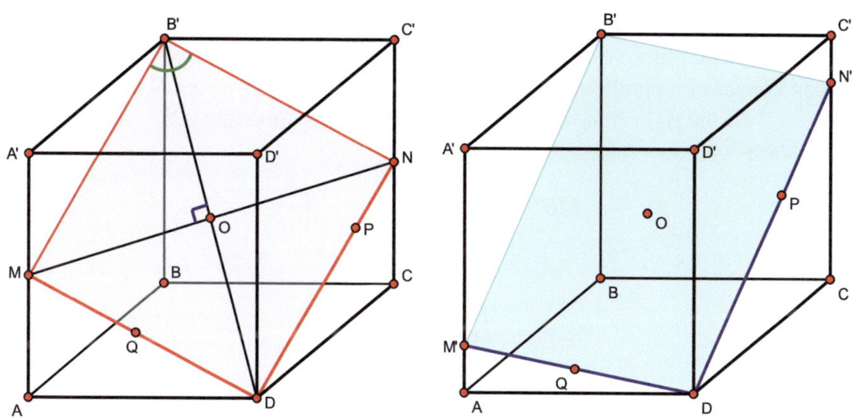

Fig. 3.2 Rhombus and parallelogram as possible cross sections of a cube

We can do it in several ways. For example, we can first prove that MN is perpendicular to $B'D$ using coordinate method. Thus, using the scalar product property and evaluating the scalar product of vectors MN and DB' we can demonstrate that it is equal to zero. For this let us introduce a coordinate system with the origin at point B. Then AB line will represent the positive direction of the X-axis, BC—the Y-axis, and BB'—the Z axis. In this system, the coordinates of points will have the following representations:

$$B' = (0, 0, 1),\ N = (0, 1, \frac{1}{2}),\ M = (1, 0, \frac{1}{2}),\ D = (1, 1, 0).$$

If the scalar product of vectors $\overrightarrow{MN}$ and $\overrightarrow{B'D}$ is zero, then the diagonals of the parallelogram $MB'ND$ are orthogonal. Obviously, all sides of it are equal that follows from Pythagorean theorem.

$\overrightarrow{MN} = (-1, 1, 0),\ \overrightarrow{B'D} = (1, 1, -1)$, then $\left(\overrightarrow{MN}, \overrightarrow{B'D}\right) = 0.$

Therefore, the diagonals of the parallelogram intersect at point O at $90°$.

Since in a square all sides are also equal and diagonals orthogonal, we need to show that the drawn cross section on the left of Fig. 3.2 is not a square but a rhombus of general type. And then we can find the angle between sides MB' and $B'N$ to show that the obtained figure $MB'ND$ is not a square by evaluating the angle between two adjacent sides of parallelogram or by using the Law of Cosines theorem applied to triangle $MB'N$:

$$\begin{aligned}&MN^2 = MB'^2 + B'N^2 - 2MB' \cdot B'N \cdot \cos\varphi,\quad \varphi = \angle MB'N\\&2 = \tfrac{5}{4} + \tfrac{5}{4} - 2 \cdot \tfrac{5}{4} \cdot \cos\varphi\\&\cos\varphi = \tfrac{1}{5} \neq 0\end{aligned}$$

Further, if we place point M anywhere below or above the midpoint of edge AA', we can construct a parallelogram through points M, B' and vertex D as it is shown in Fig. 3.2 on the right (not in scale). For simplicity, assume that $AM : MA' = 1 : 2$. In this case

$$MB' = \sqrt{1 + \left(\frac{2}{3}\right)^2} = \frac{\sqrt{13}}{3}$$

and

$$MD = \sqrt{1 + \left(\frac{1}{3}\right)^2} = \frac{\sqrt{10}}{3}$$

are the non-equal sides of the parallelogram $MB'N'D$.

All these cross sections are not difficult to imagine mentally, based on certain ideas and properties of plane figures. However, suppose that you are given simply three points on a cube through which the cutting plane α passes, and you need to construct the corresponding cross section. For example, suppose you need to construct a cross section of a cube formed by a plane passing through the points B', P and Q as shown in Fig. 3.2. B' is the vertex of the cube, and the points Q and P are just points belonging to the faces $DD'C'C$ and $AA'D'D$, respectively.

Figure 3.2 shows the results of the construction. How about constructing these cross sections based on the properties of Euclidean geometry? This task is completely different and can be difficult. It would be much easier to construct the same cross section if the condition specified that the cutting plane contains the points P and Q, but passes through the cube vertex D instead of the vertex B'.

Let's start by constructing a simple cross section and show how the resulting polygon will depend on the location of the points through which the cutting plane passes. For example, suppose we have a standard cube $ABCDA_1B_1C_1D_1$ and a plane α containing the diagonal AC and passing through the point M located on side B_1C_1. Let us show that if point M does not coincide with any of the vertices B_1 or C_1 of the cube, then the cross section will be a trapezoid. If $M = B_1$, then the cross section is the triangle AB_1C, and if $M = C_1$, then the cross section is the rectangle AA_1C_1C.

Let us draw a cube $ABCDA_1B_1C_1D_1$ and place point M on side B_1C_1 shown in Fig. 3.3.

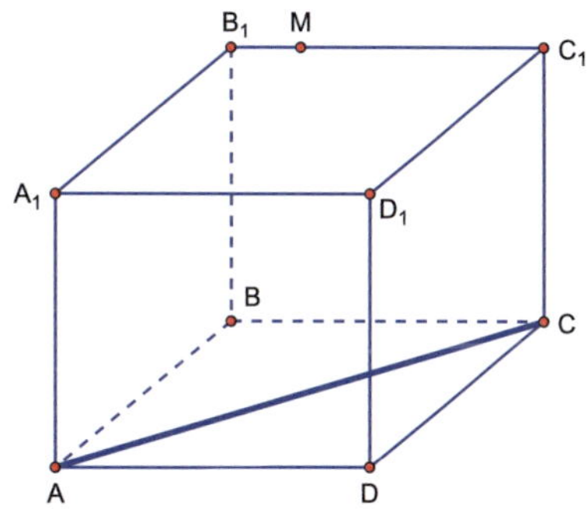

Fig. 3.3 Cube with point M and AC

Obviously, if point M coincides with vertex B_1, then we simply connect points $B_1 = M$ and C and A with M obtaining the triangle AB_1C. If we slide point M alone the edge so it becomes vertex C_1, then the cross section will be rectangle AA_1C_1C. Please, sketch these two cases on your own. If M is not one of the two vertices of the edge B_1C_1, then we will show that the cross section is going to be a trapezoid.

1. Since points M and C belong to both plane α and face BB_1C_1C, then the plane α intersects this face by segment MC, shown in Fig. 3.4. This can be written as follows.

$$\alpha \cap BB_1C_1C = MC.$$

2. We need to know how plane α intersects the face $A_1B_1C_1D_1$. Because α intersects face $ABCD$ by line AC, then by Theorem 10 it must intersect the top face by parallel to AC line that goes through point M. So, we obtain point K, such that

$$\alpha \cap AA_1B_1B = KM, \quad KM \| AC.$$

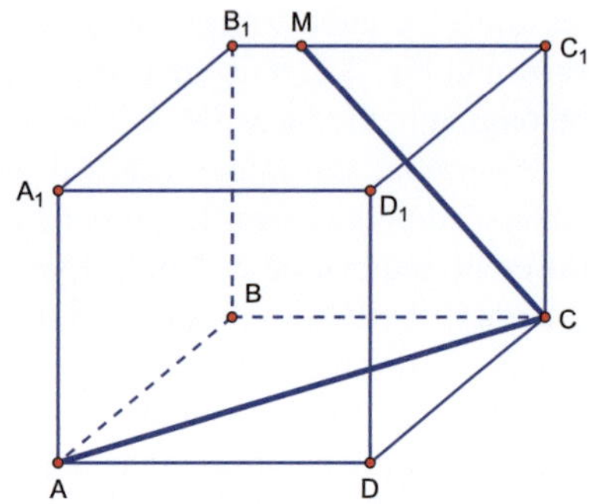

Fig. 3.4 Connecting points M and C

3. Finally, after connecting points K and A we can claim that the cross section is the trapezoid $AKMC$ (see Fig. 3.5).

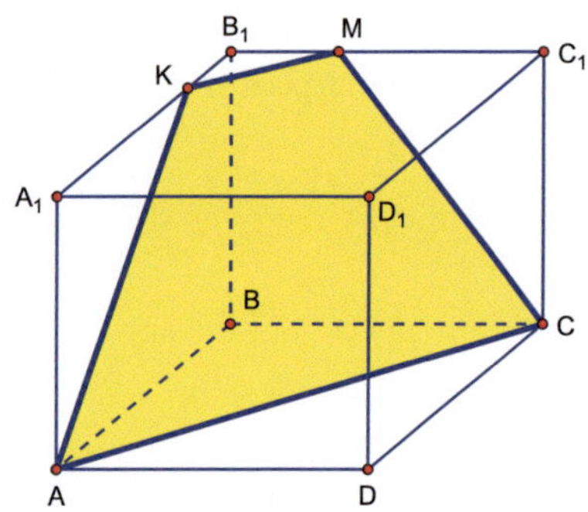

Fig. 3.5 Trapezoid is a cross section of a cube

Let us solve some interesting problems.

Problem 142 Which regular polygon can be obtained (and how) by cutting a cube with a plane?

Solution Despite the fact that cube's cross section can be a triangle, quadrilateral, pentagon, or hexagon, no plane can cut off a regular pentagon as a cross section. However, we can easily imagine an equilateral triangle, a square, and a regular hexagon as shown in Fig. 3.6. In order to cut a pentagon, the cutting plane must cross five different faces of the cube, then the two of the sides of the pentagon must lie on paralleled faces (planes) of a cube and should be parallel to each other. As we know, a regular pentagon does not have parallel edges. Therefore construction of a cross section in the shape of a regular pentagon is impossible.

In Fig. 3.6 we see a cross section bounded by the red equilateral triangle $AB'C$, with side $AB' = a\sqrt{2}$, the blue square $PEGK$, with side $PE = a$, and a green regular hexagon $ENFKHM$ with side $EN = \frac{a\sqrt{2}}{2}$.

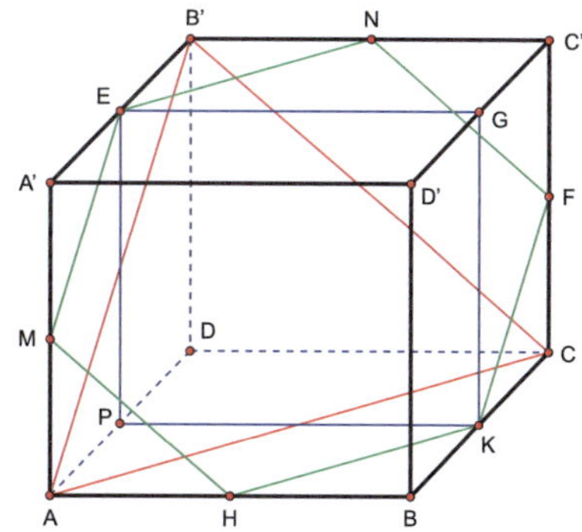

Fig. 3.6 Three regular polygons as cross sections of a cube

Problem 143 Given a cube $AA_1BB_1CC_1DD_1$ with side a. Plane α contains points L, M, and N, such that L is the midpoint of side B_1C_1, M and N are the midpoints of sides AB and CD, respectfully. Prove that the plane α divides the volume of the cube in the ratio $3 \div 1$.

Proof Since M and N are the midpoints, then MN segment is parallel to side AD and B_1C_1 and it is how plane alpha cuts the bottom face of the cube $ABCD$. Moreover, because point L of plane α belongs to side B_1C_1, then B_1C_1 is the intersection of the cube and α. Thus, the cross section of the cube by plane α is rectangle MNB_1C_1 (see Fig. 3.7).

Plane α inside the cube represents rectangle MB_1C_1N. Let us show that MB_1 is perpendicular to MN. It follows from the Theorem 20 on three perpendiculars.

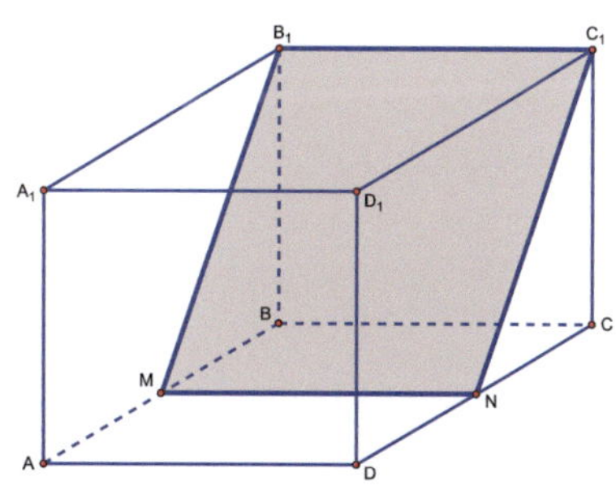

Fig. 3.7 Plane divides a cube into ratio of $3 \div 1$

$$MN \parallel BC, MB \perp MN, BC \perp BB_1$$

From our nice construction we can see that plane α cuts the cube into two convex polyhedra—one above the plane and one that is below, both bounded by the cube. The polyhedron below the cross section is the right triangular prism that has at the base a right triangle NCC_1 and the height of which is BC. The polyhedron above the cross section is the right quadrangular prism. Obviously, that if the volume of the triangular prism is V_1, then the volume of the second prism is $V_2 = V - V_1$, where V is the volume of the cube. Without loss of generality, we can assume that the side of the cube is equal to a. Thus, the volume of the triangular prism is

$$V_1 = a \cdot \frac{1}{2} a \cdot \frac{a}{2} = \frac{a^3}{4}.$$

and the volume of the quadrangular prism is

$$V_2 = a^3 - \frac{a^3}{4} = \frac{3}{4} a^3.$$

Therefore, the ratio of the volumes is $V_2 \div V_1 = 3 \div 1$. □

Problem 144 Given a cube $ABCDA_1B_1C_1D_1$. Point M is the midpoint of AB, N-midpoint of AD and K is the midpoint of side CC_1. A plane α contains points M, N, and K. Construct a cross section of the cube by this plane. Explain your constructions.

Solution First, let us sketch a cube and place the points with respect to the condition of the problem (Fig. 3.8).

Fig. 3.8 A cube with 3 points of the plane

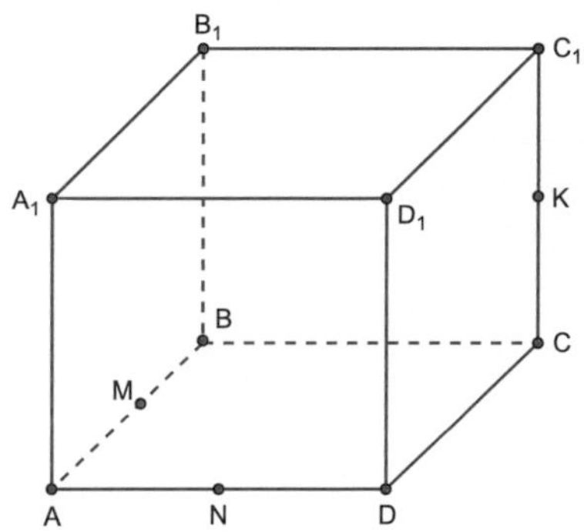

1. Since points M and N belong to both the plane and the $ABCD$ face of the cube, then the plane α crosses the bottom of the cube by line MN. Connect M and N.
2. Point K is not on plane $ABCD$, but it belongs to side CC_1 that is contained in face CC_1DD_1. So, we can find an additional point that belongs to both, plane α and plane CC_1DD_1.

Let us continue segment MN until it intersects with line DC, obtaining point P (see Fig. 3.9). This point P belongs to both plane α and plane CC_1DD_1 as well as point K. Therefore $\alpha \cap CC_1DD_1 = (PK)$. The line PK in turn, will intersect line DD_1 at point Q (Fig. 3.10)

Now, QK is the segment by which plane α intersects face CC_1DD_1. This can be written as

$$\alpha \cap CC_1DD_1 = [QK].$$

3. Moreover, since points N and Q belong to plane α and face AA_1DD_1, then $\alpha \cap AA_1DD_1 = [NQ]$ (Plane α intersects with face AA_1DD_1 by segment NQ)
Let us show these constructions, and emphasize the boundary of the polygon- cross section in red (Fig. 3.11).

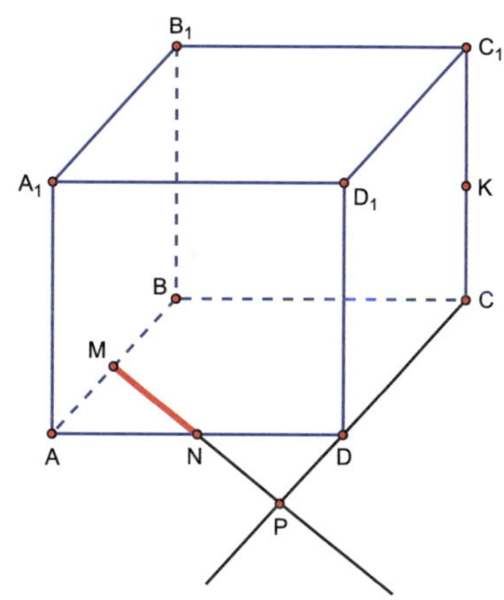

Fig. 3.9 Intersect MN and DC

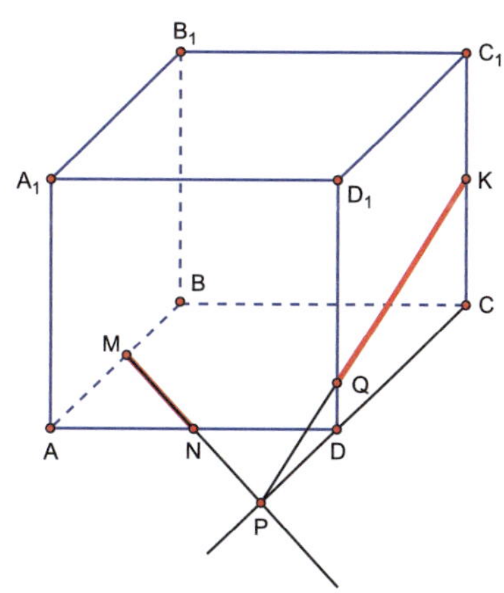

Fig. 3.10 Connect points P and K

With some experience, you would see now that the cross section is going to be a symmetric pentagon. We have already obtained its three sides MN, NQ, and QK (Fig. 3.11). Our final construction can go into two scenarios:

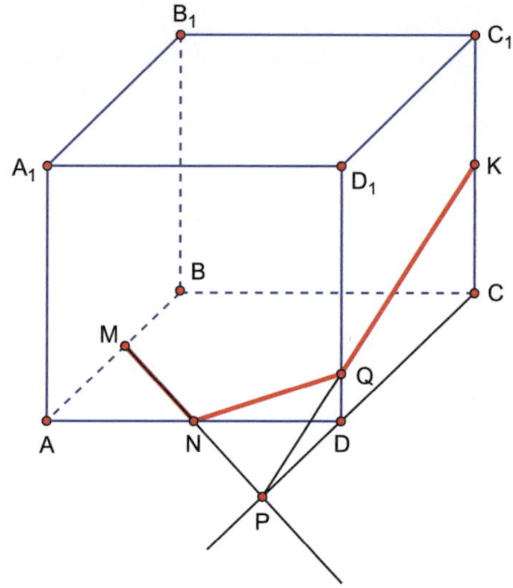

Fig. 3.11 Connect points N and Q

Method 1. We can use the property that if two planes are parallel and the 3rd plane intersects one of them, then it also intersects the second parallel plane, and the lines of intersection are parallel (see Theorem 10).

What do we have here?

The face AA_1BB_1 is parallel to DD_1CC_1 and plane α intersects face DD_1CC_1 by segment QK, then it must intersect the opposite to DD_1CC_1 face AA_1BB_1 by the line parallel to QK that goes through point M. We draw this parallel line in the plane AA_1BB_1 so it must intersect side BB_1 at some point, i.e., T (Fig. 3.12).

Point T also belongs to plane BB_1CC_1, so as point K. Therefore plane α intersects this face by segment KT. Similarly, plane α intersects face AA_1DD_1 by line NQ, since face BB_1CC_1 is parallel to AA_1DD_1, then plane α must intersect face BB_1CC_1 by the line that goes through point K and parallel to segment NQ. Obviously that this line must meet with line MT at unique point T. Our construction is done and the cross section $MNQKT$ is shown in yellow in Fig. 3.13.

Method 2. Instead of constructing parallel lines $TK \| NQ$ and $MT \| QK$, we can simply continue looking for common points of the plane α and faces of the cube. Thus, K belongs to BB_1CC_1. If we continue segment MN until it intersects line BC at point R, then point R now is contained by both plane α and face BB_1CC_1, then these two planes intersect by line RK (see Fig. 3.14).

Connecting points R and K in plane BB_1CC_1, we obtain point T of intersection of lines RK and BB_1 (Fig. 3.15). Finally, segment TK shows how α cuts face BB_1CC_1 and segment MT shows how plane α cuts face AA_1BB_1 of the cube. Pentagon $NMTKQ$ is constructed (Fig. 3.16).

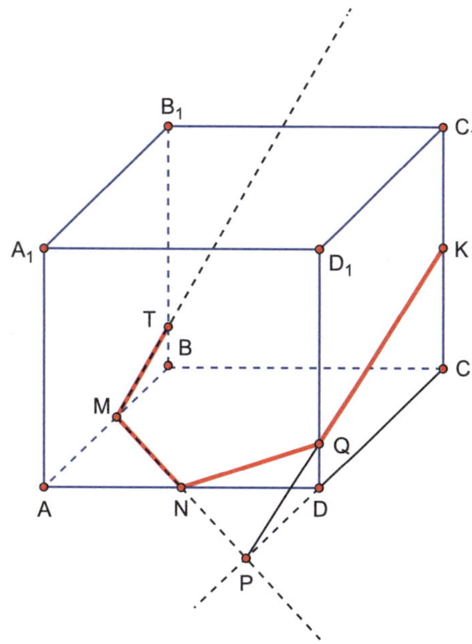

Fig. 3.12 *MT* is parallel to *QK*

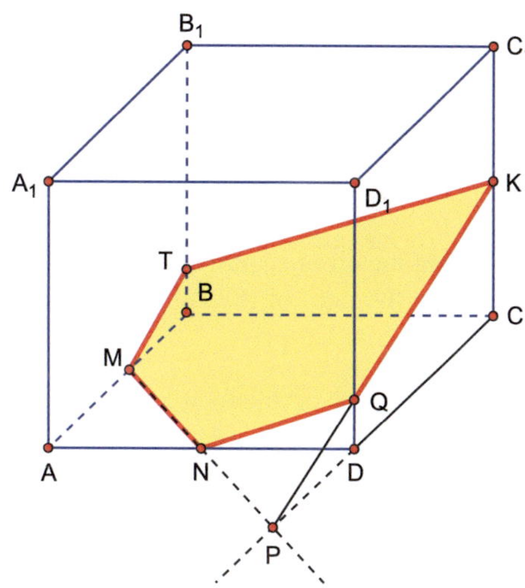

Fig. 3.13 Cross section is a symmetric pentagon

Remark. While for some of my readers construction by Method 1 may look easier than by Method 2, for many problems on cross section Method 2 produces a better visualization, the picture that is helpful for solving such problems as:

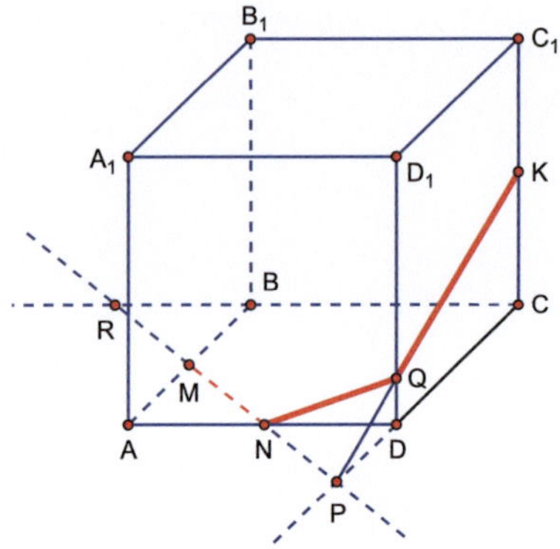

Fig. 3.14 Intersection of BC and MP

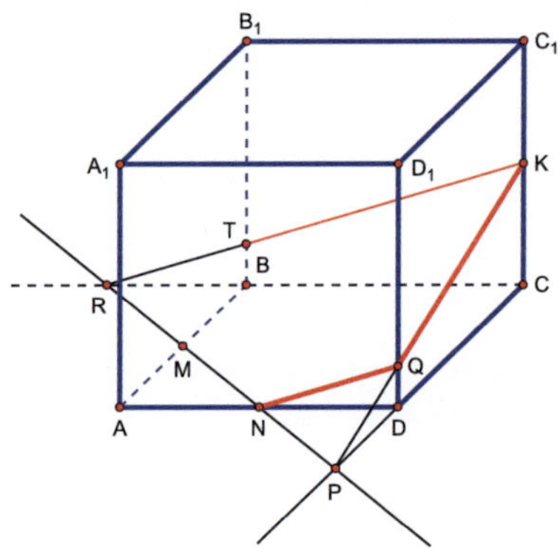

Fig. 3.15 Intersection of RK and BB_1

- Finding the cross-sectional area.
- Finding the ratio of the volumes of two spatial figures into which a given plane cuts a pyramid or prism.

Some of these problems we will solve very soon.

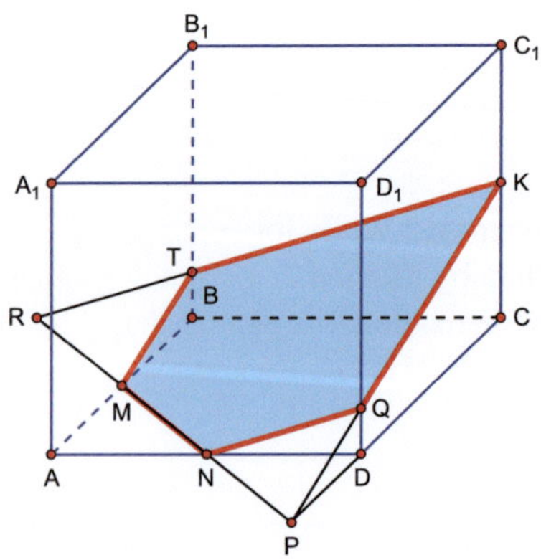

Fig. 3.16 Final construction of pentagon

Problem 145 Construct a cross section of the cube $ABCDA_1B_1C_1D_1$ by the plane that passes through points L, M and N that are the midpoints of sides C_1B_1, C_1C, and CD, respectively.

Solution By drawing a cube with the indicated points, we can immediately find how the plane intersects two adjacent faces, BB_1CC_1 and CC_1DD_1 of the cube—by simply connecting L, M and M, N (Fig. 3.17).

Next, we can continue segment LM and segment BC until they intersect at point K. Note that point K and point N both belong to two planes, α and face $ABCD$, then by connecting K and N we obtain the line by which plane α cuts the bottom of the cube and then to indicate the segment NP inside the cube (see Fig. 3.18).

Next, we can use the property of parallel planes $ABCD$ and $A_1B_1C_1D_1$ and then of AA_1BB_1 and CC_1DD_1. However, for better visualization, let us continue lines AB and PN obtaining point S. Next, we can continue lines LK and BB_1 obtaining point of intersection T. Points T and S belong to plane α and to the face AA_1BB_1 of the cube, then the plane cuts the cube by line ST as it is shown in Fig. 3.19.

Let this line intersect side A_1B_1 at point R and side AA_1 at point Q. Connecting P and Q, Q and R and then R and L, we obtain the hexagon $PHQLMN$ as the cross section of the cube by plane α (Fig. 3.20).

So far we constructed cross sections when we could connect some of the given points inside one of the faces of a polyhedron or could find some common points of a polyhedron and a secant plane right away. Sometimes three points determined a secant plane are located so that it is not possible to construct the cross section without introduction of an auxiliary plane. Let us use an **auxiliary plane method**

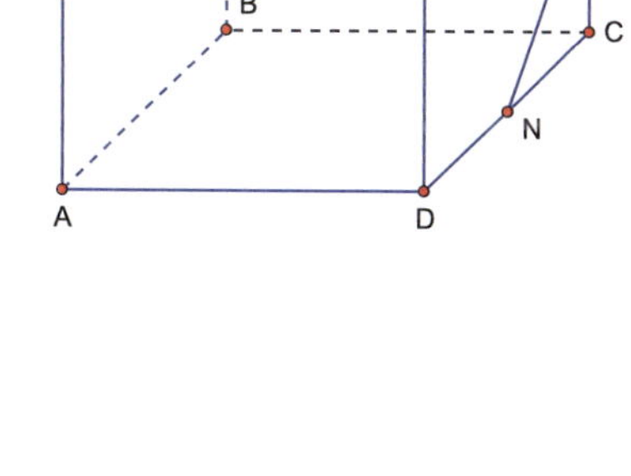

Fig. 3.17 Two sides of the cross section

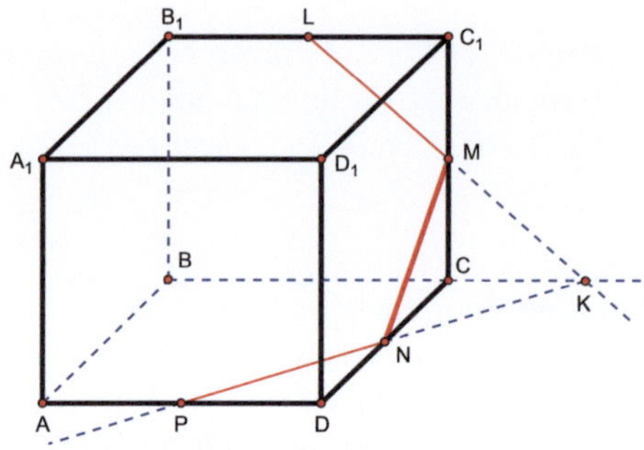

Fig. 3.18 Finding common point K and side PN

for the following problem, in which three points lie on the skew edges of a cube. You will see that one of the methods to solve this problem can be used right away for the problem described in the Introduction to this chapter.

Problem 146 Given a cube $ABCDA_1B_1C_1D_1$. A plane α contains points M, N and K that are the midpoints of the edges A_1B_1, CC_1, and AD, respectively. Construct a cross section of the cube by this plane.

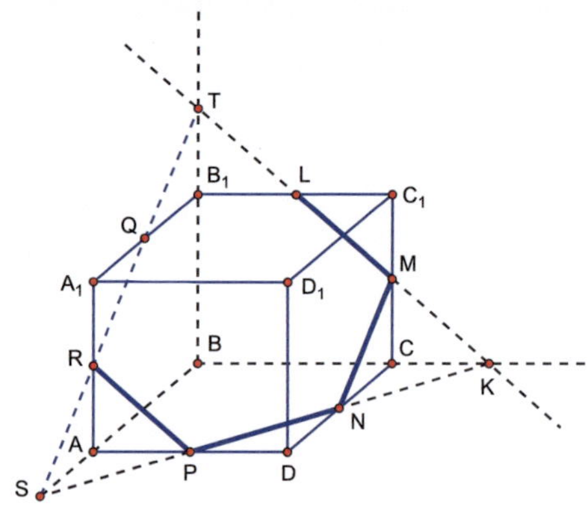

Fig. 3.19 Construction of lines ST and TK of the cross section

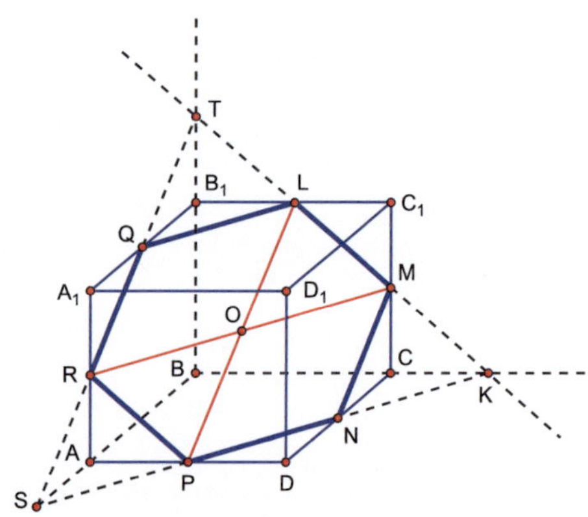

Fig. 3.20 Hexagon is the cross section

Solution Let us draw a cube and place the given points on the edges. We can see that in order to find how plane α cuts the cube, we must find an auxiliary plane that contain two of the given points. Then we could connect these two points because it would be the line of intersection of the auxiliary plane and α. Unfortunately, without auxiliary plane, such a construction will be difficult. Let us construct an auxiliary plane that contains points M and C_1. Please, follow Fig. 3.21.

1. Connect points M and C_1.

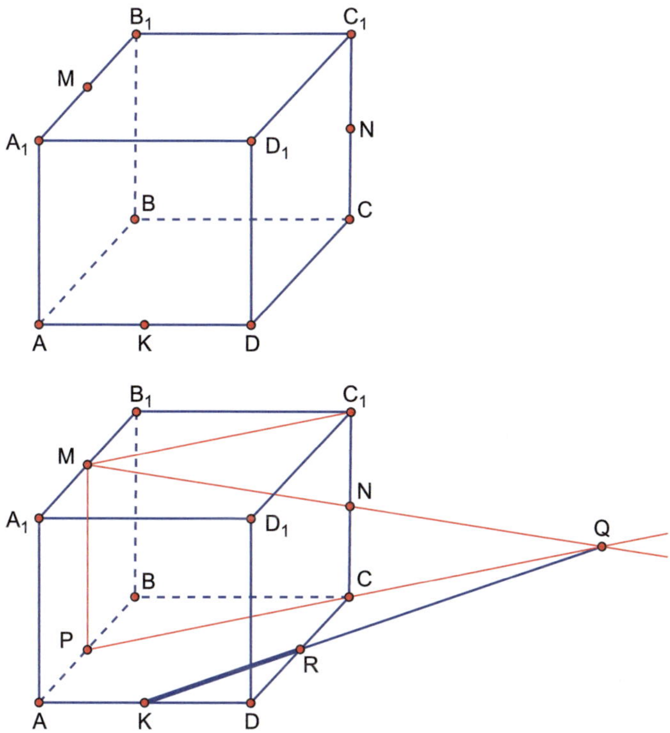

Fig. 3.21 Using auxiliary plane

2. Draw through M line parallel to C_1C. Mark the point of intersection of this plane and side AB as P. (Parallel projection)
3. Plane PMC_1C contains points M, N. Then MN intersects with line PC at some point, Q that can be written as

$$MN \cap PC = Q.$$

4. Consider points Q and K. They both belong to plane α and to plane $ABCD$ of the cube, then these planes intersect by line KQ.

$$\left.\begin{array}{l}\{K,Q\}\in\alpha\\ \{K,Q\}\in ABCD\end{array}\right\}\Rightarrow \alpha\cap ABCD=KQ$$

5. Note that $DC\cap KQ=R$. Thus, KR is the first edge of the cross-sectional polygon to be constructed.
6. Connecting points R and N we obtain segment RN, the second side of the cross section.

$$\alpha\cap DD_1C_1C=RN.$$

7. Next, we have to find how α intersects planes BB_1C_1C and $A_1B_1C_1D_1$. Continue lines RN and D_1C_1 until they intersect at point T (see Fig. 3.22).

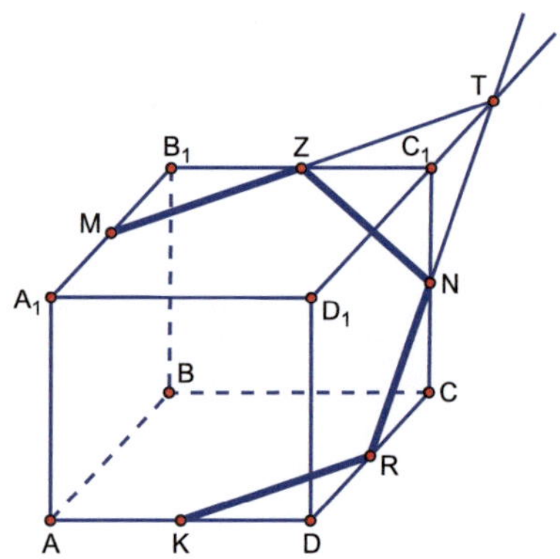

Fig. 3.22 Using auxiliary plane continued

$$RN\cap D_1C_1=T$$

8.

$$\left.\begin{array}{l}\{M,T\}\in\alpha\\ \{M,T\}\in A_1B_1C_1D_1\end{array}\right\}\Rightarrow \alpha\cap A_1B_1C_1D_1=MT$$

9. Now we can find the point of intersection of line MT and edge B_1C_1:

$$B_1C_1\cap MT=Z.$$

Next, connect points Z and N, etc. Obviously, now it is clear that the cross section is a hexagon $SMZNRK$ and I am sure that you can finish construction on your own.

For example, in the plane AA_1B_1B that is parallel to plane DD_1C_1C, we can draw a line $MS\|RN$, $S=AA_1\cap MS$, where MS is a side of the hexagon. Finally, connecting S and K, we complete our construction.

3.1.2 Cross Sections of a Pyramid

Suppose we have a triangular pyramid and a plane (α) defined by three points located somewhere on the pyramid. If the cutting plane passes through three points located on adjacent edges of the tetrahedron, then we simply connect the pairs of points and get a triangular section such as the EGF shown in the Fig. 3.23.

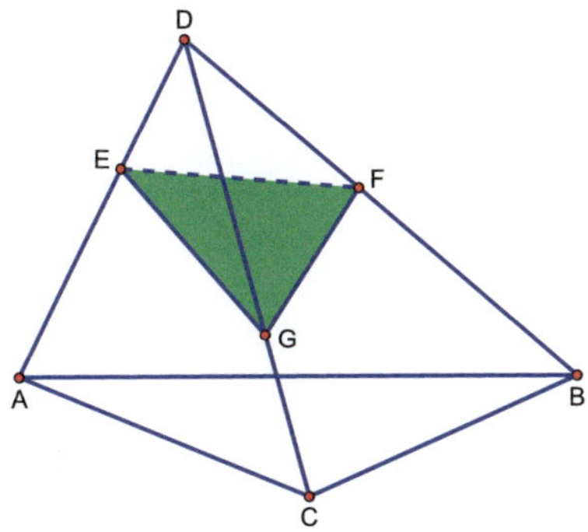

Fig. 3.23 Pyramid with three points on the edges

Let us fix the location of points E and F in Fig. 3.23 and start mentally moving point G along the edge DC. We can imagine different triangles of different sizes but if point G stays within the boundary of the segment DC, the cross section will be always a triangle. If we mentally continue the edge DC and place point G somewhere on this line below the base ABC, we will see that the line EG will intersect another side AC and hence, the cross section of the tetrahedron by this plane will be a quadrilateral. It is interesting that this case coincides with the cross section obtained when one of the points is sitting on the face ACD, that we will consider next.

If a cutting plane α goes through one point lying on one edge and two other points, each of which belongs to a different face of the pyramid (Fig. 3.24a), then we can obtain either a triangle as the cross section or a quadrilateral depending on the location of the points with respect to each other.

This situation can be written as

$$\alpha = \{M, K, L\}, \ M \in AD, \ K \in ADC, \ L \in CDB.$$

Consider the case presented by Fig. 3.24b. First, we can connect points M and K and we can see that the line MK will intersect the edge DC at some point, i.e., P.

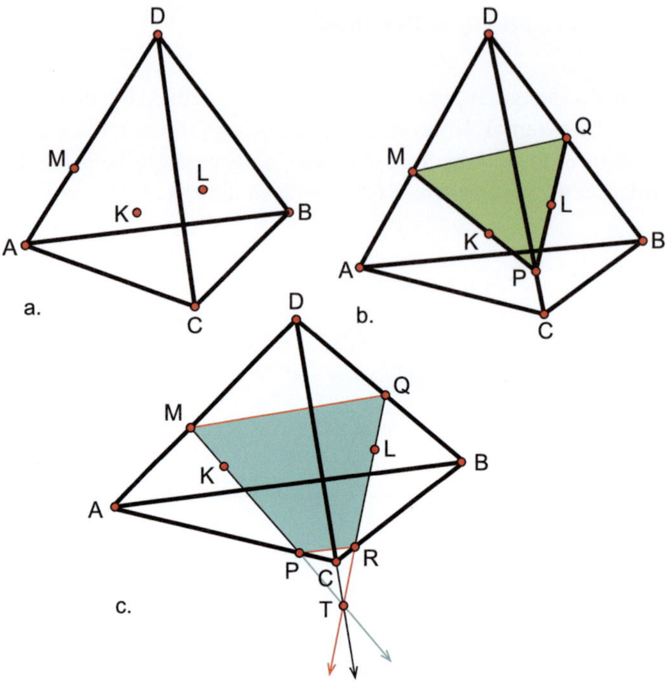

Fig. 3.24 Pyramid with two points on the faces and one on the edge

Then segment MP is the side of the cross-sectional polygon that is going to be the triangle MPQ.

$$\left.\begin{array}{l}\{M, K\} \in \alpha, \\ \{M, K\} \in ADC\end{array}\right\} \Rightarrow \alpha \cap ADC = (MK)$$

Then we can further write

$$\left.\begin{array}{l}MK \cap DC = P \\ PL \cap DB = Q\end{array}\right\} \Rightarrow \alpha \cap CDB = PQ,$$

and MPQ is the cross section.

However, if points M and K when connected make the line that intersects the edge AC instead, as shown in Fig. 3.24c, then the cutting plane will go through the base ABC and will cut the quadrilateral $MPRQ$ as the cross section. This can be written as follows.

$$\left.\begin{array}{l}MK \cap AC = P \\ MP \cap DC = T \\ TL \cap CB = R \\ TL \cap DB = Q\end{array}\right\} \Rightarrow \alpha \cap ABCD = MPRQ$$

Sometimes, in order to construct a cross section of a pyramid, we need to use auxiliary structures.

Problem 147 Construct a cross section of the triangular pyramid *DABC* by a plane α passing through the points E, F, G, if E belongs to the face *DAB*, F belongs to the face *DCB*, G belongs to the face *DAC*.

Solution Let us start from sketching a pyramid and placing the points on its lateral faces (see Fig. 3.25).

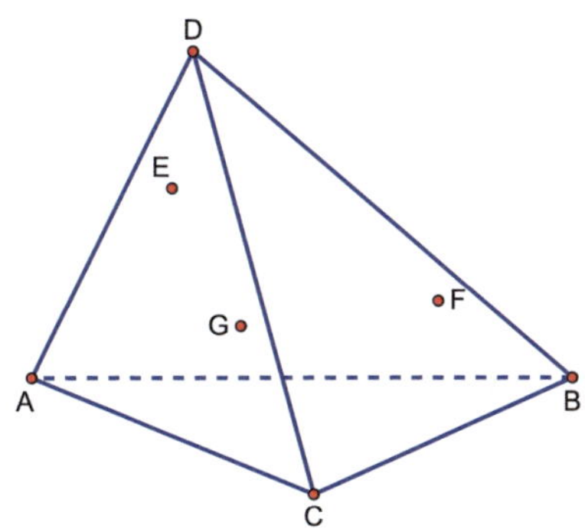

Fig. 3.25 Pyramid with three points on the lateral surface

Because all three given points lie in different planes, we need to find additional points for each face that are common for the cutting plane and the corresponding face. Let us connect points D and E, drawing line DE until it intersects with side AB at point H, then connect points D and G, making line DG until it intersects side AC at point J and finally, connect points D and F so the line DF will intersect CB at point K (see Fig. 3.26). Points H, J and K are the traces of the lines DE, DG and DF on the base ABC.

If in plane DHJ we intersect lines EG and HJ, we obtain point L shown in Fig. 3.27. The line EG intersects the basis of the pyramid, plane ABC, at this point L. This can be written as

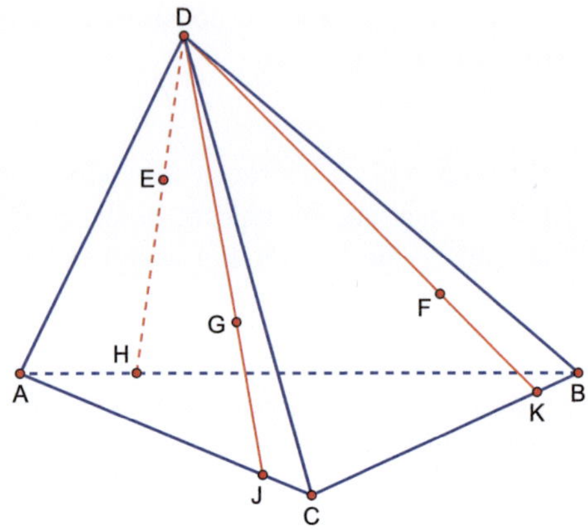

Fig. 3.26 Pyramid with traces of the three lines

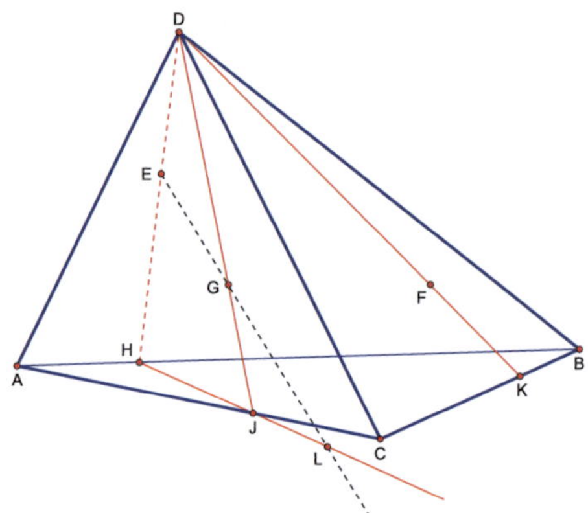

Fig. 3.27 Intersection of line HJ and EG

$$EG \cap HJ = L.$$

Consider point L. It does belong to the cutting plane because it is on line EG, $L \in \alpha$.

How can we find another point that belongs to both, plane α and the basis plane ABC? We can, for example, in plane DEF, continue lines EF and HK, they will intersect at some point M, that will belong to both plane α and to the basis plane ABC (Fig. 3.28). Next, points M and L belong to the basis plane ABC and to the cutting plane, α, then the two planes intersect by the line going through M and L and the following is true.

$$\left.\begin{array}{l}\{M, L\} \in \alpha \\ \{M, L\} \in ABC\end{array}\right\} \Rightarrow \alpha \cap ABC = (ML)$$

Next, we can find how this line ML intersects segment AC and BC. We can mark points P and Q as it is shown in Fig. 3.29.

$$ML \cap AC = P$$
$$ML \cap BC = Q$$

The first segment PQ of the cross-sectional polygon is constructed. Next, we can connect the given point G and point P. Hence plane α intersects face ACD by this line GP, line GP intersects edge AD at point N, then the second segment NP of the quadrilateral cross section is obtained (Fig. 3.29).

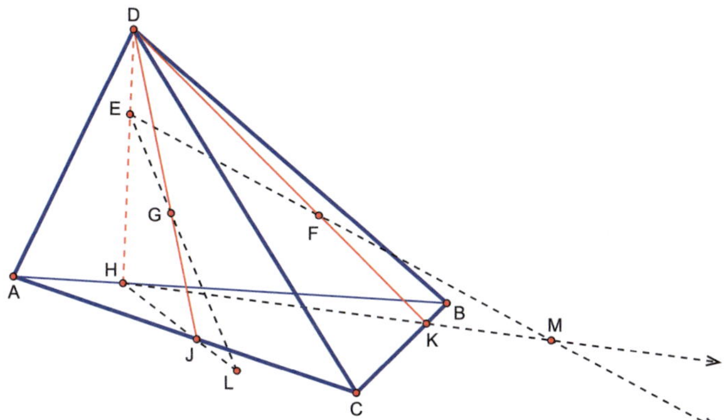

Fig. 3.28 Intersection of EF and HK

$$\left.\begin{array}{l}\{P, G\} \in \alpha \\ \{P, G\} \in ADC\end{array}\right\} \Rightarrow \alpha \cap ADC = (PG)$$

Next, $(PG) \cap (AD) = N$. As soon as we have point N, we can connect it with point E on the back face of ADB, the line NE will intersect the edge DB at point S. Then finally we connect point S and Q in face DCB to obtain the fourth side of the cross-sectional quadrilateral $PNSQ$ (Fig. 3.30).

Alternatively, we could connect points Q and F in the face DCB, line QF is the intersection of plane α and face DCB. Let S be the intersection of line FQ and segment DB. Next, we would connect points S and N to obtain the cross section $PQSN$. From the final sketch you must notice that the lines EM and SQ intersect at one point F.

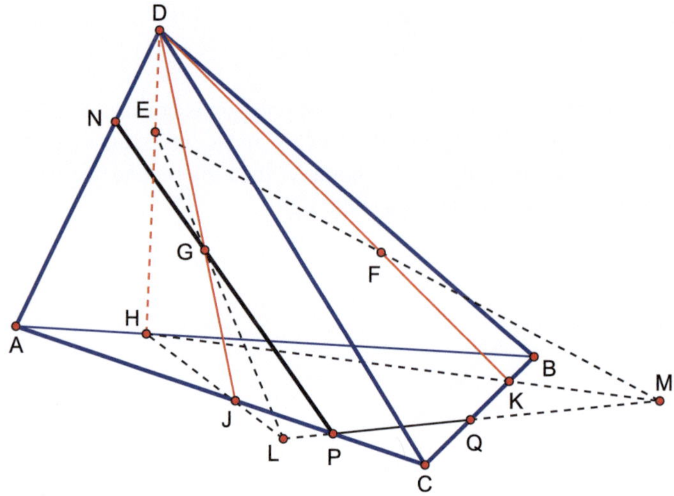

Fig. 3.29 Sides *PQ* and *NP* of the cross section

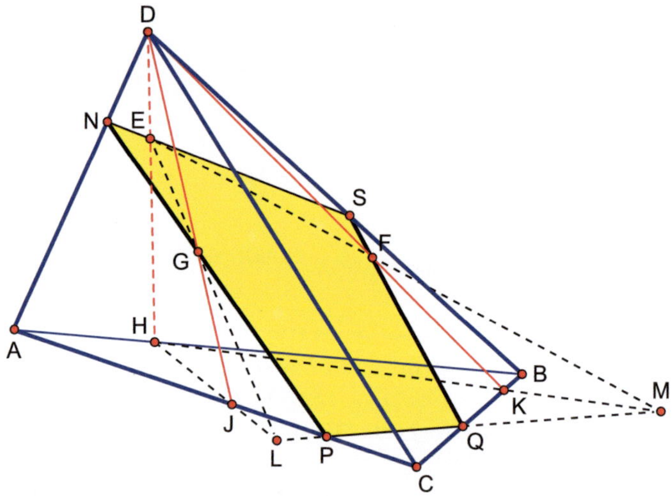

Fig. 3.30 Cross section *PQSN* is constructed

3.2 General Methods of Construction of Cross Sections

We have already constructed several cross sections of polyhedra, and perhaps without realizing it, we used different construction methods. There are two basic methods—the trace method and the projection method. If you already understand how a cross section is constructed, then you can skip this section and immediately proceed to solving problems.

3.2.1 Trace Method

Assume that the cutting plane intersects one of the faces of the polyhedron along a straight line, and this line "pierces" (intersects) the base plane at point M. This point is the trace of the line on the base plane. If another line, which is also the intersection of the cutting plane and some other adjacent face of the polyhedron, also pierces the base, but at another point N, then MN is the trace of the cutting plane on the base plane of the polyhedron. Other options are also possible. For example, we construct one point as point M, and the other can already lie on the face of the base as one of the points of the cutting plane and wait for a line to be constructed through it, along which the cutting plane will intersect the base of the polyhedron.

Trace is the line of intersection of the cutting plane and a plane of any face of the polyhedron. To construct a trace, it suffices to know two of its points, i.e., points that lie simultaneously in the cutting plane and the plane of the face under consideration. In general, a cutting plane can intersect the plane of each face of the polyhedron and each of the lines on which the edges of the polyhedron lie.

Algorithm for constructing a cross section using the Trace Method:

1. Find out if there are two points of the secant (cutting) plane in one face (if so, then the side of the cross section can be drawn through them).
2. Construct the trace of the secant plane on the base of the polyhedron.
3. Find an additional point on the edge of the polyhedron (extend the side of the base polygon on which a section point is located until it intersects with the trace).
4. Draw a line through the obtained additional point on the trace and the section point in the selected face, mark the points of its intersection with the edges of the face.
5. Repeat step 1 (if necessary).

The essence of the Trace Method lies is to construct of an auxiliary line, which is the image of the line of intersection of the cutting plane with the plane of any face of the figure. It is often convenient to find an image of the line of intersection of the cutting plane with the plane of the lower base. Using a trace, it is easy to construct images of the cutting plane points located on the edges or faces of a given figure.

Basic rules for constructing cross sections using the Trace Method:

- If two points of the secant plane (for example, α) are given (or have been constructed) on one face of the polyhedron, then the trace of the plane α on this face is a straight line passing through these two points.
- If a line of intersection of the secant plane with some face of a polyhedron is given (or constructed) and there is a point of the plane α belonging to another face of the polyhedron adjacent to this one, then it is necessary to determine the point of intersection of this trace with the common side of the two faces.
- The intersection point of a cutting plane with the base or any other face can be defined as the point of intersection of any line in the cutting plane with its projection onto the base plane.

By solving the following problem we can apply this method.

Problem 148 On the edges BB', CC' and DD' of the prism $ABCDA'B'C'D'$ the points M, N and K are respectively given. Construct the trace of the cutting plane $MNK = \alpha$ on the base face of the prism. The base of the prism is a trapezoid. Construct the cross section.

Solution Let start from sketching the situation described in the problem and obtain the first two sides *MN* and *KN* of the future cross section by connecting the given points, M, N and N, K, respectively (see Fig. 3.31).

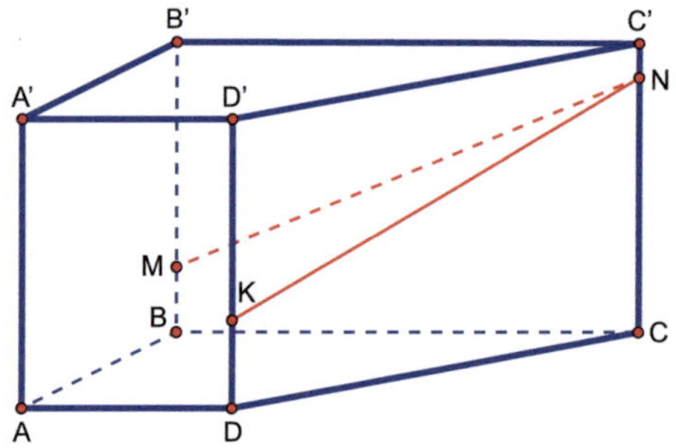

Fig. 3.31 Two sides of the cross section

In order to find out how the plane α intersects other faces of the prism, we can construct its traces, for example, on the base of the prism.

1. In the plane of face $BB'C'C$ we connect points N and M and then intersect this line with line BC (Fig. 3.32).

$$S_1 = MN \cap BC$$

   S_1 is the trace of line MN on the base of the prism.
2. In plane $DD'C'C$ connect points K and N and continue until this line will intersect with line DC at point S_2:

$$S_2 = KN \cap DC$$

   S_2 is the trace of line KN on the base of the prism.

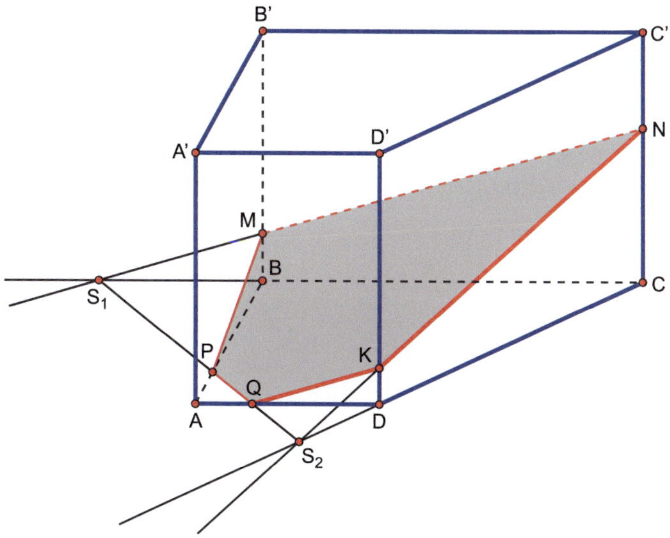

Fig. 3.32 Pentagonal cross section of a prism with two traces

3. Two points, S_1 and S_2, belong to both, the base of the prism $ABCD$ and the secant plane α, then the plane α intersects the base of the polyhedron by this line, which can be written as

$$\left.\begin{array}{l} S_1, S_2 \in ABCD \\ S_1, S_2 \in \alpha \end{array}\right\} \Rightarrow \alpha \cap ABCD = S_1 S_2$$

Therefore line S_1S_2 is the trace of the secant plane $\alpha = MNK$ on the base of the prism (see Fig. 3.32).

4. Connect points S_1, S_2 and find the intersection of line S_1S_2 with edge AB (point P) and edge AD (point Q).
5. Connect points P and M obtaining the side of the cross section.
6. The cross section of the prism (pentagon $PMNKQ$) is constructed (see Fig. 3.32).

Note 1. Because the prism has a trapezoidal base, then the faces $AA'D'D$ and $BB'C'C$ are parallel. Then QK is parallel to MN, and the entire construction of the cross section could go by a different scenario. We would still need S_1 point. We would draw KQ parallel to MN. And then since S_1, Q belong to both the secant plane and to the base $ABCD$, then we would get S_1Q by which these two planes intersect. This can be written as

$$\left.\begin{array}{l} \alpha \cap ABCD = (S_1 Q) \\ S_1 Q \cap AB = P \end{array}\right\} \Rightarrow \alpha = MNK \cap ABCD = [PQ]$$

Finally, by connecting points P and M we will get the last side of the cross section, pentagon $MNKQP$ (see Fig. 3.32).

Note 2. Analyzing different locations of points M, N, and K on the edges of the prism, we could get different cross sections. In Fig. 3.33 we can see how it is possible that the trace of the secant plane on the base (S_1S_2) can go through vertex $A = P = Q$, and then the cross section becomes a quadrilateral (trapezoid).

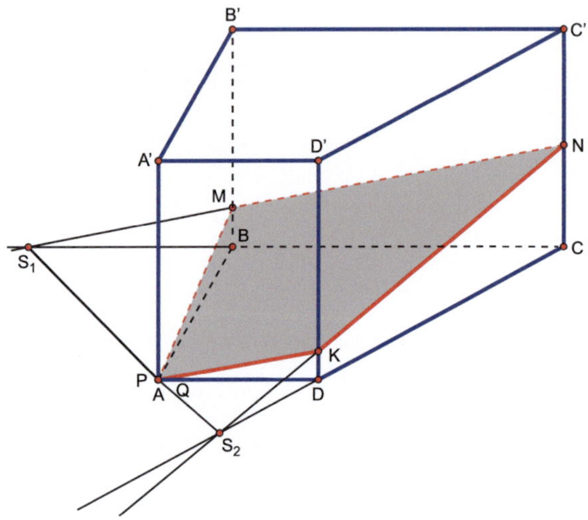

Fig. 3.33 Cross section of the prism is a trapezoid

3.2.2 Auxiliary Planes and Method of Internal Projection

Internal design projection method or the method of auxiliary planes for constructing cross sections of polyhedron is sufficiently universal. In cases where the desired trace (or traces) of the secant plane is outside the drawing, this method even has certain advantages. Additionally, in some cases the method of auxiliary plane sections turns out to be the most rational, especially when the points of the secant plane lie on the skew edges of a polyhedron (for example, as it was in the problem mentioned in the Introduction to this chapter)

When applying this method, each given point is projected onto the base plane. There are two possible types of design: central and parallel.

- **Central projection** is usually used when constructing cross sections of pyramids, while the top of the pyramid is the center of the projection. Central projection (perspective) is characterized by the fact that the projecting rays come from a single point, called projection center. The resulting image is called central projection.
- **Parallel projection** is used when constructing cross sections of prisms. Parallel projection is characterized by the fact that the projecting rays are parallel to each other. The source of the rays is located at infinity. Perpendicular projections are also called orthogonal.

If the polyhedron whose cross section is being constructed is a prism, then the projection onto the plane of the lower base is carried out parallel, in the direction determined by the lateral edge of the prism. If the polyhedron is a pyramid, then a central projection is performed onto the base plane; the center of the design is the apex of the pyramid, where all the lateral edges meet.

By parallel or central projection of points we construct auxiliary planes in which additional common points of the secant plane and one of the faces of the polyhedron can be found. This method is applied to many problems below.

The usual steps for the internal projection method can be summarized as follows

1. Construct an auxiliary plane and find the line of its intersection with the secant plane.
2. Construct a trace of the plane on the edge of the polyhedron.
3. If there are not enough cross-sectional points to complete the construction of the cross section, repeat steps 1–2.

Problem 149 In Fig. 3.34, a cube $ABCDA'B'C'D'$ is given. Points Q and P lie on the adjacent faces of the cube, such that Q and P belong to the medians dropped from the common vertex D of the triangles $AA'D$ and $DC'C$, respectively. (a) Construct the cross section of the cube by the cutting plane α that contains vertex of the cube B' and points P and Q. (b) Prove that the cross section is always the same rhombus, another vertex of which is vertex of cube D.

Solution Part (a) Construction. We can look only for the intersection of the lines that both lie in the same plane. Hence, the method of parallel projection will be beneficial. We will project points B' and Q onto the base of the cube $ABCD$ parallel to its edge BB'. Note that because point Q belong to face $AA'D'D$, its parallel projection must intersect the edge AD at some point Q'. Similarly, the internal projection of point P parallel to side BB' must intersect the edge DC of the cube, say at point P'. This can be written as follows

$$QQ' \parallel B'B \parallel PP', \quad Q' = QQ' \cap AD, \quad P' = PP' \cap DC.$$

Note. Sometimes spatial drawing is problematic and students by a mistake find the "image" of point Q on edge AB that is wrong and only an illusion since point Q belongs to face $AA'D'D$, not to $AA'B'B$.

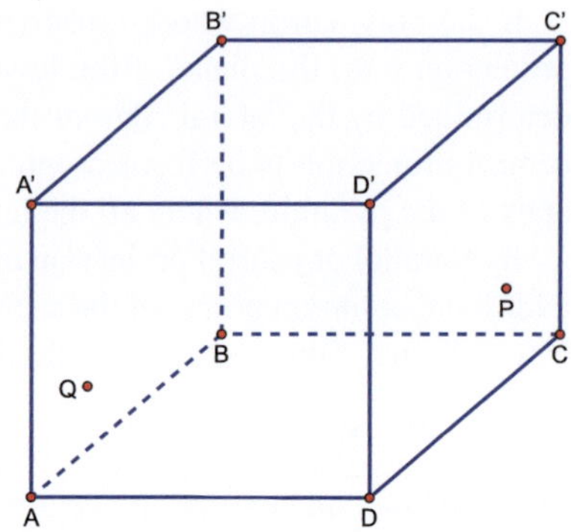

Fig. 3.34 Cube with two points on its adjacent faces

Next, in plane $QQ'BB'$ we will intersect lines $B'Q$ and BQ':

$$F = B'Q \cap BQ'$$

And in plane $BB'P'P$, we will intersect lines $B'P$ and BP:

$$G = B'P \cap BP'$$

Please, see Fig. 3.35.

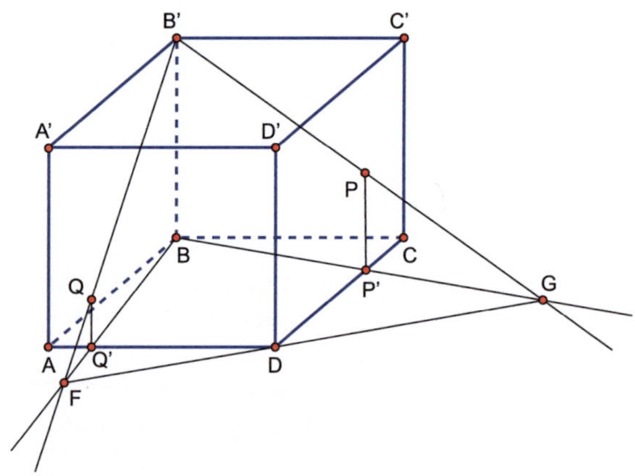

Fig. 3.35 Parallel projection is constructed

Points F and G are the traces of lines $B'Q$ and $B'P$ on the plane of the cubes' base. Moreover, the following is valid.

$$\left.\begin{array}{l}\{F, G\} \in \alpha \\ \{F, G\} \in ABCD\end{array}\right\} \Rightarrow \alpha \cap ABCD = (FG)$$

It looks like line FG indeed contains vertex D, then the cross section of the cube has D as its vertex. Connect points D and Q, this line must now intersect the edge AA' at the midpoint M. MB is the side of the cross-sectional parallelogram. Similarly, continuing line DP until in intersects the edge CC' at its midpoint point N, we will obtain the second edge of the cross section. Finally, we connect points M and B' and B' and N and obtain two other sides of the parallelogram-rhombus $MB'ND$ (see Fig. 3.36).

Part (b) Let us now prove that wherever on the medians of $\triangle AA'D$ and $\triangle CC'D$ points Q and P are located, the plane passing vertex B' and points P and Q will always make the same cross- sections inside the cube—the rhombus $B'NDM$, where M and N are the corresponding midpoints of sides AA' and CC', and D is the vertex of the cube.

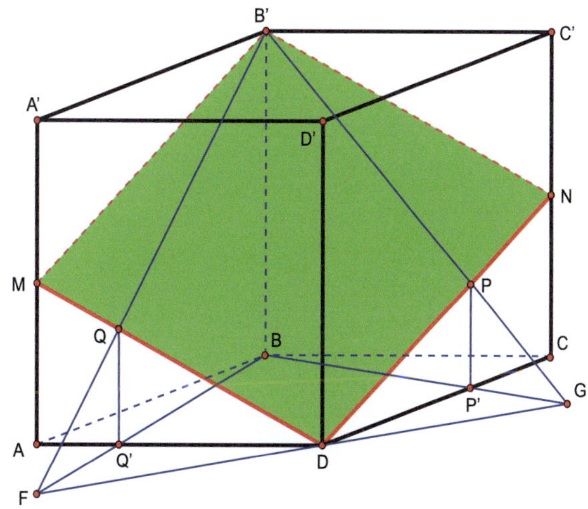

Fig. 3.36 Rhombus $MB'ND$ is constructed

To prove this let us use analytic geometry approach and introduce the coordinate system centered at point D, with the X-axis positive in direction of vector DC, the Y- axis positive in direction of vector DA and Z-axis positive in direction of vector DD' (see Fig. 3.37).

In this coordinate system, our points have the following coordinates:

$$B'(1, 1, 1),\ D(0, 0, 0),\ M(0, 1, \frac{1}{2}),\ N(1, 0, \frac{1}{2})$$

Because point $Q \in [MD]$, then it is collinear with point M and its coordinates can be written as

$$Q(0, m, \frac{m}{2}),\ 0 < m < 1.$$

Similarly, point $P \in [DN]$, then its coordinates can be expressed as follows

$$P(n, 0, \frac{n}{2}),\ 0 < n < 1.$$

In general case $n \neq m$.

Our plan can be summarized as follows

- Using formula for the line in space (1.63) passing two given points, we can now find the equations of the lines $B'Q$ and $B'P$.
- Find the coordinates of points F and G at which these lines intersect the base of the cube (plane $z = 0$)

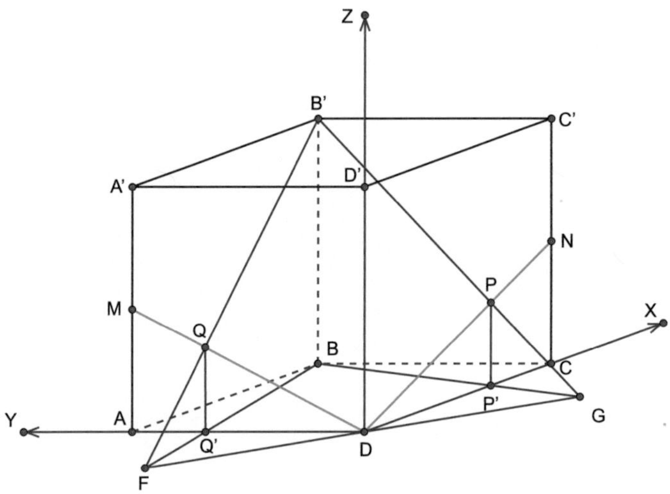

Fig. 3.37 The coordinate system is introduced

- Find the equation of line FG and show that point D belongs to this line, $D \in (FG)$.

For line $B'Q$ we obtain

$$\frac{x-1}{-1} = \frac{y-1}{m-1} = \frac{z-1}{\frac{m}{2}-1} \tag{3.1}$$

It is helpful to rewrite line (3.1) as intersection of two planes. For this, we can consider and solve separately the first and the second and the first and the third part of it.

$$\frac{x-1}{-1} = \frac{y-1}{m-1},$$
$$(x-1)(m-1) = 1 - y$$
$$(m-1)x + y = m$$

and for the first and the third terms of (3.1) we have

$$\frac{x-1}{-1} = \frac{z-1}{\frac{m}{2}-1}$$
$$(m-2)(x-1) = 2(1-z),$$
$$(m-2)x + 2z = m$$

Next, we will solve the system of three linear equations and find the point of intersection of three planes. This will find the coordinates of point F, the intersection of line $B'Q$ with the base of the cube, plane $z = 0$.

$$\begin{cases} (m-1)x + y = m \\ (m-2)x + 2z = m \\ z = 0 \end{cases} \tag{3.2}$$

We can easily solve the system (3.2), first replacing $z = 0$ into the second equation and find x, then find y coordinate from the first equation.

$$x = \frac{m}{m-2}, \quad y = -\frac{m}{m-2}, \quad z = 0.$$

Coordinates of point F are

$$F = \left(\frac{m}{m-2}, -\frac{m}{m-2}, 0\right) \tag{3.3}$$

Next, we will find the coordinates of point G as intersection of line $B'H$ with the base of the cube. For line $B'Q$ we obtain

$$\frac{x-1}{n-1} = \frac{y-1}{-1} = \frac{z-1}{\frac{n}{2}-1} \tag{3.4}$$

Considering again the first and the second term of (3.4):

$$\frac{x-1}{n-1} = \frac{1-y}{1},$$
$$(x-1) = (n-1)(1-y)$$
$$x + (n-1)y = n$$

For the first and the third terms together of (3.4), we obtain

$$\frac{x-1}{n-1} = \frac{z-1}{\frac{n}{2}-1},$$
$$\frac{x-1}{n-1} = \frac{2(z-1)}{n-2}$$
$$(n-2)x - 2(n-1)z = -n.$$

After replacing line $B'P$ by two intersecting planes, we need to intersect them with $z = 0$, the base of the cube plane:

$$\begin{cases} x + (n-1)y = n \\ (n-2)x - 2(n-1)z = -n \\ z = 0 \end{cases} \tag{3.5}$$

The solution of system (3.5) is

$$x = -\frac{n}{n-2}, \quad y = \frac{n}{n-2}, \quad z = 0.$$

Hence, the coordinates of point G are

$$G = \left(\frac{-n}{n-2}, \frac{n}{n-2}, 0\right) \tag{3.6}$$

By looking at points F and G given by (3.3) and (3.6), we notice that their coordinates are opposite numbers, so, we can state that both points lie on the line that passes the center of the coordinate system, for any defined values of parameters m and n. If we connect these two points trying to get the trace of the secant plane on the base, the line will always go the same way and will have only one intersection with the base of the cube at vertex D. Points F, G, and D are collinear. Therefore the cross section of the cube by the plane α is a always the same rhombus $MB'ND$.

Problem 150 Given a pentagonal right prism $ABCDEA_1B_1C_1D_1E_1$. Construct a cross section of the prism by the plane passing points K, L, and M, such that $K \in BB_1$, $L \in AA_1$, $M \in D_1E_1$ as it is shown in Fig. 3.38.

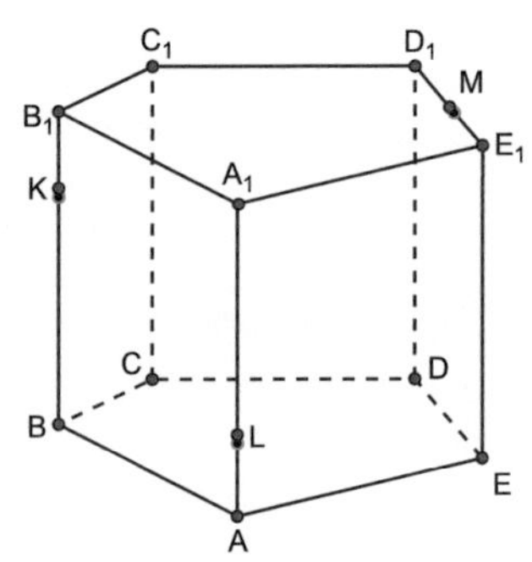

Fig. 3.38 Pentagonal prism with three given points

Solution Denote the cutting plane by α. We can easily connect points L and K in Fig. 3.38; the segment KL is the trace of plane α on face BB_1A_1A.

Let us construct an auxiliary plane BB_1M, shown in blue, in which we can intersect lines KM and FF_1 at point F_2. (Here $FF_1 = BB_1M \cap AA_1D_1D$ and where segment FF_1 is concluded between two bases of the prism.) We need this point F_2 because it

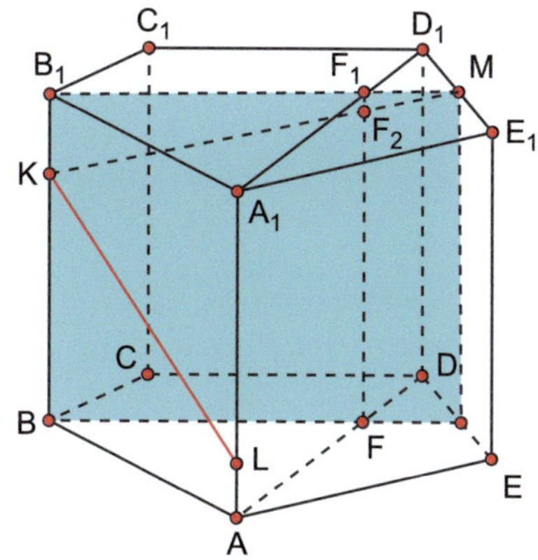

Fig. 3.39 Two auxiliary planes, BB_1M and AA_1D_1D

belongs to the secant plane α, $F_2 \in KM$ (Fig. 3.39) Moreover, the auxiliary plane AA_1D_1D also contains both these points, hence, we can find the line of intersection of two planes that can be written as

$$\left.\begin{array}{l}\{L, F_2\} \in AA_1D_1D \\ \{L, F_2\} \in KLM = \alpha\end{array}\right\} \Rightarrow \alpha \cap AA_1D_1D = (LF_2).$$

If in plane AA_1D_1D, we continue line LF_2 until it will intersect with another line DD_1, and because DD_1 lies in the face DD_1E_1E we will be able to construct the trace of plane α on the face DD_1E_1E that goes through point M of the edge D_1E_1. Please, see Fig. 3.40.

We need to find additional points. So, in the plane AA_1D_1D we will find the point of intersection of line LF_2 and DD_1 , we obtain point D_2. Point D_2 simultaneously belongs to plane α and to face DD_1E_1E, so as point M. Hence, by continuing D_2 and M until it intersects side EE_1, we obtain a new side of the cross section, MN. We can continue points L and N to obtain the trace of cutting plane α on face AA_1E_1E.

Next, we have to find how plane α will cut faces BB_1C_1C, DD_1C_1C and the top base of the prism, $A_1B_1C_1D_1E_1$. At this point we can be sure that the cross-sectional polygon will have at most six sides. Let us construct an auxiliary plane AA_1C_1C (shown in purple in Fig. 3.41) and intersect it with already constructed plane BB_1M. Within the boundaries of the top and bottom bases, these two planes will intersect by segment QQ_1. In plane BB_1M, we will find point Q_2 as $Q_2 = KM \cap QQ_1$.

Let us continue line LQ_2 in plane AA_1C_1 until it will intersect side C_1C at point P. Then the following is true

$$\left.\begin{array}{l}\{K, P\} \in BB_1C_1C \\ \{K, P\} \in \alpha\end{array}\right\} \Rightarrow \alpha \cap BB_1C_1C = KP.$$

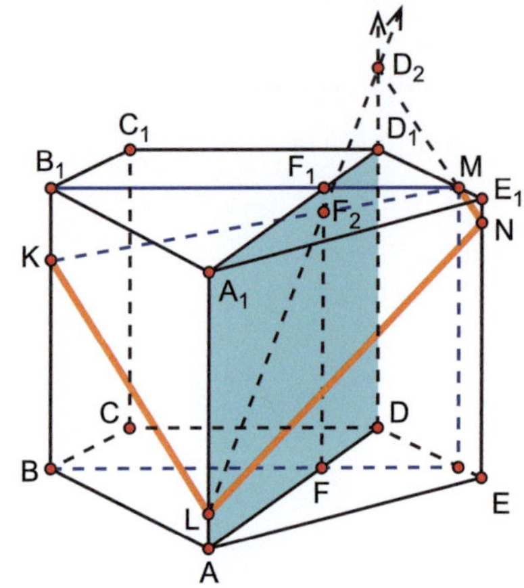

Fig. 3.40 Point D_2 is constructed

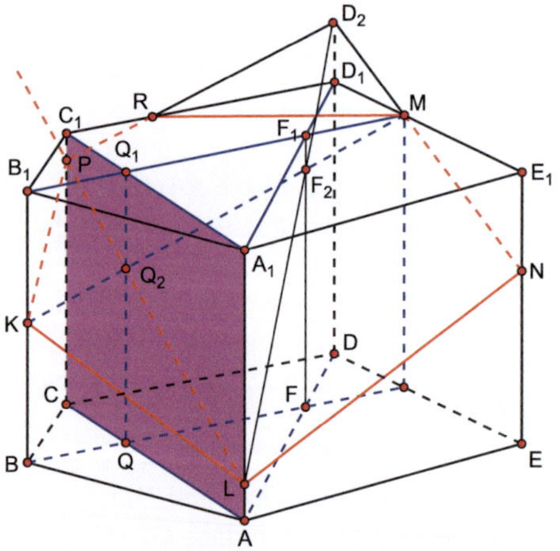

Fig. 3.41 The third auxiliary plane AA_1C_1C

Let us find the trace of plane α on face D_1C_1C. We can use the earlier constructed point D_2:

$$\left.\begin{array}{l}\{D_2, P\} \in DD_1C_1C \\ \{D_2, P\} \in \alpha\end{array}\right\} \Rightarrow \alpha \cap DD_1C_1C = (D_2P).$$

By intersecting lines (D_2P) and (C_1D_1) we obtain point R that belongs to two faces of the prism, DD_1C_1C and $A_1B_1C_1D_1E_1$, $R = C_1D_1 \cap PD_2$. Finally, connecting points P and R and then R and M we will construct the last two edges of the cross section that is a hexagon $PRMNLK$ (see Fig. 3.42).

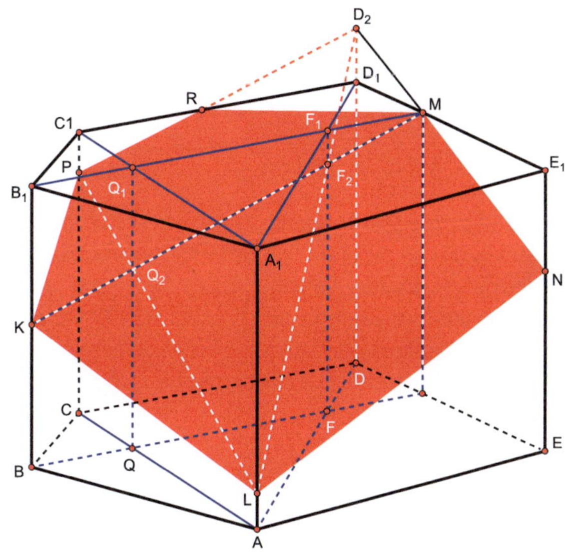

Fig. 3.42 Cross section is hexagon *PRMNLK*

3.2.2.1 Combined Construction Method

The essence of this method is the application of theorems on the parallelism of lines and planes in combination with the method of traces or with the method of auxiliary sections and internal projection.

When constructing a cross section, the following theorems are used.

1. If two planes are parallel and intersect by the 3rd plane, then the lines of intersection of parallel planes by the 3rd plane are parallel to each other.
2. If two intersecting planes are parallel to the same straight line, then the line of their intersection is parallel to this straight line.
3. If a plane and a straight line are parallel and a certain plane is drawn through this straight line and intersects the given plane, then the line of intersection of these two planes is parallel to the given straight line.

3.2.3 Cross Sections Parallel or Perpendicular to a Face or Edge of a Polyhedron

Solving the following problems, we will deal with the necessary construction of a cross section with planes running parallel or perpendicular to some edges of a pyramid. In most cases, to successfully solve such problems, we need to know the properties of perpendicular and parallel lines and planes.

Problem 151 (*Kaganov*) Prove that the perimeters of the figures obtained by intersecting a regular tetrahedron with planes parallel to two opposite edges of the tetrahedron are equal to each other.

Proof Assume that a regular tetrahedron has an edge of a. In order to prove the statement of the problem, we need to construct a cross section by the plane. Let us select two edges that belong to skew lines, they are the edges of the pyramid that are opposite. For example, in pyramid $ABCD$ we can take edge AB and DC. Next, we have to draw a cross section by a plane parallel to these two edges. We can draw segment MN in the face ADB parallel to AB:

$$MN \parallel AB.$$

Next, in the face DBC we need to draw a line parallel to side CD:

$$NL \parallel DC.$$

Next

$$KL \parallel MN \parallel AB$$

and

$$MK \parallel DC \parallel NL$$

will finish construction obtaining a parallelogram $MNLK$ (Fig. 3.43).

Perimeter of this parallelogram (P) is equal to the double sum of two sides:

$$P = 2 \cdot (MK + KL)$$

The next step is to evaluate MK and KL. How can we do it?

Remember that each face of the given pyramid is an equilateral triangle. When we randomly took a point M on the edge AD in order to draw the line parallel to AB, we finish construction by drawing segment KM parallel to edge DC. Because each face of a regular tetrahedron is an equilateral triangle with all angles of 60°, then this way we cut inside triangle ADC another equilateral triangle with each side equals AM, so $AM = MK = AK$. On the other hand, by doing our construction, we also

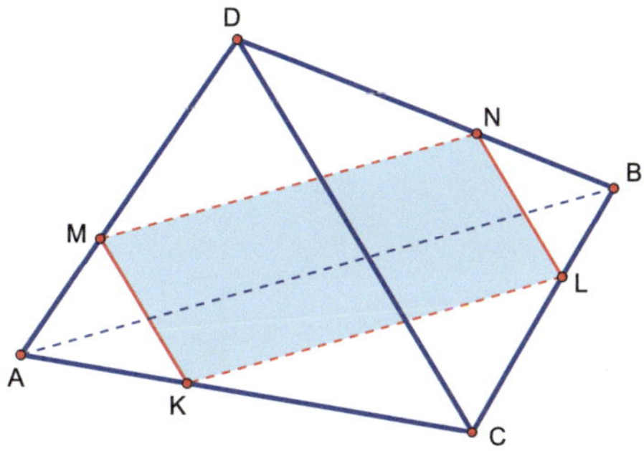

Fig. 3.43 Cross section with the same perimeter

cut another equilateral triangle KLC with each side equals KC, so $KC = LC = KL$. Finally, in the formula for the perimeter above we simply replace MK by AK and KL by KC and rewrite it as follows:

$$P = 2 \cdot (AK + KC) = 2a.$$

This formula shows that the answer does not depend on the location (initial choice) of point M in our construction. The perimeter of the cross section is always equal to twice the length of the edge of a regular tetrahedron. □

In my opinion, the best way to learn new concepts is to solve a related problem.

Problem 152 On the edges BB', CD, CC' of the prism $ABCDA'B'C'D'$, the points M, N, and L are given, respectively, and on the edge AA' the point K is given. Construct a cross section of the prism $ABCDA'B'C'D'$ by a plane passing through the point K parallel to the plane MNL.

Solution First, we will construct a cross section of the prism by the plane passing points M, N, L and then the required plane passing point K will cut the corresponding faces by the parallel lines (Fig. 3.44).

Let us connect points M and N in the face $BB'C'C$. MN is the trace of plane α on this face. Similarly, we can connect points N and L in the face $DD'C'C$. Segment NL is the trace of secant plane α on this face. We can see that point L belongs to the base $ABCD$ and to the plane α. Then we need to find the second point with the same property. For this, we will continue line MN and its projection on the base, line BC to obtain a point S (Fig. 3.45).

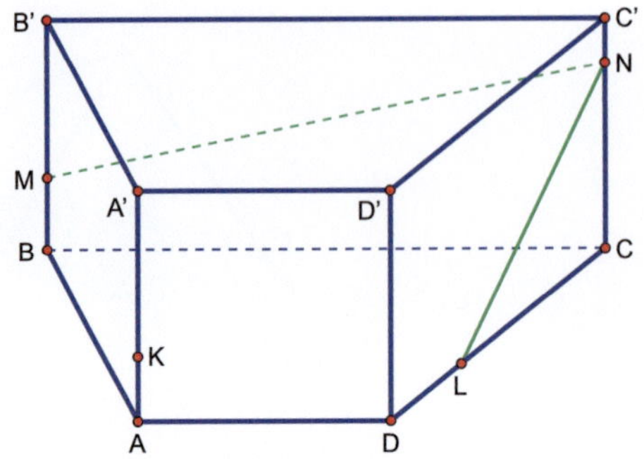

Fig. 3.44 Two sides of the auxiliary cross section

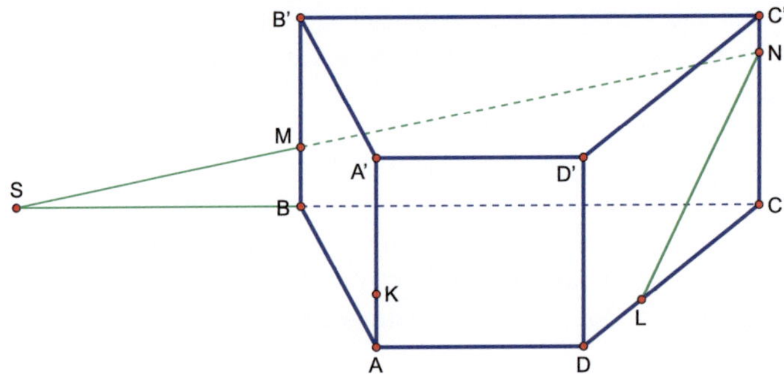

Fig. 3.45 Trace of the auxiliary cross section

$$\left.\begin{array}{l}\{L, S\} \in ABCD \\ \{L, S\} \in \alpha\end{array}\right\} \Rightarrow \alpha \cap ABCD = (SL)$$

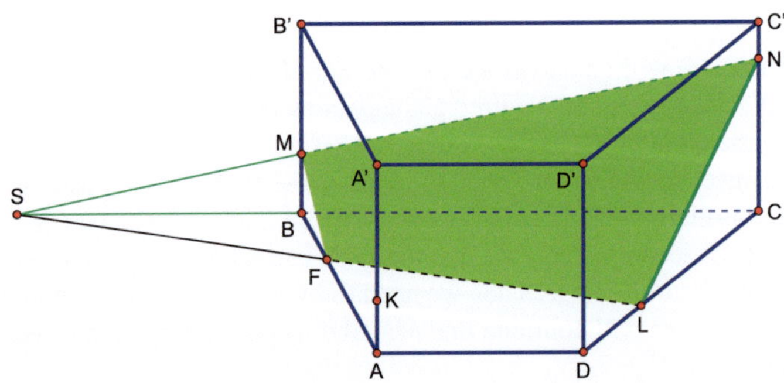

Fig. 3.46 The auxiliary cross section *MNLF*

Line *SL* is the trace of plane α on the base of the prism. Let us find its intersection with the edge *AB*:

$$(SL) \cap (AB) = F \Rightarrow \alpha \cap BB'A'A = MF.$$

The quadrilateral *MNLF* is the cross section of the prism by the plane passing through points M, N, and L (Fig. 3.46). In order to construct the plane going through point K and parallel to plane α, we first can draw line $(KP) \parallel (MF)$. Next, in the plane of face $BB'C'C$ we will draw line $(PQ) \parallel (MN)$ (Please, see Fig. 3.47)

Fig. 3.47 *PK* is parallel to plane *MNLF*

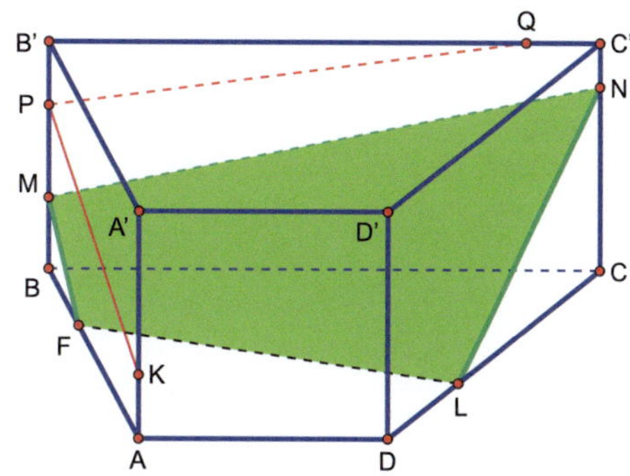

Fig. 3.48 Plane *PQEJK* is parallel to plane *MNLF*

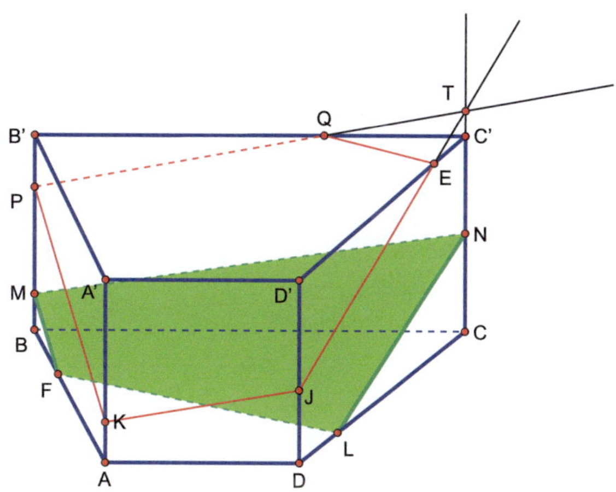

Line *PQ* will intersect with line CC' in plane $DD'C'C$. Let $(PQ) \cap (CC') = T$. Since point T belongs to the secant plane and to the face $DD'C'C$, and since plane β is parallel to α, then planes α and β intersect this face by parallel lines, hence we can draw $(TJ) \parallel (NL)$. Let E be the intersection of (TJ) and edge $D'C'$, then *JE* is another side of the cross section (see Fig. 3.48).

Finally, by connecting points K and J we complete our construction. The cross section is the pentagon $PQEJK$ in Fig. 3.48. Note that we could construct its side KJ by drawing $(KJ) \parallel (PQ)$ as well.

In the following problem you will be asked to draw a line parallel to the given edge of a pyramid.

Problem 153 Points E and G are given on the edges SB and SC of the pyramid $SABCD$, respectively. Construct a line passing through point E parallel to line AG.

Solution Let's construct a cross section of the pyramid by a plane passing through the point E and the line AG, that is, the plane defined by the three distinct points E, A, and G (see Fig. 3.49).

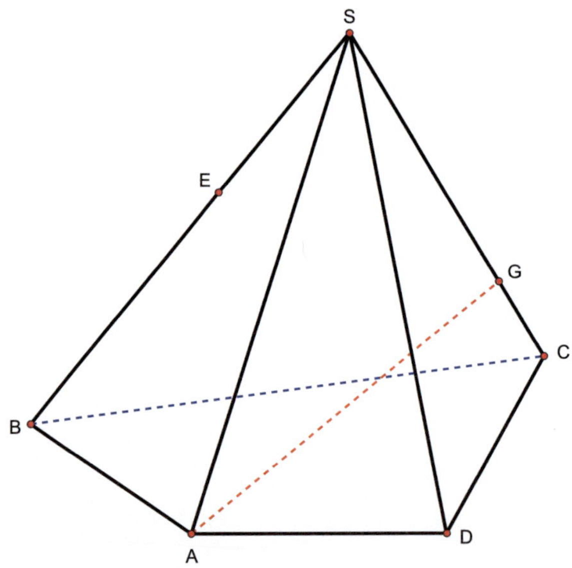

Fig. 3.49 Point E and line AG on the quadrilateral pyramid

To do this, as usual, we find the trace of the secant plane AEG. Next, we construct the cross section of the pyramid by this plane and in the plane of this section, through the point E, we draw the line EQ parallel to the line AG .

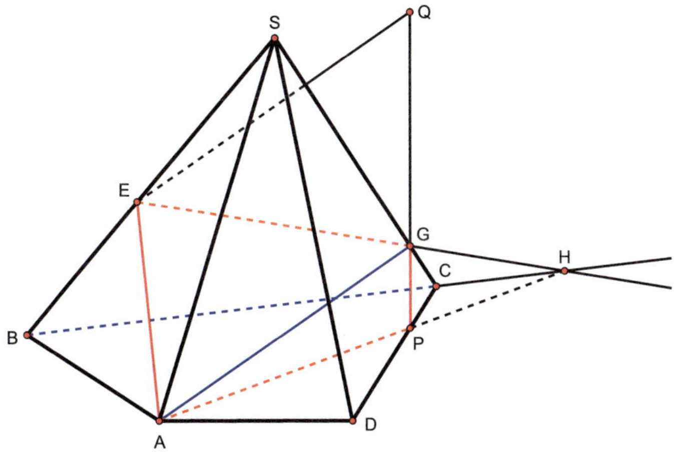

Fig. 3.50 EQ is parallel to line AG on the quadrilateral pyramid

Line EQ is the desired line.

1. Connect points E and A in face ABS, then connect points E and G in the face BSC. In order find out how the plane AEG intersects the base of the pyramid, we need a point, additional to point A.
2. Let us find the intersection of lines EG and BC and denote it by H.
3. $\left.\begin{array}{l}\{A, H\} \in ABCD \\ \{A, H\} \in \alpha\end{array}\right\} \Rightarrow \alpha \cap ABCD = (AH)$. Line AH is the trace of plane $AEG = \alpha$ on the base of the pyramid.
4. This line intersects the edge DC at point P. This can be written as $(AH) \cap (DC) = P$.
5. Connecting points P and G we obtain the fourth side of the cross section of the pyramid by plane α. It is the quadrilateral $AEGP$. Please note that AE is not parallel to GP.
6. In the plane of the cross section, we can draw line EQ parallel to the line AG (see Fig. 3.50). Done.

Problem 154 Given a regular quadrangular pyramid $ABCDS$, in which SO is its height. Point P is on the height. Construct a cross section of the pyramid by the plane α that goes through point P and parallel to the face SDC.

Solution Let us sketch the pyramid and its height and construct an axillary plane SAC shown in gray in Fig. 3.51 on the right.

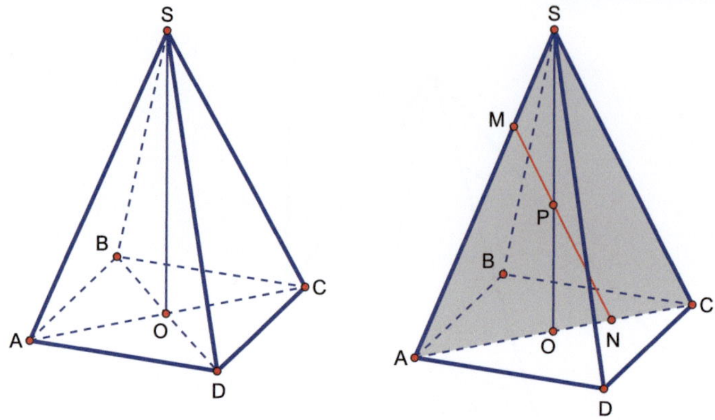

Fig. 3.51 Regular quadrangular pyramid and auxiliary plane ACS

In this plane we can draw a line MN through point P parallel to side SC:

$$P \in \alpha, \;\; P \in (MN), \;\;\; (MN) \parallel SC.$$

Note that M is on the edge AS of the pyramid, then M is one of the vertices of the cross-sectional polygon. By the condition plane α is parallel to SDC. Point N belongs to plane α since $N = (MP) \cap (AC) \;\; \Rightarrow N \in \alpha$. We know that plane SDC intersects with the base of the pyramid plane $ABCD$ by line DC. Then plane α that is supposed to be parallel to it, must intersect the base by the line parallel to DC that must go through point N.

Next, we can draw lines parallel to side DC through both points N and M in the planes SAB and $ABCD$, respectively.

In Fig. 3.52 on the left we may see our constructions, which can be described as

$$MK \cap SB = K, \;\;\; N \in (EF), \;\;\; MK \parallel EF \parallel DC.$$

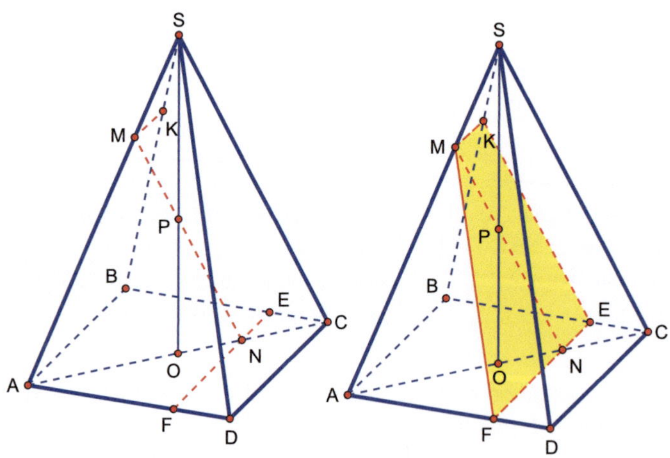

Fig. 3.52 Trapezoid $MKEF$ is the cross section of the pyramid

Plane α passes through point P and contains two intersecting lines, MN and FE, that are parallel to face SDC. Therefore trapezoid $MKEF$ is the cross section of the pyramid by plane α.

Problem 155 Given a regular quadrangular prism $ABCDA_1B_1C_1D_1$ with a square base and height equals double base length. Construct a cross section that goes through point D and perpendicular to diagonal of prism BD_1.

Solution Sketch the prism and mark all the vertices (see Fig. 3.53).

Let us construct a line through point D perpendicular to diagonal BD_1. For this we can consider an auxiliary plane BB_1D_1D shown in gray in Fig. 3.53. The question is how in this plane to construct a perpendicular to the diagonal of the prism and how this segment will be located with respect to the sides of the rectangle BB_1D_1D?

We can apply Thales's Theorem (see [2], p. 24.) If we draw a circle with side BD as its diameter and the midpoint O as its center, then it will intersect line BD_1 at the right angle. In Fig. 3.54 we marked it by G. Continuing line DG until it intersects side BB_1 we will obtain the required point M, that is one of the vertices of the cross-sectional polygon.

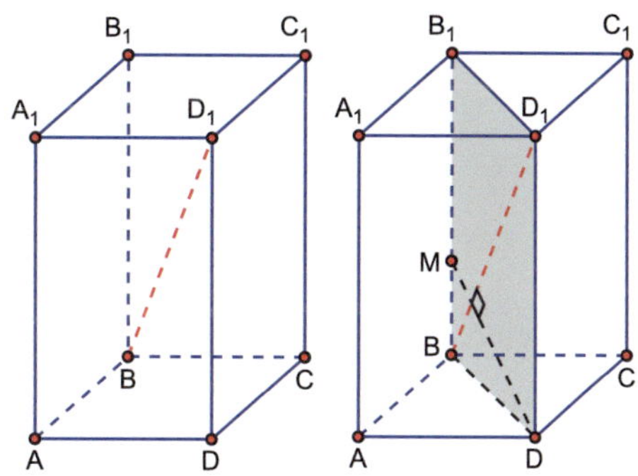

Fig. 3.53 A problem about a regular quadrangular prism

We have one line MD that belongs to the cutting plane α and perpendicular to diagonal BD_1. We need to find another one that goes through the same point D. In order to construct such a line, we can consider the base square of the prism. Its diagonals AC and BD are perpendicular. If we draw a line through point D parallel to diagonal AC, then the following will be true

$$BD \perp (LE)$$

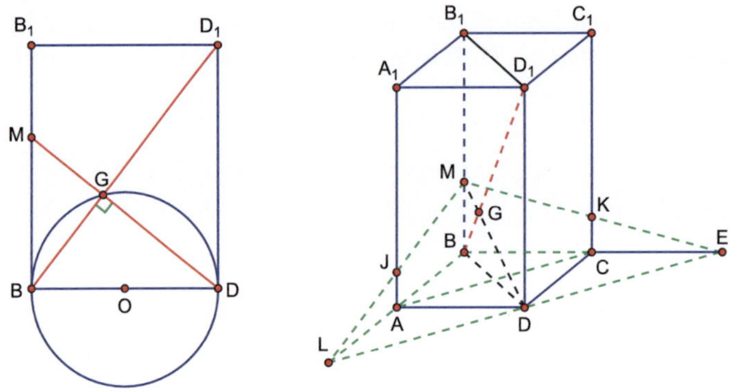

Fig. 3.54 An auxiliary plane of a regular prism

By Theorem 20 about three perpendiculars, we can state that $LE \perp BD_1$.

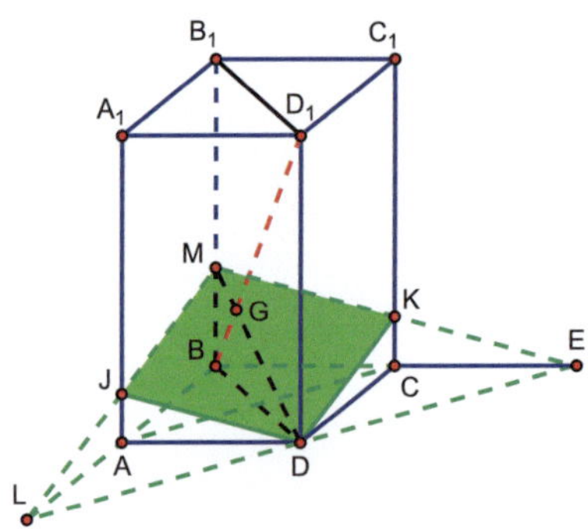

Fig. 3.55 *MKDJ* is the cross section of the given regular prism

Therefore *MKDJ* is the requested cross section of the given prism (see Fig. 3.55).

3.2.4 Construction Using Analytic Geometry

In this section we will demonstrate how two or more methods can be combined in successful solving a problem on a cross section. Let us construct a cross section of the cube given at the beginning of this chapter and that was not solved correctly by a faculty candidate.

Problem 156 Find a cross section of a cube $ABCDA'B'C'D'$ by plane α that goes through the midpoints (M, N, K) of the three skew edges (AA', $B'C'$, and CD) of the cube, respectively.

Solution Let us solve this problem using a combination of the coordinate and construction methods. Placing the coordinate system so that its center is point $B(0, 0, 0)$, the axes OZ goes as the ray BB', axes OX as the ray BA and axes OY as the ray BC, then other vertices of the cube will have the following coordinates:

$$A(1, 0, 0),\ A'(1, 0, 1),\ B(0, 0, 0),\ B'(0, 0, 1),$$
$$C(0, 1, 0),\ C'(0, 1, 1),\ D(1, 1, 0),\ D'(1, 1, 1).$$

While coordinates of the given points on the secant plane are

$$M(1.0, 0.5),\ N(0, 0.5, 1),\ K(0.5, 1, 0).$$

Using determinant formula (1.50) for the equation of the plane given by its three points and after simplification, we obtain that the secant plane α has the equation:

$$x + y + z = 1.5$$

From this equation, we can easily find how plane α intersects the coordinate planes. For this we just set $x = 0,\ y = 0$ and obtain the corresponding $z = 1.5$. This way we obtain point $P(0, 0, 1.5)$ that is sitting on the Z axes and belongs to the secant plane. Similarly by setting $x = 0,\ z = 0$ we will evaluate that $y = 1.5$ and find a new point $Q(0, 1.5, 0)$ (the secant plane cuts the Y axes at this point). When $z = 0,\ y = 0$, then $x = 1.5$ and we obtain the third point $R(1.5, 0, 0)$ at which the secant plane intersects the X axes (see Fig. 3.56 on the left).

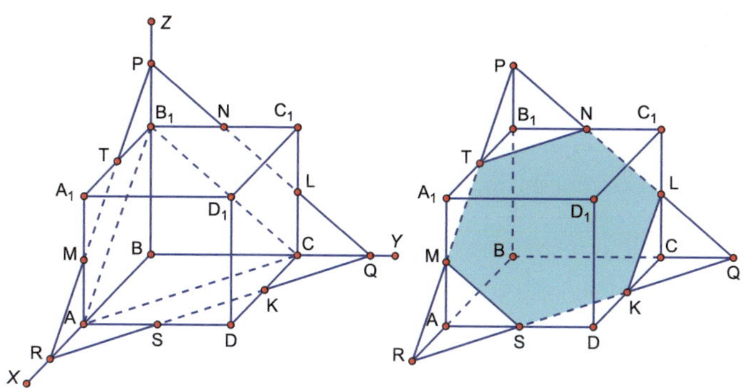

Fig. 3.56 Correct cross section through three points on the skew edges of a cube

Now we can apply the construction (trace) method. The coordinate half-plane ZOY can be seen as the plane of face $BB'C'C$. Plane α as we found has three common points with ZOY, i.e., points P , N, and Q, then these points lie on the same line (they are collinear.) Let us continue line PN in the plane $BB'C'C$ until it intersects the edge of cube $C'C$ at point L, we obtain the first side NL of the cross section. The following is true

$$\alpha \cap ZOY = (PQ) \Rightarrow \alpha \cap BB'C'C = [NL].$$

Since points $L,\ K$ belongs to both, plane α and the face of cube $DD'C'C$, and are on the edges of the cube, then we can write down

$$\alpha \cap DD'C'C = [LK].$$

Consider XOY quadrant that is also the plane of face $ABCD$ of the given cube. Points $Q,\ K$ simultaneously belong to the base of the cube and to the secant plane α, then the base face of the cube and the plane intersect by the line KQ.

$$(KQ) \cap (AD) = S \Rightarrow \alpha \cap ABCD = [KS].$$

Points $S,\ M$ belong to both, plane α and face of the cube $AA'D'D$, so, we can construct another side of the cross section, segment SM that indicates how the secant plane cuts the face $AA'D'D$:

$$\alpha \cap AA'D'D = [MS].$$

Since points R, M of plane α also belong to face $AA'B'B$, then segment MT is the side of the cross-sectional polygon and the following is valid:

$$\alpha \cap AA'B'B = [MT].$$

Finally, by connecting points T and N in the face $A'B'C'D'$ we obtain the last side of the cross section, TN. Therefore, the cross section of the cube by plane α is the hexagon $NLKSMT$ (see the right sketch in Fig. 3.56). In the following problem you will be asked to prove that this hexagon is a regular hexagon and that it contains the center of the cube.

There are different ways to construct the cross section by the given condition. For example, when we found the equation of the plane α in the given coordinate system, we could think about another plane (denote it by β) that passes through the vertices A, B' and C, that has the equation:

$$\beta :\ x + y + z = 1.$$

Comparing the equations of two planes, we know that the planes are parallel to one another.

$$\alpha \| \beta$$

Moreover, the distance between two planes is 0.5 units.

Problem 157 Prove that the cross section constructed in the previous Problem 156 is a regular hexagon. Find its side and prove that its center coincides with the center of the cube.

Proof First, let us prove that $PQ \cap B'C' = N$. Assume contradiction that $PQ \cap B'C' = V \neq N$. Because N and L are the midpoints of segments $B'C'$ and $C'C$, then NL is the mid-segment of triangle $B'C'C$ and then

$$NL = \frac{B'C}{2} = \frac{\sqrt{2}}{2}.$$

Similarly

$$LK = \frac{C'D}{2} = \frac{\sqrt{2}}{2}.$$

Using the same reasoning, we can conclude that all sides of the hexagon are equal to $\frac{\sqrt{2}}{2}$. Because the cross section of the cube is a regular hexagon, it must pass through the center of the cube. (See Theorem 42 below) □

3.3 Areas of Cross Sections and Volumes of Cutoff Solids

Many problems offered on math contests ask to find an area of a cross section or figure out how a secant plane divides the volume of a polyhedron. I would like to start from the following problem.

Problem 158 Given pyramid $SABC$. Points D and E are on the edges SA and SB, such that

$$SD \div DA = 1 \div 2, \ \ SE \div EB = 1 \div 2.$$

Plane α goes through points D and E parallel to the edge SC. How does this plane divide the volume of the pyramid?

Solution Solving this problem can be enhanced with the help of a good sketch. Following the condition of the problem, first, let us place points D and E on sides SA and SB, respectively.

The cutting plane α intersects face ASB by segment DE. Since the given points divide the edges in the same ratio, then by Thales Theorem, $DE \| AB$.

Next, we will draw through point E the line parallel to edge SC in the plane of triangle SBC. Let this line intersect BC at point M. Similarly, in the plane of face

ASC, through point D, we will draw another line parallel to edge SC that intersects edge AC at point N. ($DN \| SC$.)

Next, we will connect points N and M

$$\alpha \cap ABC = [NM].$$

Parallelogram $DEMN$ (shaded in yellow in Fig. 3.57) is the cross section of the pyramid $SABC$ by plane α.

We can see that the plane divides the solid into two parts. How can we find the ratio of the volumes? Both convex polyhedra below and above the cross section are not simple figures (neither pyramid no prism, etc.). Moreover, we are not given any information about the pyramid, so we should assume that the answer must be obtained for a general type pyramid/tetrahedron.

Let us focus on the solid located above the cross section. I want to show you how simple axillary construction can help us to find the answer to the problem. Draw $EQ\| \|BC$ in plane of face SBC. ($EQ \cap SC = Q$). Next, let us connect points Q and D (see Fig. 3.58). We understand that point Q divides edge SC in the same ratio as D and E divide the edges SA and SB, i.e., $SQ \div QC = 1 \div 2$.

Obviously, that by construction and the condition of the problem, the following is true

$$\triangle DEQ = \triangle NMC$$

Further, the planes of triangles DEQ and ABC are parallel. Hence $DEQNMC$ is a triangular prism and $SDEQ$ is a triangular pyramid. Therefore, the volume of a polyhedron located above the cross section can be found as the sum of the volumes of the corresponding prism and pyramid.

Assume that the height of the pyramid $SABC$ dropped from vertex S is equal to h. Denote by $[ABC]$ the area of the base triangle ABC. Denote the volume of the pyramid $SABC$ by V. Also assume that the volume of the small pyramid is V_1 and the volume of prism is V_2. The following is valid.

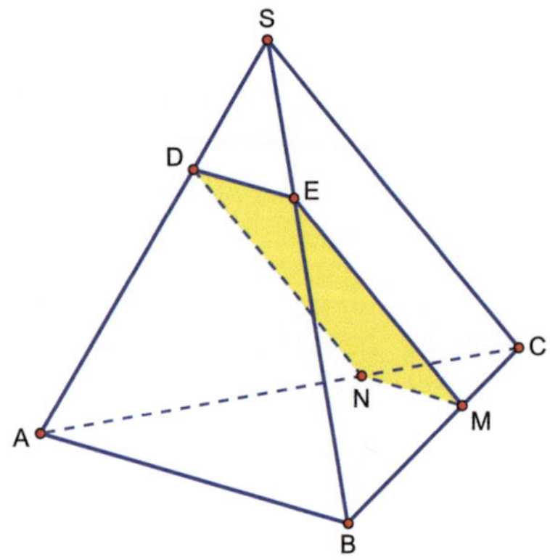

Fig. 3.57 Pyramid is cut by a plane

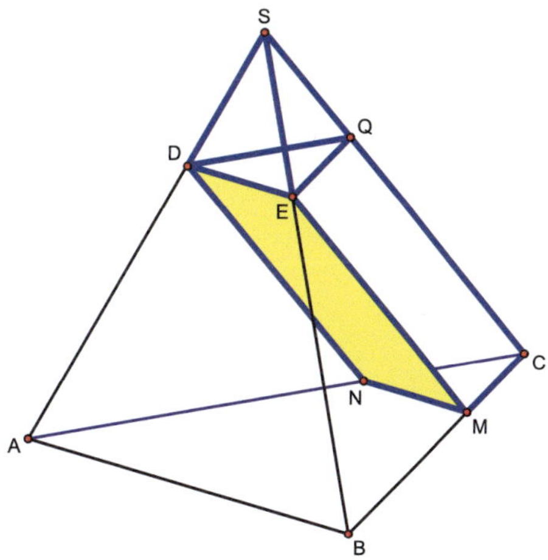

Fig. 3.58 Prism and pyramid above the cross section of a big pyramid

$$V = \frac{1}{3}h \cdot [ABC]$$

Pyramids *SDEQ* and *SABC* are similar with the coefficient of similitude $k = \frac{1}{3}$, then the following is true for their volumes.

$$V_1 = k^3 V = \frac{V}{27}$$

Next, the volume of prism *DEQMNC* is equal to the product of the height and the area of the base. Because $DN = EM = QC = \frac{2}{3}SA$, then the height of the prism h_2 is the same as the length of the segment dropped from for example, point D to the base of the prism, which can be written as follows

$$h_2 = \frac{2}{3}h$$

The base of the prism is triangle *MNC* and its area is related to the area of triangle *ABC* as

$$[MNC] = k^2 \cdot [ABC] = \frac{1}{9}[ABC]$$

Thus, the volume of prism *DEQNMC* can be evaluated as

$$V_2 = \frac{2}{3}h \cdot \frac{1}{9}[ABC] = \frac{2}{9} \cdot V$$

The volume of the solid located above the cross section *DEMN* (V_a) is the sum of two volumes,

$$V_a = V_1 + V_2 = \frac{V}{27} + \frac{2V}{9} = \frac{7}{27}V.$$

Then the volume of the polyhedron located below the cross section (V_b) can be found by subtracting the volume of the polyhedron located at the top from the volume of the pyramid,

$$V_b = V - \frac{7}{27}V = \frac{20}{27}V.$$

The ratio of the two volumes is

$$V_a \div V_b = 7 \div 20.$$

This answer does not depend on the shape of the triangular pyramid!
Answer. $7 \div 20$

In order to successfully solve variety of problems on the topic of this section, it is helpful to know some important properties and theorems.

Theorem 42 *Any plane that divides a cube into two equal parts, must pass through the center of the cube.*

Proof Assume the contradiction. Let O be the center of a cube and consider any plane π cutting the cube by half but such that O is not in plane π. Let plane π_0 pass through the center O of the cube and be parallel to π. Then $\exists$ another plane π_1 also parallel to π_0 that is at the same distance from π_0 as plane π but is located from the other side of the center of the cube. Then the parts cut off by the planes π and π_1 that do not contain point O are equal, but their volume is less than $1/2$ of the volume of the cube. Therefore plane that does not go through the center of a cube cannot divide it by half. □

Lemma 13 *A cross section of a cube by any plane passing through its center is always a polygon symmetric with respect to the center of the cube.*

Proof Let us prove this statement by considering the following three possible cases:

1. Point of a cross section belongs to a face of the cube (see Fig. 3.59 on the left)
2. Point of a cross section belongs to an edge of the cube (see Fig. 3.59 on the right)
3. Point of a cross section is one of the vertices of the cube.

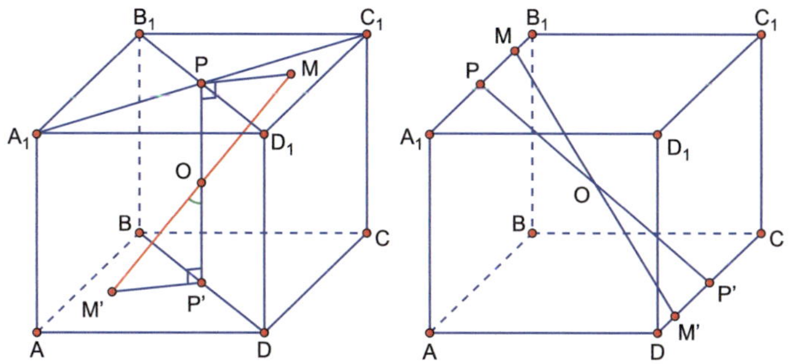

Fig. 3.59 Central symmetry of a cross section

We will only prove case 1 here, since the other two cases can be proven in a similar way. Let the point M be on the top face of the cube. Since the section passes through the center O of the cube, the line MO belongs to the plane of the section and this line will pierce the base of the cube at some point M'. Draw a line PP' through the center O perpendicular to the lower and upper faces of the cube. Next, we can connect the points P and M, and then the points M' and P'. We get two equal right triangles MPO and $M'P'O$. Therefore, $M'O = OM$ which proves that the point M is centrally symmetric to the point M' with respect to the center of the cube O. □

Theorem 43 *The area of the orthogonal projection of a polygon onto a plane ($\tilde{A}$) is equal to the product of its area (A) and the cosine of the angle (φ) between the plane of the polygon and the projection plane.*

$$\tilde{A} = A \cdot \cos\varphi \tag{3.7}$$

Proof To prove this, we first consider a triangle and its projection onto a plane passing through one of its sides, as shown in Fig. 3.60. The projection of the triangle ABC onto the plane α is the triangle ABC_1. Draw height CD of triangle ABC. By the Theorem 20 on three perpendiculars, the segment C_1D is the height of the triangle ABC_1. The angle CDC_1 is equal to the angle φ between the plane of the triangle ABC and the projection plane α, so we get the following expressions:

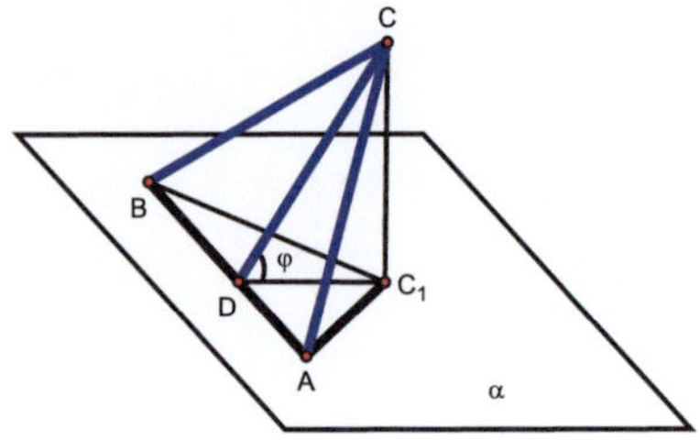

Fig. 3.60 Proof of Theorem 43

$$C_1D = CD \cdot \cos\varphi$$
$$[ABC] = \tfrac{1}{2}AB \cdot CD, \quad [ABC_1] = \tfrac{1}{2}AB \cdot C_1D$$
$$[ABC_1] = [ABC] \cdot \cos\varphi$$

Thus, in the case under consideration, the theorem is true. The theorem is also true in the case when any plane parallel to α is taken instead of the plane α. Indeed, when a figure is projected onto a parallel plane, its projections are obtained by parallel translation in the direction of the projection. Figures resulting from parallel translation are congruent.

Let us now consider the general case. Let's break this polygon into triangles. Each triangle that does not have a side parallel to the projection plane is divided into two triangles whose common side is parallel to the projection plane. Now for the areas of each triangle S_i of our partition and the area of its projection $\tilde{S}_i$ we write equalities.

$$\tilde{S}_i = S_i \cdot \cos\varphi$$

By summing all these equalities term by term, we obtain the area of the projection on the left and the area of the original polygon multiplied by cosine of the angle between two planes on the right. □

Problem 159

1. Prove that the cross section *PRQLMN* obtained in Problem 157 shown in Fig. 3.20 is a regular hexagon (it has the same side length and the central angle of 60°.) Find the area of the cross section.
2. Prove that the cross section passes through the center of the cube.

Proof Let us draw to diagonals in the hexagon, *LP* and *RM*. They (by definition) intersect at one point, the center of the regular hexagon, name it O (Fig. 3.20). On the other hand, *RM* is parallel and equal to the diagonal A_1C_1 of the cube, and *LP* is parallel and equal to C_1D, then point O is simultaneously the center of the hexagon and hence the plane divides the cube into two equal parts $1 \div 1$. (Based on Theorem 42.) The area of the cross section can be found as six areas of the equilateral triangles with side $a/\sqrt{2}$. (Please, explain why.)

$$A = 6 * (a/\sqrt{2})^2 * \sqrt{3}/4 = 3\sqrt{3}a^2/4.$$

□

Answer. $A = \frac{3\sqrt{3}a^2}{4}$

Problem 160 Construct a cross section of a cube $ABCDA'B'C'D'$ by the plane that passes through the midpoints of the edge AB and DD' and vertex A'. Find how this cross section divides the volume of the cube.

Solution Let us sketch a cube so that it would be easy to construct the cross section given the information of the problem (Please, see Fig. 3.61 on the left).

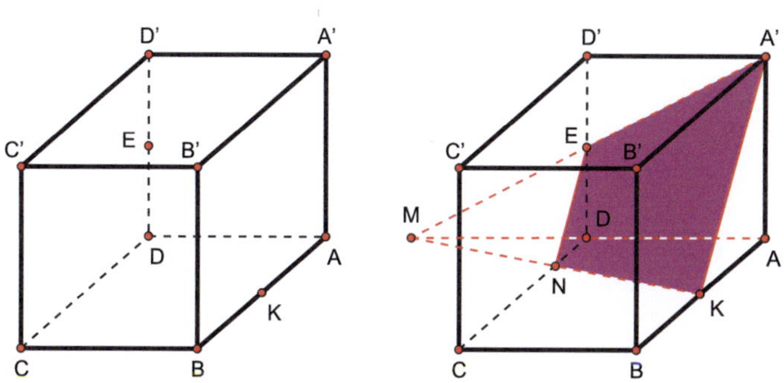

Fig. 3.61 Cross section of a cube by a plane passing the midpoints of two edges and a vertex of a cube

Let K be the midpoint of AB and E the midpoint of DD'. Denote plane $\alpha = \{K, E, A'\}$ and find how it cuts the cube. First, we will construct segment KA', such that $\alpha \cap BB'A'A = A'K$. Then we construct segment $A'E$, such that $\alpha \cap DD'A'A = A'E$. Next, we will continue line EN and the edge DA until they intersect at point M;

$$\left.\begin{array}{l}\{M, K\} \in \alpha \\ \{M, K\} \in ABCD\end{array}\right\} \Rightarrow \alpha \cap ABCD = (MK)$$

The line MK intersects the edge CD of the cube at point N, such as

$$CD \cap MK = N.$$

Connecting all points, we obtain a trapezoid $NEA'K$ as the cross section. This is shown in purple in Fig. 3.61 on the right. We see that the plane cuts off from the cube the truncated pyramid $ENDA'KA$, whose base ratio is 4 (because right triangles are similar ($\triangle\ EDN \sim \Delta A'KA$) with similarity coefficient equal to 2). Hence, we obtain

$$B_1 = [A'KA] = \frac{1}{2} \cdot \frac{a}{2} \cdot a = \frac{a^2}{4} \Rightarrow B_2 = [EDN] = \frac{1}{4} B_1 = \frac{a^2}{16}$$

Using formula (2.6) we obtain the volume of the truncated pyramid:

$$\begin{aligned} V_1 &= \tfrac{1}{3} \cdot h \cdot \left(B_1 + \sqrt{B_1 B_2} + B_2\right) \\ &= \tfrac{1}{3} \cdot a \left(B_1 + \sqrt{B_1 \cdot \tfrac{B_1}{4}} + \tfrac{B_1}{4}\right) \\ &= \tfrac{1}{3} \cdot a \cdot \tfrac{7B_1}{4} = \tfrac{a}{12} \cdot \tfrac{7a^2}{4} \\ &= \tfrac{7a^3}{48} \end{aligned}$$

Subtracting this quantity from the volume of the cube, we will obtain the volume of the second figure that is above the cross section :

$$V_2 = a^3 - \frac{7a^3}{48} = \frac{41a^3}{48}$$

Then $V_1 \div V_2 = \frac{7a^3}{48} \div \frac{41a^3}{48} = 7 \div 41$.

Answer. $V_1 \div V_2 = 7 \div 41$

Problem 161 (*MGU, physics, entrance exam, 1999*) The area of the cross section passing through the diagonal of the base of a regular quadrangular pyramid and parallel to a lateral edge that does not intersect with it is equal to s. Find the cross-sectional area of the same pyramid by a plane passing through the midpoints of adjacent sides of the base and the midpoint of the height. Express the area in terms of s.

Solution Let us sketch the cross section of the pyramid $ABCDS$ by the plane that goes through the diagonal of the base BD and parallel to the lateral edge AS that does not intersect this diagonal (see Fig. 3.62)

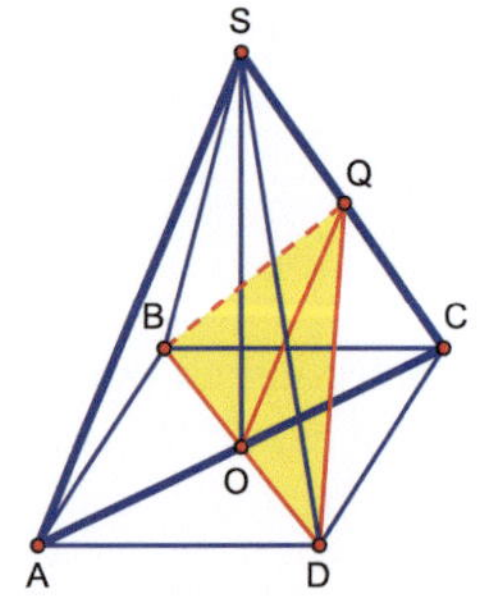

Fig. 3.62 Pyramid and its first cross section

Let O be the center of the base square. Consider triangle ASC. In this triangle draw the line OQ parallel to AS. Obviously point Q will be the midpoint of segment SC. Connecting points QD and QB we will construct the required cross section, the area of which equals s by the condition of the problem.
Next, let us construct the second cross section. By the condition of the problem, it must go through the midpoints of the adjacent sides of the base (in our Fig. 3.63

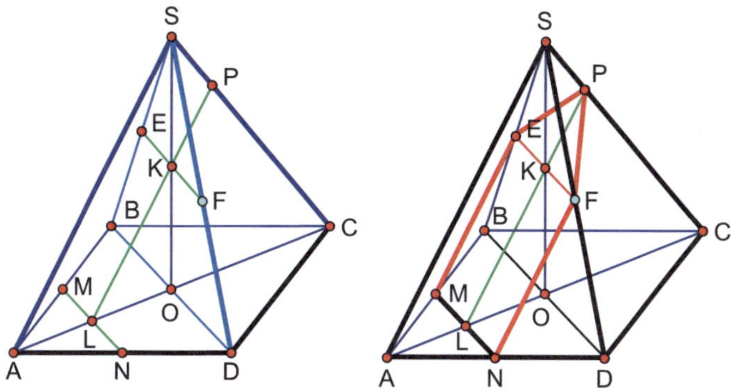

Fig. 3.63 Pyramid and its second cross section

there are points N and M) and the midpoint of the height SO (point K) Denote the cutting plane by α, we can find how it will intersect the base of the pyramid and the following is true

$$MN = \alpha \cap ABCD.$$

Let us find the point of intersection of the segment MN and the diagonal AC

$$L = MN \cap AC.$$

L is the midpoint of segment AO, K is the midpoint of segment SO. Consider triangle AOS. In this triangle LK is the midline segment, so $LK \parallel AS$. This is a very important observation! Since both points L and K lie in the cutting plane α, then the plane α must go parallel to the lateral edge AS of the pyramid. In order to finish our construction, let us consider triangle ASC shown in Fig. 3.63 in blue. Continue segment LK until it intersects the edge of SC at point P, the point of intersection of the cutting plane with the edge SC, that can be written as $P = \alpha \cap SC$.

Notice that the plane of triangle BCD is perpendicular to the base square $ABCD$. Moreover,

$$LK \perp MN, \quad MN \perp AC, \quad SO \perp AC$$

Then the cutting plane α must intersect the plane of triangle BCD by the line parallel to MN. Next, we draw line EF parallel to MN:

$$EF \parallel BD \parallel MN$$

Further, points E and F are the points of intersection of the cutting plane with the edges SB and SD, respectively. Connecting E and P, P and F, M and E, and F and N, we finally obtain the cross-sectional polygon $MNEPF$, which can be placed in the same sketch with the other cross section, the area of which equals s (see Fig. 3.64).

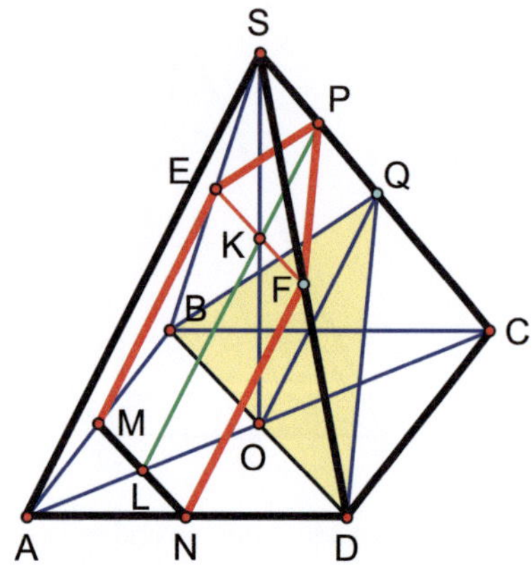

Fig. 3.64 Evaluating the area of the second cross section

Let us find the area of pentagon as the sum of the triangle EPF and the rectangle $MEFN$. From our construction, we can state that

$$\triangle\ EPF \sim \triangle\ BQD$$

with the coefficient of similitude $k = 2$ (each side of the small triangle is half the length of the corresponding side of the big triangle), then the ratio of the areas is equal to $k^2 = 4$. Hence,

$$[EPF] = \frac{1}{4} \cdot s.$$

Now, we can find the area of the rectangle. Let us show that the area of this rectangle equals the area of the yellow triangle BQD.

$$\big[BQD\big] = \left(\frac{1}{2}BD\right) \cdot QO = MN \cdot QO = MN \cdot LK = s$$

By adding two area we obtain the area of pentagon:

$$[MEPFN] = s + \frac{1}{4} \cdot s = \frac{5s}{4}.$$

Answer. $\frac{5s}{4}$.

In the following problem we will find the angle between two planes differently.

Problem 162 A regular triangular prism $ABCA_1B_1C_1$, all edges of which equal a is given

1. Construct the cross section of the prism by the plane passing the midpoints of edge AC and BB_1, and the vertex of prism A_1.
2. Find out how the cutting plane divides the edge BC of the prism.
3. Evaluate the angle between the cross-sectional polygon and the base of the prism ABC.
4. Find the area of the cross section.

Solution Part 1. Construction. We can start from drawing a regular prism and by marking the midpoints of the corresponding edges as M and E as it is shown on the left of Fig. 3.65. Denote the cutting plane by α. Next can find right away how the cutting plane intersects the face AA_1C_1C and find one of the edges of the cross-sectional polygon, A_1M. Notice that we can continue lines A_1M and C_1C until they intersect at some point T, that can be written as $A_1M \cap C_1C = T$. This point T and the midpoint E belong to both, the plane α and face BB_1C_1C, which means that the cutting plane intersects this face by the line TE:

$$\left.\begin{array}{l}\{T, E\} \in \alpha \\ \{T, E\} \in BB_1C_1C\end{array}\right\} \Rightarrow \alpha \cap BB_1C_1C = (TE)$$

In the plane BB_1C_1C we will find the intersection of the line TE and the edge of the prism CB and mark it by K ($(CB) \cap (ET) = K$.) Next, we can connect points K and E and obtain a new edge KE of the cross-sectional polygon. Finally, the following in true for the intersections of plane α with faces AA_1B_1B and ABC, respectively.

$$\left.\begin{array}{l}\{A_1, E\} \in \alpha \\ \{A_1, E\} \in AA_1B_1B\end{array}\right\} \Rightarrow \alpha \cap AA_1B_1B = AE$$

and

$$\left.\begin{array}{l}\{M, K\} \in \alpha \\ \{M, K\} \in ABC\end{array}\right\} \Rightarrow \alpha \cap ABC = MK$$

The quadrilateral MA_1EK is the cross section. Please, see our constructions on the right of Fig. 3.65.

Part 2. In order to find the ratio in which the cutting plane α divides the edge CB, we can consider two right similar triangles TCK and EBK ($\triangle\ TCK \sim \triangle\ EBK$) because the angles $\angle CKT = \angle BKE$ as vertical.

$$\frac{CK}{KB} = a \div \frac{a}{2} = 2 \div 1.$$

Moreover, $CK = \frac{2a}{3}$, $KB = \frac{a}{3}$.

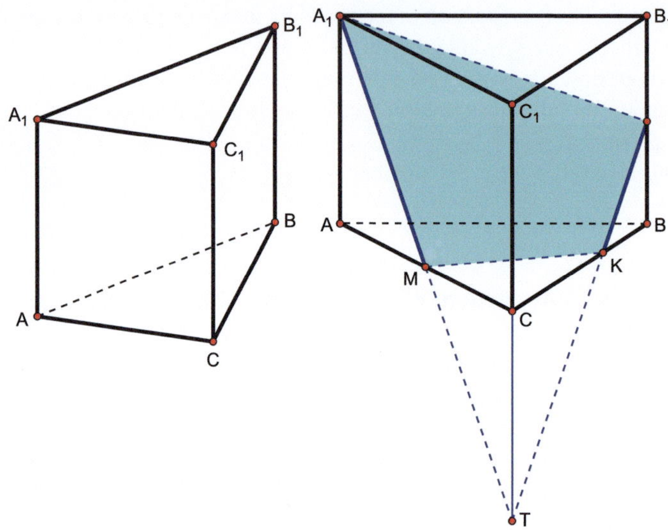

Fig. 3.65 Prism and its cross section

Part 3. Let us find and mark in our sketch the angle between the cutting plane α and the base plane ABC (see Fig. 3.66). Continue segment MK to the left and drop perpendicular AP to it from vertex A, $AP \perp MK$. Because the given prism is right, then $A_1A \perp ABC$, $A_1A \perp MK$. Hence, $A_1P \perp (MK)$ by the Theorem 20 about three perpendiculars. Therefore, the angle between the cross section and the base ABC is $\angle A_1PA$. Let us denote it by β.

How can we find the angle β? From the right triangle $\triangle A_1PA$ we find $\tan\beta = \frac{A_1A}{AP}$, $A_1A = a$, $AP =$? Then

$$\tan\beta = \frac{a}{AP} \tag{3.8}$$

Side AP can be found from the right triangle $\triangle APM$ as $AP = AM \cos\angle AMP$. Denote $\angle AMP = \varphi = \angle KMC$, then

$$AP = \frac{a}{2}\sin\varphi \tag{3.9}$$

and we can find angle φ from triangle MCK, in which we know that the angle C is $60°$. By Law of Cosines applied to $\triangle MCK$:

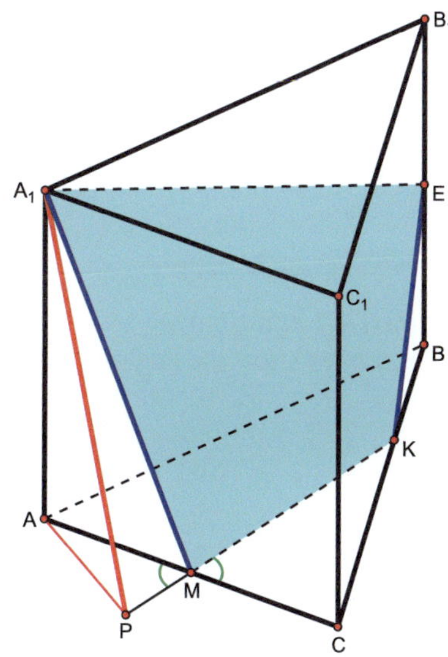

Fig. 3.66 Finding the angle between the cross section and the base ABC

$$MK^2 = MC^2 + CK^2 - 2MC \cdot CK \cos 60^\circ;$$
$$MK^2 = \frac{a^2}{4} + \frac{4a^2}{9} - \frac{a^2}{3};$$
$$MK^2 = \frac{13a^2}{36};$$
$$MK = \frac{a\sqrt{13}}{6}.$$

By the Law of Sines applied to the same triangle $\triangle MCK$,

$$\frac{CK}{\sin\varphi} = \frac{MK}{\sin 60^\circ};$$
$$\frac{2a}{3\sin\varphi} = \frac{a\sqrt{13}}{6} \div \frac{\sqrt{3}}{2}.$$

After simplification we obtain that

$$\sin\varphi = \frac{2\sqrt{3}}{\sqrt{13}} \tag{3.10}$$

Substituting (3.10) into (3.9) we obtain

$$AP == \frac{a\sqrt{3}}{\sqrt{13}} \tag{3.11}$$

Replacing AP from (3.11) into formula (3.8) we obtain the angle between the cutting plane and the base of the prism:

$$\tan\beta = \frac{a}{AP} = a \div \frac{a\sqrt{3}}{\sqrt{13}} = \frac{\sqrt{39}}{3}$$

Finally, the angle between two planes is equal to $\beta = \arctan\frac{\sqrt{39}}{3}$.

Part 4. Let us find the area of the cross section using Theorem 43. Because we know the angle, we can easily find the cosine of the same angle:

$$\cos\beta = \sqrt{\frac{1}{1+\tan^2\beta}} = \sqrt{\frac{1}{1+\left(\frac{\sqrt{39}}{3}\right)^2}} = \frac{\sqrt{3}}{4}$$

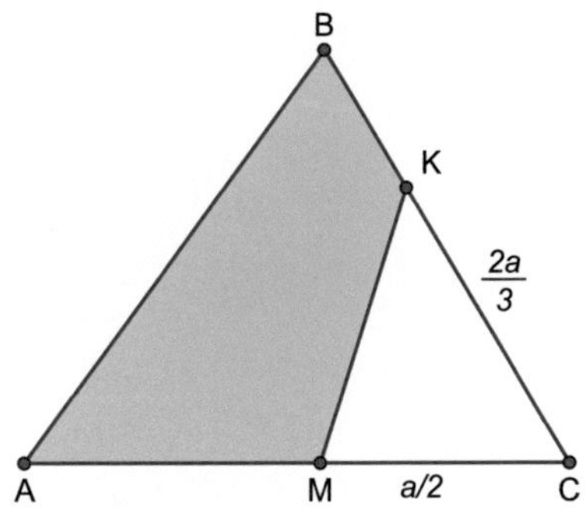

Fig. 3.67 Finding the area of the shaded region of the base

The area of the cross section is equal to the area of the shaded region $ABKM$ shown in Fig. 3.67 divided by the cosine of angle β. We can find the shaded area by subtracting the areas of triangle ABC and MKC.

The area of an equilateral triangle ABC with side a is well-known ([2]) :

$$[ABC] = \frac{a^2\sqrt{3}}{4}$$

The area of triangle MKC can be found as follows

$$\begin{aligned}[MKC] &= \tfrac{1}{2}\cdot MC\cdot KC\cdot\sin 60^\circ = \tfrac{1}{2}\cdot\tfrac{a}{2}\cdot\tfrac{2a}{3}\cdot\tfrac{\sqrt{3}}{2}\\ &= \tfrac{a^2\sqrt{3}}{12}\end{aligned}$$

Next, we will find the difference of the two areas:

$$\frac{a^2\sqrt{3}}{4} - \frac{a^2\sqrt{3}}{12} = \frac{a^2\sqrt{3}}{6}$$

Finally, the area of the cross section is equal to

$$\frac{a^2\sqrt{3}}{6} \div \frac{\sqrt{3}}{4} = \frac{a^2}{8}.$$

Answer. 1. A_1EKM; 2. $CK \div KB = 2 \div 1$; 3. $\beta = \arctan \frac{\sqrt{39}}{3}$; 4. $\frac{a^2}{8}$.

When solving problems in this section we will often use some terms from plane geometry, such as centroid, circumcenter, orthocenter, incenter of a triangle.

The **centroid** is the point of intersection of the three medians of the triangle.

The **incenter** is the point of intersection of three bisectors of the triangle. This is the center of the inscribed circle.

The **circumcenter** is the point of intersection of three perpendicular bisectors of the triangle. This is also the center of the circumscribed circle.

The **orthocenter** is the point of intersection of three altitudes of a triangle.

Please recall that only in an equilateral triangle centroid, incenter, circumcenter and orthocenter is the same point inside the triangle.

When construction of a cross section is required part of the problem, the construction itself can be complicated if the point of the cutting plane belong not to the edges of the polyhedron but to its faces. Some such problems will be considered in this section.

Problem 163 The cutting plane goes through points M, N, K, the midpoints of the edges AB, AD, and CC_1, respectively. The cross section of the cube $ABCDA_1B_1C_1D_1$ by this plane is shown in Fig. 3.68. Evaluate the area of the cross section.

Solution We will apply Theorem 43. This theorem is especially helpful because the obtained cross-sectional figure (the pentagon) is symmetric. It is not difficult to construct its orthogonal projection—we just need to find orthogonal projections of all its vertices onto the base of the cube. The pentagon $MTKQN$ projection onto the base $(ABCD)$ of the cube is pentagon $MBCDN$ (Fig. 3.68).

Fig. 3.68 Finding the area of pentagonal cross section

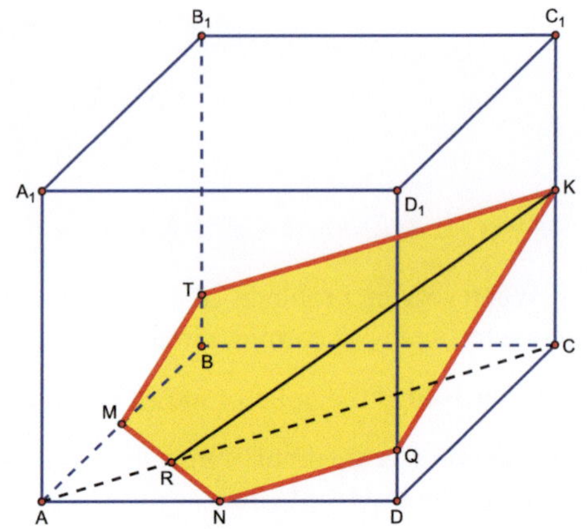

First, let us indicate the line of symmetry of the pentagon *MTKQN*. For this, we will find the intersection, *R*, of diagonal *AC* and segment *MN*. *MN* is a midline of triangle *ABD*, then *R* is the midpoint of segment *MN*. Because *KC* is perpendicular to *AC* and *AC* is perpendicular to *MN*, then *KR* is perpendicular to *MN* (by the theorem of three perpendiculars, Theorem 20) Therefore, *KR* is the line of symmetry of the pentagon *MTKQN*.

The angle between the pentagon and its orthogonal projection is $\angle KRC = \theta$.

Assume that the side of the cube is 1. Then $KC = 0.5$, $AC = \sqrt{2}$, $RC = \frac{3\sqrt{2}}{4}$, $\cos(\theta) = \frac{RC}{RK}$.

From the right triangle *KCR* we have that hypotenuse equals

$$RK = \sqrt{\frac{1}{4} + \left(\frac{3\sqrt{2}}{4}\right)^2} = \frac{\sqrt{22}}{4}$$

and cosine of angle θ is

$$\cos(\theta) = \frac{RC}{RK} = \frac{3}{\sqrt{11}}$$

The area of pentagon *MBCDN* can be found as the difference of two areas $[ABCD] - [AMN]$ or

$$1 - \frac{1}{2} \cdot \frac{\sqrt{2}}{2} \cdot \frac{\sqrt{2}}{4} = 1 - \frac{1}{8} = \frac{7}{8}$$

Finally, by dividing this projection area by the cosine of $\angle KRC$, we obtain the area of the cross -section as

$$A = \frac{7}{8} : \frac{3}{\sqrt{11}} = \frac{7\sqrt{11}}{24}.$$

Answer. $A = \frac{7\sqrt{11}}{24}$

Problem 164 Find the ratio of the volumes of the two polyhedra on which the plane α cuts the cube. For solving this problem, use the picture obtained under construction of the cross section (Problem 145, Fig. 3.16).

Solution The cross-sectional pentagon in Fig. 3.16 divides the cube into two convex polyhedra. The construction lines can give us some ideas on how to find the ratio of their volumes. Let us focus on finding the volume of the figure below the cross section. We can see that its volume can be found as the volume of the "big" pyramid with vertex K and base triangle PRC minus the volumes of two small pyramids (one with vertex Q and base triangle PDN and the other with vertex T and base -triangle MBR) All these three pyramids are right pyramids.

Once we can visualize the situation, the rest will just become routine. For example, from symmetry we know that the small pyramids are equal. Both small pyramids have right isosceles triangles with a side of 0.5 at the base. The base of the big pyramid is a right isosceles triangle with legs length 1.5. Triangles PQD and PKC are similar, $PD = 0.5, PC = 1.5, KC = 0.5$.

$$\frac{QD}{PD} = \frac{KC}{PC} \Rightarrow \frac{QD}{0.5} = \frac{0.5}{1.5}, \quad OD = \frac{1}{6} = h$$

The volume of the big pyramid is

$$V = \frac{1}{3} \cdot \frac{1}{2} \cdot \frac{1}{2} \cdot \frac{3}{2} \cdot \frac{3}{2} = \frac{3}{16}$$

The volume of the small pyramid is $v = \frac{1}{3} \cdot \frac{1}{6} \cdot \frac{1}{2} \cdot \frac{1}{2} \cdot \frac{1}{2} = \frac{1}{144}$. The volume of the figure below the pentagonal cross section and the ratio of the volumes are

$$V_1 = \frac{3}{16} - 2 \cdot \frac{1}{144} = \frac{3}{16} - \frac{1}{72} = \frac{3}{2 \cdot 8} - \frac{1}{9 \cdot 8} = \frac{25}{144}$$
$$V_2 = 1 - \frac{25}{144} = \frac{119}{144}$$
$$V_1 \div V_2 = 25 \div 119$$

Answer. $V_1 \div V_2 = 25 \div 119$

Problem 165 A regular tetrahedron $ABCS$ has side of 1. Let L be the midpoint of SB, $BK \div KC = 3 \div 1$, O is the orthocenter of the base triangle ABC. Construct a cross section of the tetrahedron by a plane passing through points K, O, and L and find how this plane divides the side AB.

Solution Let us sketch the situation described in the problem (Fig. 3.69) and construct the cross section of the pyramid.

First, we connect points L and K. Second, we find the orthocenter point O. For this, we find the midpoints of AB (denote it by J) and the midpoint of BC denoting it by T. Draw two medians and find the point of their intersection O.

$$O = (CJ) \cap (AT)$$

Next, connect O with point K, then by continuing the segment KO until it intersects side AB at point M, we obtain the segment ML and finish our construction. ΔMLK shown in Fig. 3.70) is the cross section.

Let us find the area of this triangle.

1. If $BK \div KC = 3 \div 1$, then $KC = 1/4$. Because O is the point of intersection of the medians, then it is divided by each median in a ratio of $2 \div 1$, starting from the vertex of the equilateral triangle ABC (Fig. 3.71).
   Then $AT = \frac{\sqrt{3}}{2}$ and $OT = \frac{\sqrt{3}}{6}$. Triangle OTK is the right triangle, then the following is true
$$KO^2 = OT^2 + TK^2$$
$$KO^2 = \left(\frac{\sqrt{3}}{6}\right)^2 + \left(\frac{1}{4}\right)^2$$
$$KO = \frac{\sqrt{21}}{12}$$

2. Let $x = OM$, $y = AM$. Applying Menelaus Theorem (see my book [2], Sect. 1.4.2. and Home Work Problem 19, p. 87) to the triangle MBK with a transversal AT, we obtain the following relationship:

Fig. 3.69 Sketch of the pyramid with three given points

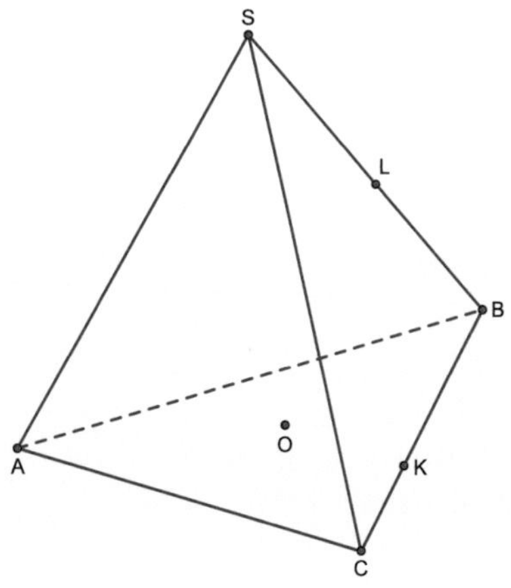

$$\frac{MO}{OK} \cdot \frac{KT}{TB} \cdot \frac{BA}{AM} = 1$$
$$\frac{x}{\frac{\sqrt{21}}{12}} \cdot \frac{\frac{1}{4}}{\frac{1}{2}} \cdot \frac{1}{y} = 1$$

Which gives us that

$$6x = \sqrt{21}y$$

Or if we square both sides:

$$12x^2 = 7y^2 \tag{3.12}$$

3. From the right triangle JMO we have that $JO = \frac{\sqrt{3}}{6}$. Since $AJ = 1/2$ and $AM = y$, then

$$MJ = \frac{1}{2} - y.$$

Using Pythagorean Theorem, we obtain that

$$\frac{3}{36} + \left(\frac{1}{2} - y\right)^2 = x^2,$$

which can be simplified to

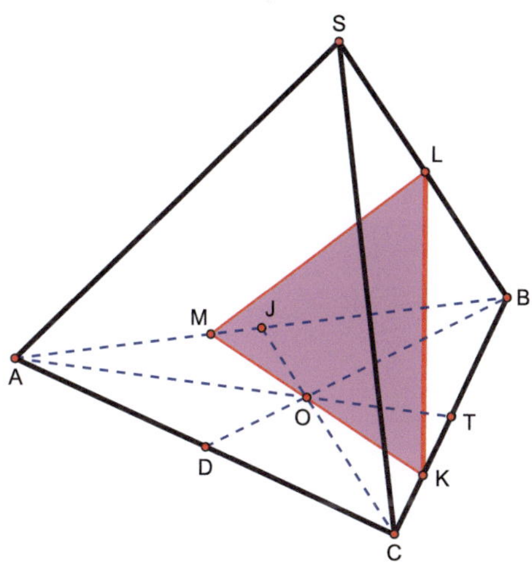

Fig. 3.70 Cross section of a triangular pyramid

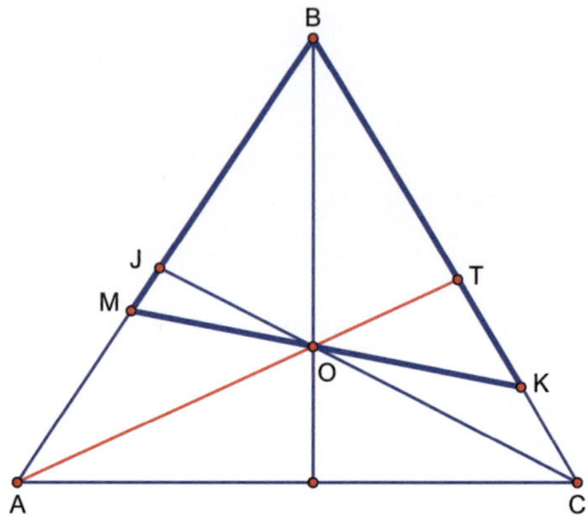

Fig. 3.71 View of the base of a triangular pyramid

$$4 - 12y + 12y^2 = 12x^2 \tag{3.13}$$

Substituting (3.12) into (3.13), we obtain a quadratic equation for y:

$$5y^2 - 12y + 4 = 0$$
$$y = 2, \;\; y = \tfrac{2}{5}$$

4. Since only $y = AM = 2/5$ satisfies the condition of the problem ($y < 1/2$), we get from (3.12) that

$$x = \frac{\sqrt{21}}{15}, \quad MK = OK + x = \frac{3\sqrt{21}}{20}.$$

5. Now we know that point M divides side AB in a ratio of

$$AM \div MB = \frac{2}{5} \div \frac{3}{5} = 2 \div 3.$$

Answer. $AM \div MB = 2 \div 3$

Problem 166 Using the condition of the previous Problem 169, find the lengths of all sides of the triangle MKL obtained in the cross section of the tetrahedron.

Solution 1. MK was found in the previous problem as $MK = \frac{3\sqrt{21}}{20}$.

2. Next, in order to evaluate KL, we in the plane of triangle CSB will drop perpendicular from vertex L onto side CB, obtaining point F. Note that triangle CSB is an equilateral triangle ($CS = SB = 1$), then its height equals $\frac{\sqrt{3}}{2}$. Since L is the midpoint of SB, then $LF = \frac{\sqrt{3}}{4}$. $KF = \frac{1}{2}CB = 1/2$,
3. Applying the Pythagorean theorem to the right triangle LFK, we find $KL = \frac{\sqrt{7}}{4}$.
4. Similarly, let us drop perpendicular from L onto segment AB in the plane of triangle ASB (see Fig. 3.72). Similarly, $LN = \frac{\sqrt{3}}{4}$

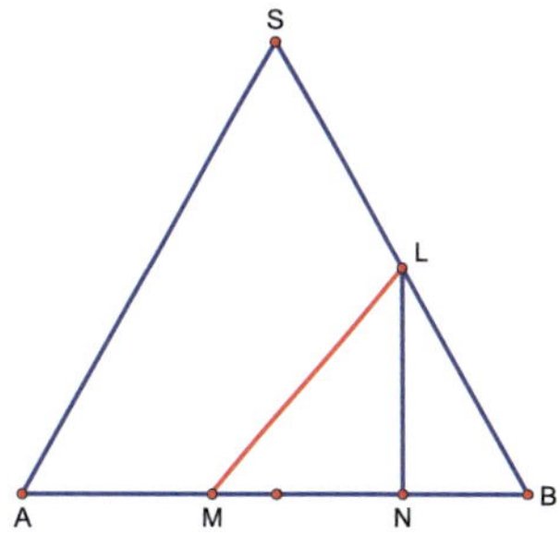

Fig. 3.72 Finding side ML

5. From the right triangle LNB we find that

$$NB^2 = LB^2 - LN^2$$
$$NB = \frac{1}{4}$$
$$AN = 1 - 1/4 = 3/4$$
$$MN = 3/4 - 2/5 = 7/20$$
$$ML = \sqrt{\frac{3}{16} + \frac{49}{400}} = \frac{\sqrt{31}}{10}$$

Therefore, $ML = \frac{\sqrt{31}}{10}$, $MK = \frac{3\sqrt{21}}{20}$, and $KL = \frac{\sqrt{7}}{4}$.

Answer. $ML = \frac{\sqrt{31}}{10}$, $MK = \frac{3\sqrt{21}}{20}$, and $KL = \frac{\sqrt{7}}{4}$.

Problem 167 In a regular quadrangular pyramid $MABCD$ with vertex M, the sides of the base are equal to 1, and the lateral edges are equal to 2. The point N belongs to the edge MC, and $MN \div NC = 2 \div 1$. (1) Construct the cross section of the pyramid by the plane α passing through points B and N parallel to line AC. (2) Evaluate its area.

Solution Part 1. (Construction)

1. Draw pyramid $ABCDM$ and place point N on edge CM. Since the cutting plane goes parallel to diagonal AC, then show this diagonal on your original sketch (Fig. 3.73)

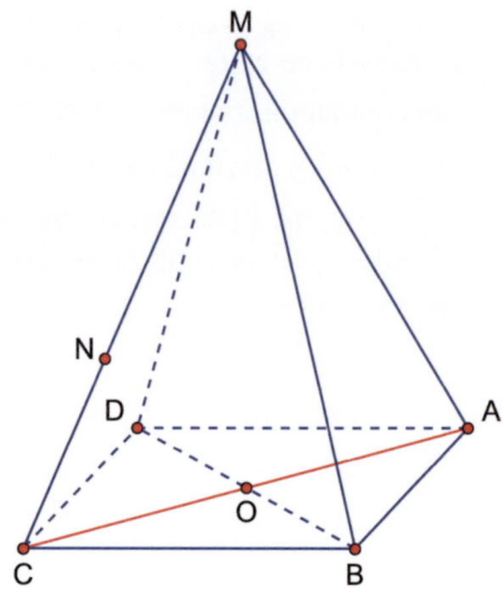

Fig. 3.73 Regular square pyramid

2. Let us draw in the triangle plane ACM, segment NK parallel to AC (point K belongs to edge MA). Let NK intersect MO at point P (O is the center of the base of the pyramid), and $MP : PO = MN : NC = 2 : 1$ (by Thales) then the point P is the centroid- the point of intersection of the medians of the triangle MBD. Our cutting plane goes through points B, N and K (Fig. 3.74)

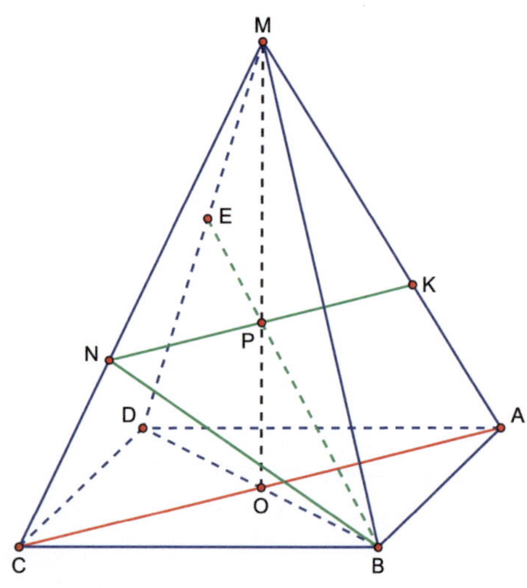

Fig. 3.74 Drawing line NK parallel to AC in the square pyramid

3. How does it cut the faces CDM and ADM?

Line BP intersects edge MD at point E. This line belongs to the cross section.

$$\left.\begin{array}{l}\{E, N\} \in \alpha \\ \{E, N\} \in MDC\end{array}\right\} \Rightarrow \alpha \cap MDC = EN$$

4. Connect E and K and K and B. Quadrilateral $BNEK$ is the required cross section (Fig. 3.75).

Part 2. Segment BE is the median of triangle MBD, so we can evaluate its length from this triangle:

$$BE = \frac{\sqrt{2BD^2 + 2MB^2 - MD^2}}{2} = \frac{\sqrt{4AB^2 + MB^2}}{2} = \frac{\sqrt{4 \cdot 1 + 4}}{2} = \sqrt{2}.$$

In the isosceles triangle AMC:

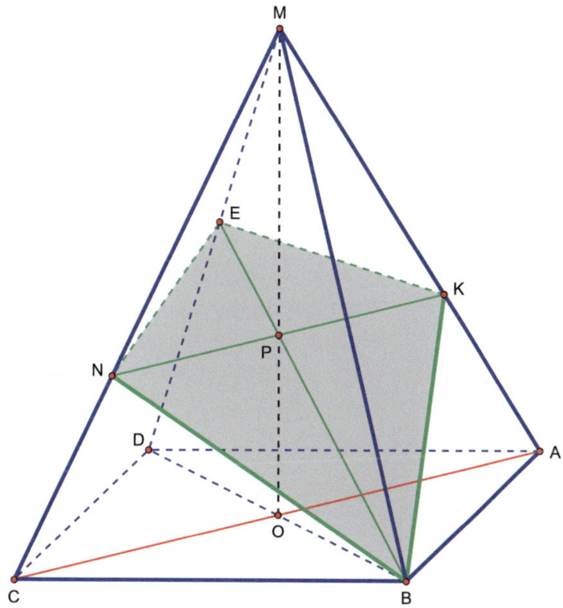

Fig. 3.75 Cross section *NEKB* of square pyramid

$$\left.\begin{array}{l} NK \parallel CA \\ MO \perp CA \end{array}\right\} \Rightarrow NP = PK, \quad BE \perp NK.$$

We can evaluate NK as $NM = \frac{2}{3}MC, \Rightarrow NK = \frac{2}{3}AC = \frac{2\sqrt{2}}{3}$ Since the line BD is perpendicular to the plane MAC, the diagonals BE and NK of the quadrilateral $BNEK$ are perpendicular. Therefore, the cross section is a kite, and its area equals half the product of the diagonals:

$$[NEKB] = \frac{1}{2} \cdot BE \cdot NK = \frac{1}{2} \cdot \sqrt{2} \cdot \frac{2\sqrt{2}}{3} = \frac{2}{3}.$$

Answer. $\frac{2}{3}$.

Problem 168 Find the area of the cross section of the cube $ABCDA'B'C'D'$ with the edge of 1 by plane α that passes through points $L \in AB,\ N \in AD,\ M \in CC'$. It is know that $AL \div LB = 2 \div 1,\ AN \div ND = 2 \div 1,\ CM \div MC' = 2 \div 1$.

Solution Part 1. (Construction) 1. First, we can connect points L and N in the face of $ABCD$ and construct the first segment of the cross section (Fig. 3.76)

$$\left.\begin{array}{l}\{L, N\} \in \alpha \\ \{L, N\} \in ABCD\end{array}\right\} \Rightarrow \alpha \cap ABCD = (LN)$$

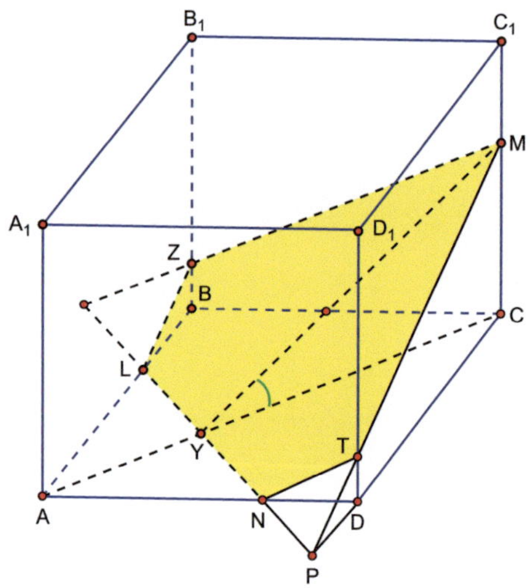

Fig. 3.76 The pentagonal cross section of a cube

2. Second, in order to find out how the plane cuts the face $DD'C'C$, we need to find an additional point of these two planes. For this, we can continue line LN and intersect it with line CD in the plane of the face $ABCD$ to obtain point P. We can state the following

$$\left.\begin{array}{l}(LN) \cap (CD) = P \\ \{P, M\} \in \alpha \\ \{P, M\} \in DD'CC'\end{array}\right\} \Rightarrow \alpha \cap DD'CC' = (PM).$$

3. Let the line PM cut the edge DD' at point T. Then segment TM is the second edge of the cross section.
4. If we continue the line LN inside the face $ABCD$ to intersect with the line BC, we obtain point W (not shown in Fig. 3.76) Both W and M belong to the cutting plane

and to the face $BB'CC'$, then two planes intersect by the line WM, and the face—by segment ZM. By connecting points Z and L we obtain the last segment LZ and finish construction of the cross section, pentagon $NTMZL$.
Alternatively, the edge LZ can be obtained by drawing parallel line to MT, because the cutting plane must cut two parallel faces of the cube by parallel lines (see Theorem 10).
5. The cross section is the pentagon, $NTMZL$.

Part 2. We can find the area of the cross section using the Theorem 43. Let us find cosine of the angle between this plane and the base of the cube $ABCD$. Y is the intersection of the diagonal AC and LN, Y is the midpoint of LN. Moreover, since the pentagon is symmetric with respect to MY, then $MY \perp LN$.Next, $YC \perp LN$, $(LN \parallel BD)$. Hence, using the Theorem 20 on three perpendiculars, the angle $\gamma = \angle MYC$ is the angle between the cutting plane and the face $ABCD$.

$$\cos\gamma = \frac{YC}{YM}.$$

By Thales Theorem, point Y divides diagonal AC in a ratio of $2 \div 1$ as point N divides edge AD, then $YC = \frac{2}{3}\sqrt{2}$. From the right triangle $\triangle\, MCY$ we have

$$YM = \sqrt{YC^2 + MC^2} = \sqrt{\left(\tfrac{2\sqrt{2}}{3}\right)^2 + \left(\tfrac{2}{3}\right)^2} = \tfrac{2\sqrt{3}}{3}.$$
$$\cos\gamma = \tfrac{2\sqrt{2}}{3} : \tfrac{2\sqrt{3}}{3} = \tfrac{\sqrt{2}}{\sqrt{3}}$$

The projection area (purple region in Fig. 3.77) equals the area of the square with side 1 minus the area of the right triangle LAN with legs of $2/3$, then

$$[LBCDN] = 1 - \frac{1}{2}\cdot\left(\frac{2}{3}\right)^2 = \frac{7}{9}.$$

Finally, the pentagon area can be found as follows

$$[LZMTN] = \frac{7}{9} : \cos\gamma = \frac{7}{9}\cdot\frac{\sqrt{3}}{\sqrt{2}} = \frac{7\sqrt{6}}{18}.$$

Answer. $\frac{7\sqrt{6}}{18}$.

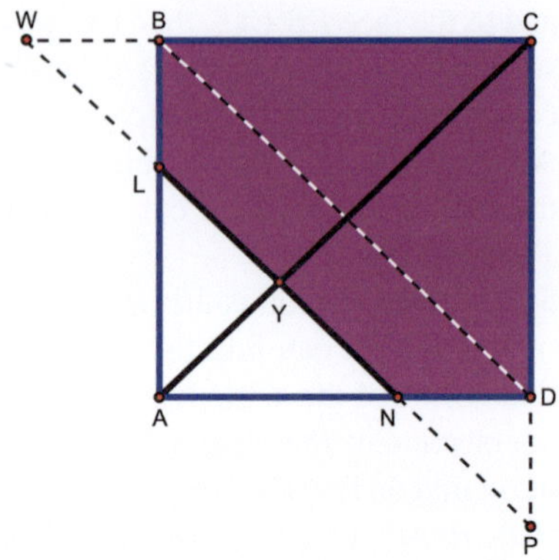

Fig. 3.77 Projection of the cross section onto face *ABCD*

Problem 169 In a regular quadrangular pyramid *SABCD* with base *ABCD*, a cross section is drawn through the midpoints of the edges *AB* and *BC* and the vertex *S*. Find the area of this cross section if all the edges of the pyramid are equal to a.

Solution Let M be the midpoint of AB and N the midpoint of BC. Then the area of the cross section is equal to the area of the triangle SMN. Find sequentially SM, MN and SN (see Fig. 3.78).

Here SM and SN are the medians of triangles SAB and SBC, respectively. Since these triangles are equilateral (because all the edges of the pyramid are of the same length), then

$$SM = SN = \sqrt{SB^2 - BN^2} = \frac{a\sqrt{3}}{2}$$

Let us now find the length of MN. Since it is the midline segment of AC, then it is half of AC, $MN = \frac{a\sqrt{2}}{2}$. Now let's find the area of the isosceles triangle SMN. To do this, we draw the height SH, which, according to the Pythagorean Theorem, is equal to $h = SH = \sqrt{\left(\frac{a\sqrt{3}}{2}\right)^2 - \left(\frac{a\sqrt{2}}{2}\right)^2} = \frac{a}{2}$, and calculate the area:

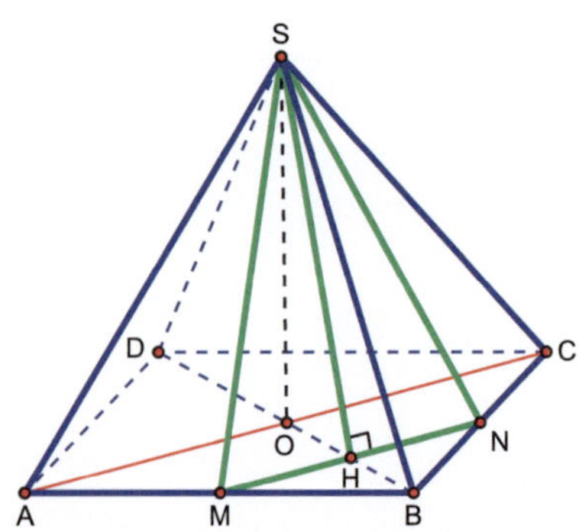

Fig. 3.78 The cross section of pyramid *SABCD*

$$[SMN] = \frac{1}{2} \cdot \frac{a\sqrt{2}}{2} \cdot \frac{a}{2} = \frac{a^2\sqrt{2}}{8}$$

Answer. $[SMN] = \frac{a^2\sqrt{2}}{8}$

Problem 170 Point E is the midpoint of edge BB_1 of cube $ABCDA_1B_1C_1D_1$. 1. Find the cross-sectional area of the cube cut by the plane D_1AE, if the edges of the cube are equal to 4 units. 2. How does the cutting plane divide the volume of the cube?

Solution 1. (Construction) The straight line AE intersects the straight line A_1B_1 at the point K. The straight line $D_1K \cap C_1B_1 = F$ intersects the edge at its middle point F. Thus, $AEFD_1$ is the cross section of the cube by the plane D_1AE (see Fig. 3.79).

In an isosceles triangle AKD_1 we have that

$$D_1K = AK = 2AE = 4\sqrt{5}$$

and the height of the triangle is equal to

$$h = \sqrt{AK^2 - \left(\frac{AD_1}{2}\right)^2} = 6\sqrt{2}$$

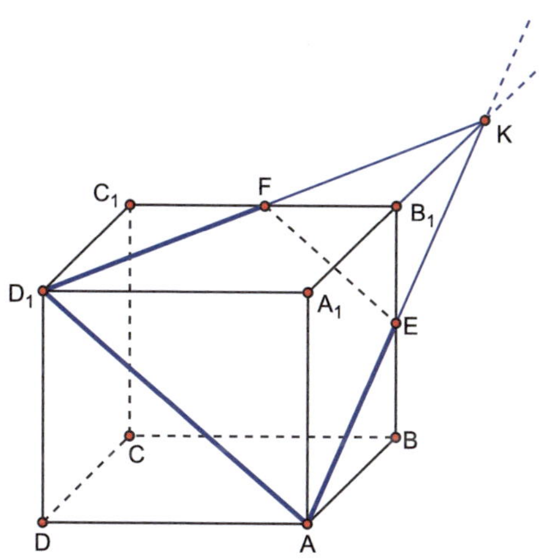

Fig. 3.79 The cross section of a cube

Since EF is the midline segment of the triangle AKD_1, we get for the areas:

$$[KEF] = \frac{1}{4} \cdot [AKD]$$

Next, $[AEFD_1] = [AKD_1] - [KEF] = \frac{3}{4} \cdot [AKD_1] = 18.$

2. Let us now find how this cross section divides the volume of the cube. The sketch in Fig. 3.79 is perfect if we saved and did not erase our construction of vertex K shown by dotted lines. We can see that the cross section AD_1FE divides the cube into the truncated pyramid with volume V_1 to be determined and another body below this pyramid with the volume $V_2 = V - V_1 = a^3 - V_1$. The truncated pyramid has the upper base B_1, the lower base of B_2, and the height h, such that

$$B_1 = [FB_1E], \quad B_2 = [D_1A_1A], \quad h = A_1B_1$$

Using formula (2.6), the volume of the truncated pyramid $FB_1ED_1A_1A$ can be found as follows

$$V_1 = \frac{1}{3} \cdot h \cdot \left(B_1 + B_2 + \sqrt{B_1B_2}\right),$$

where

$$B_1 = \frac{1}{2}\left(\frac{a}{2}\right)^2 = \frac{a^2}{8}$$

$$B_2 = \frac{1}{2}a^2 = \frac{a^2}{2}$$

$$\sqrt{B_1B_2} = \frac{a^2}{4}$$

Next, we have for the two volumes:

$$V_1 = \frac{1}{3} \cdot a \cdot \left(\frac{a^2}{8} + \frac{a^2}{2} + \frac{a^2}{4}\right) = \frac{7a^3}{24}, \quad V_2 = a^3 - \frac{7a^3}{24} = \frac{17a^3}{24}$$

Hence, the ratio of the volumes is $V_1 \div V_2 = 7 \div 17$.

Answer. 1. $A = 18$ *units*^{2}
2.$V_1 \div V_2 = 7 \div 17$

Problem 171 In a regular quadrangular pyramid, the dihedral angle at the base is α. A plane is drawn through its edge forming with the base the angle of β. The side of the base is equal to a. Evaluate the cross-sectional area.

Solution Let us sketch the situation and denote the angle of inclination of one of the lateral faces of the pyramid to the base as α. (This angle is marked by two arcs in Fig. 3.80) Let M be the midpoint of segment AD, and K be the midpoint of BC. Then $MK \perp AD$ and $SM \perp AD$, then $\angle SMK = \alpha$. A secant plane is drawn through the edge AD is forming with the base $ABCD$ angle $\angle NMK = \beta = \angle IDC$. (This angle is marked by one arc)

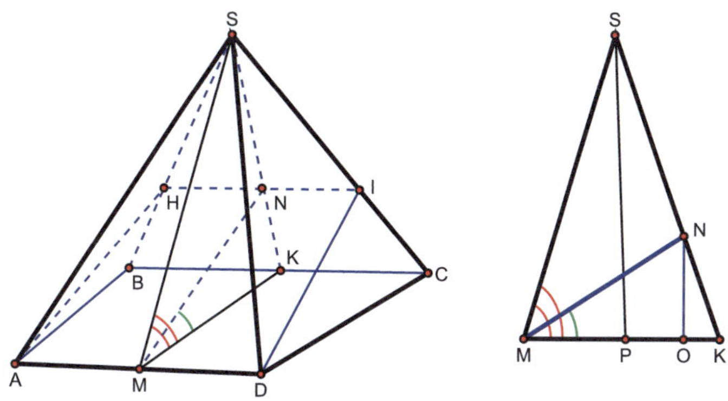

Fig. 3.80 Problem 175 on quadrangular pyramid

For convenience, let us consider triangle MSK shown on the right of Fig. 3.80. Mark point N and drop perpendiculars NO and SP to the base of the triangle MKS.

Denote $MO = x,\ NO = y$. Using trigonometry, we have from $\triangle MNO$:

$$y = x \tan \beta$$

Triangle MSK is an isosceles triangle, then $\angle SMK = \alpha = \angle SKM$. From triangle NOK:

$$\left.\begin{array}{l} y = NO = OK \tan \alpha \\ OK = MK - MO \\ OK = a - x \end{array}\right\} \Rightarrow y = (a - x) \tan \alpha$$

Now we can equate two formulas for y:

$$y = x \tan \beta = (a - x) \tan \alpha$$
$$x = \frac{a \tan \alpha}{\tan \alpha + \tan \beta},\ y = \frac{a \tan \alpha \tan \beta}{\tan \alpha + \tan \beta}$$

From triangle MON, we can evaluate the side of the rectangle in the cross section $b = DI = MN = \frac{x}{\cos\beta}$. Substituting here x we obtain

$$b = \frac{a\tan\alpha}{\tan\alpha\cos\beta + \sin\beta}$$

The area of the cross section is

$$A = ab = \frac{a^2\tan\alpha}{\tan\alpha\cos\beta + \sin\beta}$$

Using formula of trigonometry, we can rewrite the right-hand side of this formula as follows:

$$\frac{a^2\sin\alpha}{\cos\alpha\cdot\left(\frac{\sin\alpha\cdot\cos\beta}{\cos\alpha}+\sin\beta\right)} = \frac{a^2\sin\alpha}{\sin\alpha\cos\beta+\sin\beta\cos\alpha}$$

Let us apply the sine of the sum of two angles formula to the denominator. Finally, the area of the cross section is

$$A = \frac{a^2\sin\alpha}{\sin(\alpha+\beta)}.$$

Answer. $A = \frac{a^2\sin\alpha}{\sin(\alpha+\beta)}$

Problem 172 The height H of a regular quadrangular pyramid forms an angle α with its lateral face. Through the side of the base, a cross section is drawn perpendicular to the opposite face. Find the volume of the pyramid cut off by the plane.

Solution Part 1. Construction. Let us sketch a pyramid and drop the height SO from its apex. Here $SO = H$ and $\angle OSP = \alpha$. Assume that our plane goes through the edge BC of the pyramid $ABCDS$ shown in Fig. 3.81. Then this plane must be perpendicular to face ADS because this face is opposite to edge BC.

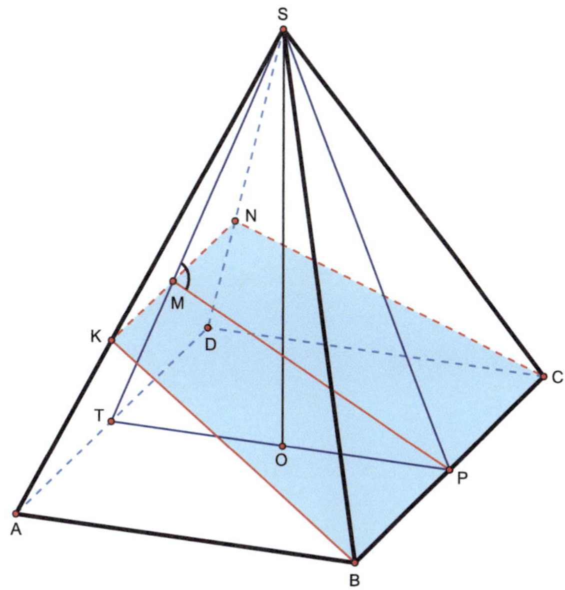

Fig. 3.81 Cross section perpendicular to face *ADS*

How can we proceed?

Consider triangle BSC. It is isosceles. If P is the midpoint of BC, then SP is the median and height of triangle BSC. Let us draw line through P parallel to side AB:

$$PT \parallel AB.$$

Connect points T and S , then

$$TS \perp AD$$

The segment KN of the cross section must be parallel to BC and intersect ST at such point (for example, denoted by M) that the angle

$$\angle SMP = 90^\circ.$$

It is hard to get an exact sketch for this problem, but we need to sketch something similar to what I have in Fig. 3.81 and finish construction by connecting points N and C, and B and K ($KN \parallel BC$). The cross section is the trapezoid $KNCB$.

Part 2. Pyramid $SKNCB$ is cut off by this cross section and its volume we need to evaluate. Denote $h = SM$ as its height dropped to its base $KNCB$.
It is very important to mention that since triangle TSP is isosceles, $ST = SP$ and the following is true.

$$\angle TSP = 2\angle OSP = 2\alpha.$$

Next, from the right triangle *SMP*, we can easily evaluate the length of the side *MS*, which is the height of the new pyramid *SKNCB* that has been cut off by the cross section from the original pyramid.

$$MS = SP\cos(2\alpha).$$

SP in turn, can be found from the right triangle *SOP* as

$$SP = \frac{SO}{\cos\alpha} = \frac{H}{\cos\alpha}.$$

We now find that

$$MS = \frac{H\cos(2\alpha)}{\cos\alpha};\ MP = 2H\sin\alpha.$$

The triangles $\triangle SKN \sim \triangle SAD$, then the following is true

$$\frac{KN}{AD} = \frac{SM}{ST} = \frac{SM}{SP} = \cos 2\alpha,$$

from which we obtain that

$$KN = AD \cdot \cos 2\alpha$$

Next, $AD = 2OP$ and

$$AD = 2H\tan\alpha = BC.$$

Then

$$KN = 2H\tan\alpha\cos(2\alpha).$$

Let us find the area of the trapezoid *KNCB*.

From the right triangle *SMP* we obtain that $MP = SP \cdot \sin(2\alpha)$. Since $SP = \frac{H}{\cos\alpha}$, then $MP = 2H\sin\alpha$. The area of the trapezoid is

$$[KNCB] = \frac{1}{2}(KN + BC) \cdot MP = \frac{1}{2}h(KN + BC)$$

where

$$KN = 2H\tan\alpha\cos 2\alpha.$$

and

$$BC = 2h\tan\alpha.$$

Using trigonometric formula for cosine of double angle, we rewrite

$$1 + \cos 2\alpha = 2\cos^2\alpha$$

Also we know that $\tan\alpha = \frac{\sin\alpha}{\cos\alpha}$ Then after simplification

$$\frac{1}{2}(KN + BC) = H\tan\alpha(1 + \cos 2\alpha) = 2H \cdot \frac{\sin\alpha}{\cos\alpha}\cos^2\alpha = H\sin 2\alpha.$$

The area of the trapezoid is

$$[KNCB] = \frac{H\cos 2\alpha}{\cos\alpha} \cdot H\sin 2\alpha = 2H^2\cos 2\alpha\sin\alpha.$$

$$[KNCB] = 4H^2\sin^2\alpha \cdot \cos\alpha.$$

The volume of the pyramid is

$$V(SKNBC) = \frac{1}{3} \cdot SM \cdot [KNCBS] = \frac{1}{3}2H\sin\alpha 2H^2\cos\alpha\sin\alpha.$$

After simplification we finally obtain

$$V = \frac{4}{3}H^3\sin^2\alpha\cos 2\alpha.$$

Answer. $V = \frac{4}{3}H^3\sin^2\alpha\cos 2\alpha.$

Next, we will consider a problem given at 1965 International Math Olympiad.

Problem 173 (*1965 International Math Olympiad*) Given the tetrahedron $ABCS$ whose edges AB and CS have lengths a and b, respectively. The distance between the skew lines AB and CS is d, and the angle between them is ω. Tetrahedron $ABCS$ is divided into two solids by plane ε, parallel to lines AB and CS. The ratio of the distances of ε from AB and CS is equal to t. Compute the ratio of the volumes of the two solids obtained.

Solution 1. Let us construct the cross section of the tetrahedron $ABCS$ by plane $\alpha = \varepsilon$. This plane is parallel to the pair of the skew edges of the tetrahedron, and its construction is similar to one explained earlier in this section and given in Figs. 3.57 and 3.58.

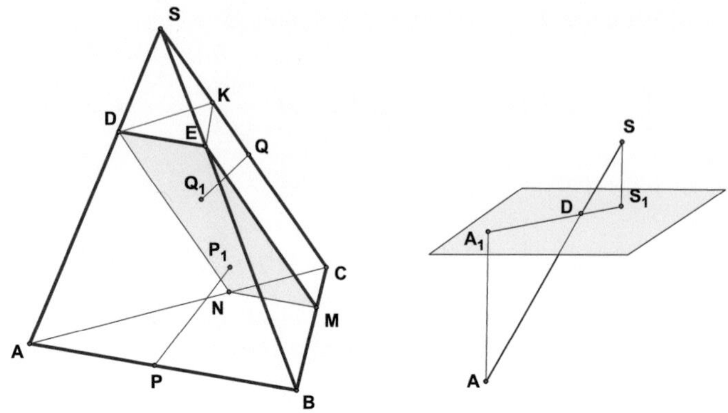

Fig. 3.82 International Math Olympiad 1965 Problem 3

2. Thus, plane α makes the cross section that is parallelogram $DEMN$, inside the tetrahedron. Given the condition of the problem, we must find how this cross section divides the tetrahedron, so this problem should be like one associated with Figs. 3.57 and 3.58. Can we use the ideas of that problem?
3. In that problem we were given the information about how point D divides the edge AS. We do not know this here; however, we know that the ratio of the distances from edge AB and SC to the plane α is equal to t. How can we use this information? Take point P on the edge AB and let $PP_1 \perp \alpha$, then take a point Q on edge SC and drop a perpendicular QQ_1 to plane α, such that $QQ_1 \perp \alpha$. By the condition of the problem we have

$$\frac{PP_1}{QQ_1} = t$$

and since two perpendiculars to the same plane are parallel, then $PP_1 \parallel QQ_1$.
4. Consider line AS going through plane α and intersecting it at point D as it is shown on the right of Fig. 3.82. Let us drop perpendiculars to this plane from both ends of segment AS. Denote them $AA_1,\ SS_1$. Since $AB \parallel \alpha$and $SC \parallel \alpha$, then $AA_1 = PP_1,\ SS_1 = QQ_1$.
5. Consider similar triangles $\triangle AA_1D \sim \triangle SS_1D$. Their corresponding sides are proportional, then the following is true

$$\frac{AA_1}{SS_1} = \boxed{\frac{AD}{DS} = \frac{PP_1}{QQ_1} = t} \tag{3.14}$$

Therefore, the cross section divides the edge of the tetrahedron in the same ratio as the ratio of the distances between the plane and two skew edges of the tetrahedron. Now we see that the solution to our problem does not depend on

the length of the edges, nor on the distance between two skew edges, nor on the angle between them.
It follows from Eq. (3.14) that $DS = \frac{1}{t+1} \cdot AS$ and that $AD = \frac{t}{t+1} \cdot AS$.

6. By drawing a plane parallel to the base ABC through the side of the cross section DE, we obtain the cross-sectional triangle DEK (see the left sketch in Fig. 3.82). Hence, the volume of the polyhedron above the cross section can be obtained as the sum of the volumes of pyramid $SDEK$, V_{pyr} and prism $DEKNMC$, V_{prism}.
7. Pyramids $SDEK$ and the original tetrahedron are similar with the coefficient of similitude of k:

$$k = \frac{1}{t+1}, \tag{3.15}$$

then the volume of the small tetrahedron can be written in terms of the volume of the given tetrahedron (V) as follows

$$V_{\text{pyr}} = \frac{1}{(t+1)^3} \cdot V \tag{3.16}$$

8. Assume prism $DEKNMC$ has height h, then

$$h = \frac{t}{t+1} \cdot H \tag{3.17}$$

where H is the height of the original tetrahedron.
The base of the prism is triangle NMC that is similar to triangle ABC of the base with the same coefficient of similitude as (3.15), so the area of the base of the prism $[NMC]$ can be written in terms of the base area $[ABC]$ of triangle ABC as follows:

$$B = [NMC] == \frac{1}{(t+1)^2} \cdot [ABC]. \tag{3.18}$$

Using (3.17) and (3.18) we obtain the formula for the volume of prism $DEKNMC$.

$$\begin{aligned} V_{\text{prism}} &= hB = 3 \cdot \tfrac{1}{3} \cdot \tfrac{t}{t+1} \cdot H \cdot \tfrac{1}{(t+1)^2} \cdot [ABC] \\ &= \tfrac{3t}{(t+1)^3} \cdot V \end{aligned} \tag{3.19}$$

If we add (3.16) and (3.19) we obtain the volume of the polyhedron above the cutting plane.

$$V_1 = \frac{1}{(t+1)^3} \cdot V + \frac{3t}{(t+1)^3} \cdot V = \frac{3t+1}{(t+1)^3} \cdot V$$

The volume of the polyhedron below the plane is

$$V_2 = V - \frac{3t+1}{(t+1)^3} \cdot V = \frac{t^2\,(t+3)}{(t+1)^3} \cdot V$$

9. Finally, the ratio of the two volumes is

$$V_2 \div V_1 = \frac{t^2(t+3)}{3t+1}.$$

Answer. $V_2 \div V_1 = \frac{t^2(t+3)}{3t+1}$.

Remark. You probably noticed that we solved the problem without using any information about the lengths of the edges. Neither the distance between the two skewed edges nor the angle between them was used. Perhaps the creators of the problem wanted to confuse the participants, or, most likely, they planned to ask them to find the area of the cross section, but decided not to. In the following problem we will find the area of the cross section.

Problem 174 Given the tetrahedron $ABCS$ whose edges AB and CS have lengths a and b respectively. The distance between the skew lines AB and CS is d, and the angle between them is ω. Tetrahedron $ABCS$ is divided into two solids by plane ε, parallel to lines AB and CS. The ratio of the distances of ε from AB and CS is equal to t. Find the area of the cross section.

Solution Let us use the same Fig. 3.82. The area of the cross section (parallelogram $NMED$) can be evaluated as follows

$$A = NM \cdot EM \cdot \sin\omega, \ \angle EMN = \omega$$

Because $MN \| AB$ and $EM \| SC$, then the lengths of MN and EM can be found from the corresponding similar triangles. Let d_1 and d_2 be the distances from AB and SC to the cross section. Then by the condition of the problem

$$\frac{d_1}{d_2} = t = \frac{AD}{DS} = \frac{MB}{MC}.$$

Consider $\triangle ACB \sim \triangle NCM$, then

$$\frac{AB}{MN} = \frac{CB}{CM}$$

Since $BM = t \cdot MC$, and $t \cdot MC + MC = CB$, then $MC(1+t) = CB$, then substituting $AB = a$, we obtain

$$\frac{AB}{MN} = 1 + t \Rightarrow MN = \frac{a}{1+t}$$

Next, $\triangle BSC \sim \triangle BEM$, then

$$\frac{SC}{ME} = \frac{BC}{BM} = (t+1)/t$$

Because $SC = b$, then

$$ME = \frac{bt}{1+t}$$

Finally, the area of the cross section is

$$A = \frac{abt}{(1+t)^2} \sin \omega.$$

Answer. $A = \frac{abt \sin \omega}{(1+t)^2}$.

Remark. We see that when finding the cross-sectional area, we did not use the distance between the two skew edges of the tetrahedron, which for some reason was also specified in the problem statement. Can you create a problem with the same condition where you could use this distance? It turns out that all the given parameters in the IMO 1965 problem can be used if you need to find the volume of the tetrahedron. This problem is given to you as Problem 31 in the homework after Chap. 3.

Problem 175 Points N and M and K and L are taken on two pairwise skew edges of a tetrahedron $ABCD$ shown in Fig. 3.83. Does NM intersect with KL?

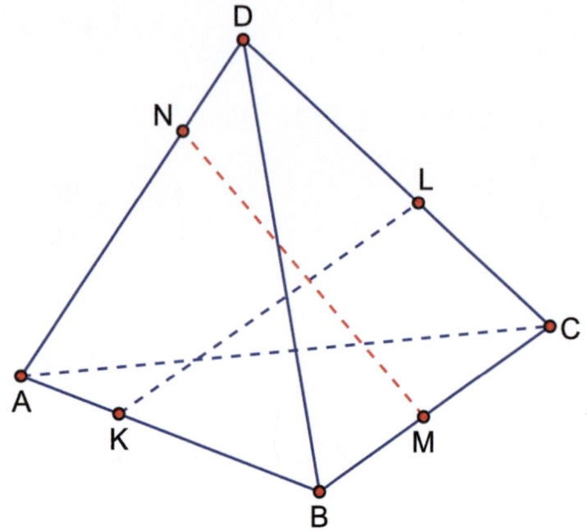

Fig. 3.83 Does *MN* intersect *KM*?

Solution Consider Fig. 3.83. If $NM \cap KL$, then N, L, M, and K lie on the same plane.

If we connect N and L, $NL \cap AC = P$, then $PM \cap AB = Q$ (This is shown in Fig. 3.84).

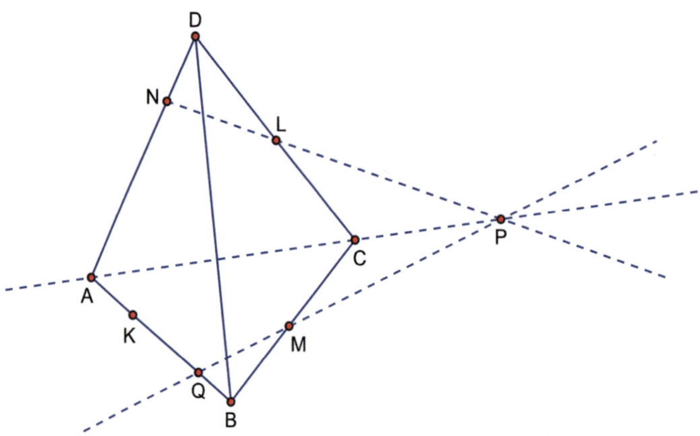

Fig. 3.84 *MN* does not intersect *KL*

If $K = Q$, then four points belong to the same plane and $NM \cap LK = T$. Otherwise, lines are skew. In the case demonstrated in Fig. 3.83, the given lines NM and KL are skew lines.

3.4 Some Applications of Cross Sections and Construction Methods

Stereometry is used in architecture, geodesy, animated movies, mechanical engineering, molecular physics, chemistry and in many other fields.

3.4.1 Construction of Shadows

The legend tells that Thales (sixth century BC) when he was in Egypt, struck the Pharaoh Amasis by being able to accurately determine the height of the Cheops pyramid in Giza. He waited for the moment when the shadow of the stick becomes equal to its height and measured the length of the shadow of the pyramid.

Because Sun is so huge, its ray fall go parallel by the same vector to all the objects on Earth at the same time of the day. We also understand that shadow from the sun goes in the same direction. The angle at which the sun rays will fall on the stick would be the same as they fall on the pyramid. In Fig. 3.85 we reconstructed possible experiment of Thales. The shadow of Cheops pyramid $ABCDS$ will be triangle DS_1C, the vertex of which S_1 is the projection of the pyramid vertex S onto line OH. Simultaneously the shadow from the stick S_1G will be segment S_1G_1. The height of the pyramid, SO will have projection OS_1. Triangles SOS_1 and GS_1G_1 are similar triangles.

Fig. 3.85 Finding the height of Cheops pyramid

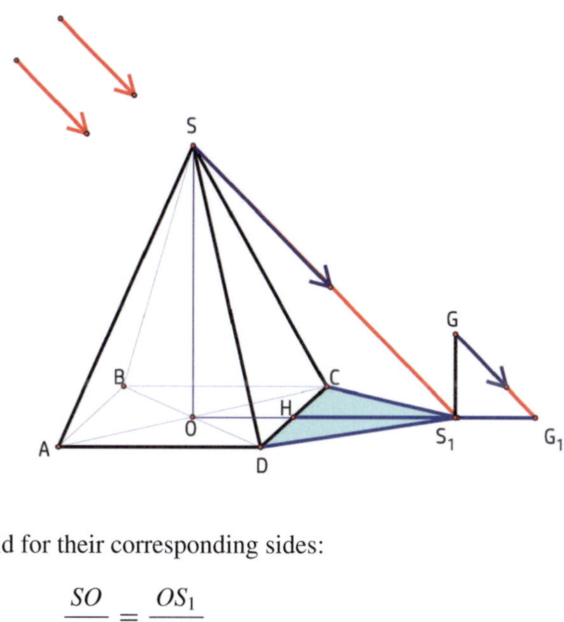

Then the following is valid for their corresponding sides:

$$\frac{SO}{GS_1} = \frac{OS_1}{S_1G_1}$$

Obviously, when the length of the shadow from the stick equals its height, then the height of Cheops pyramid is equal to the length of segment S_1O. Further, $h_1 = SO = OS_1 = HS_1 + OH$, where $OH = \frac{1}{2}AD$, $AD = 230$ m.

Another story tells that Thales did not wait until the shadow from the stick would be equal to its height, but could find its height at any time of a day. He knew that the length of the base of the great Pyramid is approximately 230 meters, but he wanted to evaluate the height, $h_1 = SO$, of the Pyramid. At certain time of the day, the pyramid cast a shadow, HS_1, so that $L_1 = OS_1 = 223$ meters long. Thales placed a vertical stick at the end of this shadow and found that his $h_2 = 1.8$ m stick made a shadow of $L_2 = S_1G_1 = 2.75$ m long. Using the properties of the similar triangles, Thales obtained:

$$\frac{L_1}{L_2} = \frac{h_1}{h_2} \quad \Rightarrow \quad h_1 = \frac{223 \cdot 1.8}{2.75} = 146 \text{ m}.$$

It is amazing but the original height of Cheops pyramid before its cap was removed for building masques in Cairo was indeed about 146 m!

Problem 176 Find the shape of shadow cast produced by a regular quadrangular pyramid under sun rays. Assume that there is a screen placed between the pyramid and the vertex of the shadow triangle, just in the middle. The base of the pyramid and its height are equal to a. Find the area of the natural shadow cast and the composite shadow of the pyramid on the ground and on the screen. Assume that the sun rays form with its projection on the ground at the moment angle α.

Solution Let us sketch a regular pyramid $ABCDS$ and select direction of the sun rays given by red vector and projection of the rays on the horizontal plane given by dark blue vector. Let us find the image of the apex of the pyramid S on the horizontal plane, point S_1. See Fig. 3.86.

Part 1. If we do not have the screen, then triangle BS_1C will be the shadow. We do not see the entire triangle in our picture because part of the shadow is hidden by the pyramid. If we have the screen, then if some points do not get to land on the ground plane, they would stop on the screen forming an additional shadow on the perpendicular screen shown as pink triangle WS_2X.)

Let us see how shadow can be constructed. For example, let us take point $L \in SC$, first, and drop perpendicular to the base of the pyramid. Its foot will be on the diagonal AC of the base. Let denote it by point M. Second, through point M we must draw line parallel to the direction given by vector OT and intersect it with another line passing through point L parallel to the sun rays (red vector) The intersection of these two rays will be point L_1 In our sketch this point belongs to the boundary of the shadow if there is not screen or it will be a point on the boundary of the pink triangle WS_2K (this is the point at which the ray emanating from L pierces the plane of the screen

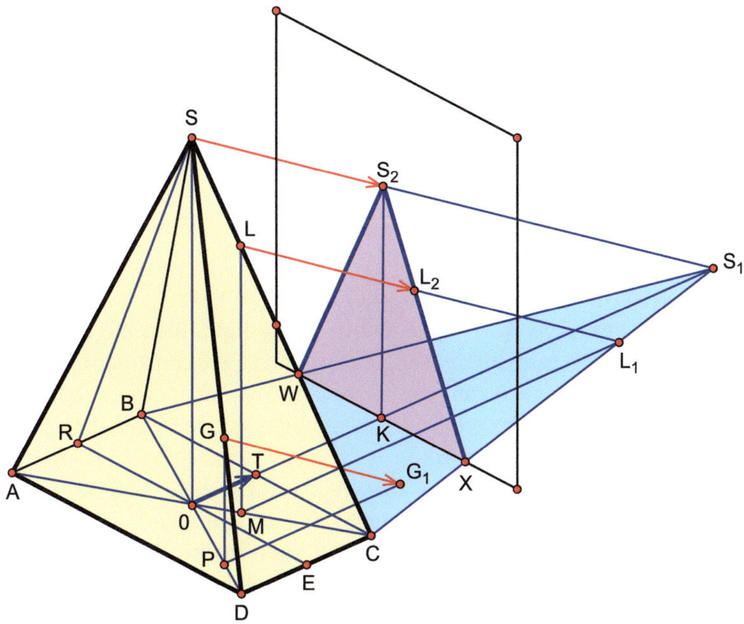

Fig. 3.86 Shadow of pyramid on the ground and on the screen

before it would fall to the ground otherwise.) Consider point $G \in SD$. Similarly, drop perpendicular from this point onto the base of the pyramid obtaining point P, then draw though P line parallel to vector OT and intersect it with another ray coming from point G parallel to the sun rays. We obtain point G_1. Point G_1 will land on the ground plane before reaching the plane of the screen.

Part 2. If point K is the midpoint of OS_1, then triangle WS_2X is similar to triangle RSE with the coefficient of similitude $k = 1/2$, and then the area of the pink shadow is $1/4$ of the area of triangle RSE, or $\frac{1}{2}a^2$. If there is no screen, then the area of the blue shadow is the area of triangle BS_1C. Its height $h = OS_1$ can be found from the right triangle SOS_1 as follows

$$h = \frac{SO}{\tan\alpha} = \frac{a}{\tan\alpha}$$

The area of the blue shadow is

$$[BS_1C] = \frac{a^2}{2\tan\alpha}$$

In case of the screen, there will be a piece of blue shadow before the screen. For this we can find the area of the trapezoid $BWXC$, the area of which is $3/4$ of the area of the entire blue shadow area ($\triangle WS_1X \sim \triangle BS_1L$ with coefficient of similitude of

1/2, so its area is 1/4 of the area of the big triangle, then the trapezoid area is the difference of the areas of the big and small triangles)

Finally, the area of the composed shadow cast is

$$\frac{3a^2}{8\tan\alpha} + \frac{3}{8}a^2.$$

Answer. (1) $\frac{a^2}{2}$; (2) $\frac{3a^2(1+\tan\alpha)}{8\tan\alpha}$

Next, let us consider shadow from a single source of light.

If the light is produced by a single source of the light, then the light rays will approach different objects at different angles. In general, the construction of shadows is carried out using the same techniques as the construction of cross sections; in this case, it is necessary to use the projection of the light source on the plane on which the shadow falls as an auxiliary point.

For example, let's build a shadow cast by a vertical match AB on the α plane (the end of the match B rests on the plane), if the light source is located at the point L, and L_1 is the projection of the point L onto the plane α. Draw two lines (AL), (BL_1) intersecting at point A_1. Segment A_1B will be the shadow of match AB. See constructions in Fig. 3.87 on the left.

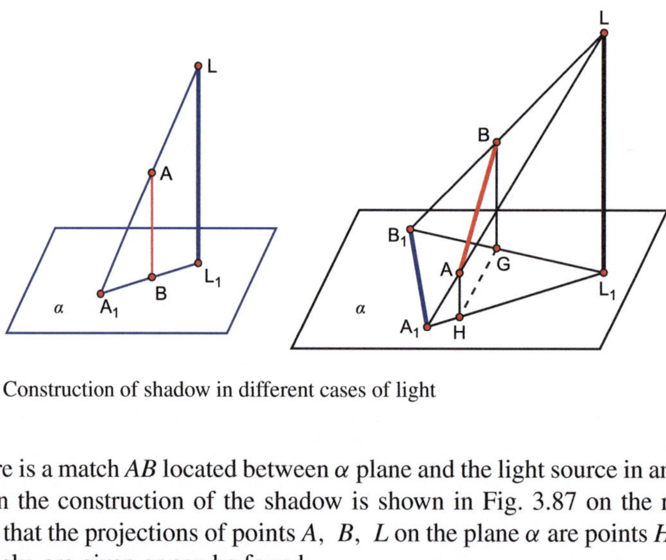

Fig. 3.87 Construction of shadow in different cases of light

If there is a match AB located between α plane and the light source in an arbitrary way, then the construction of the shadow is shown in Fig. 3.87 on the right. It is assumed that the projections of points A, B, L on the plane α are points H, G, L_1, respectively, are given or can be found.

Instead of constructing the shadow of the match AB, we construct the shadows A_1H and B_1G of the vertical matches AH and BG, and then by connecting the points A_1 and B_1 we get the desired shadow. The projection of the match AB onto the plane α is actually given as a segment HG. The shadow cast by this match on the α plane, if the light source is located at point L, will be the segment A_1B_1 in Fig. 3.87 on the right.

Let's solve the following problem.

Problem 177 The floor lamp is located behind the cube, directly above the base plane of the cube. The height of the floor lamp is twice the edge of the cube. The light bulb forms an isosceles triangle with neighboring vertices of the top face of the cube. Construct the shadow cast by the cube on the plane of its bottom base.

Solution Of course, one could construct separately the shadow cast by each vertical edge of the cube, and then connect the corresponding vertices. However, here it is easier to use the fact that the edges of the upper base of the cube are parallel to the plane of the lower base, therefore the shadow cast by the upper base of the cube will be a square. Since LL_1 is twice the edge of the cube, then the side of this square will be equal to 2 times side of the cube. If we draw a line connected the centers of the lower and upper bases (OO_1) in the cube and construct the shadow cast by it, we can obtain point M as the intersection of line LO and L_1O_1. M is also the center of the square $PQST$. Then it will not be difficult to draw the shadow cast by the entire upper base, and then by the entire cube (see Fig. 3.88).

Fig. 3.88 Shadow cast from the cube

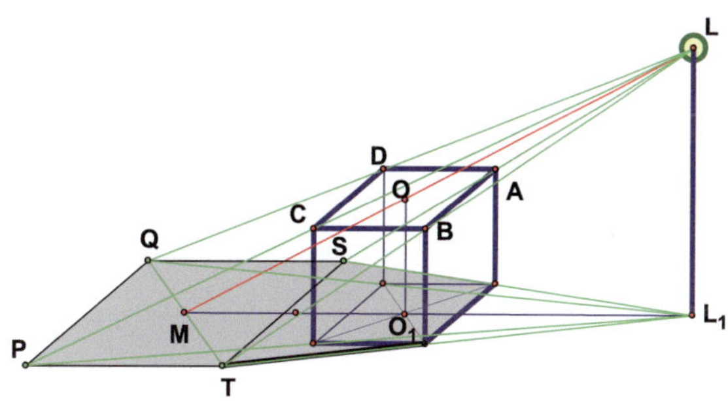

Problem 178 A lamp attached to the ceiling hangs exactly in the middle of the top edge of the table, which has the shape of a cube with side a. The distance from the light bulb to the middle of the center of the top face of the cube is equal to twice the side of the cube. Plot the shadow cast by the cube on the floor plane and find the size of the shadow as a function of the distance between the top surface of the table and the ceiling lamp in the room.

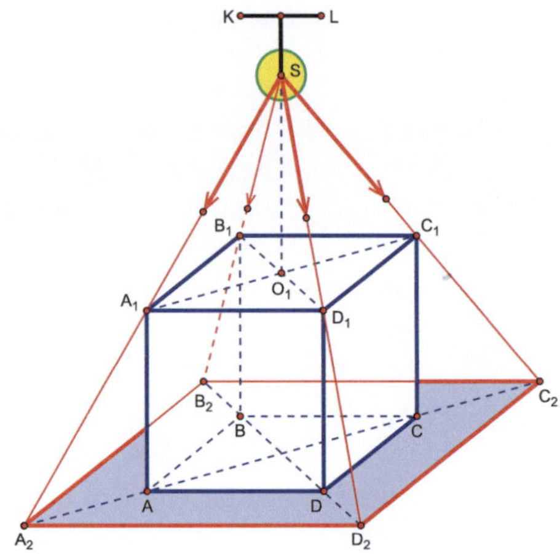

Fig. 3.89 Shadow of the cube with the ceiling lamp on the top

Solution Let us construct the shadow cast by finding the image of each vertex of the cube's top, and then connect the corresponding vertices. For example, we connect point S that represents the ceiling lamp with vertex A_1 and intersect this line with line AC until two lines intersect at point A_2 on the floor plane. Similarly we will obtain points B_2, C_2, and D_2, forming the boundary of the shadow cast that is a square. Such construction is shown in Fig. 3.89.

The shaded region between the cube and the boundary of the square $A_2B_2C_2D_2$ is the shadow cast. By the condition of the problem $\Delta A_1SD_1 \sim \Delta A_2SD_2$ with the coefficient of similitude $k = \frac{2}{3}$, then the side of the shadow cast is equal to $\frac{3a}{2}$. Then the area of the shaded region is

$$\frac{9a^2}{4} - a^2 = \frac{5a^2}{4}.$$

Answer. $A = \frac{5a^2}{4}$.

3.4.2 *Construction of New Polyhedra*

We have seen that one cutting plane divides a spatial figure into two parts. This allows us to construct new convex polyhedra from the known polyhedra. A plane that cuts a solid does not have to be parallel to any of the faces of that solid. We say that a plane truncates the solid. Think about a cross-sectional polygon as a new face of a new solid figure formed under this operation.

Truncation. We take a vertex of a trihedral angle and truncate a solid at this vertex. If we have a pyramid, then this can be continued indefinitely. By truncating we obtain three trihedral angles from each truncated vertex.

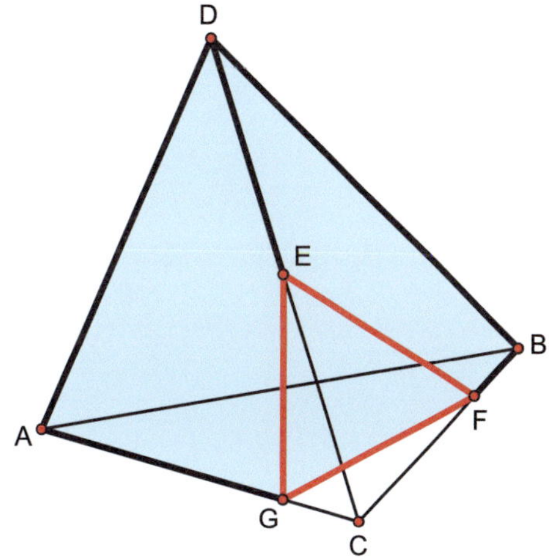

Fig. 3.90 Pyramid truncated by plane FEG

We can start with a triangular pyramid (a tetrahedron) because it has $F = 4$ faces, $V = 4$ vertices and $E = 6$ edges (Fig. 3.90). Assume that the pyramid is made of solid wood.

If we take points on three sides of the same trihedral angle with vertex C, (points E, G and F on edges DC, AC, and BC) this plane cuts the tetrahedron into two parts—convex polyhedron $ADBFEG$ and tetrahedron $GEFC$. The tetrahedron also has fours faces, four vertices and six edges. However, convex polyhedron $ADBFEG$ obtained after removing $GEFC$ will have three additional edges, two additional vertices and one additional face (the cross-sectional polygon GEF). Under such truncation we have the following relationships between the new and old number of the vertices, edges, and faces.

$$\begin{aligned} E' &= E + 3 \\ V' &= V + 2 \\ F' &= F + 1 \end{aligned} \tag{3.20}$$

If we additionally truncate the given tetrahedron $ABCD$ at a different vertex A, after removing two small tetrahedra, we again obtain a convex polyhedron concluded between two cross-sectional polygons shown in Fig. 3.91. This polyhedron $HIJDBFEG$ has six faces, eight vertices, and twelve edges ($F = 6, E = 12$ and $V = 8$). It is made of two triangles, two quadrilateral and two pentagons. Euler's formula holds (check). The second iteration (truncation) produces three more edges, two more vertices, and one more face that can be written as follows.

$$\begin{aligned} E'' &= E' + 3 = E + 6 = 6 + 6 = 12 \\ V'' &= V' + 2 = V + 4 = 4 + 4 = 8 \\ F'' &= F' + 1 = F + 2 = 4 + 2 = 6 \end{aligned}$$

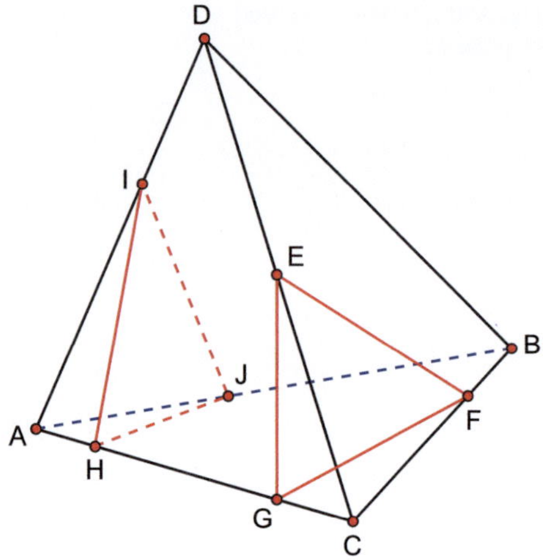

Fig. 3.91 Pyramid *ABCD* truncated by two planes

Next, let us truncate our tetrahedron *ABCD* not at vertex *C* and then at *A*, but differently. Let us truncate it at one already obtained new vertex (*G*), then the new convex polyhedron *ADBFEKLH* shown in Fig. 3.92 will have the same number of vertices, faces and edges ($E = 12,\ V = 8,\ F = 6.$) as the polyhedron truncated at vertex *A*. We can see that Euler's formula works.

$$F + V = E + 2$$
$$6 + 8 = 12 + 2$$

Our convex polyhedron *ADBFEKLH* has a different shape from convex polyhedron *HIJDBFEG* but it also consists of 2 triangulars, 2 quadrilateral and 2 pentagonal faces. In all cases, if we truncate a tetrahedron, all obtained convex polyhedra will have only trihedral angles, so there will be three edges meeting at one vertex.

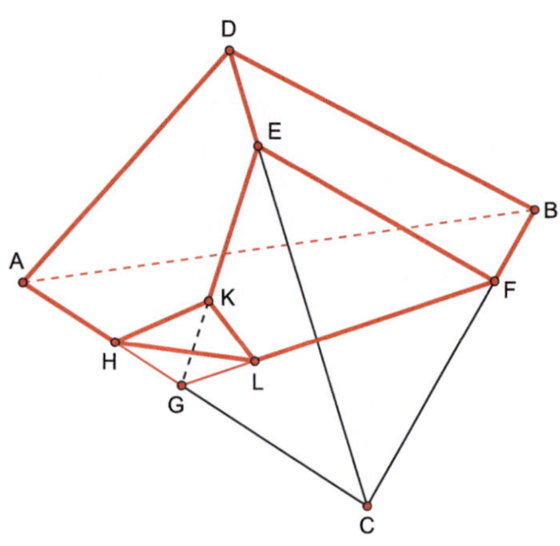

Fig. 3.92 Convex polyhedron *ADBFEKLH* with $E = 12,\ V = 8,\ F = 6$ is shown in red color

Recall that an n-gonal pyramid has $n+1$ faces ($F=n+1$), $n+1$ vertices ($V=n+1$), and $2n$ edges ($E=2n$). The following lemma is true.

Lemma 14 *Truncating a n-gonal pyramid at any vertex of its base leads to obtaining a tetrahedron and a convex polyhedron that has one additional face, two additional vertices and three additional edges than the original pyramid. The new convex polyhedron will have the following number of faces, edges, and vertices.*

$$F' = F+1 = n+2,\ E' = E+3 = 2n+3,\ V' = V+2 = n+3.$$

Now let us see what will happen if we truncate a tetrahedral or pentahedral angles. We start from a pyramid $ABCDS$. We cut it by a plane parallel to the base ABC and obtain truncated pyramid $ABCDA_1B_1C_1D_1$ (see Fig. 3.93). Pyramid $ABCDS$ had five faces (four triangular faces and one quadrilateral), five vertices, and eight edges ($F=5,\ V=5,\ E=8$) that satisfy Euler's formula $F+V=E+2\quad (5+5=8+2)$.

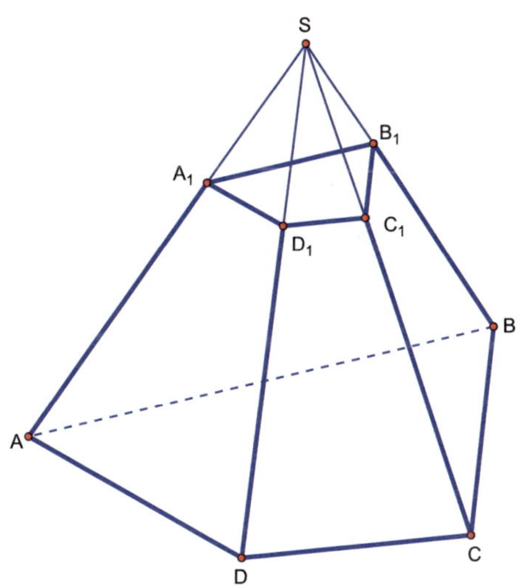

Fig. 3.93 $ABCDA_1B_1C_1D_1$ is a convex polyhedron with $E'=12,\ V'=8,\ F'=6$

By cutting the top of the pyramid by parallel plane, we replace the tetrahedral angle $\angle SABCD$ by four trihedral angles at the new vertices $A_1,\ B_1,\ C_1,\ D_1$. The number of faces is increased by one, the number of the vertices by three and the number of edges by four. Thus, we obtain

$$E' = E + 4$$
$$V' = V + 3$$
$$F' = F + 1$$

After such truncation we will have $F' = 6,\ V' = 8,\ E' = 12$. Now, all the six faces are quadrilaterals.

Note. We do not need to cut the original pyramid by a plane parallel to the base. We could take any plane that makes a cross section in the form of a quadrilateral and intersects all the edges of the tetrahedral angle. For us it is important that points $A_1,\ B_1,\ C_1,\ D_1$ be coplanar.

Next, let us truncate a pentagonal pyramid that has six vertices ($V = 6$), six faces ($F = 6$) and ten edges ($E = 10$). After truncation, the pentahedral angle is replaced by five trihedral angles (see Fig. 3.94).

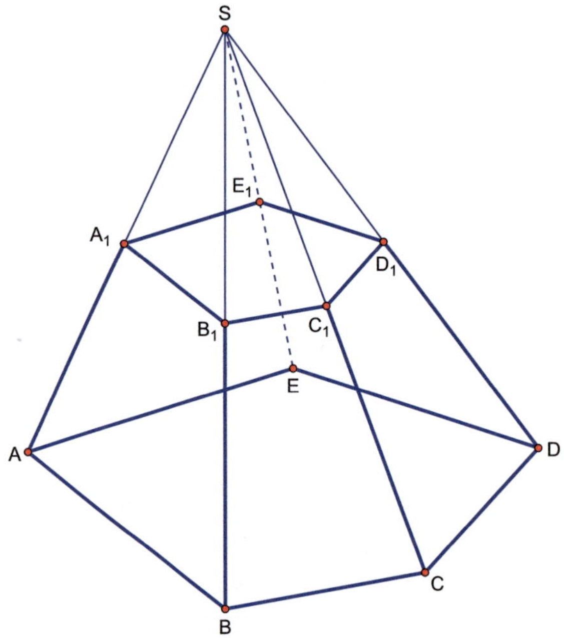

Fig. 3.94 $ABCDEA_1B_1C_1D_1E_1$ is a convex polyhedron with $E' = 15,\ V' = 10,\ F' = 7$

$$E' = E + 5$$
$$V' = V + 4$$
$$F' = F + 1$$

So, the truncated pyramid $ABCDEA_1B_1C_1D_1E_1$ has 10 vertices (all associated with trihedral angles), seven faces and fifteen edges.

Lemma 15 *Truncating a n-gonal pyramid at its apex by a plane parallel to its base leads to obtaining a similar pyramid and a convex polyhedron that has* $(n-1)$ *additional vertices, one more face and n more edges as the original pyramid. The new convex polyhedron will have the following number of faces, edges, and vertices.*

$$F' = n + 2,\ E' = 3n,\ V' = 2n.$$

Let us solve some problems.

Problem 179 Construct a convex polyhedron with three edges at each vertex and which has eighteen edges, twelve vertices, and eight faces. The answer may vary.

Solution If we truncate tetrahedron at each of the four vertices, we obtain the fourth iteration of the formula (3.20):

$$\begin{aligned} E^{(4)} &= E + 4 \cdot 3 \\ V^{(4)} &= V + 4 \cdot 2 \\ F^{(4)} &= F + 4 \cdot 1 \end{aligned}$$

This will produce convex polyhedron shown in Fig. 3.95 that has four triangular and four hexagonal faces.

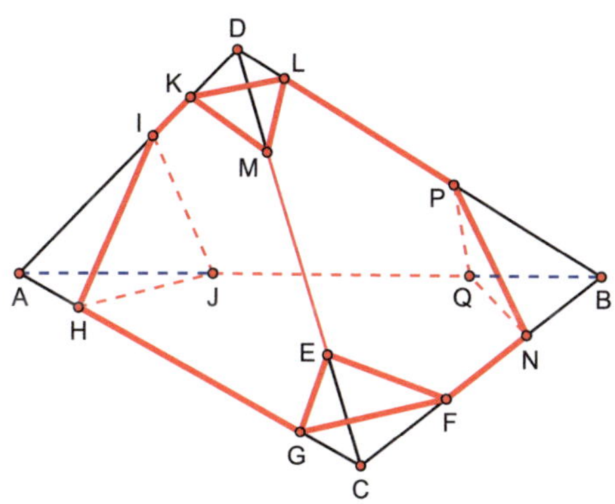

Fig. 3.95 *HIJKLMPQNFEG* has $E = 18,\ F = 8,\ V = 12$

Since each vertex of the polyhedron associates with a trihedral angle, we obtain a simple convex polyhedron. Therefore this red convex polyhedron *HIJKLMPQNFEG* has $E = 18$ (eighteen edges), $V = 12$ twelve vertices, and $F = 8$ eight faces as requested.

Problem 180 A tetrahedron *ABCD* is truncated by plane α containing points E, F, and G, such that $E \in DC$, and $DE \div EC = 2 \div 1$, $G \in AC$ and $AG \div GC = 4 \div 1$, and $F \in CB$ and $BF \div FC = 1 \div 2$. The volume of a tetrahedron *ABCD* is V, find the volume of the convex polyhedron *ADBFEG*.

Solution First, let us sketch the tetrahedron and cut it by plane *GEF*. We will obtain a sketch similar to one in Fig. 3.90. Second, denote the volume of the convex polyhedron by V_1 and the volume of the small tetrahedron *GEFC* by V_2. It is clear that

$$V_1 = V - V_2.$$

So, we need to evaluate the volume of the small tetrahedron V_2. This volume is equal to $1/3$ of the product of the base (the area of $\triangle GCF$) and the height h (the perpendicular dropped from the vertex E onto base GCF.

We know that the volume of the tetrahedron is equal to

$$V = \frac{1}{3} \cdot [ABC] \cdot H.$$

Because E is the midpoint of segment DC, then the height of the small pyramid h is $1/2$ of the height of the tetrahedron H:

$$h = \frac{1}{2} \cdot H.$$

Next, let us find the relationship between the area of triangles ABC and GFC. They are not similar. However, we know the location of points G and F.

Denote the height of the triangle ABC dropped from vertex B as h_B and the height of triangle FGC dropped from vertex F by h_F. Since

$$FC = \frac{2}{3}BC,$$

then

$$h_F = \frac{2}{3}h_B.$$

Similarly, we can find the relationships between the bases AC and GC of the triangles, ABC and GFC:

$$GC = \frac{1}{5}AC.$$

Next, the areas of the two triangles are related as follows

$$[FGC] = \frac{1}{2} \cdot GC \cdot h_f = \frac{1}{2} \cdot \frac{1}{5}AC \cdot \frac{2}{3}h_B = \frac{2}{15} \cdot [ABC].$$

Now we can evaluate the volume of the small tetrahedron:

$$V_2 = \frac{1}{3} \cdot \frac{2}{15} \cdot [ABC] \cdot \frac{1}{2} \cdot H = \frac{V}{15}.$$

Finally, the volume of the convex polyhedron that is cut off by the plane is

$$V_1 = V - \frac{V}{15} = \frac{14V}{15}.$$

Answer. $V_1 = \frac{14V}{15}$.

3.4.3 Applications to Plane Geometry

We all know how to construct a triangle through given three non collinear points in a plane-just connect these points. What if we are given not the vertices of the triangle, but simply three distinct points that are not the vertices of a triangle but lie on the sides of a triangle, one point on each side? Can you construct such a triangle? Would it be unique? I want to offer you some problems to solve.

Problem 181 Given three non-collinear points A, B, and C. Construct a triangle such that each given point lies on one of the sides of the triangle, but is not its vertex.

Solution One thing that comes to mind is the unfolding of a tetrahedron with all equal faces. Remember how we constructed it?

We took a triangle, the face of the future tetrahedron and then drew a line through each vertex parallel to the side opposite that vertex. At the intersection we obtained a large triangle similar to the original triangle. Each of the vertices of the small triangle (in our case ΔABC) belongs to a side of the large triangle $\Delta D_1D_2D_3$, but is not its vertex. You can look at our drawing in Fig. 2.99 given in Problem 130 (Chap. 2). As a result of our construction we obtain that $\Delta ABC \sim \Delta D_1D_2D_3$, and point $A \in D_2D_3$, point $B \in D_1D_3$ and point $C \in D_1D_2$. Moreover, A, B, C are the midpoints of the corresponding sides of the large triangle. Thus, we have constructed a triangle similar to the original triangle ABC, which satisfies the condition of the problem.

Are there other solutions? Let us find out by solving the following problem.

Problem 182 Three different non-collinear points M, K and N are given. Construct a triangle ABC such that $M \in AB$, $N \in AC$ and $K \in BC$. Make sure triangle ABC is not similar to triangle MNK.

Solution If there are no other restrictions, then it is obviously possible to construct a set of triangles ABC whose sides contain one of three different points. Sometimes the problem "can be solved from the end" by drawing a triangle ABC and placing the points M, N and K on its sides AB, AC and CB respectively. Next, suppose that vertex B moves to another location and becomes B_1. We fix the location of our points M, N and K, but we want to draw a second triangle $A_1B_1C_1$, each side of which will also contain one of the given points. Then we definitely need to connect K and B_1, M and B_1. If point N is inside the angle MB_1K, then to construct a triangle we just need to draw through point N any straight line that will intersect two rays B_1K and B_1M. Moreover, each such line will give us two more vertices of the triangle: A_i and C_i. As you can see, in order to construct at least one triangle with the required properties, we need to select vertex B_1 so that there is point N inside the angle MB_1K.

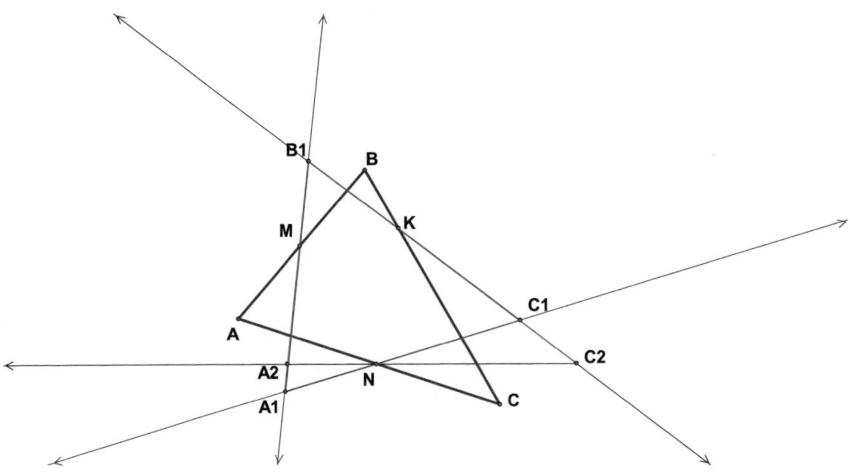

Fig. 3.96 Given three points on the sides of a triangle

Thus, we select point C_1 on line B_1K, then connect C_1 and N; the intersection of rays B_1M and C_1N will give us the third vertex A_1: $C_1N \cap B_1M = A_1$ (see triangle $\triangle A_1B_1C_1$ in Fig. 3.96).

We could also take a different point C_2 of the future triangle on the ray B_1K and then connect it with N and then again intersect C_2N and B_1M obtaining the new vertex A_2, such that $C_2N \cap B_1M = A_2$, etc. (see green triangle $\triangle A_2B_1C_2$ in Fig. 3.96). We equally could start taking a new point on the ray B_1M instead, and continue by a similar scenario. Therefore in our construction we have two degrees of freedom, and each time we get a new triangle.

Problem 183 Three rays emanating from one point of the plane divide the plane into three regions, and the angles formed by adjacent sides of the rays have the measure of α, β and γ; each angle is less than $180°$. Inside each angle there is a point, for example M, K and N, such that these points do not lie on one straight line. It is required to construct a triangle ABC, each side of which contains one of the given points, and vertices A, B and C lie on the sides of the angle (one vertex on each ray).

Solution Why don't we use the previous scenario and, for example, take some arbitrary point C on one of the rays and try to see what happens? That is, now we must connect point C with point K. We soon see that the situation is changing. The fact is that a new vertex, say A, should now be obtained at the intersection of the second ray and line CK, but then if we connect vertex A and N, we must obtain the third vertex B at the intersection of ray AN and the third ray. Moreover, BC segment must contain point M. As it follows from the sketch, (Fig. 3.97) we may not succeed at all, that is, it turns out that the choice of point C (the first of the vertices) is not free.

We understand that we need to look for other ways to solve this problem.

Let's solve the problem using stereometry and imagine that the rays form in space the edges of a trihedral angle with the vertex at the point O. Let's represent the points M, N and K on the faces of the trihedral angle $OXYZ$ (Fig. 3.98). These points define a certain plane, let's denote this plane by β. (The future triangle ABC will be in this plane as well.) Plane β will intersect the faces of the trihedral angle $OXYZ$ along the desired sides BC, CA and AB.

Let N be the intersections of the plane β with the plane XOY of the trihedral angle. Let us find the second point of intersection of these planes. To do this, we project the points M and K onto the plane of the trihedral angle XOY with respect to the line OZ, that is, we draw parallel lines

$$MM_1 \parallel KK_1 \parallel OZ.$$

Then we draw the plane MKK_1M_1, that goes through these two parallel lines. If MK and M_1K_1 are not parallel, they will intersect at some point P in the plane MKK_1M_1, such that $MK \cap M_1K_1 = P$ (see Fig. 3.99).

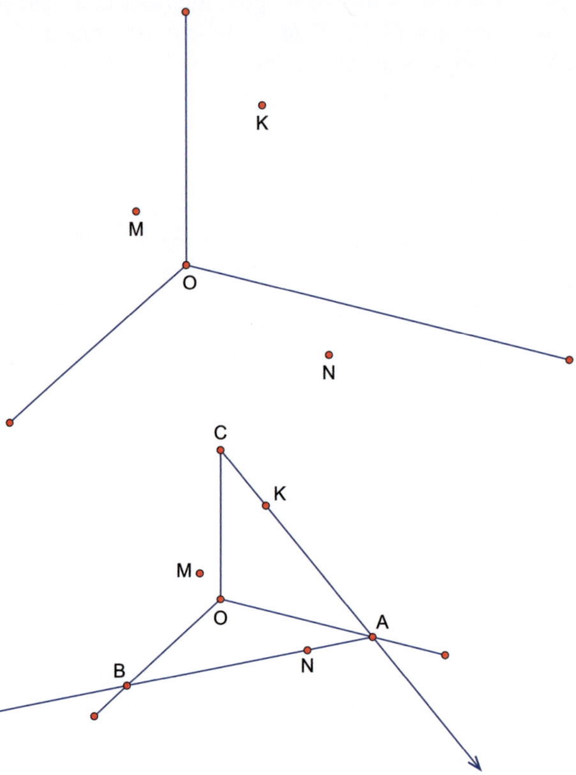

Fig. 3.97 Attempt to construct a triangle

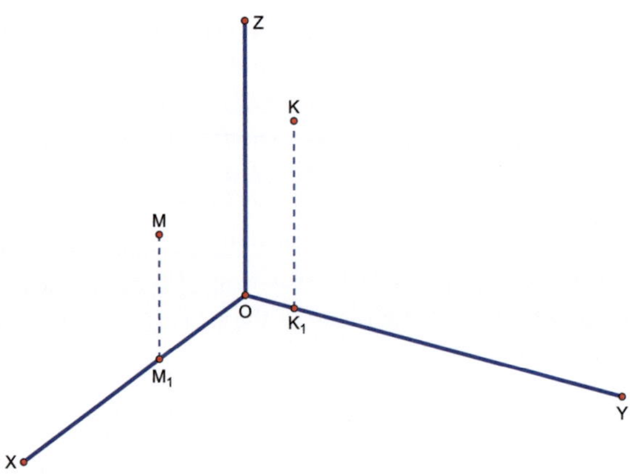

Fig. 3.98 Three points in the plane

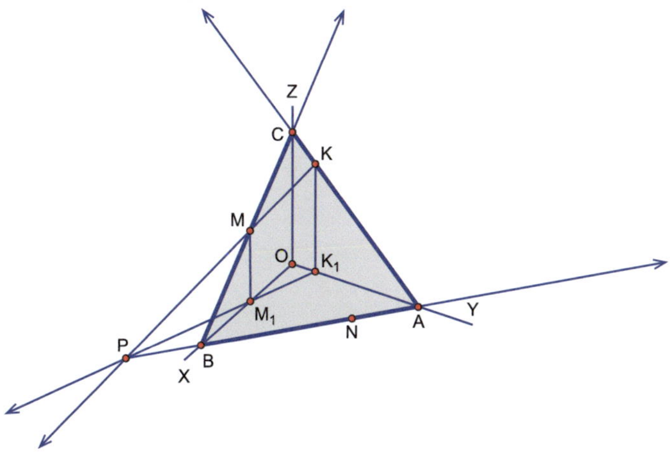

Fig. 3.99 Construction of triangle ABC

In other words, we get the second point of intersection of the β plane with the XOY face of the trihedral angle. Since this point P and the given point N both belong to plane β and to the face XOY, then these two planes will intersect by the line connecting these two points. Line PN will cut on the rays XO and OY the vertices of the triangle, B and A, respectively. Finally, if we connect B and M and separately points A and K, then the rays BM and AK will intersect at the unique point C, the 3rd vertex of the triangle ABC.

$$
\begin{aligned}
&XOY \cap \beta = PN \\
&XOY \cap \beta = BA \\
&BM = \beta \cap XOZ, \quad BC = \beta \cap XOZ \\
&KA = \beta \cap YOZ, \quad \beta \cap YOZ = CA \\
&BM \cap AK = C
\end{aligned}
$$

Triangle ABC is constructed.

Remark. Let's figure out under what conditions the construction of such a triangle using the method offered above is impossible. Firstly, we immediately discard the case when the segment connecting any two given points, for example M and K, runs parallel to the segment M_1K_1, where points $M_1 \in OX$ and $K_1 \in OY$ are obtained by projection of points M and K parallel to OZ. In this case we will obtain that $KM \| K_1M_1$ and we will not be able to construct $\triangle ABC$. See the left sketch in Fig. 3.100.

The second case occurs if any point, for example, M, lies on the projection of a segment connecting two other points (K and N) on the plane containing this point, M. For example, consider the segments KN and K_2N_2, such that $KK_2 \| NN_2 \| OY$

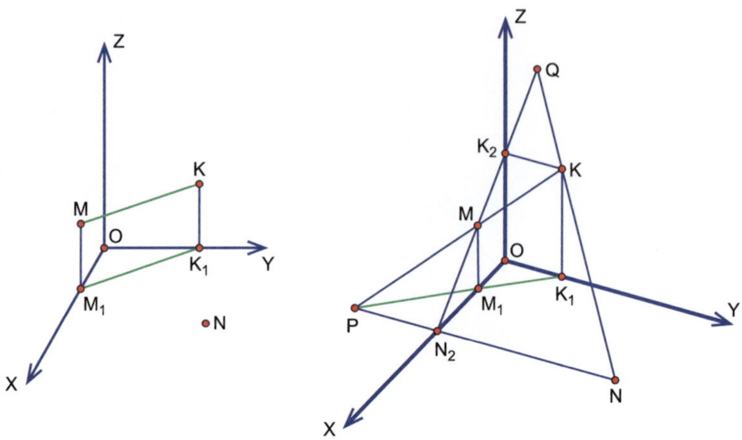

Fig. 3.100 Cases when construction of $\triangle ABC$ is impossible

(projection parallel to the OY axis). Assume that line K_2N_2 is not parallel to KN and intersects it at some point Q. Point M, $Q \in ZOX$. However, the third of these points M lies on K_2N_2, then constructing triangle ABC is also impossible.

The third case occurs when MK is not parallel to M_1K_1, so we can connect points M and K, construct their parallel projections and intersect segment MK with M_1K_1 at point P. However, when we connect point P with the third point N, then this line PN goes parallel to axis OY, so there is no intersection. On the right in Fig. 3.100 we can observe this case together with case 2 in one drawing. It can be shown that if the second case is valid, and point M lies on line K_2N_2, then point N connected with P will make the line parallel to line OY, so further construction of the triangle would not be possible.

In the following problem, we will see how 2D-geometry can help solving 3D-geometry problems.

Problem 184 The base of the radio tower is located at distances of 300, 600, and 900 m, respectively, from three points standing on a horizontal plane. The angles at which the tower is visible from these points were measured and their sum is 90°. What are these angles and what is the height of the radio tower?

Solution Let us make a sketch for the problem. In Fig. 3.101 we show the tower DE and points A, B, C of the plane. In this figure, $AE = 300$, $BE = 600$, and $CE = 900$. The angles α, β and γ at which the tower is observed from each of the three points are marked by one, two, and three arcs.

Fig. 3.101 Radio tower

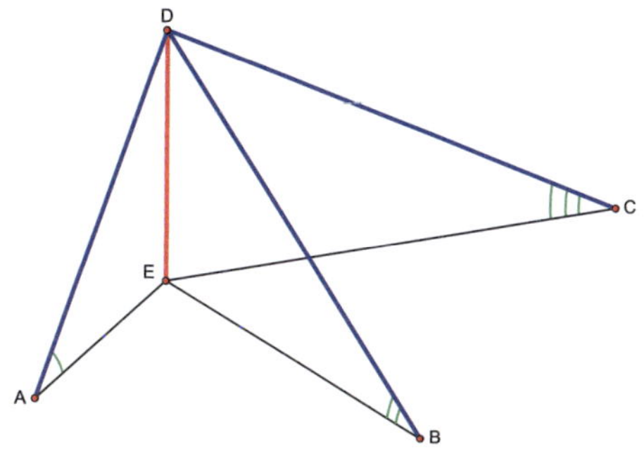

Let us introduce

$$\alpha + \beta + \gamma = \frac{\pi}{2} \tag{3.21}$$

Let the height of the tower be $DE = h$. The following is true:

$$\begin{aligned} \tan\alpha &= \frac{DE}{AE} = \frac{h}{300} \\ \tan\beta &= \frac{DE}{BE} = \frac{h}{600} \\ \tan\gamma &= \frac{DE}{CE} = \frac{h}{900} \end{aligned} \tag{3.22}$$

Denote

$$x = \frac{h}{300}. \tag{3.23}$$

Then from (3.21)–(3.23) we have

$$\arctan x + \arctan\frac{x}{2} + \arctan\frac{x}{3} = \frac{\pi}{2} \tag{3.24}$$

Solving (3.24) we find x and then the angles and height of the tower.

The function on the left-hand side (LHS) of (3.24) is monotonically increasing function (this can be checked if we take the first derivative of the LHS). Hence, Eq. (3.24) must have only one unique solution.

Let us show that $x = 1$ is the solution. There is a known fact that

$$\arctan 1 = \frac{\pi}{4},$$

so we must prove that

$$\arctan\frac{1}{2} + \arctan\frac{1}{3} = \frac{\pi}{4}. \tag{3.25}$$

Consider a rectangle 2×3 units and the right isosceles triangle MNK inscribed in the rectangle (see Fig. 3.102).

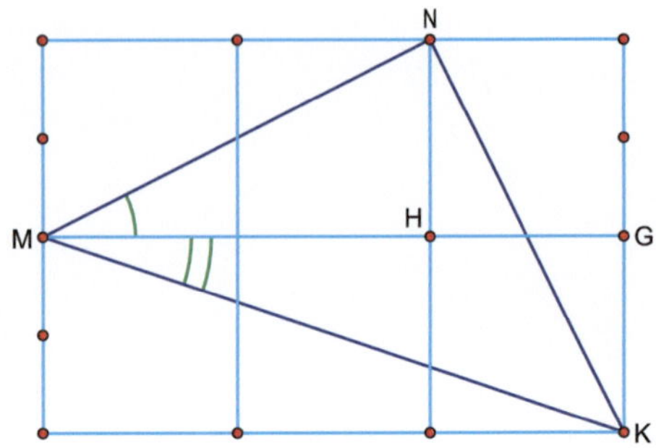

Fig. 3.102 Proof that $\arctan\frac{1}{2} + \arctan\frac{1}{3} = \frac{\pi}{4}$

Because $MN = NK$, then $\angle NMK = 45°$ or $\angle NMK = \frac{\pi}{4}$.
It is obvious that the following is true

$$\angle NMH + \angle GMK = \angle NMK = \frac{\pi}{4}$$

From right triangle NHM, we can find $\angle NMH$:

$$\tan\angle NMH = \frac{NH}{MH} = \frac{1}{2}$$

From right triangle KGM we have:

$$\tan\angle GMK = \frac{GK}{MG} = \frac{1}{3}$$

Moreover, $\arctan 1 + \arctan\frac{1}{2} + \arctan\frac{1}{3} = \frac{\pi}{2}$

Hence, we can state that Eq. (3.24) indeed has the solution $x = 1$.
From (3.23) we can find the height of the radio tower and the three angles:

$$h = 300, \quad \alpha = \frac{\pi}{4}, \quad \beta = \arctan\frac{1}{2}, \quad \gamma = \arctan\frac{1}{3}.$$

Answer. $h = 300, \quad \alpha = \frac{\pi}{4}, \quad \beta = \arctan\frac{1}{2}, \quad \gamma = \arctan\frac{1}{3}$.

3.5 Homework on This Chapter. Problems to Solve

1. Given pyramid $SABC$. Points D and E are on the edges SA and SB, such that

$$SD \div DA = 1 \div 3, \; SE \div EB = 1 \div 3$$

Plane α goes through points D, E and parallel to the edge SC. How does this plane divide the volume of the pyramid?

Hint. See problem 158.
Answer. $5 \div 27$.

2. Let L be the midpoint of SB, $BK \div KC = 2 \div 1$, O is the centroid of the base triangle ABC. Construct a cross section that passes through K, O, and L and evaluate its area, if pyramid is a tetrahedron with side 1.

   Hint. See Problem 165.
   Answer. 1/6 sq. units.

3. $ABCDEFA_1B_1C_1D_1E_1F_1$ is a regular hexagonal prism. Points M, N, K are the midpoints of E_1D_1, BC and AA_1, respectively. Construct the trace of the cutting plane MNK onto plane of (a) ABC; (b) AA_1B_1B; (c) $A_1B_1C_1D_1E_1F_1$; (d) DD_1E_1E; (e) CC_1D_1D. Find the cross-sectional polygon.

   Hint. Use orthogonal projection method.

4. The base of the parallelepiped $ABCDA_1B_1C_1D_1$ is the square $ABCD$, the vertices of which are equidistant from the vertex A_1 of the upper base. The angle between the edge AA_1 and the base plane ABC is 60°. Construct a cross section of the parallelepiped by a plane perpendicular to edge AA_1 and passing through vertex C.

   Hint. The equidistance of the vertices of the square $ABCD$ from point A_1 means that the perpendicular A_1O, dropped from this point to the plane of the lower base, falls into the center of the square, point O. From the condition of the problem we know that the angle between the edge AA_1 and the plane ABC (angle A_1AC) is 60°, it is easy to prove that the triangle A_1AC is equilateral.

   Solution. In our Fig. 3.103 the equal angles are marked by one arc.
   The height CH of the equilateral triangle AA_1C falls into the middle of the edge AA_1. Next, on the faces AA_1D_1D and AA_1B_1B we construct, respectively HK and HM - perpendiculars to the line AA_1. The points of their intersection with the edges B_1B and D_1D (points K and M) are connected by line segments with the point C. The quadrilateral $HKCM$ is the required cross section. Prove on your own that this quadrilateral is a parallelogram (Fig. 3.104).
5. In a regular tetrahedron, a plane is drawn through the apothem, so that the resulting cross section has the smallest possible area. Find the ratio of the volumes of the parts into which the plane divides the tetrahedron.

   Answer. $V_1 : V_2 = 3 : 19$.

6. A rectangular parallelepiped $ABCDA_1B_1C_1D_1$ has the square base with side a. Through the midpoints of the edges AA_1 and CC_1 and point K on the edge BB_1, a cutting plane is drawn. Find the distance from the point K to the midpoint of the edge BB_1, if it is known that the cross-sectional area of the parallelepiped by this plane is $2a^2$ and the length of the side edge is $4a$.

   Answer. $\sqrt{6}/2$.

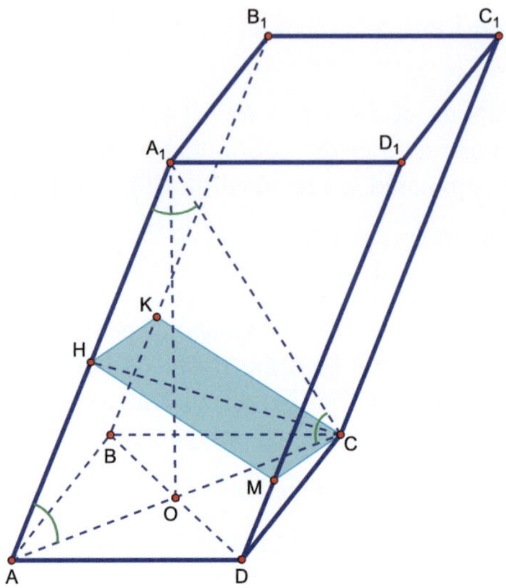

Fig. 3.103 Cross section of an oblique parallelepiped

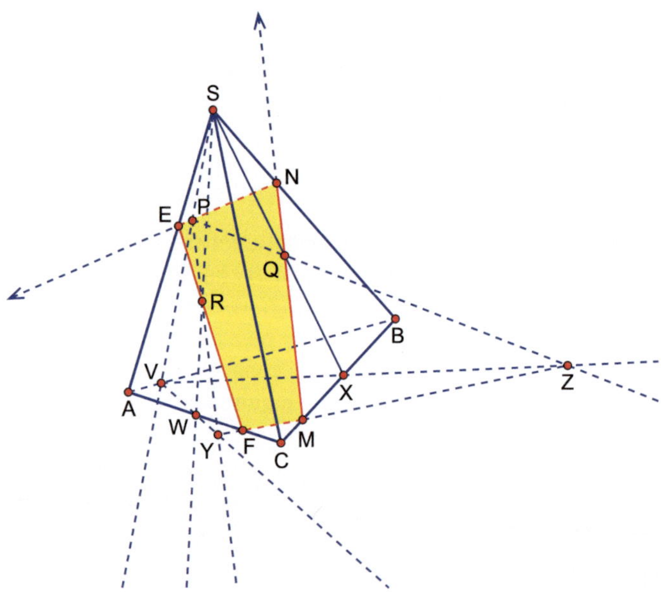

Fig. 3.104 ENMF is the cross section of a pyramid

7. The base of pyramid *TABC* is triangle *ABC* with the angle *A* equals 60°. The lateral edge *TA* is the height of the pyramid and it equals *h*. The edge *TC* is perpendicular to the base side *BC* and the angle between the edge *TB* and the

bisector of the base AD equals 60° What is the smallest area of a pyramid's cross section by a plane passing through the bisector AD and intersecting the edge TB?

Answer. $\frac{h^2}{6\sqrt{6}}$.

8. In a cube $ABCDA_1B_1C_1D_1$ with edge a, a plane P is drawn through a point A parallel to line BD, forming an angle equal to $\arcsin \frac{1}{2\sqrt{2}}$ with line AB. Find the cross-sectional area of the cube by plane P.

   Answer. $S = \frac{2a^2}{\sqrt{3}}$.

9. In a regular triangular pyramid $SABC$ the side edge makes with the plane of the base angle α, $\sin \alpha = \frac{\sqrt{7}}{3}$. The plane that passes through side BC and goes perpendicular to the edge SA intersects this edge at point M. The volume of pyramid $MABC$ equals $1/2$. Find the volume of pyramid $SMBC$.

   Answer. 1.

10. In a regular quadrangular pyramid $SABCD$, a cross section is drawn through the base diagonal BD perpendicular to the edge SC. Find the volume of the pyramid, if the cross-sectional plane forms an angle φ, $\sin \varphi = \sqrt{7}/4$ with the base plane, and the cross-sectional area is 189.

    Answer. 3024 cubic units.

11. The volume of pyramid $SABC$ equals V. A plane is drawn through points M and N of the edges AS, AB, respectively, and the internal point P of the face ABC. This plane intersects the line CS at point L. Let D and E be the points of intersection of lines AP and BP with the edges BC and AC, respectively. It is known that $AN = NC, AM = MS, AP = 3PD$, and $BP = 2PE$. Find the volume of pyramid $ACLN$.

    Answer. The volume of pyramid $ACLN$ can take any values from the interval $[0; \frac{3V}{2}]$ (for the corresponding pyramid $SABC$)

12. Points E and F are taken on the sides AA' and CC' of the cube $ABCDA'B'C'D'$ where $AE = 2A'E$ and $CF = 2C'F$. A plane through points B, E, and F is drawn that divides the cube into two parts. Find the ratio of the volume of the body containing point B' to the entire volume of the cube.

    Answer. $25/72$.

13. The triangular pyramid is dissected by a plane so that the medians of the side faces are divided by intersection points in the ratio $2 \div 1$, $3 \div 1$ and $4 \div 1$, counting from the top of the pyramid. In what ratio, counting from the top of the pyramid, are the side edges divided?

    Answer. $12 \div 7$, $12 \div 5$, $12 \div 1$

14. The base of the quadrangular pyramid *SABCD* is the parallelogram *ABCD*. Points K, L, M belong to the edges *SA*, *SB*, *SC*, respectively and divide these edges in ratios , $2 \div 1$, $1 \div 2$, and $3 \div 1$, counting from the top of the pyramid. Does the plane *KLM* intersect the edge *SD*? If so, in what ratio does it divide this edge?

    Answer. The plane does not intersect the edge *SD*.

15. Construct a cross section of the triangular pyramid *SABC* by a plane passing through the points P, Q, R , if *P* belongs to the face *SAB*, *R* belongs to the face *SAC*, and *Q* belongs to the face *SCB*.

    Solution Let us sketch the pyramid and place all necessary points on its faces. Connecting points *P*, *Q*, and *R* would not do anything because we have to find out how the cutting plane through these points cuts each of the faces and edges. We can use the trace method here.
    Let's draw straight lines through the given points and the top of the pyramid. In this case, the straight line *SP* lies in the back, invisible, face, and is shown by a dotted line. Let us mark the points of intersection of the edges AC, *CB*, and *AB* by these straight lines as W, X, V, respectively. The line *PR* belongs to the secant plane since both points—*P* and *R*—belong to the plane. Let's draw it and draw its trace in the plane of the base ABC. We obtain VW. Then the point of intersection of lines *PR* and VW is the *Y* point, and it belongs to both planes: the secant plane and the plane of the base of the pyramid. We will do exactly the same with the points *P* and *Q*: we draw a line through them, and then it is traced through the points *V* and X. The point *Z* is the intersection of the lines *PQ* and VX.
    Now we have two points (*Y* and *Z*) that belong to the cross section and are conveniently located in the plane of the base, *ABC*, which makes it possible to draw a straight line through them, which will certainly intersect the edges *AC* and *BC*, which also lie in the plane of the base. This line will intersect edge *BC* at point M and edge *AB* at point F. Points M and *Q* belong to face *CSB*, so we can connect them with a line. This line will intersect the edge *SB* at the point N. Similarly, we can now draw a line through the points *N* and *P* belonging to the face ABS. The line *PN* will intersect the edge *AS* at the point E. It can be connected to the point *R* since the points are located in the same plane—the plane of the face ASC. Let us draw a line through the points *E* and *R*. The line *ER* intersects the edge AC at the point F. We have already obtained this point as an intersection of line *YZ* and edge AC. The point *F* can be connected to the point M since both of them lie in the plane of the base. Finally, *ENMF* is the cross section.

16. Construct a cross section of the parallelepiped $ABCDA'B'C'D'$ that goes through edge AB' and parallel to the edge $A'D'$.

17. Construct a cross section of the parallelepiped $ABCDA'B'C'D'$ that goes through edge $A'D'$ and parallel to the edge AC.

18. Given a pyramid $SABC$. Draw a cross section of the pyramid by a plane that goes through the point of intersection of the medians of face ABC parallel to face ASB.

19. Construct a cross section of the pyramid $SABCD$ that contains its height SO and goes parallel to the edge AB.

20. Construct a cross section of the pyramid $SABCD$ that contains its height SO and goes perpendicular to face ABS.

21. Construct the cross section of the cube $ABCDA'B'C'D'$ by the plane that passes the midpoint of edge AB and goes parallel to the plane $BC'D$.

22. Given a cube $ABCDA'B'C'D'$. Find the line of intersection of planes $AB'C$ and $BC'D$.

23. Given a cube $ABCDA'B'C'D'$ with side 1. Find the area of a hexagon that is obtained in the cross section of this cube by a plane passing through the center of the cube and the midpoints of the edges AB and BC.

    Hint. Since the plane goes through the center of the cube, then prove that the cross section is a regular hexagon with side $\frac{1}{\sqrt{2}}$

    Answer. $\frac{3\sqrt{3}}{4}$

24. A beam of parallel rays falls on a plane α perpendicular to it. How to place a cube with edge a above the plane so that the shadow it casts has the maximum area?

    Hint. If the base of the cube $ABCDA_1B_1C_1D_1$ is parallel to the plane, then the cast shadow will be equal to the area of the base or a^2. Therefore, to get a larger cast shadow, we need to place one of its vertices, A on the plane α, and then tilt the cube by an angle φ relative to the plane so that the lower base of the cube forms an angle φ with the plane α. In this case, the shadow will have the shape of a hexagon. Applying Theorem 43, the area of this hexagon can be evaluated as the sum of the areas of the projections onto the plane α of ΔBCD ($\frac{1}{2} \cdot a^2 cos\varphi$), $\Delta A_1B_1D_1$ (the same as for ΔBCD) and of the diagonal cross section BB_1D_1D ($a \cdot a\sqrt{2} \cdot \cos(\frac{\pi}{2} - \varphi)$.) Next, add these areas of the shadow cast to find A, shadow's area and then find its maximum.

    Solution. The area of the shadow cast is

$$A = 2 \cdot \frac{1}{2} a^2 \cos\varphi + a^2\sqrt{2}\sin\varphi = a^2\left(1 \cdot \cos\varphi + \sqrt{2}\sin\varphi\right)$$

    Since $\sqrt{1^2 + (\sqrt{2})^2} = \sqrt{3}$, by multiplying and dividing the right hand side of this expression by $\sqrt{3}$, we can introduce an auxiliary angle $\beta = \arcsin\frac{1}{\sqrt{3}}$ and rewrite this expression as the sine of the sum of two angles.(See my book [3], page 192):

$$A = a^2 \cdot \sqrt{3}\left(\frac{1}{\sqrt{3}}\cos\varphi + \frac{\sqrt{2}}{\sqrt{3}}\sin\varphi\right) = a^2\sqrt{3}\cdot\sin(\beta+\varphi) \le a^2\sqrt{3}.$$

The inequality follows from the fact that the maximum value for the sine is 1. **Answer**. $A_{max} = a^2\sqrt{3}$.

25. The base of pyramid *SABC* is right triangle *ABC*, its lateral edge *SC* is perpendicular to the plane *ABC* and $SC = AC = BC$. Construct a cross section of the pyramid by a plane perpendicular to line *SA* and passing through point *E*, the midpoint of the edge *AC*.

    Hint. To construct a cross section by a plane perpendicular to a given line, we, as always, need to drop two intersecting perpendiculars to this line. Here, three faces of the pyramid are equal isosceles right triangles, and one face (*ASB*) is an equilateral triangle. Let us use the plane drawings of these triangles, so you can clearly see where the directions and ends of the perpendiculars will be constructed. Because triangle *SCA* is isosceles and right triangle, then we can construct the midpoint of the hypotenuse *AS* and then connect points *K* and *C*. The plane will intersect face *ASC* by segment $FE \| KC$. Next, in the equilateral triangle *ASB*, we can find the midpoint of *AS*, then connect points *P* and *B* and draw through *F* line FG parallel to *PB*. We construct $EF \perp SA$, then $FG \perp SA$, *F* and *G* are connected (see Fig. 3.105).

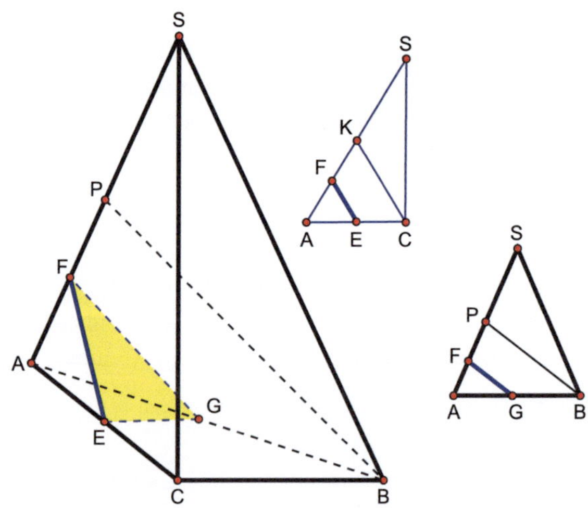

Fig. 3.105 Cross section EFG of a right triangular pyramid

The cross section ($\triangle EFG$) is constructed.

26. A pyramid *ABCDE* has a parallelogram *ABCD* at its base. Point *E* is the apex of the pyramid. The point *F* is the midpoint of the edge *DE*, and the point *G* divides the edge *BE* in the ratio $3 \div 1$, starting from the vertex.

(a) Construct the cross section of the pyramid by plane α containing the points A, F and G.
(b) Let point H be the intersection of the line EC and the plane α. Find the ratio in which the point H divides the edge EC.

Hint. See Problem 14 of Chap. 1.

Answer. $EH \div HC = 3 \div 4$.

27. The base of the quadrangular pyramid $SABCD$ is the parallelogram $ABCD$. Points K, L, M belong to the edges SA, SB, SC, respectively, and divide these edges in ratios $2 \div 1$, $1 \div 2$, and $3 \div 1$, counting from the top of the pyramid. Construct the cross section of the pyramid with the plane KLM. Does this plane intersect the edge SD? If so, in what ratio does it divide this edge?

    Hint. Check Problem 52 of Chap. 1.

    Answer. The plane does not intersect SD.

28. The plane passes the vertex B_1 of the cube $ABCDA_1B_1C_1D_1$ and the points M and N on the edges C_1C and AD respectively. It is known that $C_1M \div MC = 2 \div 1$, and N is the midpoint of the edge AD. Construct the cross section of the cube and evaluate the area of the cross section.

    Hint. Let us sketch the cube and construct the cross-sectional pentagon B_1MSNQ. Please see a solution to Problem 39 of the Homework section of Chap. 1.

    Solution. Because the secant plane cuts the edge DC by the midpoint S, then

$$NS \| AC,$$

where AC is the diagonal of the base square $ABCD$. Since the diagonals of the square are orthogonal, then $BC \perp NS$, then by Theorem 20 on three perpendiculars, $B_1P \perp NS$, where $BD \cap NS = P$.
The angle between the cross section and the base of the cube $ABCD$ can be found from the right triangle B_1BP, in which $BP = \frac{3\sqrt{2}}{4}$, $BB_1 = 1$ and $B_1P = \frac{\sqrt{17}}{2\sqrt{2}}$ and so,

$$\cos\phi = \cos\angle B_1PB = \frac{BP}{B_1P} = \frac{3}{\sqrt{17}}.$$

The orthogonal projection of the cross section is a pentagon $NABCS$, the area of which is equal to the difference between the areas of the unit square and the right isosceles triangle NDS with sides $1/2$. Then the cross-sectional area A can be found using Theorem 43 and formula (3.7):

$$A = \frac{7}{8} \div \frac{3}{\sqrt{17}} = \frac{7\sqrt{17}}{24}.$$

Answer. $A = \frac{7\sqrt{17}}{24}$

29. The plane passes the vertex B_1 of the cube $ABCDA_1B_1C_1D_1$ and the points M and N on the edges C_1C and AD respectively. It is known that $C_1M \div MC = 2 \div 1$, and N is the midpoint of the edge AD. Construct a cross section of the cube and find the ratio in which the cross section divides the volume of the cube.

    Solution. Assume the cube is a unit cube. Let's continue the construction started in the previous problem and extend the line NS until it intersects the line AB at point R. Continue line RB_1 to the intersection of AA_1 at point Q. The cross section of the cube is a pentagon B_1MSNQ. This pentagon divides the volume of the cube into two figures: one above the pentagon, the other below it. If you follow the problems solved in this chapter, then you will be able to find the volume of the figure above the cross section similarly as the difference in volume of the big pyramid B_1RBT and two small equal pyramids, $QRAN$ and $MSCT$, obtaining $25/72$. Then the volume of the spatial figure below the cross section is equal to $1 - 25/72 = 47/72$. The ratio of the two volumes is $25 \div 47$.

    Answer. $V_1 \div V_2 = 25 \div 47$

30. A regular triangular pyramid $ABCD$ has an equilateral triangle ABC at its base, and all lateral edges of the pyramid are equal to 1 one. The trihedral angle at vertex D is $270°$. The plane intersects the pyramid, passing through the points M, N, K belonging to edges AD, AB and DC such that $AM \div MD = 1 \div 2$, $AN \div NB = 2 \div 1$ and KM is parallel to edge AC. Construct a cross section by the plane and find the ratio in which the cross section divides the edge CB.

    Hint. Place the pyramid on one of its lateral faces, making the triangle CDB as its base. Note that all the lateral faces are isosceles right triangles, and all the plane angles of the trihedral angle at the vertex D are right angles (see solution in Fig. 3.106).

Solution. Method 1. (Construction)

- Construct $MK \| AC$.
- $MN \cap DB = T$. In plane CDB we can connect points T and K to obtain $L = KT \cap CB$.
- $KMNL$ is the cross section (see Fig. 3.106).
- $AD = DB = DC = 1$; $MN = \frac{2}{3}$. Because triangle NPB is right isosceles triangle ,then $NP = PB = \frac{1}{3}$, $NB = \frac{1}{3}AB = \frac{\sqrt{2}}{3}$.
- $\triangle MDT \sim \triangle NPT$, then denote $BT = x$ and the following is valid

$$\frac{\frac{2}{3}}{\frac{1}{3}} = \frac{1+x}{\frac{1}{3}+x},$$

  which gives us $x = 1/3 = BT$.
- Since $MK \| AC$,then $MK = \frac{2}{3}AC = \frac{2\sqrt{2}}{3}$, and $CK \div KD = 1 \div 2$. Hence, in triangle KDT, L is the midpoint of side KT.

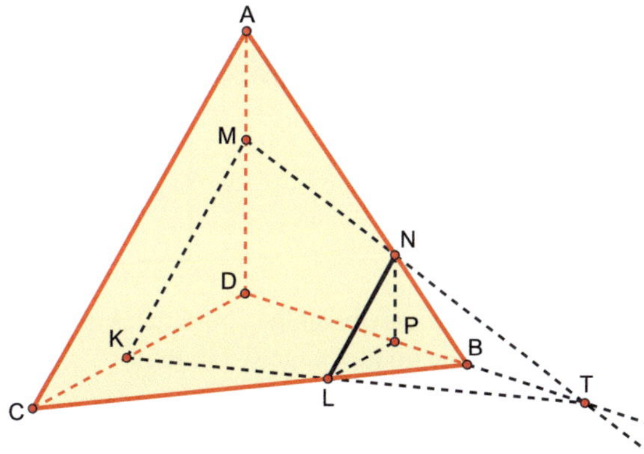

Fig. 3.106 Cross section *KMNL* of a triangular pyramid

- Finally, $BL \div LC = 1 \div 2$.

Method 2. (Vector-coordinate method) The idea is to find the equation of plane ABC and plane MNK and find how these two planes intersect with edge CB. Let us introduce the coordinate system with the origin at D and the X-axis in the direction of vector DC, Y-axis in the direction of vector DB and the Z axis in the direction of DA. The points of the plane MNK have the following coordinates:

$$M = (0, 0, \frac{2}{3}),\ N = (0, \frac{2}{3}, \frac{1}{3}),\ K = (\frac{2}{3}, 0, 0).$$

Using formula (1.50), the equation of plane MNK can be found as follows

$$\begin{vmatrix} x & y & z - \frac{2}{3} \\ 0 & \frac{2}{3} & -\frac{1}{3} \\ \frac{2}{3} & 0 & -\frac{2}{3} \end{vmatrix} = 0,$$

which gives us

$$6x + 3y + 6z = 4.$$

Obviously the plane passing the vertices $C,\ B,\ A$ has the equation

$$x + y + z = 1$$

To solve the system of two linear equations

$$\begin{cases} 6x + 3y + 6z = 4 \\ x + y + z = 1 \end{cases}$$

we can multiply the second equation by 6 and subtract it from the first equations, obtaining $y = 2/3$. Next, we can find the equation of the line of intersection of these planes as

$$\begin{cases} x = \frac{1}{3} - t \\ y = \frac{2}{3} \\ z = t \end{cases} \Leftrightarrow \begin{bmatrix} x \\ y \\ z \end{bmatrix} = \begin{bmatrix} \frac{1}{3} \\ \frac{2}{3} \\ 0 \end{bmatrix} + t \cdot \begin{bmatrix} -1 \\ 0 \\ 1 \end{bmatrix}$$

Replacing $t = 0$, we obtain the coordinates of point L as $L = (\frac{1}{3}, \frac{2}{3}, 0)$. Therefore the cutting plane divides edge CB in ratio

$$BL \div LC = 1 \div 2.$$

Answer. $1 \div 2$

31. Given the tetrahedron $ABCS$ whose edges AB and CS have lengths a and b respectively. The distance between the skew lines AB and CS is d, and the angle between them is ω. Tetrahedron $ABCS$ is divided into two solids by plane α, parallel to lines AB and CS. The ratio of the distances of α from AB and CS is equal to t. Find the volume of the tetrahedron.

    Hint. We can reuse and modify the left sketch in Figure 3.82. Erase sides DK, KE and perpendiculars PP_1 and QQ_1 and connect points A and E, A and M, and S and M. Now our tetrahedron is divided into four pyramids (two triangular and two quadrangular), i.e. $ADEMN$, $EABM$, $SDEMN$, and $SNMC$. Evaluate their volumes and add together.

    Solution. Let A be the area of the cross section found in Problem 174. Let d_1 be the distance from point A to the plane ε and d_2 be the distance from point S to this plane. Denote the volume of the tetrahedron by V, the height of the tetrahedron dropped from S by H and the height of pyramid $EABM$ dropped from vertex E by h. We have the following for the volume of the tetrahedron:

$$\frac{1}{3} \cdot d_1 \cdot A + \frac{1}{3} \cdot h \cdot [MBA] + \frac{1}{3} \cdot d_2 \cdot A + \frac{1}{3} \cdot H \cdot [NMC] = V,$$

    where $[MBA]$ and $[NMC]$ are the areas of triangles MBA and NMC, respectively. Because $d_1 + d_2 = d$ and $V = \frac{1}{3} H \cdot [ABC]$ and using information of the problem and the solution of Problem 174, we have

$$\frac{1}{3} \cdot A(d_1 + d_2) + \frac{1}{3} \cdot \frac{t}{t+1} \cdot H \cdot (\frac{t}{t+1})^2 \cdot [ABC] + \frac{1}{3} \cdot H \cdot \frac{1}{(t+1)^2} \cdot [ABC] = V$$
$$\frac{1}{3} A \cdot d + (\frac{t}{t+1})^3 \cdot V + \frac{1}{(t+1)^2} \cdot V = V.$$

In the last equation we can move all terms with V to the RHS and factor V. After simplification we obtain:

$$\frac{1}{3}A \cdot d = V \cdot \left(1 - \frac{t^3}{(t+1)^3} - \frac{1}{(t+1)^2}\right) = V \cdot \frac{(3t^2 + 2t)}{(t+1)^3}$$

After substitution here the formula (3.20) for the cross sectional area, A, obtained in solving Problem 174, this can be simplified to

$$\frac{3t+2}{t+1} \cdot V = \frac{abd \sin \omega}{3},$$

which gives us the answer.

Answer. $V = \frac{abd \sin \omega \cdot (t+1)}{3(3t+2)}$

Chapter 4
Cylinders, Cones, Spheres, and More

It's difficult to say when people felt the need to describe something as being a cylinder, cone, or sphere. Perhaps they were observing stars in the sky and looked at the full Moon and thought about its shape. They might have considered that an orange, peach, apple, or apricot does not exactly look like a polyhedron, because they all have curvature. Having contemplated nature, they might have been artistically inspired to create complex objects themselves. In any case, as civilization progresses, there appears to be an increasing occurrence of artifacts with greater geometric sophistication in the historical record. When making something one has to understand the properties of the object so it seems that this geometric sophistication in art goes hand-in-hand with practical advances in geometric knowledge.

For instance, in any museum of ancient cultures, you can find interesting assortments of household items. From 2000 BC, containers shaped like modern bowls were used for food. Perhaps with the invention of pottery, plates began to take the form of objects obtained by rotating a flat figure around an axis. It is not unlikely that this idea of rotational generation of cylindrical symmetries might have arisen from pottery production. Sometime much later, it was noticed that the surface of the Earth is not flat. This was evidenced by the following observation: when a ship appears on the horizon, the tops of the masts are first visible, and only then the entire ship appears. The assumption about the spherical shape of the Earth arose shortly before Pythagoras. Experimental confirmation of the hypothesis came from Aristotle, who studied the boundary of the shadow cast during a lunar eclipse.

Formulas for the volume and surface area of most spatial figures, including those with curvature, were found by the great ancient Greek scientist and mathematician Archimedes (287–212 BC), who came close to the concept of a definite integral, ahead of the rest of humanity by almost two thousand years. All scientists know-how Archimedes exclaimed, "Give me a reliable fulcrum and I will turn the world around!" By rotating a segment around the axis of a parabola, Archimedes obtained a paraboloidal segment and determined its volume to be equal to half the volume of a cylinder of the indicated dimensions. He was the first known to estimate the bounds

E. Grigorieva, *Methods of Solving Solid Geometry Problems*,
https://doi.org/10.1007/978-3-031-86969-3_4

of the irrational number π. However, Archimedes seems to have been most proud of his formulas for the volume and surface area of a sphere. In his work he stated that

- The volume of a ball is equal to $\frac{2}{3}$ the volume of the cylinder circumscribed around the ball.
- The surface area of the ball is equal to the area of the lateral surface of the cylinder circumscribed around the ball.
- The surface area of the ball is four times the area of the great circle of the ball.

In memory of these great achievements, a ball and cylinder marked the tomb of Archimedes. In this chapter we will trace some of Archimedes's brilliant ideas and touch a little on his methods of calculation.

The main goal of this chapter is to learn to solve challenging problems about cylinders, cones, spheres, bodies of revolution and their combinations. We will learn how formulas for surface area and volume of these figures were derived and see the connections between the modern calculus approach to evaluating the volume of a figure of revolution and how it was done by Archimedes and Cavalieri (1598–1647). The Pappus–Guldin theorems will also be presented. This chapter is devoted to the study of various combinations of figures with curvature and to solving Olympiad-type problems associated with inscribing and circumscribing polyhedra.

4.1 Cylinders

According to the geometric definition, **cylindrical surface** is a surface formed by parallel movement of a straight line L (**ruling**) along some generating curve C (**directrix**). Line L in not parallel to the plane of C.

Right cylindrical surfaces are formed if the rulings are perpendicular to the plane containing C, the directrix. In Fig. 4.1, we observe a right cylindrical surface with the rulings parallel to the Z-axis and the directrix curve $y = \sin x$ perpendicular to the ruling. Let us select a circle with radius R for C located in XY-plane. If the ruling line is parallel to the Z axis, then the surface formed will be a familiar right circular cylindrical surface shown in Fig. 4.2.

Oblique cylindrical surfaces are formed if the rulings are not perpendicular to the plane containing the directrix, C.

A cylinder has two bases that are parallel to each other and connected by a cylindrical surface. If the directrix is a circle and the ruling is perpendicular to it, then the resulting cylinder is called round and straight. The line passing through the centers of the two bases is called the axis of the circular cylinder. All points of the figure are at the same distance from this line, which is equal to the radius of the base. The right circular cylinder generated by a vertical line (the ruling) moving around the circle of radius R with the algebraic equation $x^2 + y^2 = R^2$ in the XY-plane is shown in Fig. 4.2. The circle is the **generating curve** for the right cylinder.

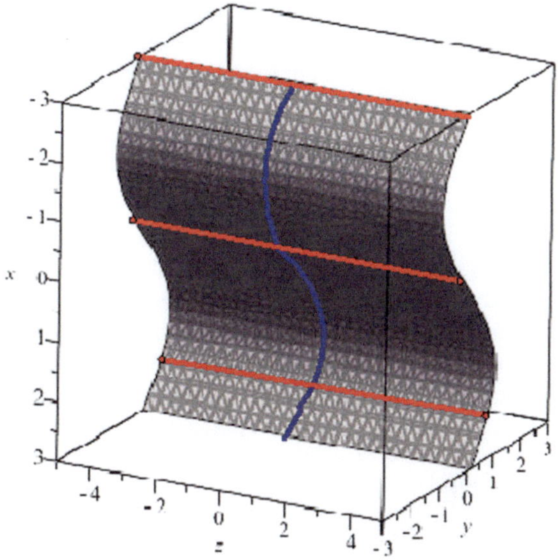

Fig. 4.1 Cylindrical surface with directrix C $y = \sin(x)$ and rulings parallel to the Z axis

In solid geometry, often cylindrical surface is surface of revolution that is traced by a straight line (the ruling) that always moves parallel to itself or some fixed line or direction (the axis). The path, to be definite, is directed along a curve (the directrix), along which the line glides. Such cylinders will be discussed further in detail.

A generalized cylinder is a body that consists of two identical plane figures lying in the parallel planes that can be combined by translation of all segments connecting the corresponding points of these figures. These identical figures are

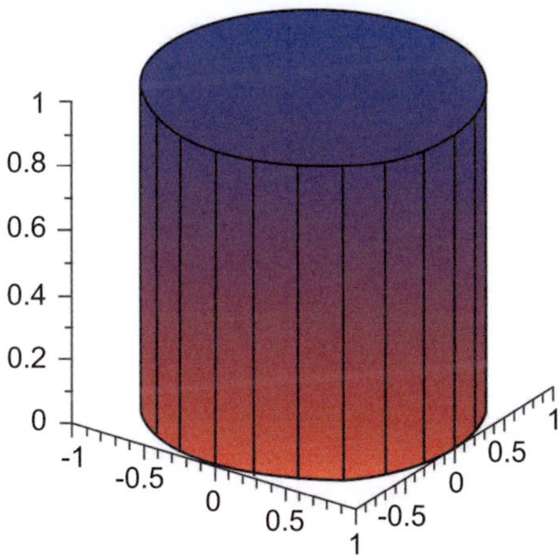

Fig. 4.2 Right cylindrical surface

called the bases of the cylinder. If the base is a circle, then such a cylinder is called **a circular cylinder**. The circles are called the bases of the cylinder, and the segments connecting the corresponding points of the circumference of the circles, are called the forming segments or simply sides of a cylinder. Since translation is a rigid motion, the bases of the cylinder are equal. By the translation the plane moves onto a plane moves onto a parallel plane, then the bases of the cylinder lie in parallel planes. The points move along parallel or coinciding lines by the same distance, so the forming segments of the cylinder are parallel.

A circular cylinder is called **right** if its sides are perpendicular to the planes of the bases. The radius of a cylinder is the radius of its base, the height of a cylinder is the distance between the planes of its bases. The axis of a cylinder is a straight line passing through the centers of the bases. A right circular cylinder can be obtained by rotating a rectangle around one of its sides. For example, in Fig. 4.3, the right cylinder is obtained by rotating rectangle ABO_2O_1 around axis O_1O_2.

Fig. 4.3 Right cylinder obtained by revolution of rectangle ABO_2O_1 about O_1O_2 axis

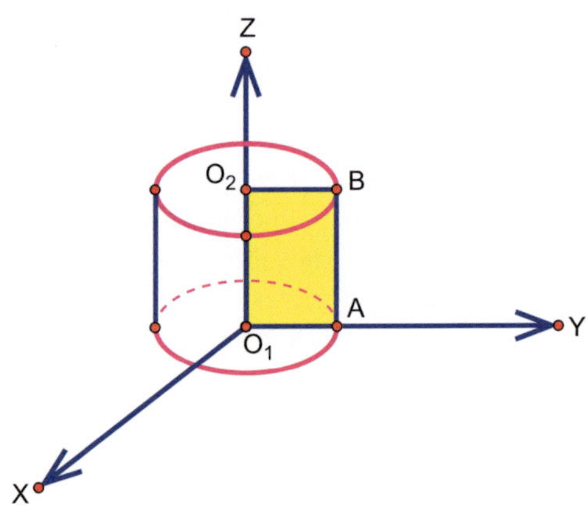

A plane parallel to the plane of the base of the right circular cylinder intersects its lateral surface in a circle with a perimeter equal to the circumference of the base. The cross section of a cylinder by a plane parallel to its axis is a rectangle, with two of its sides are the rulings of the cylinder, and two other sides are parallel chords of the bases. The cross sections of a circular cylinder parallel to the bases have the same areas as the base.

Oblique circular cylinder. Let us select a circle with radius R for C located in XY plane. If the ruling lines are not parallel to the Z axis and form with the base an acute angle, then the surface formed will be the surface of the oblique circular cylinder. An oblique circular cylinder has the lateral edge (side) not orthogonal to the base circles that forms an angle α with the base. The axis O_1O_2 of an oblique cylinder connects the centers of the basis circles and is not perpendicular to the bases, forming with the base the same angle α as the lateral edge (see Fig. 4.4). The height (altitude) of an oblique cylinder h is the length of the perpendicular dropped from one circle to the other or simply the distance between two bases of the cylinder.

Fig. 4.4 Oblique cylinder

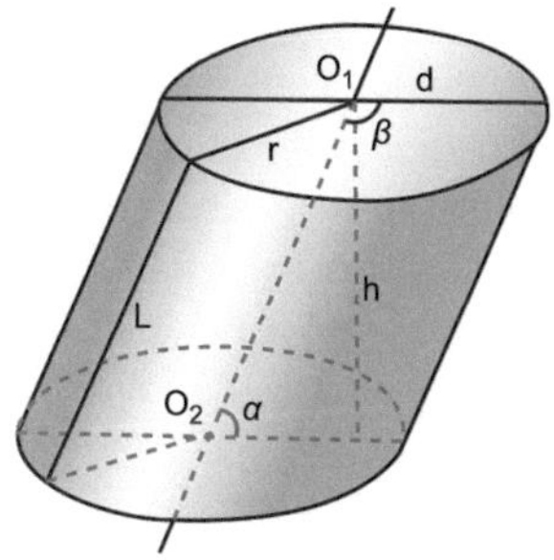

Theorem 44 *The volume of a right cylinder is equal to the product of the area of its base, Q, and height, h, i.e.,*

$$V = Q \cdot h,$$

where Q is the area of the base and h is the height of the cylinder.

Proof Since the base area of the cylinder is Q, there are sequences of circumscribed and inscribed polygons with areas Q_n and Q'_n such that

$$\lim_{n \to \infty} Q_n = \lim_{n \to \infty} Q'_n = Q.$$

Let us construct sequences of prisms whose bases are the circumscribed and inscribed polygons considered above and whose lateral edges are parallel to the ruling of the given cylinder with length h. These prisms are circumscribed and inscribed for the given cylinder. Their volumes are $V_n = Q_n h$ and $V'_n = Q'_n h$, respectively. Hence, the volume of the cylinder is

$$V = \lim_{n \to \infty} Q_n h = \lim_{n \to \infty} Q'_n h = Qh.$$

□

If the cylinder is circular, then $Q = \pi r^2$ so the volume of a circular cylinder is

$$V = \pi r^2 h. \tag{4.1}$$

The volume of an oblique cylinder with height h and base circle of radius r can be evaluated also by (4.1). In this case, the axial cross sections are parallelograms. The area of the lateral surface of a right cylinder is given by $A = 2\pi rh$, where the radius of the base of the cylinder r and the height h. Think of cutting the surface of the cylinder by a line parallel to its axis and then unwrap it. It is a rectangle with the height that is equal to the height of the cylinder, h, and the width equals the circumference of its circular base, $2\pi \cdot r$ (Fig. 4.5).

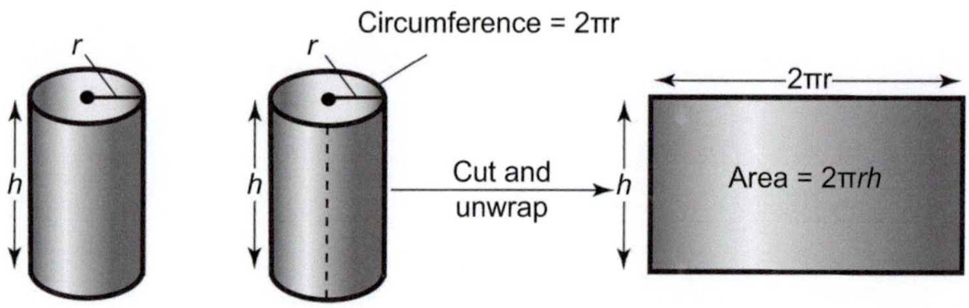

Fig. 4.5 Cylinder and its lateral surface

Fig. 4.6 I am near the cylindrical leaning Tower of Pisa

The total surface area of a right circular cylinder must include two areas of the top and bottom circles of the radius r,

$$S = 2\pi rh + 2\pi r^2. \tag{4.2}$$

The surface area of oblique cylinder cannot be calculated by formula (4.2). If the cylinder is oblique, then the lateral surface area must be equal to the product of the length of its generatrix (lateral edge) and the perimeter of the cross section, which is perpendicular to this generatrix. Unfortunately, there is no simple formula for expressing the lateral surface area of an oblique cylinder in terms of its height and its base parameters. Many famous buildings have the shape of a cylinder. For example, the cylindrical leaning Tower of Pisa is shown in Fig. 4.6.

4.2 Cones

A **generalized cone** is a body that consists of a plane figure—the base of the cone, a point not lying in the plane of this circle, the apex of the cone, and all segments connecting the apex with the points of the base. If the base of the cone is a circle, it is a circular cone. If the apex of a cone is not projected at the center of the base circle, then such a cone is an oblique cone (Fig. 4.7).

Fig. 4.7 Generalized cone

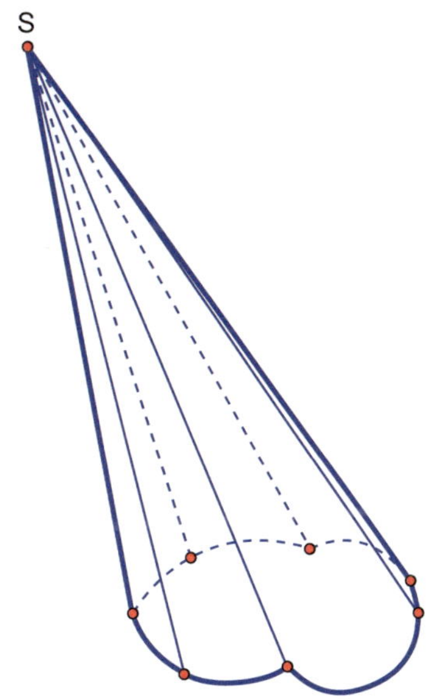

A cone is called a **right cone** if the line connecting the apex (vertex) of the cone with the center of the base is perpendicular to the plane of the base. Some things familiar to us that have conic shapes include: ice cream cones, Christmas trees, and teepees among the Indians of North America. Two options for a circular cone (right on the left and oblique on the right) are shown in Fig. 4.8.

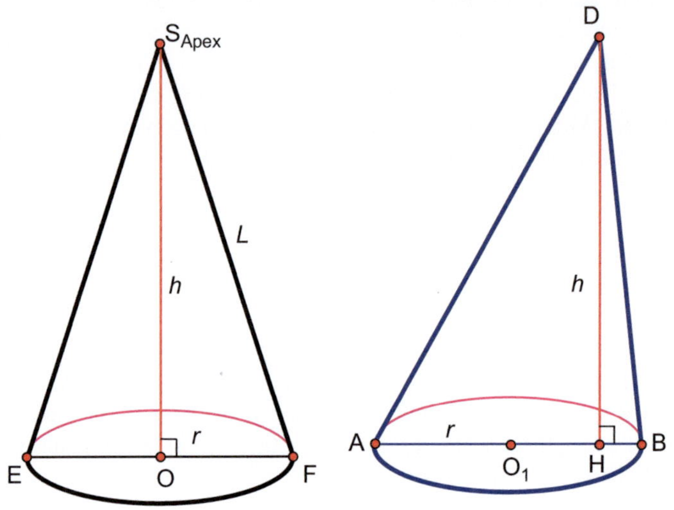

Fig. 4.8 Right and oblique circular cones

A right circular cone can be considered as a solid body obtained by rotating a right triangle about one of its leg, as an axis. For example, when rotating a right triangle with legs 3 units and 4 units about the large leg, you get a cone of 4 units high and with the base radius of 3 units, and if you rotate the same triangle around the smaller side, you will obtain the cone with height of 3 units and base radius of 4 units. Additionally, a right cone can be considered as a body obtained by rotating an isosceles triangle around its height as an axis. For example, the right cone on the left of Fig. 4.8 can be obtained by rotating $\triangle ESF$ around the axis SO.

A right cone in which the lateral edge L is equal to the diameter of the base ($L = 2r$) is called an **equilateral cone**. Such a cone is shown in Fig. 4.9. We can see the cross section of the cone by a plane passing the diameter and the height of the cone is an equilateral triangle. If you glue a circular sector along the two radii bounding it, you will get the lateral surface of a right circular cone, the vertex of which is the center of the circle from which the sector was cut, and the lateral edge of which is the radius of this circle. The volume of a cone is

$$V = \frac{1}{3}Qh, \tag{4.3}$$

Formula (4.3) can be also proven similar to Theorem 44 for a cylinder. We can inscribe and circumscribe polygons about the base circle. If we connect the apex of the cone with each vertex of such polygons, we will obtain the cone sandwiched between the inscribed and circumscribed n-gonal pyramids with the following volumes: $V_n = \frac{1}{3}Q_n h$ and $V_n' = \frac{1}{3}Q_n' h$, respectively. Hence, the volume of the cone is

$$V = \frac{1}{3}\lim_{n\to\infty} Q_n h = \frac{1}{3}\lim_{n\to\infty} Q_n' h = \frac{1}{3}Qh.$$

Here Q is the area of the base of the cone. If the cone is circular, then $Q = \pi r^2$ and its volume is calculated as

$$V = \frac{1}{3}\pi r^2 h, \tag{4.4}$$

where r is the radius of the circle of the base and h is the height of the cone shown in Fig. 4.9. So the area of the lateral surface of a right circular cone with radius r and lateral edge L is

$$S_{\text{lateral}} = \pi r L \tag{4.5}$$

and the full surface area of a right circular cone is

$$S = \pi r L + \pi r^2. \tag{4.6}$$

Let us solve the following problem that will help us to derive formulas (4.5) and (4.6) for the general case of a right circular cone.

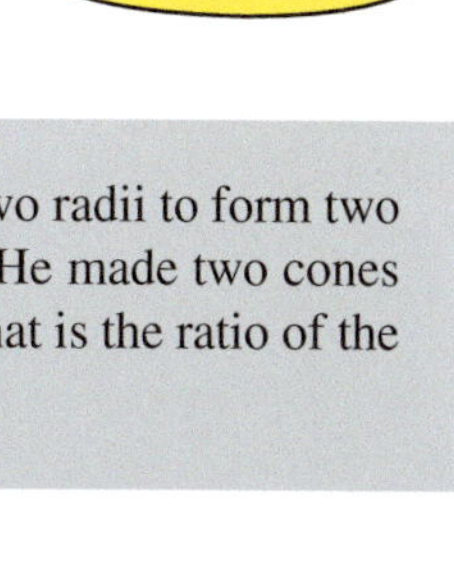

Fig. 4.9 Right equilateral cone

Problem 185 A student cut a circular paper disk along two radii to form two sectors, the smallest having a central angel of $\alpha = 120°$. He made two cones using each sector to form the lateral surface of a cone. What is the ratio of the volume of the smaller cone to that of the larger?

Solution Assume that the circle has radius r. Please, find Fig. 4.10 and make two cones with scissors by cutting the circle along the blue line and then gluing each sector separately along the blue line.It would be obvious to you that the radius of the original circle, r, will become the slant height of the both cones and that each sector forms a cone with the lateral edge r. The red arc EDB will became the circumference C_1 of the base of the cone made out of the 120° sector. Hence, its length equals $1/3$ of the circumference of the original circle,

$$C_1 = 2\pi r_1 = \frac{1}{3}2\pi r = 2\pi \cdot \frac{r}{3}.$$

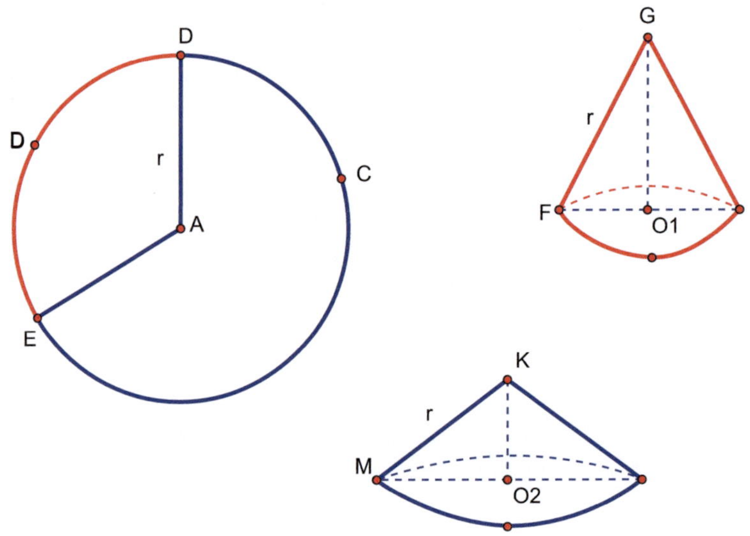

Fig. 4.10 Two cones made of a circle

Therefore its radius is $r_1 = \frac{r}{3}$ and its height obtained using Pythagorean theorem is

$$h_1 = \sqrt{r^2 - \left(\frac{r}{3}\right)^2} = \frac{2\sqrt{2}}{3}r.$$

Similarly, we can find that the bigger cone has the radius $r_1 = \frac{2r}{3}$ and its height is

$$h_2 = \sqrt{r^2 - \left(\frac{2r}{3}\right)^2} = \frac{\sqrt{5}}{3}r.$$

The volume of a cone is $1/3$ height times base. The bases can be expressed as $B_1 = \pi(r/3)^2$, $B_2 = \pi(2r/3)^2$. Then after simplification, the ratio of the volumes is

$$\frac{V_1}{V_2} = \frac{2\sqrt{2}}{4\sqrt{5}} = \frac{1}{\sqrt{10}}.$$

Note that this ratio does not depend on the radius of the original circle. However, in order to make a cone out of a circular sector, the central angle α of the sector must be greater than zero and strictly less than 180°.

Answer. $V_1 \div V_2 = \frac{1}{\sqrt{10}}$.

Remark. In the problem above, the central angle was 120°, which means that the sector from which one of the cones was formed, had the area of $1/3$ of the area of the original circle. It also means that the circumference of the base of the corresponding cone is also $1/3$ of the circumference of the big circle.

Let C_1 be the circumference of the base of the right cone obtained from a sector and C be the circumference of the big circle. Then the following must be true

$$C_1 = x \cdot C, \qquad 0 < x < 1.$$

Here x indicates the fraction. Let the big circle have radius $R = L$ and let r be the radius of the base of the cone. The parameter x can be evaluated from the equation $2\pi \cdot r = x \cdot 2\pi \cdot L$ so, $x = \frac{r}{L}$. It is understood that the lateral surface of the cone (A_1) shown in Fig. 4.10 in red is equal to the area of the segment $EDBA$ from which it is created. Let A be the area of the original circle. The following is valid

$$A_1 = x \cdot A = x \cdot \pi L^2 = \pi \frac{r}{L} \cdot L^2 = \pi r L.$$

This matches formula (4.5).

The part of a cone concluded between the base of the cone and the cutting plane that goes parallel to its base is called a **truncated cone**. A truncated cone can be obtained by rotating a right trapezoid around its side edge that is perpendicular to the base. The formula for the volume of a truncated cone is very similar to that of the truncated pyramid, i.e.,

$$V = \frac{1}{3}\pi h \left(r^2 + rR + R^2\right). \tag{4.7}$$

The lateral surface of a right truncated cone with lateral edge L and bases R and r is

$$S = \pi L(r + R). \tag{4.8}$$

The volume of an oblique cone with height h and radius of the base r is also calculated by formula (4.3). Note that a lateral surface of an oblique cone cannot be simply expressed in terms of its height and radius of the base. Recall that formula (4.4) for the volume of a right circular cone was derived by Archimedes. Let us look at some of his ideas.

As a child, I had a pyramid toy that consisted of a rod and a set of rings of different radii like one in Fig. 4.11 on the left. It was necessary to string these rings onto the rod, placing the rings on top of each other, starting with the largest and ending with the smallest. The result was a cone-shaped structure—a kind of pyramid resembling a cone. This toy is very interesting. Apparently, it developed creativity in kids and for me as an adult because reminiscing about this toy I see that we can derive the formula for the volume of a cone.

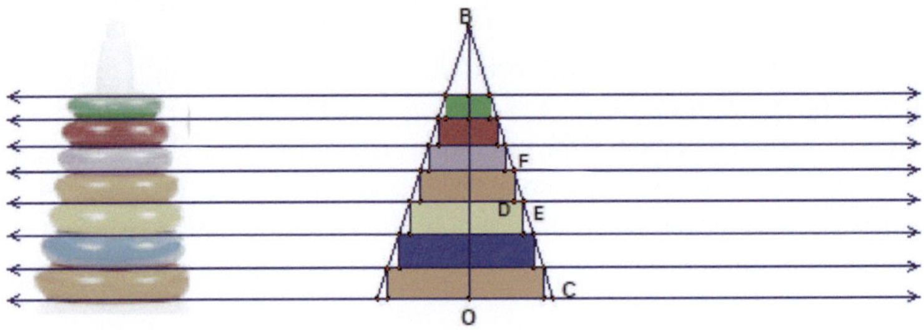

Fig. 4.11 Kid's toy pyramid and making a cone

Let a cone with the radius of the base $r = OC$ and height $h = OB$ be given. Consider its axial cross section and slice the entire cone by the rings of the same very little height of $\epsilon = DF$. Assume that we have n such slices, then

$$\epsilon = \frac{h}{n}. \tag{4.9}$$

Now let the radii of two consecutive rings differ by the same value of $\delta = DE$. Thus, the radius of the first ring is $r - \delta$, of the second is $r - 2 \cdot \delta$, etc., so the radius of the kth ring is $r - k \cdot \delta$. Because $\triangle COB \sim \triangle EDF$, then the following is true:

$$\frac{h}{\epsilon} = \frac{r}{\delta},$$

from which we obtain that

$$\delta = \frac{\epsilon r}{h}. \tag{4.10}$$

The volume of the kth ring can be evaluated using the formula for the volume of the cylinder

$$V_k = \pi \epsilon \cdot (r - \delta \cdot k)^2. \tag{4.11}$$

Substituting in (4.11) formula (4.10) for δ and (4.9) for ϵ, we get

$$V_k = \pi\left(r - k \cdot \frac{\epsilon r}{h}\right)^2 \epsilon = \frac{\pi r^2 h}{n}\left(1 - \frac{k}{n}\right)^2.$$

This formula can be rewritten as

$$V_k = \frac{\pi r^2 h \cdot (n-k)^2}{n^3}. \tag{4.12}$$

This way Archimedes found the volume of the kth ring.

Let us put together all the rings. The tinier the rings are, the more our construction resembles the actual cone. If $n \to \infty$, then as the limit the volume of our construction will be the volume of the cone. At this point let us use modern methods of summation.

$$\lim_{n\to\infty} \frac{\pi r^2 h}{n^3} \cdot \sum_{k=1}^{n} (n-k)^2$$

In order to find this limit, first, we can find the sum

$$\frac{1}{n^3} \cdot \sum_{k=1}^{n} (n-k)^2 = \frac{1}{n^3} \cdot \sum_{k=1}^{n} \left(n^2 - 2nk + k^2\right),$$

which can be broken into three sums:

$$\frac{1}{n^3} \cdot \left(n^2 \cdot \sum_{k=1}^{n} 1 - 2n \cdot \sum_{k=1}^{n} k + \sum_{k=1}^{n} k^2\right),$$

and then evaluated as

$$\begin{aligned}
&\frac{1}{n^3} \cdot \left(n^3 - \frac{2n \cdot n(n+1)}{2} + \frac{n(n+1)(2n+1)}{6}\right) \\
&= \frac{n}{n^3} \cdot \left(n^2 - n(n+1) + \frac{(n+1)(2n+1)}{6}\right) \\
&= \frac{-6n + 2n^2 + 3n + 1}{6n^2} \\
&= \frac{2n^2 - 3n + 1}{6n^2} \\
&= \frac{1}{3} - \frac{1}{2n} + \frac{1}{6n^2}.
\end{aligned}$$

If the number of the rings is increasing, then this sum gets closer and closer to $\frac{1}{3}$. Therefore, the volume of the cone is approaching the value of

$$V = \lim_{n\to\infty} \frac{\pi r^2 h}{n^3} \cdot \sum_{k=1}^{n} (n-k)^2 = \frac{1}{3}\pi r^2 h.$$

We know this formula as the formula (4.4) for the volume of a circular cone having radius of the base r and height h.

4.3 Balls and Spheres

A **ball** is a body that consists of all points in space located not greater than a certain distance from a given point (**center of the ball**). The distance from the center to the boundary of the ball is the **radius** of the ball. The boundary of the ball is a spherical surface (**sphere**). All points of the sphere are equidistant from the center of the ball at a distance equal to the radius. A ball is a body of revolution—it is the rotation of a semicircle around an axis of its diameter. For example, the rotation of a circle with a radius of 3 units around its diameter will be a ball whose radius is 3 units.

A plane passing through the center of a ball is called a **diametral plane**. A cross section of a ball by a diametral plane is called a **great circle**. For example, in Fig. 4.12 two perpendicular planes passing through diameter AB form great circles. Another great circle with the radius OD is shown in red. Any section of a ball by a plane is a circle the center of which is the foot of the perpendicular dropped from the center of the ball to the secant plane. All the planes equidistant from the center dissect the ball into equal circles. Any plane that does not pass through the center of the ball will cross it by a circle of a smaller radius. One of such smaller circles can be also seen in Fig. 4.12. Here O_1 is the center of the small circle with the radius $r = O_1D, \quad r < R = OD$.

The boundary of a ball is called a **sphere**. In this book we will often deal with a sphere so we need to find either its surface area, volume or both, given its radius. The volume of a sphere of radius R is

$$V = \frac{4}{3}\pi R^3 \tag{4.13}$$

and its surface area is

$$S = 4\pi R^2. \tag{4.14}$$

Both formulas (4.13)–(4.14) were obtained by Archimedes (circa 287–212 BC).

Consider the toy pyramid described in the previous section and let's use it to derive formula (4.13). We need to make not one conical pyramid, but three different ones: cylindrical (when the rings of radius 1 are very thin but add to have form a cylinder with height 1), conical (from the same thin rings, but different radii, from which you can assemble an "almost" cone of height and base radius equal to 1), and

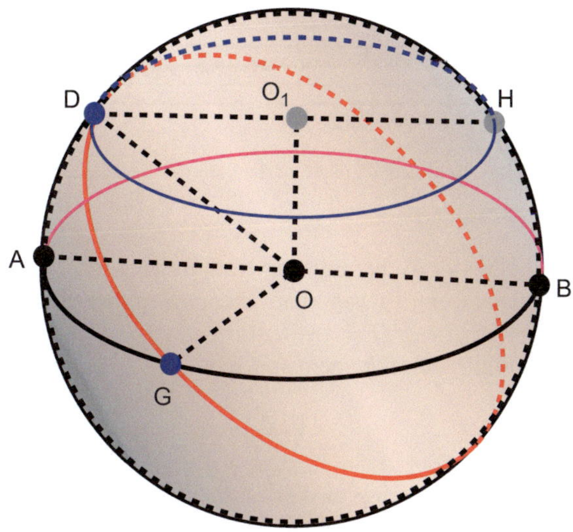

Fig. 4.12 A ball dissected by several planes

"hemispherical" (again, from the same thin rings, from which you can assemble an "almost" hemisphere of radius 1). In this case, all rings must be made of the same material. Following Archimedes, let's take the very precise scales with flat bowls and put a cylinder toy assembled from rings on one bowl and a cone and a hemisphere on the other, placing the cone with its base on the scales and the hemisphere "on the head," so that the flat base of the hemisphere is on top and located horizontally. Let the heights of the rings be the same and equal to δ, where δ is a very small number.

Now let's calculate the volumes of rings located at the same height h for each of the three figures. We can refer to Fig. 4.13 where each figure is represented by orthogonal cross sections of a plane passing through the axes of the cylinder, cone, and the center of the hemisphere. The location of the rings at height h is shown in red. For the cylinder, the radius of the base is $R = FD = 1$. For all three figures, we calculate the volumes of tiny rings at the height $h = AH = JL = PQ$. The volume of the cylindrical ring of radius 1 is

$$V = \pi\delta.$$

It is easy to find the radius of the tiny ring from the cone. Because $SL = LI = 1$, triangle SLI is a right isosceles triangle with equal legs. Then $\triangle SJK$ is similar to it and the radius of the tiny cylindrical ring coming from the cone will have radius $r_1 = JK = SJ = 1 - h$, so its volume is

$$V_1 = \pi \cdot r_1^2 \cdot \delta = \pi(1-h)^2\delta.$$

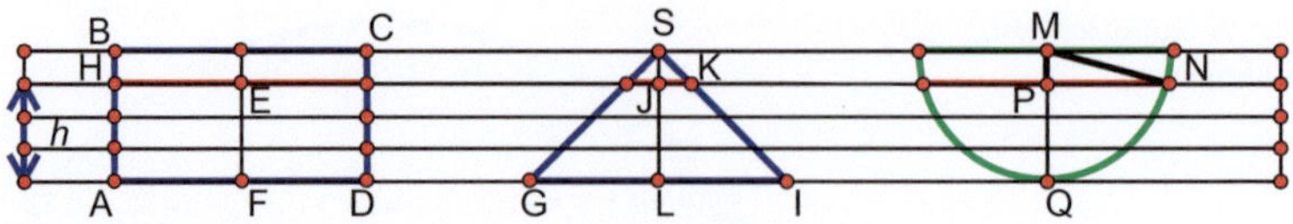

Fig. 4.13 Archimedes's proof of the volume of a sphere formula

The radius r_2 of the tiny ring from the hemisphere (in Fig. 4.11, $r_2 = PN$) can be evaluated using Pythagorean Theorem applied to triangle MPN, where $MN = 1$, $MP = MQ - PQ = 1 - h$, so $r_2 = \sqrt{1-(1-h)^2}$. Hence, the volume of the tiny ring coming from the hemisphere is

$$V_2 = \pi r_2^2 \delta = \pi(1-(1-h)^2)\delta.$$

If we put the ring made of a cylinder on the left pan of the scale, and the rings from a cone and hemisphere, respectively, on the right pan of the scale, then we will see that the scales will balance. Algebraically, we obtain the equality,

$$V = V_1 + V_2.$$

Let us assume that the thickness δ of the rings is very small, then a conical toy made from such rings will be practically indistinguishable from a real cone, a hemispherical toy will be almost a hemisphere, and a cylindrical toy will always be a cylinder. In the limit, we find that the volume of a hemisphere of radius 1 is equal to the volume of a cylinder with a base radius and height equal to 1 minus the volume of a cone with a base radius and height also equal to 1.

$$V_{\text{hemisphere}} = \pi - \frac{1}{3} \cdot \pi = \frac{2}{3} \cdot \pi$$

Hence, the volume of the sphere is twice that quantity. Thus, the volume of a hemisphere of radius R is $\frac{4}{3} \cdot \pi R^3$. Archimedes formulated it as

The volume of the ball is equal to $\frac{2}{3}$ the volume of the cylinder circumscribing the ball:

$$\frac{2}{3}\pi R^2 \cdot (2R) = \frac{4}{3}\pi R^3.$$

Next, let us see how Archimedes derived formula (4.14) for the surface area of a sphere. Consider a sphere of radius R inscribed into a cylinder. Then the radius of the base of this cylinder must be equal to the radius of the sphere and its height h must be equal to the diameter of the sphere. Let us dissect the cylinder and sphere by two planes parallel to the base of the cylinder at height h and $h - \Delta h$, respectively. These two planes will cut a circular strip of radius R and thickness Δh on the surface of the cylinder and a strip of radius r and thickness Δl on the surface of the sphere.

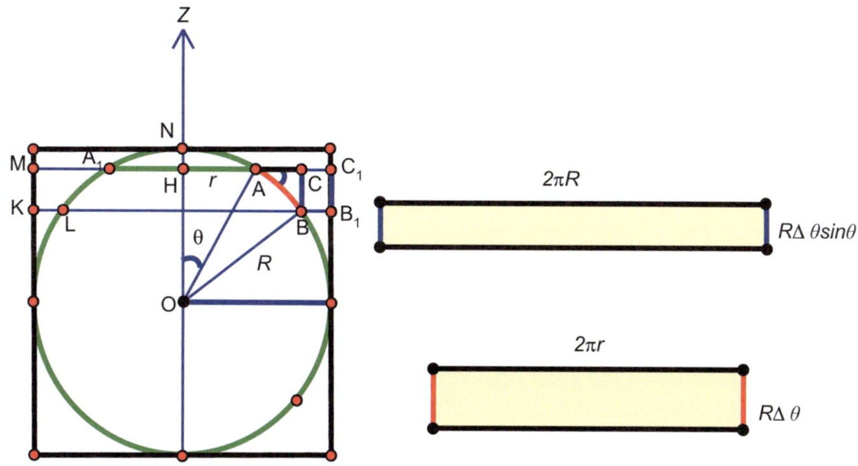

Fig. 4.14 The lateral surface areas of a cylinder and the sphere inscribed in this cylinder are the same

Now, let us show that the areas of both strips are equal (Fig. 4.14). Denote $\angle NOA = \theta$, $OB = OA = R$, and $HA = r$. Then the strip cut from the cylinder will have length $2\pi R$ and width $\Delta h = C_1B_1$. The corresponding strip cut from the sphere will have length $2\pi r$, where $r = R \sin\theta$ and width $AB = \Delta l = R\Delta\theta$. Because the change of the central angle $\Delta\theta \to 0$, then $AB \approx AT$, the length of the segment of tangent drawn at point A to the great circle. In the right triangle ACB: $\angle CAB = \angle AON = \theta$, then $\Delta h = C_1B_1 = CB = AB\sin\theta = R\Delta\theta\sin\theta$. Archimedes showed that if $\Delta h \to 0$, then the areas of cylindrical and spherical strips are the same.

If we unfold each of the strips, we would see that their areas are equal. It is shown on the right of Fig. 4.14.

$$\Delta A_{\text{cylinder}} = 2\pi R \cdot R\sin\theta\Delta\theta = 2\pi R^2\sin\theta\Delta\theta$$
$$\Delta A_{\text{sphere}} = 2\pi r \cdot R\Delta\theta = 2\pi R^2\sin\theta\Delta\theta.$$

Since we know that any cylinder circumscribed around a sphere of radius R must have height $2R$, its lateral surface is equal to $2\pi R \cdot 2R = 4\pi R^2$ which is actually the surface area of the sphere of radius R.

Archimedes was the first to derive formula (4.14). He formulated it as,

> The surface area of the ball is equal to the area of the lateral surface of the cylinder circumscribed around the ball, $2\pi R \cdot (2R) = 4\pi R^2$, and four times the area of the great circle of the ball, $4 \cdot (\pi R^2) = 4\pi R^2$.

Formula (4.14) for the surface area of a sphere can come from the fact that it is equal to the derivative of the volume with respect to R because the total volume inside of a sphere can be considered as the sum of the surface areas of an infinite number of spherical shells of infinitesimal thickness, concentrically stacked on each other from radius 0 to radius R. Let's assume that Earth is round and has the shape of a ball of radius R. And now consider a fairly thin layer of the atmosphere above Earth with thickness δR so that it is much less than the radius of the Earth ($\delta R \ll R$). Subtracting the two volumes, we will obtain the following:

$$\begin{aligned}\delta V &= \frac{4}{3}\pi(R+\delta R)^3 - \frac{4}{3}\pi R^3 = \frac{4}{3}\pi(R^3 + 3R^2\delta R + 3R(\delta R)^2 + (\delta R)^3 - R^3)\\ &= 4\pi R^2\delta R + 4\pi R(\delta R)^2 + \frac{4}{3}\pi(\delta R)^3\\ &= \left(4\pi R^2 + 4\pi R\delta R + \frac{4}{3}\pi(\delta R)^2\right)\delta R.\end{aligned}$$

For any given radius of the additional volume (δV), the surface area at the radius R denoted by $A(R)$ is measured by the additional thickness (δR) as $\delta V \approx A(R)\cdot\delta R$. If δR is infinitesimal, we obtain the formula for the surface area of the ball,

$$A(R) = \lim_{\delta R\to 0}(4\pi R^2 + 4\pi R\delta R + \frac{4}{3}\pi(\delta R)^2) = 4\pi R^2.$$

Formula (4.14) for the surface area of a sphere also can be visualized without using calculus. Imagine that you are trying to peel an orange, and the peeling of the orange occurs along the "meridians" from top to bottom. My dad used to brag about how beautifully he peeled oranges. He called it "plain cut" and "fashionable cut." (see pictures on the left and right in Fig. 4.15, respectively). Of course, my dad, showing it as his own invention, tried to interest us in geometry, although many people peel an orange in this way. Suppose an orange is spherical. To cut an orange with a knife along arcs of great circles means to obtain sections of the orange with planes passing through the center of the orange. You can also say that we get a cut of an orange along the meridians, like a globe, and then we separate the peel. That's what we need! Each piece of peel is not flat, but a three-dimensional object. Try to flatten it or imagine that it is stretchable and very elastic. Unwrapping (expanding the peel) helps to calculate the area of a sphere!

Fig. 4.15 Peeling an orange along the "meridians"

In Fig. 4.15 we cut the orange peels into 7 and 17 pieces, respectively. You can make many cuts to get several identical pieces and then lay them out in a horizontal line, as shown in Fig. 4.16. Obviously, the more pieces we produce, the more we comply with the conditions of equality, which means the more accurately we can measure the surface area of the sphere! Our orange peel unwrapping in the limit will be a rectangle with a length equal to $2\pi r$ and a width of $2r$, where r is the radius of the sphere. The area of such a rectangle is $A = 2\pi r \cdot 2r = 4\pi r^2$ that coincides with the surface area of a sphere with radius r (4.14).

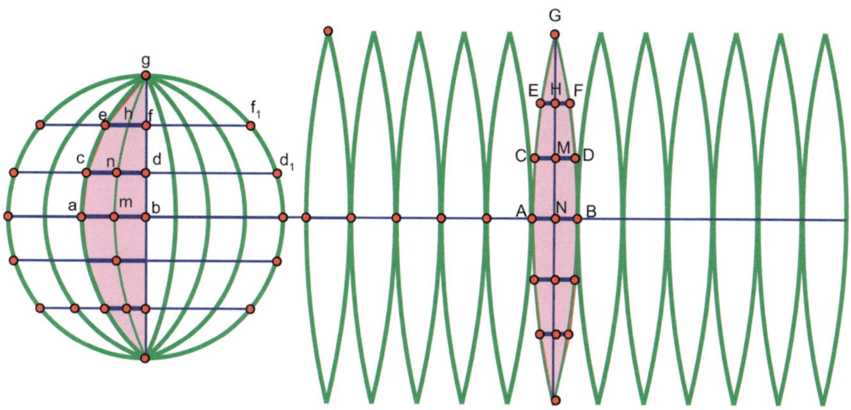

Fig. 4.16 Unfolding a spherical orange alone the meridians

Let us find the volume of a ball using calculus. If we take 1/4 of the circle centered at the origin and of radius R (located in the first quadrant) and rotate it about the Y-axis, then each point of the arc will make a circle of radius $r = \sqrt{R^2 - x^2}$. The area of such a circle will be $S(x) = \pi \cdot (R^2 - x^2)$. The volume of the hemisphere is

$$V = \int_0^R \pi(R^2 - x^2)dx = \pi R^3 - \frac{1}{3}\pi R^3 = \pi R^2 \cdot R - \frac{1}{3}\pi R^2 \cdot R = \frac{2}{3}\pi R^3.$$

Therefore, the volume of a hemisphere is equal to the difference between the volume of the cylinder with radius of the base the same as its height and the volume of the equilateral cone. Obviously, the volume of a sphere is twice the volume of the hemisphere of the same radius.

A spherical segment is the part of a ball cut off from it by a plane. Any cutting plane divides a ball into two **spherical segments**. If such a plane goes through the center of the sphere, then we obtain a **hemisphere**. It is known that Archimedes found the volume and lateral surface area of a spherical segment:

$$V_{\text{segment}} = \pi h^2 \left(R - \frac{h}{3}\right). \tag{4.15}$$

$$S_{\text{segment}} = 2\pi Rh. \tag{4.16}$$

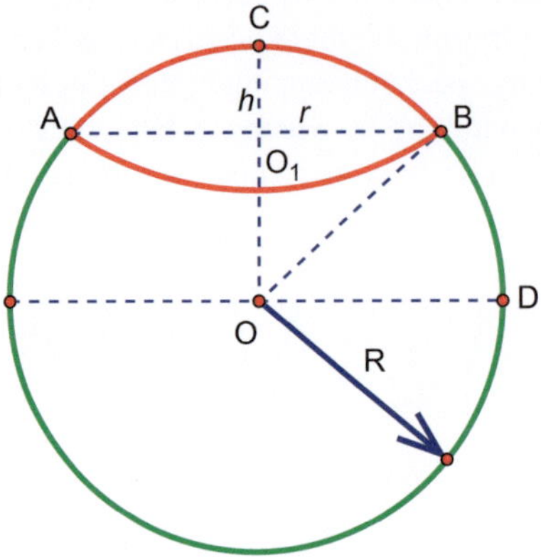

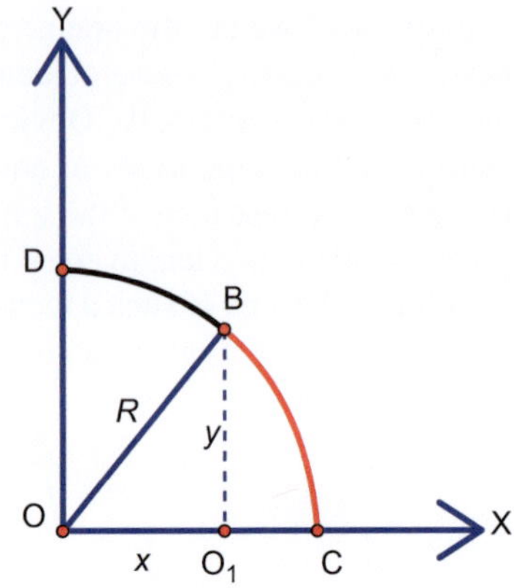

Fig. 4.17 A ball and a segment

The proof of these formulas can be performed by a student who has mastered integration. Consider one of the segments of a general type. It is determined by the parameters $R,\ h$ in Fig. 4.17. Let a ball of radius R be given and a segment of height h be cut off from the ball. Consider the axis passing through the center of the ball and the center of the cutting plane. It is clear that it suffices to consider the case of a smaller segment, where, as in Fig. 4.17 on the left, the height of the segment is not more than the radius of the ball. Construct in the coordinate plane a circular arc with the equation $y = \sqrt{R^2 - x^2}$, such that $R - h \le x \le R$, and rotate it around the X-axis.

$$V = \int_{R-h}^{R} \pi y^2 dx = \pi \left(R^2 x - \frac{1}{3} x^3 \right) \Big|_{R-h}^{R} = \left(R^3 - \frac{R^3}{3} \right) - \left[R^2 (R-h) - \frac{1}{3} (R-h)^3 \right].$$

After collecting like terms we obtain the required formula (4.15) for the volume of a spherical segment.

Next, let us evaluate the area of the segment,

$$S = 2\pi \int_{R-h}^{R} y \sqrt{1 + \left(\frac{dy}{dx} \right)^2} dx.$$

Since $\frac{dy}{dx} = -\frac{x}{\sqrt{R^2-x^2}}$, then $1 + \left(\frac{dy}{dx}\right)^2 = \frac{R}{\sqrt{R^2-x^2}} = \frac{R}{y}$. Hence, the expression under the integral is equal to R, and it is very easy to evaluate the surface area for the segment,

$$S = 2\pi Rx|_{R-h}^{R} = 2\pi R\,(R - (R-h)) = 2\pi Rh.$$

Remark. If $h = R$, then formulas (4.15) and (4.16) will express the volume and surface area of a hemisphere, respectively. Multiplying both formulas by 2 we obtain formulas for the volume (4.13) and surface area (4.14) of a ball of radius R.

Let's apply this knowledge to solving a couple problems.

Problem 186 The hemisphere-shaped bowl is filled with water and then tilted at an angle of 45°. How much water will pour out?

Solution When the bowl is tilted by 45° (the position is indicated by a red dotted line FB in Fig. 4.18), water will not spill out of the spherical segment ABC, the volume of which is equal to

$$V = \frac{\pi h^2}{3}(3R - h)$$

where the height of the segment is

$$h = DC = OC - OD = R - \frac{R\sqrt{2}}{2}.$$

Here the radius of the hemisphere is $R = OB = OC$. Therefore,

$$V_{\text{segment}} = \frac{\pi}{3}\left(R - R\frac{\sqrt{2}}{2}\right)^2\left(2R + R\frac{\sqrt{2}}{2}\right) = \frac{\pi R^3\left(16 - 5\sqrt{2}\right)}{12}.$$

Because the volume of the hemisphere is equal to

$$V_{\text{hemisphere}} = \frac{2}{3}\pi R^3.$$

We can divide the volumes of the segment and hemisphere to obtain the ratio,

$$\frac{V_{\text{segment}}}{V_{\text{hemisphere}}} = \frac{2\pi R^3\left(8 - 5\sqrt{2}\right)}{3 \cdot 8} \div \frac{2\pi R^3}{3} = \frac{\left(8 - 5\sqrt{2}\right)}{8} \approx 0.116$$

that is, approximately 12% of the water will stay in the cup and hence approximately 88% will pour out.

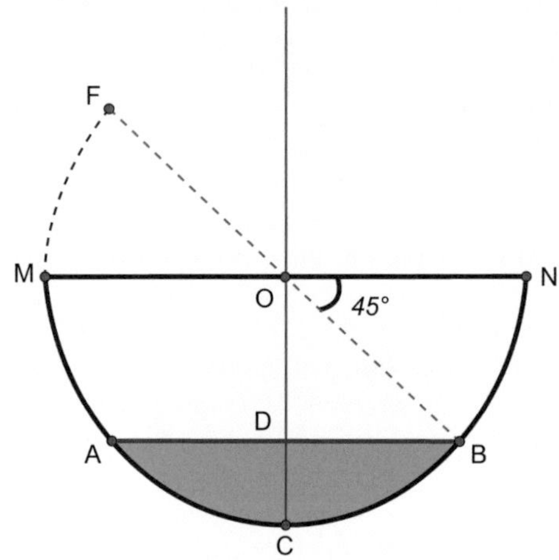

Fig. 4.18 A hemispherical bowl filled with water and then tilted

Answer. Approximately 88% of water will pour out the bowl.

Problem 187 A mathematician ordered a gold ring of a given width from a jeweler, so that the ring would have the shape of a body enclosed between a sphere and a cylinder (its axis passes through the center of the sphere). When he received the order, it turned out that the ring was too small—it was difficult to put on his finger. The customer wanted to increase the diameter of the cylindrical part of the ring, leaving its width h the same. The jeweler demanded that gold be added, which would increase the price of the ring. The mathematician responded with a decisive refusal and explained his reason. Prove that the mathematician was correct.

Proof The ring is obtained from a ball by removing a cylinder with two segments at its ends shown in Fig. 4.19. The volume of the cylinder is

$$\pi r^2 h = \pi (R^2 - \frac{h^2}{4})h.$$

The volume of each spherical segment is found by formula (4.15) with the replacement of h by the value $H = R - \frac{h}{2}$.

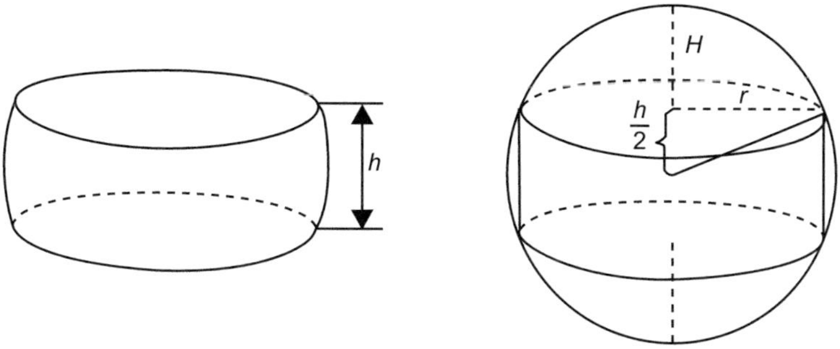

Fig. 4.19 Ring problem

The volume of the ring is

$$
\begin{aligned}
V &= \frac{4}{3}\pi R^3 - \pi\left(R^2 - \frac{h^2}{4}\right)h - 2\pi\left(R - \frac{h}{2}\right)^2 \cdot \frac{2R + \frac{h}{2}}{3} \\
&= \frac{4}{3}\pi R^3 - \pi h \frac{\left(4R^2 - h\right)}{4} - \frac{\pi\left(R - \frac{h}{2}\right)^2 \cdot (4R + h)}{3} \\
&= \frac{16\pi R^3 - 3\pi h \cdot 4R^2 + 3\pi h^3 - 4\pi\left(R^2 - Rh + \frac{h^2}{4}\right)(4R + h)}{12} \\
&= \frac{16\pi R^3 - 12\pi h R^2 + 3\pi h^3 - \pi\left(4R^2 - 4Rh + h^2\right)(4R + h)}{12} \\
&= \frac{16\pi R^3 - 12\pi h R^2 + 3\pi h^3 - 16\pi R^3 + 12\pi h R^2 - \pi h^3}{12}
\end{aligned}
$$

After cancelling opposite terms, we finally obtain

$$V = \frac{\pi h^3}{6}.$$

From this formula it follows that the volume of the ring (and therefore the consumption of gold, which affects the price) does not depend on the radius of the ring, but only on its width h, which does not change with the increase in the size of the ring. Therefore, the mathematician was right. □

Lemma 16 (Principle of Cavalieri) *If two spatial objects Φ_1 and Φ_2 can be arranged in space such that a plane parallel to the given plane cuts them making so that the cross sections of the objects F_1 and F_2 have equal area, then the volumes of the original spatial objects are equal.*

The formula for the volume of a sphere (4.13) can also be proven using Cavalier's principle.

Let a hemisphere of radius R be given, the great circle of which is located on the plane α. Consider a cylinder whose base is a circle of radius R, located in the same plane, and whose height is also R. In Fig. 4.20 we show our solid figures by cross sections, in two projections, when cutting the objects by the plane perpendicular to plane α that goes through the diameters of both cylinder and the hemisphere, and by the plane β parallel to plane α at the distance x from α. We inscribe a cone into the cylinder, the base of which will be the upper base of the cylinder, and the vertex will be the center of the lower base of the cylinder.

Let us prove that the volumes of the hemisphere and of the object obtained from removing the cone from the cylinder are equal. The object on the right of Fig. 4.20 consists of the points of the cylinder that do not fall inside the cone. Imagine that we have not just the cylindrical surface, but a cylindrical rod. Let's mentally cut the cone out of it and take it away from the cylinder.

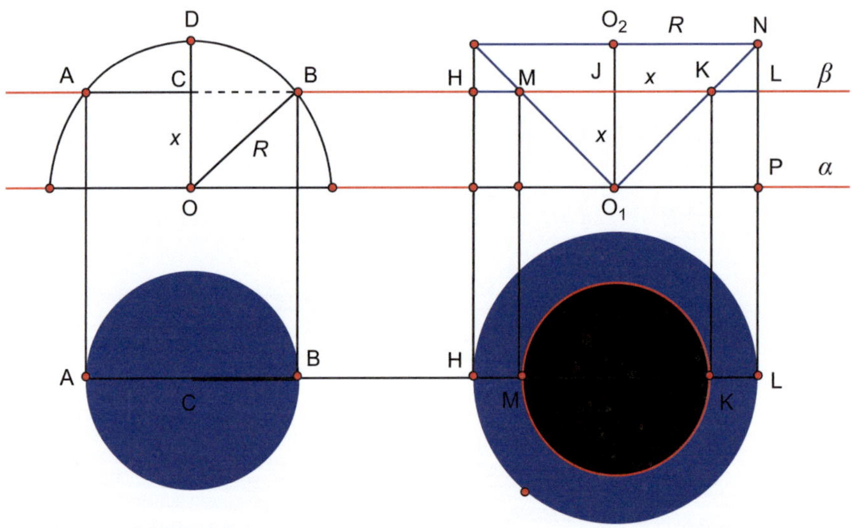

Fig. 4.20 Hemisphere versus cylinder with the upside cone taken out of the cylinder

To use Cavalieri's principle, we need to compare the cross-sectional areas of both spatial figures with a plane parallel to the given one. At the bottom of Fig. 4.20 we see the cross sections by the plane β, parallel to the plane α, at a distance x from it, $0 < x < R$. In the section of the hemisphere by this plane we obtain a circle of radius $r = \sqrt{R^2 - x^2}$ and area πr^2. It can be calculated from the right triangle OCB, since $OB = R$ and $OC = x$, so the area of the blue circle is $\pi \cdot CB^2 = \pi \left(R^2 - x^2\right)$. In the cross section of the other object, we obtain a ring. Because the cylinder has the radius of the base equal to its height and since any plane β parallel to the plane of the base α at distance x from α will cut the cone at circles of radius x. Note that by

Fig. 4.20 we can find that the right triangle O_1JK is an isosceles triangle. In the ring, the radius of the inner circle is equal to x and the radius of the outer circle is equal to R. The area of this ring is equal to the difference of the areas of two circles, i.e., $\pi R^2 - \pi x^2$ and it is showing in blue. We can see that it is equal to the cross-sectional area of the hemisphere (this cross section is also shown in blue).

It follows from Cavalieri's principle that the hemisphere and the constructed object (with the cone upside down taken out of the cylinder) have equal volumes. Let's calculate this volume. It is equal to the difference between the volumes of the cylinder and the cone, i.e.,

$$V = \pi R^2 R - \frac{1}{3}\pi R^2 R = \frac{2}{3}\pi R^3.$$

The volume of a sphere is twice the volume of a hemisphere and, therefore is expressed by the formula, $V = \frac{4}{3}\pi R^3$.

In the modern understanding of comparing two areas or volumes using Cavalieri's principle, each point x of the numerical axis corresponds to the functions $m_1 = m_1(x)$ and $m_2 = m_2(x)$, defining the areas or lengths of "indivisibles." Using the definite integral allows you to prove Cavalieri's principle regardless of whether you are calculating area or volume. Therefore, the functions $m_1 = m_1(x)$ and $m_2 = m_2(x)$ determine the areas or lengths of "indivisibles" (both can be called a "measure" by height x, where $0 \le x \le H$, H is the upper boundary for x). In the case of lengths, products of the form $f(x)dx$ are the areas, and in the case of areas, they are the volumes. However, in both cases the integrals $\int_0^H m_1(x)dx = \int_0^H m_2(x)dx$, which proves the statement called Cavalieri principle.

Remark. In my opinion, a study of how Archimedes estimated the volume of a sphere or cone, or an application of Cavalieri's principle to derive some volume formulas for spatial objects can be of great interest to students. The derivation of some formulas using Cavalieri's principle may not be as rigorous as one would like in a modern geometry course, however, it has the advantage in that it builds geometric intuition which is helpful for more complex topics.

4.3.1 Spheres in a Coordinate System

A sphere of radius R centered at point $M = (x_0, y_0, z_0)$ has the following equation:

$$(x - x_0)^2 + (y - y_0)^2 + (z - z_0)^2 = R^2. \tag{4.17}$$

Problem 188 Solve the system of nonlinear equations:

$$\begin{cases} x^2 + y^2 + z^2 = 14 \\ x + 2y + 3z = 14 \end{cases}$$

Solution If we substitute $x = 1,\ \ y = 2,\ \ z = 3$, this triple will satisfy both equations. We can look at this problem from a geometric point of view. The system geometrically represents the intersection of the sphere centered at the origin with radius of $\sqrt{14}$ with the plane orthogonal to vector $\overrightarrow{n}\,(1, 2, 3)$. Point (1,2,3) is on the sphere and also on the plane. The distance from the origin to the plane is

$$\rho = \frac{|ax_0 + by_0 + cz_0 - d|}{\sqrt{a^2 + b^2 + c^2}} = \frac{|1 \cdot 0 + 2 \cdot 0 + 3 \cdot 0 - 14|}{\sqrt{1^2 + 2^2 + 3^2}} = \frac{14}{\sqrt{14}} = \sqrt{14}.$$

Therefore the plane is tangent to the sphere and such a point is unique.

Answer. $x = 1,\ \ y = 2,\ \ z = 3.$

4.4 Figures of Revolution

You can easily imagine a surface of revolution, for example, using ordinary means at hand. Each of us knows how to use a compass, but let's get a little creative and use a compass to draw not just a circle of a certain radius, but put the compass on a sheet of paper, fix the center of the future circle and fix a point that will form the radius of the future circle. Now let's look at the plane that unites the side of the compass, which will rotate around the other side of the compass (the so-called the axis of rotation).

The axis of rotation of the compass will not rotate when we draw a circle, but at each moment of time the second side of the compass, together with the axis of rotation, describes a right triangle, which we rotate around this axis (see Fig. 4.21). As a result, we get a circle on a sheet of paper. At each moment of time, we can fix the position of a right triangle rotating around this axis; its hypotenuse AB will form a generatrix of the future cone, leg AO will be the height and leg OB the radius of the cone.

Let's imagine that we take a rectangular trapezoid and begin to rotate it around one of the sides that is perpendicular to the parallel bases; as a result of this rotation, we will get a truncated cone. Take a regular pencil, cut out any triangle from paper and now glue one of its sides to the surface of the pencil. Now let's rotate the pencil around its axis. At each moment of time, the triangle will be in some other position, describing a three-dimensional figure. If you think about it, this figure will represent two cones that have a common base, and the radius of this cone will be equal to the height of the triangle drawn from the vertex opposite the base glued to the pencil.

In Fig. 4.22 we take $\triangle MDK$ and rotate it around one of its sides KM. If DN is the height of the triangle dropped to side MK, then this height is equal to the radius of the base of each cone, $R = MN$. The height DN splits the original triangle into two right triangles (blue and yellow). In the first cone the height is the length KN and in the second cone, the height is the length MN.

Now take the right triangle MND and rotate it around the line that goes through its leg MN. You will see that the obtained spatial figure is the right circular cone with

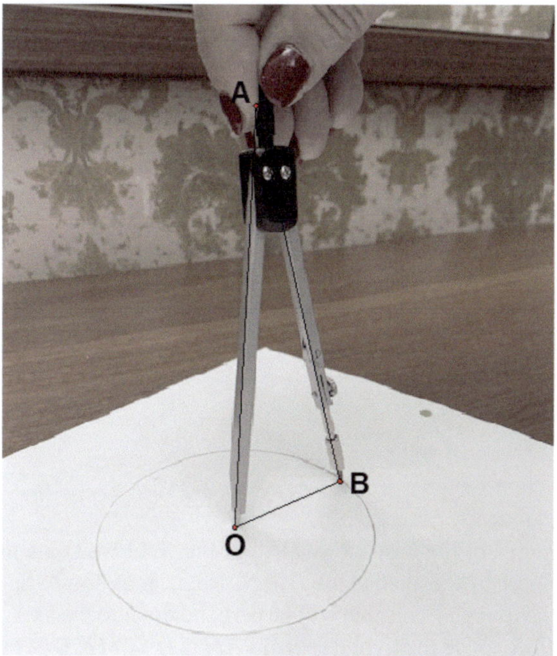

Fig. 4.21 Using compass to imagine a circular cone

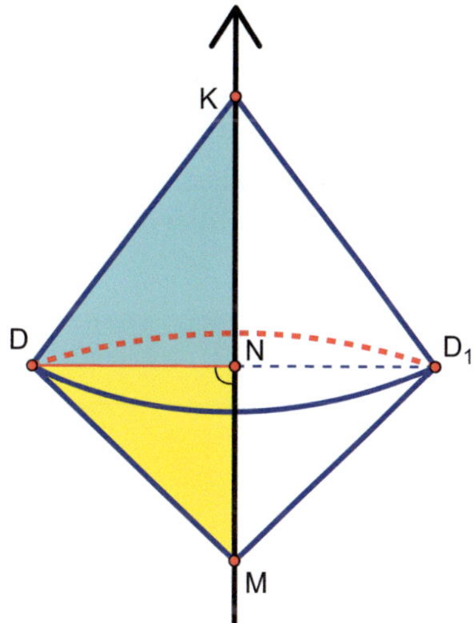

Fig. 4.22 Rotating a triangle

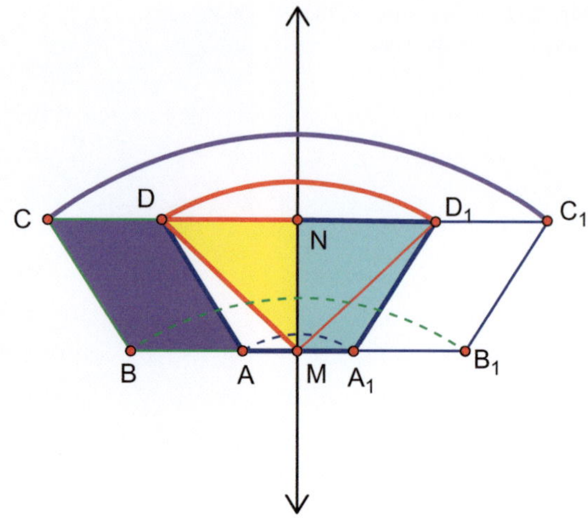

Fig. 4.23 Spatial figures obtained by rotation

radius of the base, $R = DN$. In Fig. 4.23 we show the cross section of the rotated figure by the axial plane DMN. Next, let us draw a line (MB) parallel to side DN of the triangle. This line will be perpendicular to the axis of rotation MN ($MB \perp MN$). If we rotate the right trapezoid $AMND = MND_1A_1$ with respect to MN, we obtain a truncated cone, with the radius of big circle of $R = DN$ and the radius of the small circle $r = AM$. Both the cone and the truncated cone are solid figures (no holes).

If we want to get a depression in a three-dimensional figure, this can also be achieved by modifying the same picture. Thus, we can, for example, rotate another triangle DAM around axis MN. As a result of this rotation we will obtain the figure with the cavity inside that can be filled with something. This figure is obtained as if we cut off the cone from the truncated cone.

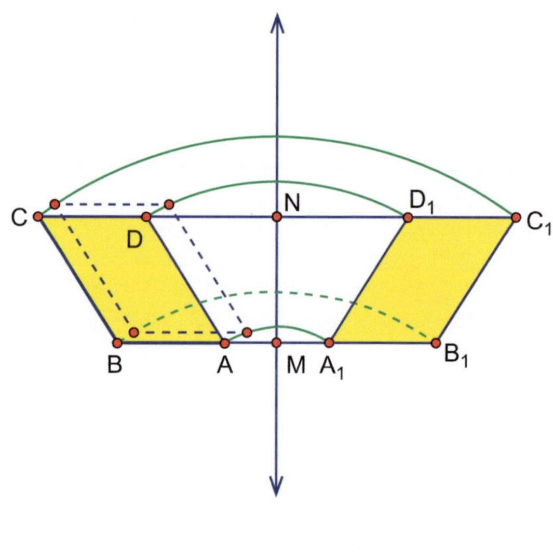

Fig. 4.24 Pot with a hole

You can imagine all kinds of shapes by rotating different plane figure around the same axis. For example, let's take a parallelogram $ABCD$ and draw a perpendicular through one of its vertices (i.e., vertex A) and rotate it around this axis (see the purple parallelogram in Fig. 4.23). We will get a three-dimensional figure which will look like something in the form of a bowl with conical cavity inside.

For another example, we can draw the axis of rotation not through one of the vertices of the parallelogram, but a little further, for example, using the MN axis again. Then we will form an object that can be modeled as a truncated cone with base radii $R = CN$ and $r = BM$, from which we take out a smaller truncated cone with radius $R_1 = DN$ and $r_1 = AM$ (Fig. 4.24). Since this figure has a through hole, it cannot model any container for storing the solution, because you can't pour anything into such a container. However, this problem can be easily fixed. Simply patch the bottom with a piece that fits.

It is interesting to calculate the volume of such a container using the properties of the volumes of known figures. We can rotate a half-circle (a semicircle) with radius $R = OK$ around axis MN and obtain a sphere of the same radius. If we rotate a piece of the curve DKA around the same axis MN, we will obtain a symmetric circular figure of the shape reminding us part of a hyperboloid (Fig. 4.25).

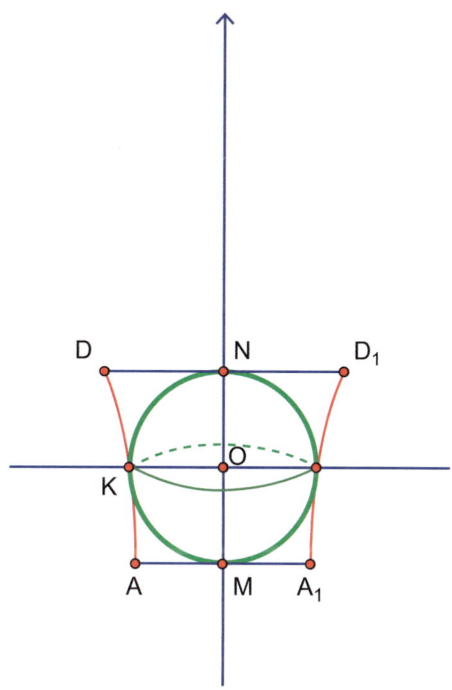

Fig. 4.25 Sphere and hyperboloid

4.4.1 Volumes and Surface Areas of Figures of Revolution

Let the function $y = f(x)$ be defined on the interval $[a, b]$ and continuous on (a, b). Determine the volume of the body obtained by rotating a curved trapezoid formed by function $y = f(x)$ around the segment $[a, b]$.

If a function is rotated around the X-axis, then we need to calculate the integral,

$$V = \pi \int_a^b f^2(x)dx. \tag{4.18}$$

If a function $x = g(y)$ is rotated over the Y-axis between points c and d, then the volume of the obtained figure is

$$V = \pi \int_c^d g^2(y)dy. \tag{4.19}$$

Let us recall another problem solved by Archimedes. He showed that *the volume of a paraboloid of revolution inscribed in a right cylinder is half the volume of the corresponding cylinder*. Consider such a paraboloid where the radius of the base is R and height H. It is clear that if we take parabola $y = f(x) = x^2$, that by rotating it (or actually the curve $x = g(y) = \sqrt{y}$) around the Y-axis between $y = 0$ and $y = H$, we obtain the paraboloid of the revolution whose greatest cross section is a circle of radius $\sqrt{H}$. Hence, it can be inscribed only in the cylinder with the base radius of $R = \sqrt{H}$. If $x = g(y) = \sqrt{y}$, then the volume of paraboloid will be equal to

$$V_{\text{paraboloid}} = \pi \int_0^H y dy = \frac{\pi H^2}{2}.$$

At height $y = H$, the corresponding horizontal coordinate of the figure is equal to $x = g(H) = \sqrt{H}$. This is equal to the radius of the base of the cylinder of height H in which the paraboloid is inscribed. The volume of the cylinder is

$$V_{\text{cylinder}} = \pi \left(\sqrt{H}\right)^2 H = \pi H^2.$$

If we compare the two formulas we see that the volume of the circumscribed cylinder is twice as much as the volume of the paraboloid.

Now it is time to start solving some interesting problems.

Problem 189 The figure is obtained by rotating a circle with center at point O and radius R relative to a straight line a lying in the plane of the circle and not intersecting it. Point E is the base of the perpendicular drawn from point O to line a, and the distance $OE = d$ $(d > R)$. Find the volume of the figure of revolution.

Solution Let's think about what happens when rotating a circle around a given axis. It takes a little imagination to understand that the resulting figure resembles a donut or bagel; in mathematics such a figure is called a torus. The larger d, the larger the hole in the torus we get. For example, a torus obtained by rotating a circle is shown in Fig. 4.26 using Maple 17.

Imagine that we have a torus made of elastic material and that we can change its shape. Let's cut it along the arc indicated in red in Fig. 4.26 and unwind the torus. You will get a cylindrical "sausage" with a radius of the original circle. By applying the Cavalieri principle to the torus and the cylinder, with a cut at level x along a horizontal plane perpendicular to line a, we can find the volume of the torus.

Let's draw a plane α perpendicular to the straight line a at a distance x from the point E $(0 \leq x < R)$. Then in the cross section of the torus by this plane we get a ring, the radius of the outer circle of which is equal to $R_1 = d + \sqrt{R^2 - x^2}$ (In Fig. 4.27, it is distance $AD = AC + CD$, where $AC = d$ and CD is the leg of the right triangle OCD, $CO = x$, $OD = R$), and the inner circle, the radius of which is equal to $R_2 = d - \sqrt{R^2 - x^2}$ represents the distance $AB = AC - BC$.

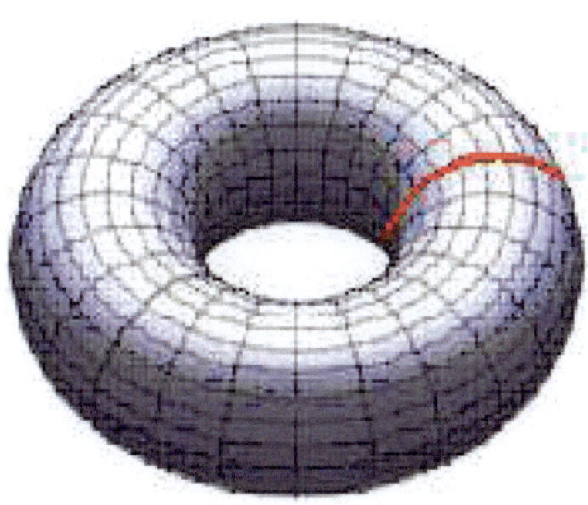

Fig. 4.26 Torus with one of the rotating circles shown in red

Sometime we do not want to eat an entire bagel but want to have only a slice of it. Many breakfast places cut bagels into two equal parts, which can help our imagination as well. Before putting cream cheese on it, let us look at the half of the bagel from the top. You will see the ring—a plane figure. If you cut a thin slice of one half of the bagel, you will get a new ring with a smaller hole inside. The projection of this slice gives an inner circle of radius $R_2 = AB$ (Fig. 4.28) and an outer circle of radius $R_1 = AD$. All you can eat of this slice is between the boundaries of the inner and outer circles. So, the area of the ring is $\pi R_1^2 - \pi R_2^2$, i.e.,

$$\text{Area}_{\text{ring}} = \pi\left(d + \sqrt{R^2 - x^2}\right)^2 - \pi\left(d - \sqrt{R^2 - x^2}\right)^2 = 4\pi d\sqrt{R^2 - x^2}. \tag{4.20}$$

Consider again the right cylinder shown on the right in Fig. 4.27, whose axis is the straight line going through the center of the circle O_1, radius of the base is $GO_1 = R$, and height of $2\pi d$. Let us show that the torus and cylinder satisfy the Cavalieri principle. In the cross section of the cylinder by the plane α, a rectangle with sides $2\sqrt{R^2 - x^2}$ and $2\pi d$ is obtained. Therefore the area of the rectangle is

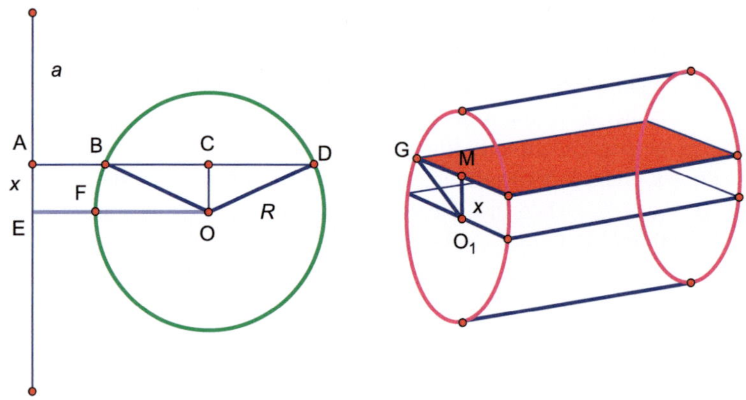

Fig. 4.27 Volume of a Torus

$$\text{Area}_{\text{rectangle}} = 4\pi d\sqrt{R^2 - x^2} \tag{4.21}$$

By Cavalieri's principle (Lemma 16), since the cross-sectional areas are equal, the volume of the torus is the volume of the cylinder. The volume of this cylinder is

$$V_{\text{cylinder}} = \pi R^2 \cdot 2\pi d = 2\pi^2 R^2 d$$

so the volume of the torus is

$$V_{\text{torus}} = 2\pi^2 R^2 d.$$

Now suppose that a plane curve is rotated about an axis external to the curve. The following theorems named in honor of two scientists, Pappus, a fourth-century Greek geometer known as Pappus of Alexandria, and Paul Guldin (1577–1843), a Swiss mathematician and astronomer, handle this situation.

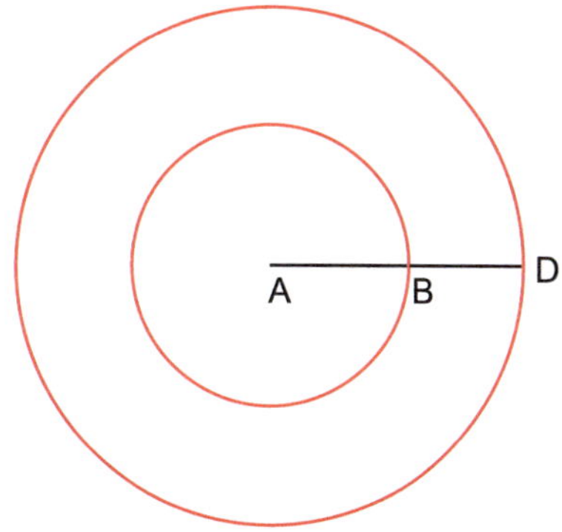

Fig. 4.28 Ring as a horizontal slice of a Torus

Theorem 45 (Pappus–Guldin Theorem 1) *"Each solid of revolution is equal to a right cylindrical figure whose base is the figure out of rotation of which the solid is produced and whose altitude is the circumference of a circle in which the centroid of the figure is revolved" ([25]).*

For a closed curve, the resulting volume of revolution is equal to the product of the plane area enclosed by the curve and the displacement of the centroid of the area.

Theorem 46 (Pappus–Guldin Theorem 2) *The surface area of a rotating body is equal to the product of the length of generatrix of the revolving figure and the circumference of a circle in which the centroid of the figure is revolved.*

To find the surface area of the torus in the previous problem, we could evaluate it as the lateral surface area of the corresponding cylinder,

$$A_{\text{torus}} = 2\pi R \cdot 2\pi d = 4\pi^2 Rd. \tag{4.22}$$

Next, let's look at some calculus techniques to find the volume and surface area of the torus obtained by rotating the circle described in the previous problem.

Problem 190 The figure is obtained by rotating a circle with center at point O and radius R relative to a straight line a lying in the plane of the circle and not intersecting it. Point E is the base of the perpendicular drawn from point O to line a, and the distance $OE = d$ $(d > R)$. Find the volume of the figure of revolution.

Solution

Method 1. We can find the volume of our torus using formula (4.19) as a difference of two figures of revolution. Let us place our circle in a standard coordinate system, shown in Fig. 4.29.

The equation of the circle with center at $(d, 0)$ and radius of R is

$$(x-d)^2 + y^2 = R^2,$$

from which we obtain the equation of the "red" curve as

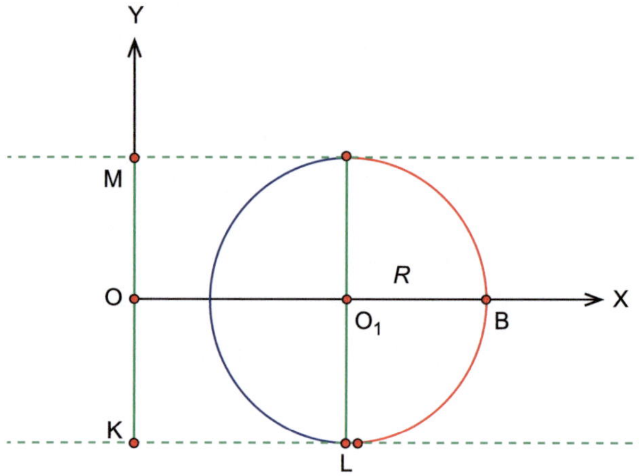

Fig. 4.29 Torus as a difference of two figures of revolution

$$x_2(y) = d + \sqrt{R^2 - y^2}, \quad -R \le y \le R$$

and the "blue" curve as

$$x_1(y) = d - \sqrt{R^2 - y^2}, \quad -R \le y \le R.$$

The circle is revolved around the Y-axis so the volume of the torus is evaluated as the difference of two figures of revolution obtained by $x_2(y)$ and $x_1(y)$,

$$V = \pi \int_{-R}^{R} \left(x_2^2(y) - x_1^2(y)\right) dy = \pi \int_{-R}^{R} \left(d + \sqrt{R^2 - y^2}\right)^2 dy - \pi \int_{-R}^{R} \left(d - \sqrt{R^2 - y^2}\right)^2 dy$$

Using the difference of squares formula, we can simplify our integral as

$$V = \pi \int_{-R}^{R} 2d \cdot 2\sqrt{R^2 - y^2} dy = 4\pi d \cdot \int_{-R}^{R} \sqrt{R^2 - y^2} dy. \qquad (4.23)$$

Fortunately the integral on the right of (4.23) can be calculated as the half area of the circle centered at the origin and of radius R. The expression simplifies as

$$V_{\text{torus}} = 4\pi d \cdot \frac{\pi R^2}{2} = 2\pi^2 R^2 d.$$

Method 2. Let's divide our circle into n thin rectangles with straight lines parallel to the X-axis. For example, one such rectangle $MNPK$ is shown in blue in Fig. 4.30. If we rotate this rectangle around the Y-axis, we get a ring, the thickness of which will depend on the number of rectangles dividing the circle. The thinner the rings are, the better the approximation. If the number of the rings $n \to \infty$, then in the limit, the sum of the volume of the rings approaches the volume of our torus.

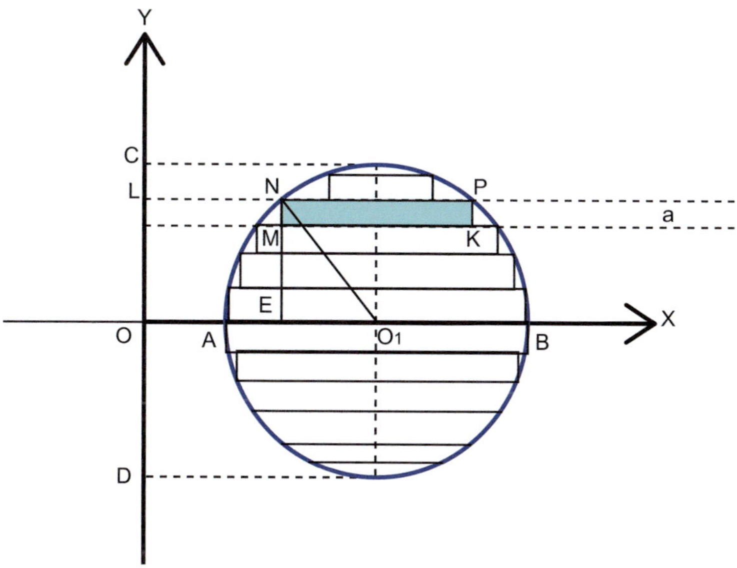

Fig. 4.30 A torus made up of thin rings

In our sketch points $OO_1 = d$, $AO_1 = R$, points N and P belonging to the selected rectangle are at level y and have the coordinates $N = (x_1, y)$ and $P = (x_2, y)$. Let us find the volume of the small ring obtained under rotation of the blue rectangle. The ring will have height dy and its volume is

$$dV = \pi x_2^2 dy - \pi x_1^2 dy = \pi (x_2^2 - x_1^2) dy$$

Here x_1 and x_2 can be found from the corresponding right triangles. From triangle NEO_1 we obtain that $EO_1 = \sqrt{NO_1^2 - NE^2} = \sqrt{R^2 - y^2}$, then

$$\begin{aligned} x_1 &= d - \sqrt{R^2 - y^2} \\ x_2 &= d + \sqrt{R^2 - y^2} \end{aligned} \tag{4.24}$$

The volume of the torus is

$$V = \lim \sum dV = \int_{-R}^{R} \pi \left(x_2^2 - x_1^2\right) dy = \pi \int_{-R}^{R} (x_2 + x_1)(x_2 - x_1)\, dy \quad (4.25)$$

Above we used the difference of squares formula again. Substituting formulas (4.24) into (4.25) we get,

$$V = \pi \int_{-R}^{R} 2d \cdot 2\sqrt{R^2 - y^2} dy = 4\pi d \int_{-R}^{R} \sqrt{R^2 - y^2} dy. \quad (4.26)$$

We could use trigonometric substitution. However, it is not difficult to notice that the integral $\int_{-R}^{R} \sqrt{R^2 - y^2} dy$ represents the area of a semicircle with center at the origin and radius R so its value is $\frac{1}{2}\pi R^2$. Substituting this into (4.26), we obtain

$$V_{\text{torus}} = 4\pi d \cdot \frac{1}{2}\pi R^2 = 2\pi^2 R^2 d. \quad (4.27)$$

Method 3. Now, cut the circle by n equidistant lines parallel to the Y-axis (Fig. 4.31). The volume of a small cylinder can be found as $dV = 2\pi x\,(y_2 - y_1)\, dx$, where $x = OP,\ \ y_1 = -\sqrt{R^2 - (x-d)^2},\ \ y_2 = \sqrt{R^2 - (x-d)^2}$, so

$$y_2 - y_1 = 2\sqrt{R^2 - (x-d)^2}$$

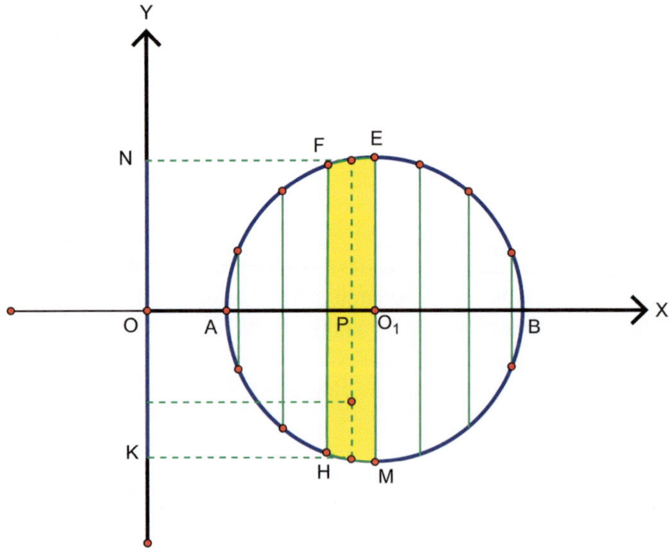

Fig. 4.31 A torus made up of thin cylinders

If we add all the cylinders and let $dx \to 0$, then the volume of the torus is

$$V = \int_{d-R}^{d+R} 2\pi x \,(y_2 - y_1)\, dx = 4\pi \int_{d-R}^{d+R} x\sqrt{R^2 - (x-d)^2}dx$$

Or, using the substitutions:

$$\begin{array}{l} u = x - d, \;\; x = u + d, \;\; du = dx \\ u_1 = -R, \;\; u_2 = R \end{array},$$

we can rewrite that integral as

$$V = 4\pi \int_{-R}^{R} (u+d)\sqrt{R^2-u^2}du = 4\pi \left(\int_{-R}^{R} u\sqrt{R^2-u^2}du + d\int_{-R}^{R} \sqrt{R^2-u^2}du \right)$$

The first integral inside parentheses is zero ($g(u) = u\sqrt{R^2-u^2}$ is an odd function). The second integral $\int_{-R}^{R} \sqrt{R^2-u^2}du = \frac{1}{2}\pi R^2$ because it is the area of the semicircle centered at the origin and radius R. Finally, the volume of the torus is

$$V = 4\pi d \cdot \frac{1}{2}\pi R^2 = 2\pi^2 d R^2.$$

Answer. $V_{\text{torus}} = 2\pi^2 R^2 d$.

In the next problem, we will rely on Calculus for the solution.

Problem 191 The figure is obtained by rotating a circle with center at point O and radius R relative to a straight line a lying in the plane of the circle and not intersecting it. Point E is the base of the perpendicular drawn from point O to line a and the distance $OE = d$ ($d > R$). Find the surface area of the figure of revolution.

Solution Let us evaluate the surface area the torus using Calculus. Consider again a circle centered at point $O_1 = (d, 0)$ (Fig. 4.32) where

$$(x-d)^2 + y^2 = R^2. \tag{4.28}$$

The torus is obtained by rotating this circle around the Y-axis. To find the surface area of the torus, we "cut" the circumferential curve into many small arcs ds, so that when rotated around the Y-axis, the arc forms a strip of area,

$$dA = 2\pi x ds, \tag{4.29}$$

where ds can be found as

$$ds = \sqrt{1 + \left(\frac{dy}{dx}\right)^2}\, dx. \tag{4.30}$$

We can find the area formed by rotating half a circle and then multiply the answer by two. From Eq. (4.28) we find y as

$$y = \sqrt{R^2 - (x - d)^2}. \tag{4.31}$$

Next, we will find its derivative

$$\frac{dy}{dx} = -\frac{(x - d)}{\sqrt{R^2 - (x - d)^2}}$$

and then substitute it in (4.30) and in (4.29):

$$dA = 2\pi x \sqrt{1 + \frac{(x - d)^2}{R^2 - (x - d)^2}} dx.$$

After simplifying the expression under the square root, we have a new formula,

$$dA = 2\pi R \cdot \frac{x}{\sqrt{R^2 - (x - d)^2}} dx.$$

Since the x coordinate changes between $x_1 = d - R$ and $x_2 = d + R$, we can find the area of half a torus,

$$A = 2\pi R \cdot \int_{d-R}^{d+R} \frac{x}{\sqrt{R^2 - (x - d)^2}} dx. \tag{4.32}$$

The integral in (4.32) can be found rewritten by the substitution:

$$\begin{array}{lll} u = x - d, & x = u + d & \\ du = dx & u_1 = x_1 - d = -R, & u_2 = x_2 - d = R \end{array}$$

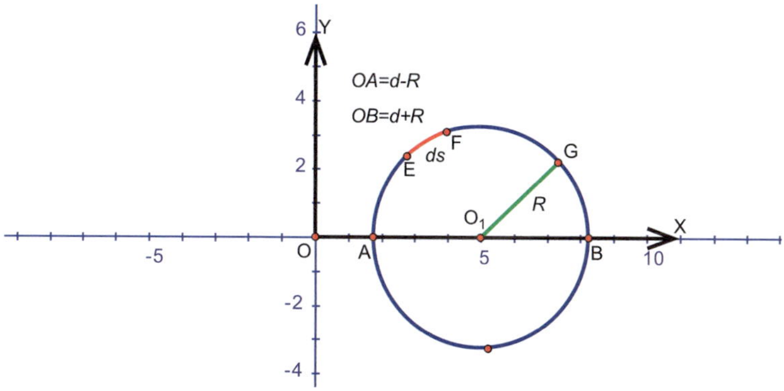

Fig. 4.32 Circle that makes a torus

as,

$$\begin{aligned} A &= 2\pi R \cdot \int_{-R}^{R} \frac{u+d}{\sqrt{R^2-u^2}} du \\ &= 2\pi R \cdot \left(\int_{-R}^{R} \frac{u}{\sqrt{R^2-u^2}} du + \int_{-R}^{R} \frac{d}{\sqrt{R^2-u^2}} du \right). \end{aligned} \tag{4.33}$$

The first integral in (4.33) is equal to zero because $f(u) = \frac{u}{\sqrt{R^2-u^2}}$ is an odd function so (4.33) can be simplified as

$$A = 2\pi Rd \int_{-R}^{R} \frac{du}{\sqrt{R^2-u^2}}. \tag{4.34}$$

The integral $\int_{-R}^{R} \frac{du}{\sqrt{R^2-u^2}}$ can be found with the trigonometric substitution,

$$\begin{aligned} &u = R\sin\theta, \quad du = R\cos\theta d\theta \\ &\theta_1 = -\tfrac{\pi}{2}, \quad \theta_2 = \tfrac{\pi}{2}. \end{aligned}$$

Under the integral we will have $\frac{du}{\sqrt{R^2-u^2}} = \frac{R\cos\theta d\theta}{R\cos\theta} = d\theta$ and hence the area in (4.34) is

$$A = 2\pi Rd \int_{-\frac{\pi}{2}}^{\frac{\pi}{2}} d\theta = 2\pi^2 Rd. \tag{4.35}$$

Multiplying (4.35) by 2 we will obtain the total surface area of the torus,

$$A_{\text{torus}} = 4\pi^2 Rd. \tag{4.36}$$

and see that the formulas (4.22) and (4.36) are identical, as expected.

Answer. $A_{\text{torus}} = 4\pi^2 Rd$

Problem 192 The quarter circle AOB rotates around its radius OB. At what distance OP from the center of the circle O must the line PD be drawn parallel to the line OA so that the ring formed by the rotating segment CD will have a given area πm? Find the maximum area of this ring when the distance OP changes from zero to $OB = a$.

Solution Let us sketch the problem (see Fig. 4.33).
Let $OP = x$, where $0 < x \leq a$. Then from the right triangle POD we have

$$PD^2 = a^2 - x^2.$$

Next, we can find the area of the ring as the difference of the areas of two circles,

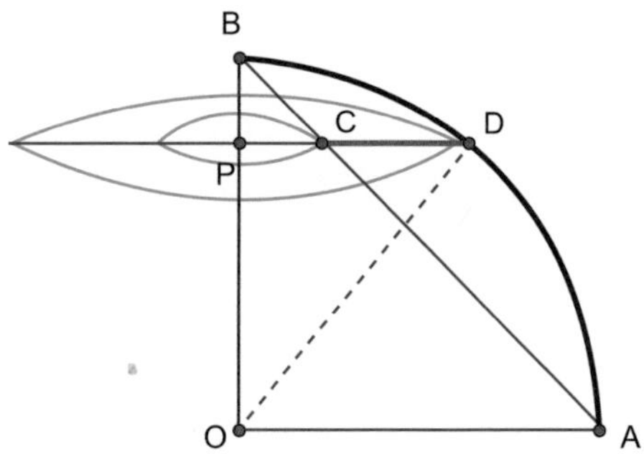

Fig. 4.33 Ring and quarter of the circle

$$\pi PD^2 - \pi PC^2 = \pi m \tag{4.37}$$

From (4.37) we understand that the larger m, the larger the ring area. Since the triangle BPC is an isosceles so $PC = PB = a - x$, and the formula simplifies as

$$a^2 - x^2 - (a - x)^2 = m,$$

which in turn can be rewritten as a quadratic equation in x,

$$2x^2 - 2ax + m = 0. \tag{4.38}$$

This equation has one ($x = \frac{a}{2}$) or two real roots ($x = \frac{a \pm \sqrt{a^2 - 2m}}{2}$) if the discriminant of (4.38) is nonnegative, i.e.,

$$a^2 - 2m \geq 0 \Leftrightarrow \boxed{0 < m \leq \frac{a^2}{2}}.$$

The maximum value of $m = \frac{a^2}{2}$ corresponds to $x = \frac{a}{2}$.

Answer. $x = \frac{a}{2}$. The maximum area $\frac{a^2}{2}$ is at $x = a/2$, the midpoint of OB.

Problem 193 Let the volume of the body obtained as a result of rotation of a right triangle around the hypotenuse be equal to V, and around the legs, respectively, V_1 and V_2. Prove that

$$\frac{1}{V^2} = \frac{1}{V_1^2} + \frac{1}{V_2^2}$$

Proof Let us sketch the cones obtained in all three cases and find the volume of the first figure when rotation is about the hypotenuse. Figure 4.34 represents the sum of two cones with radius r and heights $h_1 = OA$ and $h_2 = OB$, respectively. Therefore

$$V = \frac{1}{3}\pi r^2 h_1 + \frac{1}{3}\pi r^2 h_2 = \frac{1}{3}\pi r^2 (h_1 + h_2) = \frac{1}{3}\pi r^2 c.$$

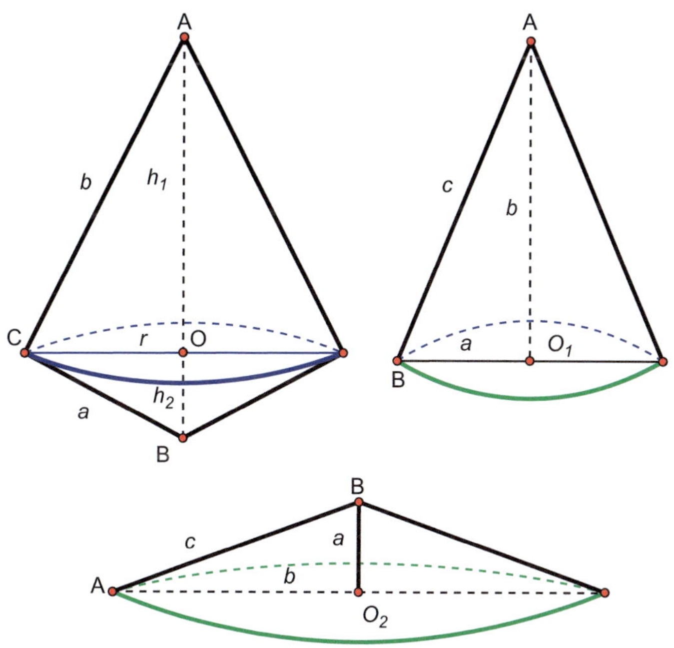

Fig. 4.34 Rotation of a right triangle about its hypotenuse and each of its legs

The radius of the base ($r = OC$) can be found from equating the areas of triangle ABC found it two different ways: $\frac{r \cdot c}{2} = \frac{a \cdot b}{2}$, so $r = \frac{ab}{c}$, where $a = CB$, $b = AC$, and $c = AB$. The volume of the figure can be found as, $V = \frac{1}{3}\pi r^2 c = \frac{1}{3}\pi \frac{a^2b^2}{c}$ so $\frac{1}{V^2} = \frac{9c^2}{\pi^2 a^4 b^4}$.

Next, we can find the volumes of the cones by rotation around each leg of the right triangle,

$$V_1 = \frac{1}{3}\pi b^2 a \Rightarrow \frac{1}{V_1^2} = \frac{9}{\pi^2 a^2 b^4}$$

and

$$V_2 = \frac{1}{3}\pi a^2 b \Rightarrow \frac{1}{V_2^2} = \frac{9}{\pi^2 b^2 a^4}.$$

Adding the last two formulas, we obtain the desired relationship,

$$\begin{aligned} \frac{1}{V_1^2} + \frac{1}{V_2^2} &= \frac{9}{\pi^2 a^2 b^4} + \frac{9}{\pi^2 b^2 a^4} = \frac{9}{\pi^2 a^2 b^2}\left(\frac{1}{b^2} + \frac{1}{a^2}\right) \\ &= \frac{9}{\pi^2 a^2 b^2}\left(\frac{a^2+b^2}{a^2 b^2}\right) = \frac{9c^2}{\pi^2 a^4 b^4} = \frac{1}{V^2} \end{aligned}$$

□

4.5 Circumscribed Spheres

4.5.1 Spheres Circumscribed About Polyhedra

A sphere is called **circumscribed** about a polyhedron if it passes through all the polyhedron's vertices. The polyhedron itself is said to be inscribed in a sphere. In Fig. 4.35 a sphere is circumscribed about a tetrahedron, triangular right prism, and a pentagonal right prism.

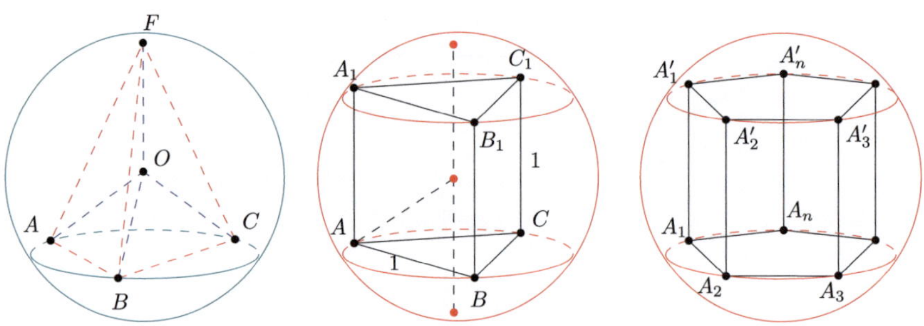

Fig. 4.35 Spheres circumscribed about different polyhedra

In order for a sphere to be circumscribed about a polyhedron, it is necessary but not sufficient that a circle can be circumscribed about any of its faces. The center of the circumscribed sphere (if any exists) lies in the planes perpendicular to the edges of the polyhedron passing through their midpoints (these planes are called perpendicular bisectors of the edges) and also it lies on the straight lines perpendicular to the faces of the polyhedron passing through the centers of the circles circumscribed about the corresponding faces.

Consider Fig. 4.36. We can circumscribe a circle about any face of the inscribed cuboctahedron. For example, the circle in which the square $ABCD$ is inscribed (shown in green). Its center O_1 that is the midpoint of diagonal BC is also shown. The center of the circumscribed sphere lies on the line perpendicular to a face that goes through the center of the circumscribed circle. Assume that we draw such perpendicular, $OO_1 \perp BC$. If this line is continued until it intersects the "North" and "South" poles of the sphere (points G and H, respectively), the center of the sphere O will be the midpoint of diameter GH and point O_1 will be collinear with them. Let R be the radius of the sphere and r be the radius of the circle circumscribed about square $ABCD$ in Fig. 4.36. The sketch is not to scale. However, we understand that $OB = OG = OH = R$ and that $O_1B = r$. If we connect points O and C, then the triangle BOC is an isosceles, in which the base $BC = 2r$ and $BO = OC = R$. We also can express diagonal BC in terms of the edge of the square or cuboctahedron, $AB = a$, $2r = a\sqrt{2}$. Hence, in this triangle we do not know only the angle $\angle BOO_1$.

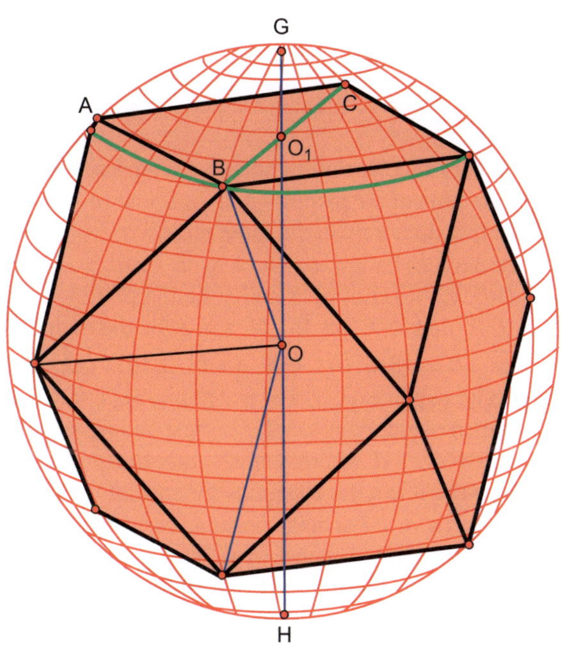

Fig. 4.36 A sphere circumscribed about a cuboctahedron

In this section you will be given an opportunity to find the radius of the sphere R in terms of the side of the cuboctahedron, a. However, you will see that we really do not need to find this angle and that the idea of our interesting solution is based on a construction of this polyhedron. Because all vertices of a polyhedron must lie on the circumscribed sphere, then the center of the circum-sphere is equidistant from any four vertices of this polyhedron that are not on the same plane (not coplanar) This equal distance is the radius of the circumscribed sphere.

You can circumscribe a sphere about any triangular pyramid. A sphere can be circumscribed about an n-gonal pyramid if and only if a circle can be circumscribed about its base. A sphere can be circumscribed about a prism if and only if the prism is right and if it is possible to circumscribe a circle about its base.

We can see in Fig. 4.36 that cuboctahedron has 8 triangular and 6 square faces and obviously a circle can be circumscribed about any of its faces. This circle is the intersection of the circumscribed sphere and the plane of the face of the polyhedron. Thus, the fact that circles can be circumscribed about the faces of a polyhedron is the necessary condition for a sphere to be circumscribed about a polyhedron. In some cases, this condition is also sufficient.

Problem 194 Find the coordinates of the center of a sphere and its radius if the sphere is circumscribed about a tetrahedron with vertices $(0, 0, 0)$, $(8, 0, 0)$, $(0, -2, 0)$ and $(0, 0, -6)$.

Solution

Method 1. Using coordinate methods of analytic geometry.

Recall that the center of the sphere circumscribed about the tetrahedron is the point of intersection of the planes—perpendicular bisectors of any three edges of the tetrahedron that do not lie in the same plane.

Let $ABCD$ be a tetrahedron with vertices $A(0, 0, 0)$, $B(8, 0, 0)$, $C(0, -2, 0)$ and $D(0, 0, -6)$. Let α, β and γ be the planes—perpendicular bisectors of edges AB, CB, and AD. Let M, N, and K be the midpoints of these edges. We can find vectors $\overrightarrow{AB} = (8, 0, 0)$, $\overrightarrow{CB} == (8, 2, 0)$, $\overrightarrow{DA} = (0, 0, 6)$. These vectors are perpendicular to planes α, β and γ at points $M(4, 0, 0)$, $N(4, -1, 0)$, and $K(0, 0, -3)$, that are the midpoints of the corresponding edges. The equation of a plane is $ax + by + cz = d$, where vector $\mathbf{n}(a, b, c)$ is normal to the plane. Using this equation we can write down the equations for each of the three planes.

Plane $\alpha : x = 4$. For plane β we can write that $8x + 2x + 0 \cdot z = d$. Since the plane goes through point N, then $8 \cdot 4 + 2(-1) = d$, and we find $d = 30$ and $8x + 2y = 30$ or $4x + y = 15$. Plane γ: $z = -3$. By solving the system of equations:

$$\begin{cases} x = 4 \\ 4x + y = 15 \\ z = -3 \end{cases}$$

we can obtain $x = 4$, $y = -1$, $z = -3$, the coordinates of the center of the circumscribed sphere $O(4, -1, -3)$. The radius of the sphere can be found as the magnitude of a vector connecting the center with any of the vertices of the tetrahedron. For example, we can use vertex A:

$$\overrightarrow{OA} = (4, -1, -3)$$
$$R = \left\| \overrightarrow{OA} \right\| = \sqrt{4^2 + 1^2 + 3^3} = \sqrt{26}$$

Method 2. Without using analytic geometry.

Let us sketch the tetrahedron in a coordinate system and then simply work with the tetrahedron after understanding its shape. Consider our sketch of the tetrahedron $ABCD$ in Fig. 4.37. Here $AB = 8$, $AC = 2$, and $AD = 6$. From the condition of the problem, three faces of the tetrahedron are the right triangles ($\triangle CAB$, $\triangle DAB$, and $\triangle DAC$). Then their hypotenuses are the diameters of the circles respectively circumscribed about them. In our sketch, points E, F, and G are the centers of the circles circumscribed about triangles $\triangle CAB$, $\triangle DAC$, and $\triangle DAB$, respectively.

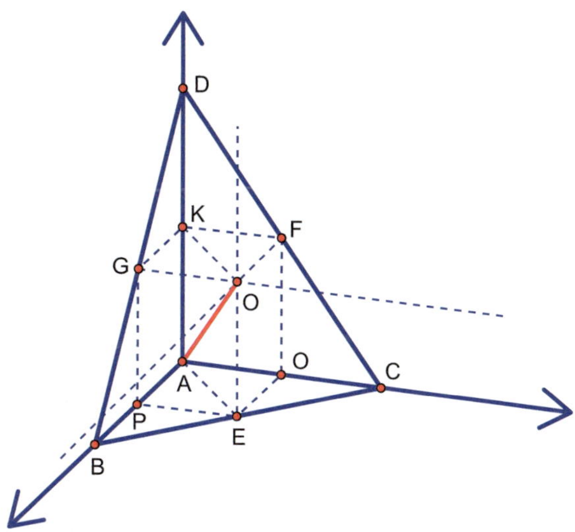

Fig. 4.37 A tetrahedron $ABCD$ inscribed in a sphere with radius $R = OA$

We need these points because we know that the center of the circumscribed sphere (i.e., point O) lies on the straight lines perpendicular to the faces of the tetrahedron passing through the centers of the circles circumscribed about the faces. Therefore, the center of the circum-sphere must lie on the line perpendicular to the base ABC that goes through point E. This line must be parallel to DA as shown in Fig. 4.37. Similarly, the circumcenter of the sphere must lie on the line that goes through point G perpendicular to face DAB (so parallel to AC). The point of intersection of three lines is the circumcenter, O, and $OA = R$ is its radius. Continuing our construction, we can build up a rectangular parallelepiped $PAQEGKFO$, where $KA = \frac{1}{2}AD$, $PA = \frac{1}{2}AB$, and $PE = \frac{1}{2}AC$. In Fig. 4.37 the center of the sphere is one of its vertices, outside of the face BDC and its main diagonal AO is the requested radius, R. It can be evaluated as,

$$AO^2 = PE^2 + AP^2 + KA^2, \quad R = \sqrt{4^2 + 1^2 + 3^2} = \sqrt{26}.$$

Answer. $O(4, -1, -3)$, $R = \sqrt{26}$

A sphere can be circumscribed about a right prism if and only if a circle can be circumscribed about the base of this prism. Its center will be point O, which is the midpoint of the segment connecting the centers of the circles circumscribed about the bases of the prism. The radius of the sphere R is calculated by the formula,

$$R = \sqrt{\left(\frac{h}{2}\right)^2 + r^2}, \tag{4.39}$$

where h is the height of the prism and r is the radius of the circle circumscribed about the base of the prism.

Problem 195 Find the radius of a sphere circumscribed about a regular hexagonal prism with all edges equal to 1.

Solution Let us find the radius of the sphere using our sketch in Fig. 4.38. Each base of the regular prism is a regular hexagon with side of 1. Denote the center of the bottom hexagon $ATSRQP$ by E. Then $AT = AI = AE = 1$. Let O be the center of the sphere. It must lie on the midpoint of the segments connecting the centers of both hexagons. Then $OE = \frac{1}{2}DE = \frac{1}{2}$. Next, from the right triangle OEA we have,

$$R = \sqrt{AE^2 + OE^2} = \sqrt{1 + \frac{1}{4}} = \frac{\sqrt{5}}{2}$$

or, we could just use formula (4.39). Here h is the length of the side edge $AI = 1$ and r is the radius of the circle inscribed in the hexagon, i.e., $r = AE = 1$.

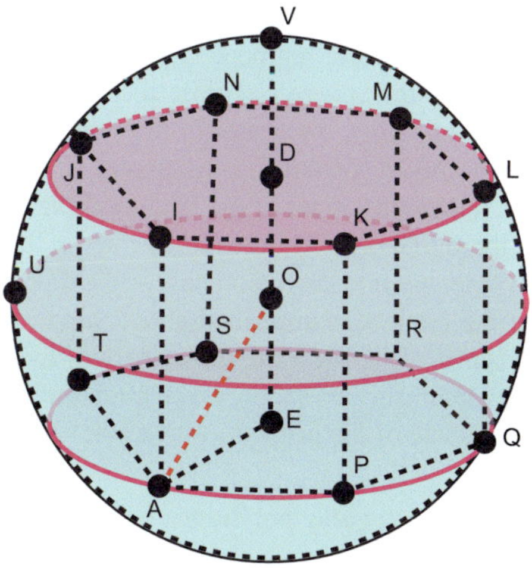

Fig. 4.38 Hexagonal prism inscribed in a sphere

Answer. $R = \frac{\sqrt{5}}{2}$

Problem 196 Is it possible to circumscribe a sphere about:

(a) a cuboid (right parallelepiped) with unequal edges a, b, and c?
(b) an oblique parallelepiped with axial section as rhombus?
(c) an oblique prism?

Solution

(a) A sphere can be circumscribed about any rectangular parallelepiped. Its radius is equal to half the main diagonal (2.33), $R = \frac{1}{2}\sqrt{a^2 + b^2 + c^2}$.
(b) No sphere can be circumscribed about an oblique parallelepiped with an axial section as a rhombus.
(c) No sphere can be circumscribed about an oblique prism.

Problem 197 Can the center of a sphere circumscribed about a prism be outside the prism? Justify your answer.

Solution Yes, it is possible. The prism must be a right triangular prism and its bases must be obtuse triangles. Consider such a triangular right prism. We can circumscribe circles about any triangle, the base of a prism. Let O_1 and O_2 be the centers of the circles. Then the segment O_1O_2 is the height of the prism and its midpoint is the center of the sphere circumscribed about this prism. Depending on the location of segment O_1O_2, the center of the sphere will be either inside the prism, on the face of the prism, or outside the prism. If a triangle is acute (all angles are less than 90°), then the center of the circum-circle will be inside the triangle. If the triangle is right, by Thales's Theorem the center of the circum-circle will be the midpoint of the hypotenuse and the hypotenuse is the diameter. If a triangle is obtuse, then the center of the circle circumscribed about this triangle will lie always outside the triangle. Assume O_1 and O_2 are the centers of the bottom and top bases, then the center of the circumscribed sphere is the midpoint of segment O_1O_2. This entire segment will be outside of the prism as well as the center of the circum-sphere.

All five regular polyhedra can be inscribed in a sphere so that all their vertices are on the circumscribed sphere. Furthermore:

- For each regular polyhedron, all the dihedral angles are equal
- There is a transformation (rotation) that maps a figure onto itself, with each vertex being mapped onto its neighboring vertex.

Let us consider a regular tetrahedron shown in Fig. 2.10. Rotation around axis DO_1 (this axis passes through the vertex and the center of the opposite face ABC) in 120° angle will not change the figure. Each vertex under this rotation will move to the other vertex. There are four vertices in a regular tetrahedron, hence, there are four such lines of symmetry. There are three pairs of skew edges (opposite) in a regular tetrahedron. For example, using Fig. 2.10, these pairs are $AD - BC$, $DC - AB$, $DB - AC$. In the same figure points E, F, and H are the midpoint of edges BC, AB, and AC. Let us draw perpendicular bisectors DFC, AED and DBH. Each of these three planes is the plane of reflection. Reflecting the tetrahedron $ABCD$ with respect to each of these planes will not change the figure (its shape).

For example, if we reflect $ABCD$ with respect plane of symmetry DAE, points B and C will be the reflecting images of one another, but vertices A and D will not move. Therefore a regular tetrahedron has three planes of reflection. A regular triangular pyramid also has the same three reflection planes. From chemistry we know that methane molecule looks like a regular tetrahedron and a molecule of ammonia looks like a regular triangular pyramid. These molecules have three planes of reflection and belong to a so-called group of symmetry C_{3v}.

Using a sphere of radius 1, we can construct a regular octahedron inscribed in a sphere. Draw three mutually perpendicular diameters of the sphere and connect the ends of the diameters with line segments. We will obtain eight equilateral triangles—the faces of the octahedron—where each edge has length of $\sqrt{2}$.

Problem 198 Find the radius of a sphere circumscribed about a unit icosahedron.

Solution If an icosahedron is inscribed in a sphere, then all its vertices must lie on the sphere. If the icosahedron like one in Fig. 4.39 is inscribed in a sphere of radius R, then AC must be its diameter. For the rectangle $ABCD$, $AB = CD = 1$ so that BC and AD are the diagonals of regular pentagon with unit sides. The diagonal BC of the regular pentagon $FKBNC$ can be calculated in the same way as we did in Problem 116 of Chap. 2. The length of the diagonal is determined by the formula (2.67).

If the side of pentagon is equal to 1 ($a = 1$), then

$$BC = AD = \frac{1 + \sqrt{5}}{2}.$$

Once we have found the diagonal of the regular pentagon, we can find the diameter of the sphere circumscribed about the icosahedron. Applying the Pythagorean theorem to the right triangle ABC in Fig. 4.39, we have

$$AC^2 = AB^2 + BC^2$$

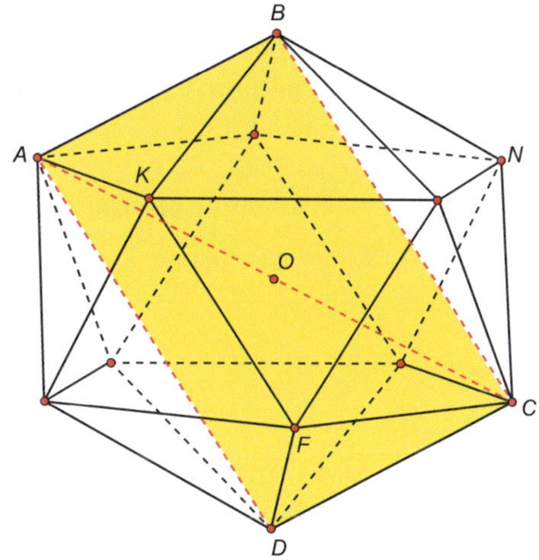

Fig. 4.39 If an icosahedron is inscribed in a sphere

$$AC^2 = 1^2 + \left(\frac{1+\sqrt{5}}{2}\right)^2 = \frac{4+1+2\sqrt{5}+5}{4} = \frac{10+2\sqrt{5}}{4},$$

so

$$\frac{AC^2}{4} = \frac{10+2\sqrt{5}}{16}.$$

Since the desired radius is equal to half of this diagonal, i.e., $R = AO = OC = \frac{AC}{2}$, it can be found as the square root from the expression above,

$$R = \frac{\sqrt{10+2\sqrt{5}}}{4}. \tag{4.40}$$

The radius of the sphere circumscribed about the icosahedron with side a,

$$R = \frac{a\sqrt{10+2\sqrt{5}}}{4}. \tag{4.41}$$

Answer. $R = \frac{\sqrt{10+2\sqrt{5}}}{4}$.

Problem 199 Find the radius of a sphere circumscribed about a unit cuboctahedron.

Solution A cuboctahedron is a semi-regular polyhedron whose faces consist of different types of regular polygons—triangles and squares. We can visualize a cuboctahedron using Maple (Fig. 4.40) where the faces are six green squares (as a cube) and eight red triangles (as an octahedron).

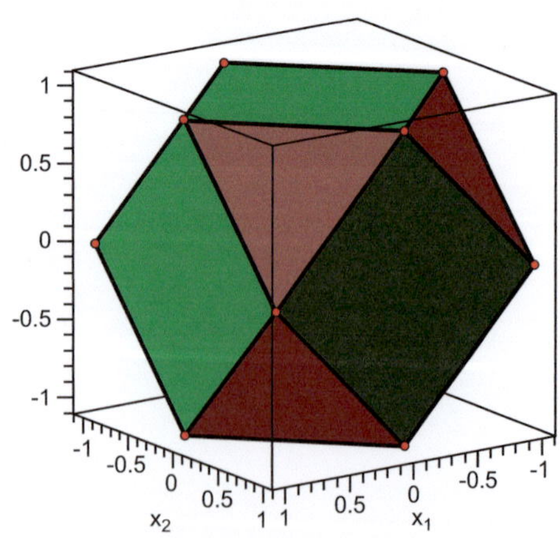

Fig. 4.40 Cuboctahedron constructed using Maple software

Recall that a cuboctahedron is obtained from a cube by cutting off regular triangular pyramids with vertices at the cube and side edges equal to half the edge of the cube (Chap. 2). It is interesting that constructing a cuboctahedron from a cube will help us find the desired radius of the sphere circumscribed about this polyhedron (Fig. 4.41).

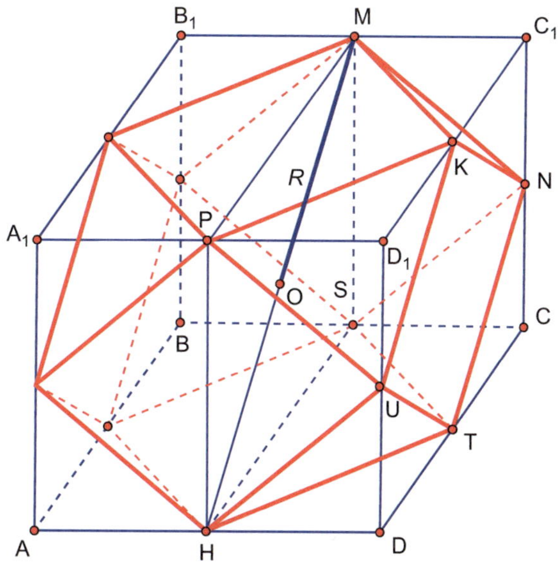

Fig. 4.41 Cuboctahedron constructed from the cube by removing 8 pyramids

Consider just one of the regular triangular pyramids, at vertex D (pyramid $DHUT$). Denote the side of its base triangle by $a = HT = HU = UT$, that is also the side of the cuboctahedron. It is easy to find the side of the cube, b, since $HT = \frac{b\sqrt{2}}{2} = a$, so $b = a\sqrt{2}$. If the edge of the cuboctahedron is $a = 1$, then the edge of the corresponding cube is $b = \sqrt{2}$. The diameter of the circumscribed sphere is equal to the distance HM shown in our sketch. Let us draw an auxiliary plane $HPMS$ that crosses the original cube. The radius of the sphere $R = OM$ is equal to the distance from the center of the cube O to the midpoint of its edge, M, i.e., is equal to

$$R = \frac{1}{2}b\sqrt{2} = \frac{b}{\sqrt{2}} = \frac{a\sqrt{2}}{\sqrt{2}} = a.$$

If $a = 1$, then the radius of the sphere is also 1.

Answer. $R = a = 1$.

4.5.2 Spheres Circumscribed About Cylinders and Cones

A sphere is said to be circumscribed about a cylinder or a cylinder is said to be inscribed in a sphere if the circles of the bases of the cylinder lie on the sphere. It is always possible to circumscribe a sphere about a right cylinder. For example, in Fig. 4.42 a sphere is circumscribed about a right cylinder.

Fig. 4.42 A sphere circumscribed about a right cylinder

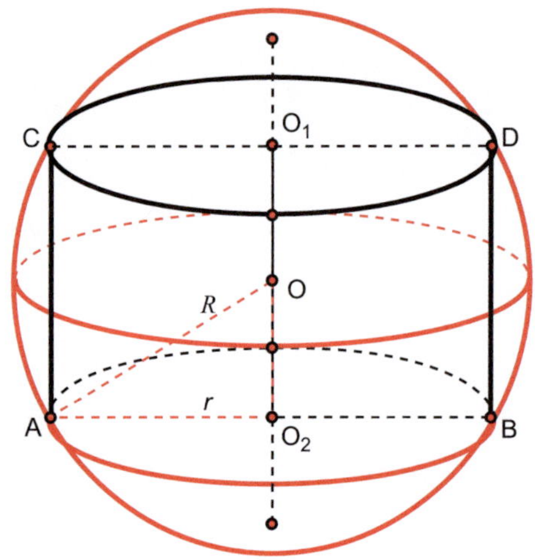

In any inscribed right cylinder, the center of the circumscribed sphere, (point O in Fig. 4.42) will be the midpoint of the segment connecting the centers of the bases O_1 and O_2 of the cylinder. The sphere's radius R is calculated by the formula,

$$R = \sqrt{\left(\frac{h}{2}\right)^2 + r^2} \tag{4.42}$$

where h is the height of the right cylinder and r radius of the base circle.

Let us use Fig. 4.42 and recall how the mathematician from Problem 187 of this chapter ordered a gold ring. Imagine that we have a golden ball, from which the cylinder previously inscribed in this ball has been removed and also two sectors located above and below the cylinder have been cut off from the ball. The gold ring has been "created."

An oblique cylinder can be "inscribed" in a sphere under certain conditions. Thus, in Fig. 4.43 a sphere is not really "circumscribed" about an oblique cylinder, but contains it. Consider an oblique cylinder with the base radius r and height h placed in a sphere. To find the minimal radius R of the sphere, we need to know the angle, α, that the generatrix of the cylinder makes with the diameter of the base of the cylinder. Now consider the cross section of the sphere that passes through the axial plane of the cylinder as shown in Fig. 4.44. Because AC is the diameter of the sphere, then

$AK \perp CK$ and this means that $CK = h$ (the height of the cylinder). From the right triangle CKA and applying the Pythagorean Theorem, we obtain $AC^2 = CK^2 + AK^2$ or since $AC = 2R$, $4R^2 = h^2 + AK^2$ where segment $AK = 2r + BK$. BK can be found using the right triangle CKB as $BK = CK \cot \alpha = h \cot \alpha$. Thus, we have $4R^2 = h^2 + (2r + h \cot \alpha)^2$ which can be rewritten as,

$$R = \frac{1}{2}\sqrt{h^2 + (2r + h \cot \alpha)^2} \tag{4.43}$$

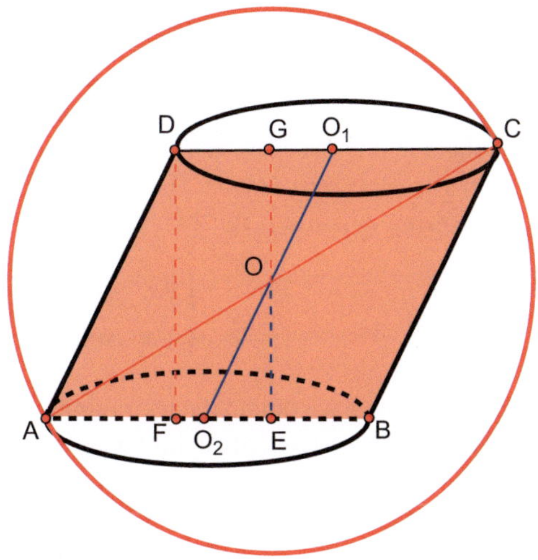

Fig. 4.43 A sphere circumscribed about oblique cylinder

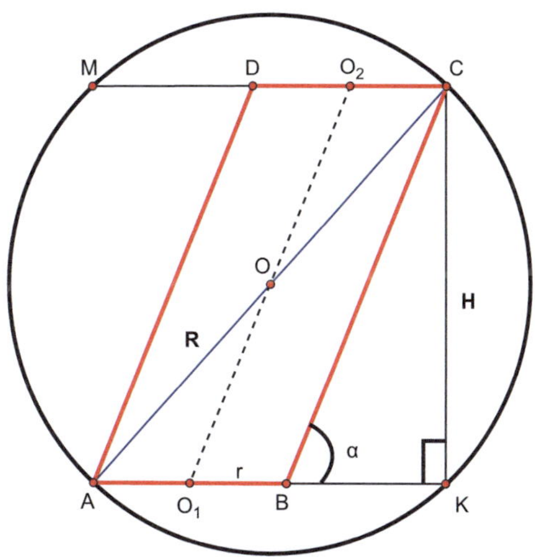

Fig. 4.44 Finding the radius of the sphere circumscribed about an oblique cylinder

In formula (4.43) the angle α satisfies the inequality, $0 < \alpha \leq 90^\circ$. If the angle is $\alpha = 90^\circ$, then the cylinder becomes a straight cylinder and the formula (4.43) becomes the formula (4.42).

Problem 200 Can an oblique cylinder with a base radius of 1 and generatrix length of 2 "fit" into a sphere of radius $R = \sqrt{3}$?

Solution

Method 1.
Assume that an oblique cylinder with the given characteristics is shown in Fig. 4.44. Denote the side of the cylinder by l, the height of the cylinder by h, and the radii of the sphere and cylinder by R and r, respectively. Also denote the length of segment $BK = x = h \cot \alpha = l \cos \alpha$. If an oblique cylinder can be placed in a sphere, then the following conditions must be valid:

$$\begin{cases} 4R^2 = h^2 + (2r + x)^2 \\ x^2 = l^2 - h^2. \end{cases}$$

From this system we obtain the restrictions similar to (4.43) on the radius of a sphere in case when the side of the cylinder and its inclination angle, α is given

$$4R^2 = l^2 + 4r^2 + 4rl \cos \alpha. \tag{4.44}$$

The last condition (4.44) can be rewritten as $12 = 4 + 4 + 8 \cos \alpha$, or as $\cos \alpha = \frac{1}{2}$, $\alpha = 60^\circ$. Therefore, the oblique cylinder of radius $r = 1$ with side of 2 can be inscribed in a sphere of radius $R = \sqrt{3}$.

Method 2.
Let's use Fig. 4.44 again and draw the second diagonal DB of the parallelogram $ABCD$. If the generatrix is equal to 2, which is equal to the diameter of the oblique cylinder, then the parallelogram $ABCD$ in Fig. 4.44 will be a rhombus. Then the diagonals AC and BD are perpendicular. In the right triangle COB, $OC = \sqrt{3}, CB = 2$, then $OB = 1$, $\angle CBO = 60^\circ$ and $\angle OCB = 30^\circ$. Applying the Pythagorean theorem to triangles AKC and CKB, we find $BK = 1$. $\Delta OCB = \Delta KCB$ and then triangle OCK is an equilateral triangle with side $\sqrt{3}$. In this case, $OM = OK = \sqrt{3}$ and the rectangle $AMCK$ is inscribed in a circle of radius $R = OK = \sqrt{3}$, a great circle of a sphere of the same radius. Also, $\angle CBK = 60^\circ$ and the formula (4.43) is valid. Therefore such a cylinder can be inscribed in a given sphere.

Answer: Yes. This oblique cylinder can be inscribed in the sphere of radius $\sqrt{3}$.

Sphere is circumscribed about a cone if the apex of the cone and the circumference of the base of the cone lie on the sphere. A sphere can be always circumscribed about a cone; its radius is equal to the radius of the circle circumscribed about an axial cross section of the cone. Please, look at the sketch in Fig. 4.45 for a right cone. Similar rules are applicable to an oblique cone.

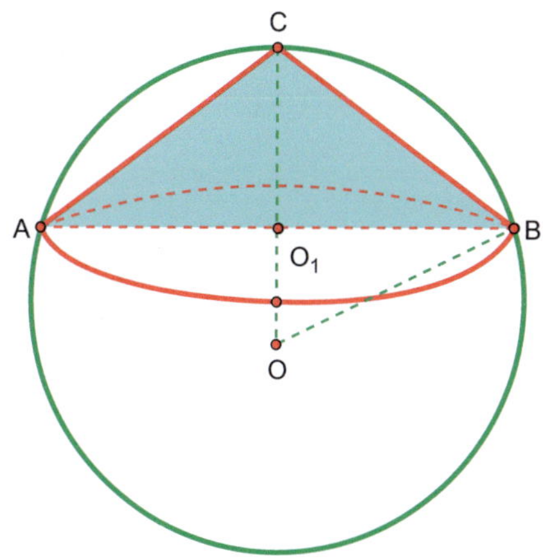

Fig. 4.45 Right cone inscribed in a sphere

When solving problems about inscribed cones, it is nice to remember the formula for the radius R of a circle circumscribed about a triangle with sides a, b and c:

$$R = \frac{abc}{4S}, \tag{4.45}$$

where S is the area of the inscribed triangle (shaded area in Fig. 4.45). In our sketch, it would be the area of triangle ABC, in which $a = AB$, $b = AC$, and $c = AB$ is the diameter of the base of the cone.

4.6 Inscribed Spheres

4.6.1 Spheres Inscribed in Polyhedra

A sphere is inscribed in a polyhedron if it is tangent to all planes containing the faces of the polyhedron at the interior points of the faces (Fig. 4.46). A sphere can always be inscribed in a triangular pyramid. A sphere can be inscribed in an n-gonal pyramid if and only if the bisector planes of all dihedral angles of the pyramid have a common point. In order to check whether a sphere can be inscribed in a pyramid, it is enough to check whether there is a point of intersection of the bisector planes of all internal dihedral angles at the base of the pyramid. If such a point exists, then it will be equidistant both from the base of the pyramid and from each of the lateral (side) faces.

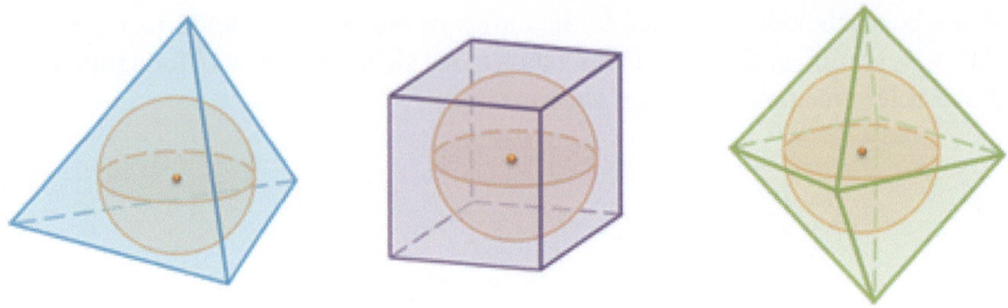

Fig. 4.46 Spheres inscribed in polyhedra

Often, when solving problems of planimetry, the relation is used that the area of a triangle is equal to its semi-perimeter multiplied by the radius of the inscribed circle. (You can, for example, find it in my book [2], p. 125.) Using Fig. 4.47, we have the obvious equality,

$$A = \frac{ar}{2} + \frac{br}{2} + \frac{cr}{2} = \frac{1}{2}r\,(a + b + c) = p \cdot r.$$

It is easy to verify that this formula is also valid for any polygon, in which a circle is inscribed; in this case, we must replace the area in the formula with the area of the polygon, and the semi-perimeter with the semi-perimeter of the polygon. Interestingly, a similar formula can be obtained in stereometry.

Fig. 4.47 Circle inscribed in a triangle

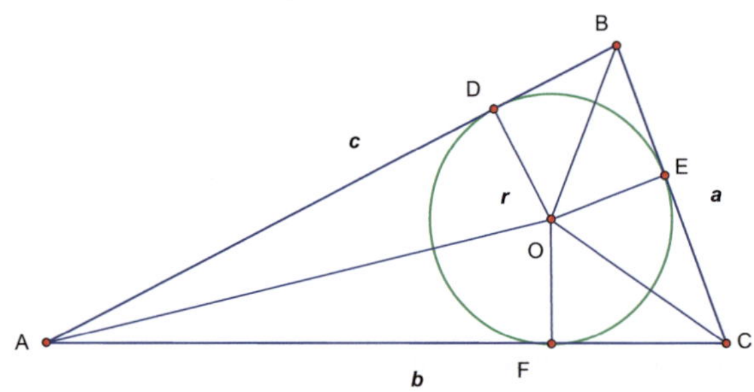

Lemma 17 *If a sphere of radius r is inscribed in a pyramid with n faces, then its volume can be calculated by the formula*

$$V_{\text{pyramid}} = \frac{1}{3}S \cdot r, \quad S = \sum_{i=1}^{n} A_i, \tag{4.46}$$

where A_i is the area of each face of the pyramid.

Proof Similar to a circle inscribed into a triangle, the volume of a pyramid is equal to one-third of the product of the total surface area of the pyramid, S, and the radius of the inscribed sphere, r. The proof of this formula is also similar to the plane case.

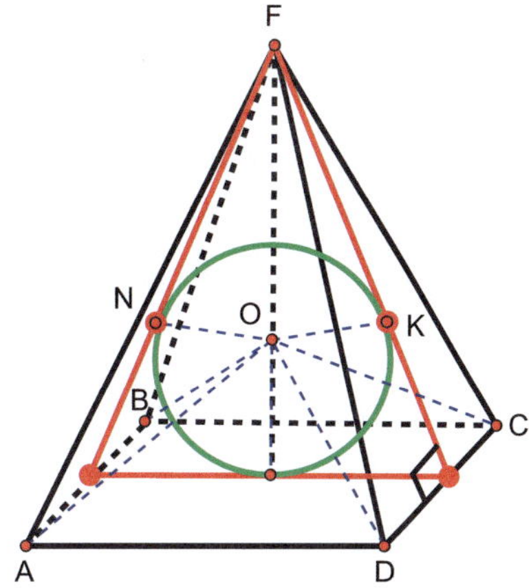

Fig. 4.48 Sphere inscribed in a regular square pyramid

Let us connect the center of the sphere with all the vertices of the pyramid, after which the pyramid can be imagined as divided into several small pyramids (Fig. 4.48). Note further that the radius of the inscribed sphere in this case will be the height of each small pyramid, and therefore we can calculate the volume of this pyramid as the sum of the volumes of these small pyramids. Here, a sphere of radius $r = OK$ is inscribed into a regular square pyramid $ABCDF$. If we connect the center of the sphere O with all five vertices of the original pyramid, we divide it into five smaller pyramids ($OABF$, $OFBC$, $OADF$, $OADF$, and $OABCD$), each with vertex at O and with the height r, that is the radius of the inscribed sphere. □

Problem 201 Given a regular square pyramid with base area, A, and $60°$ angles between the base and lateral faces, find the radius of the sphere inscribed in the pyramid.

Solution Let's use the pyramid $ABCDF$ as shown in Fig. 4.48. Assume that $FK \cap DC = P$, where P (not shown in the sketch) is the midpoint of DC. If we project any of the four lateral faces onto the base plane, the area of its projection will be $1/2$ of the area of the square of the base, A. Next, let us find the area of the lateral face A_1 using Theorem 43 from Chap. 3 as $A_1 \cos 60° = \frac{1}{2} \cdot A$, from which $A_1 = A$.

Therefore, all faces of the square pyramid are equal to A, so their sum is equal to $5A$. Substituting this in formula (4.46) we obtain,

$$V = \frac{1}{3} \cdot 5A \cdot r. \tag{4.47}$$

On the other hand, the volume of the pyramid can be found differently. Consider the right triangle FMP in Fig. 4.48. In this triangle $\angle FPM = 60°$, side of the base $AD = \sqrt{A}$, then $MP = \frac{\sqrt{A}}{2}$ and then the height of the pyramid $h = FM = MP \tan 60° = \frac{\sqrt{A}}{2\sqrt{3}}$. Hence,

$$V = \frac{1}{3} A \cdot \frac{\sqrt{A}}{2\sqrt{3}} \tag{4.48}$$

By equating the right-hand sides of (4.47) and (4.48), we have $5A \cdot r = A \cdot \frac{\sqrt{A}}{2\sqrt{3}}$ from which we obtain the radius of the inscribed sphere, $r = \frac{A}{10\sqrt{3}}$.

Answer. $r = \frac{A}{10\sqrt{3}}$

Problem 202 Find the radius of a sphere inscribed in a regular triangular pyramid with base side of a. All dihedral angles are 60°.

Solution First, we will sketch the pyramid with the ball inside. Second, we will use the fact that the center of the inscribed sphere O is the point of intersection of the plane bisectors of the dihedral angles at the base of the pyramid. Moreover, it is located on the height of the pyramid, SE. Since OD is the bisector of angle ADS, then $\angle ODA = 30°$. It is clear that one of the big circles of the ball is inscribed into plane angle ADS.

In Fig. 4.49, segment $OE = OF = r$, radius of the ball. From the right triangle OED we can find this radius as, $OE = ED \tan 30°$. Point E of triangle ABC is the point of intersection of the medians, so it divides AD in the ratio $2 \div 1$ starting from the vertex A. Hence, segment ED is $1/3$ of AD, then $AD = \frac{a\sqrt{3}}{2}$, then $ED = \frac{a\sqrt{3}}{6}$. Thus, the radius of the sphere is $r = OE = \frac{a\sqrt{3}}{6}\frac{\sqrt{3}}{3} = \frac{a}{6}$.

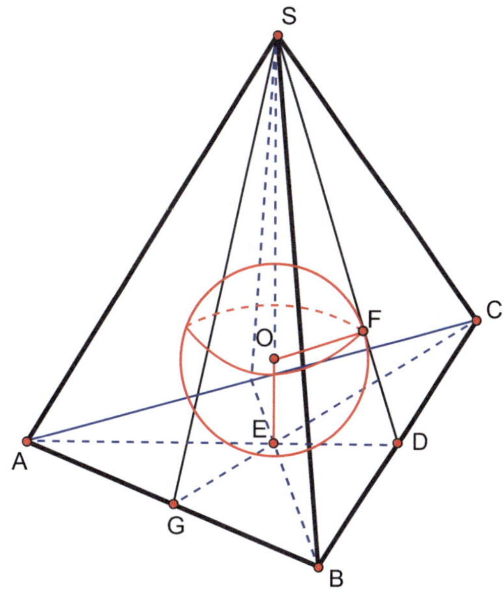

Fig. 4.49 Sphere inscribed in a triangular pyramid

Answer. $r = \frac{a}{6}$.

A sphere can be inscribed in a right prism if and only if the height of the prism is equal to the diameter of the circle inscribed in the base. A sphere can be inscribed in a prism if and only if its height is equal to the diameter of a circle that can be inscribed in its orthogonal cross section. A sphere inscribed in a cube is shown in Fig. 4.50.

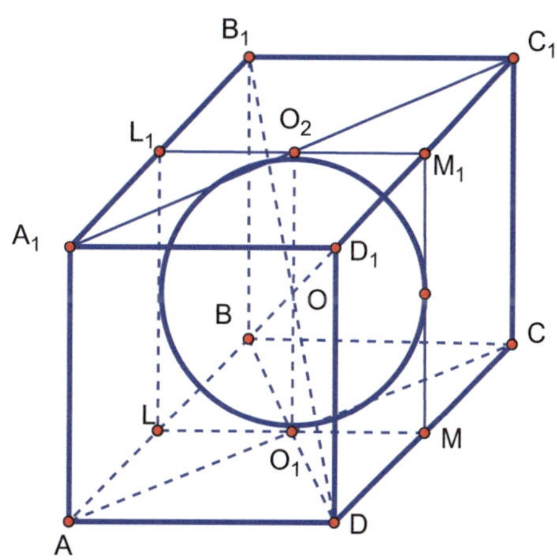

Fig. 4.50 Sphere inscribed in a cube

Problem 203 Find the full surface area of a right triangular prism circumscribed about a sphere if the base area of the prism is equal to B.

Solution Since a sphere is inscribed in the prism, the radius of its great circle must be equal to the radius of the circle inscribed in the polygon of the base of the prism. Suppose that the triangle of the base has sides a, b and c. Then the area of the base of the prism B is equal to

$$B = \frac{1}{2}Pr = \frac{1}{2}r \cdot (a + b + c),$$

where P is the perimeter of the base of the prism and r is the radius of the ball. Additionally, if H is the height of the prism, then it is equal to the diameter of the ball, $H = 2r$. The total surface area of the right triangular prism is equal to the sum of the areas of two triangular bases and of three rectangular lateral faces,

$$\begin{aligned} S &= 2B + H \cdot a + H \cdot b + H \cdot c = 2B + H(a + b + c) \\ &= 2B + 2r \cdot \tfrac{2B}{r} = 6B. \end{aligned}$$

Answer. $S = 6B$.

4.6.2 *Spheres Inscribed in a Cylinder or Cone*

4.6.2.1 A Sphere Inscribed in a Cylinder

A sphere is said to be inscribed in a cylinder if it is tangent to the bases of the cylinder and its cylindrical lateral surface. If a sphere is inscribed in a cylinder, then it must be tangent to both its bases and each generatrix of the cylinder. A sphere can be inscribed in a cylinder if the height of the cylinder is equal to the diameter of its base. Its center will be the point O, which is the midpoint of the segment connecting the centers of the bases O_1 and O_2 of the cylinder. The radius of the sphere R will be equal to the radius of the circle of the base of the cylinder.

It was precisely this case with a sphere inscribed in a cylinder that Archimedes considered when he proved and derived formulas for the volume and surface area of a sphere. Figure 4.51 shows the axial cross section of the cylinder, rectangle AA_1B_1B. This rectangle must be circumscribed about the great circle of the sphere. Therefore this rectangle must be a square with sides equal to the diameter of the inscribed sphere,

$$2R = O_1O_2 = AA_1 = B_1B = AB = A_1B_1.$$

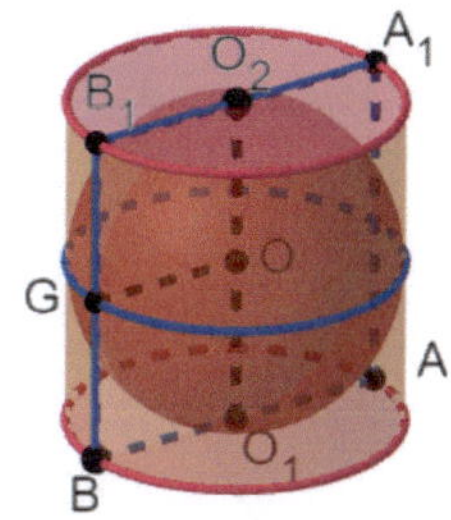

Fig. 4.51 A sphere inscribed in a cylinder

Problem 204 Is it possible to inscribe a sphere in a cylinder whose axial section is a rhombus?

Solution No, it is impossible. Only if a rhombus is a square.

A sphere cannot be inscribed in an oblique cylinder, but a sphere can be placed inside an oblique cylinder! Imagine a ball of radius R, which in real life you want to place in a bucket shaped like an inclined cylinder. The "bucket" can be a glass container, so the colorful ball can be seen from the outside. The ball does not have to be tangent to all surfaces of the cylinder. It just needs to fit inside the cylinder.

Problem 205 Find the condition under which a ball of radius R can be placed inside an oblique circular cylinder with base radius r, the lateral edge of which has length a and forms an angle α with the diameter of the base.

Solution Consider the axial sections of two different oblique cylinders shown in Fig. 4.52. Assume that they both have the same base radius $AO_1 = r$ and that the ball that we are trying to place inside each cylinder has radius R. For both cylinders we will introduce angle $\alpha = \angle DAB$ that their generatrix $AD = a$ forms with the base of the cylinder. If the ball can be placed inside either cylinder, its center will be the midpoint of the axis O_1O_2. Let us focus on the taller cylinder on the right of Fig. 4.52. The height of this cylinder $h = DE$ is $h = AD \cdot \sin\alpha = a \cdot \sin\alpha$.

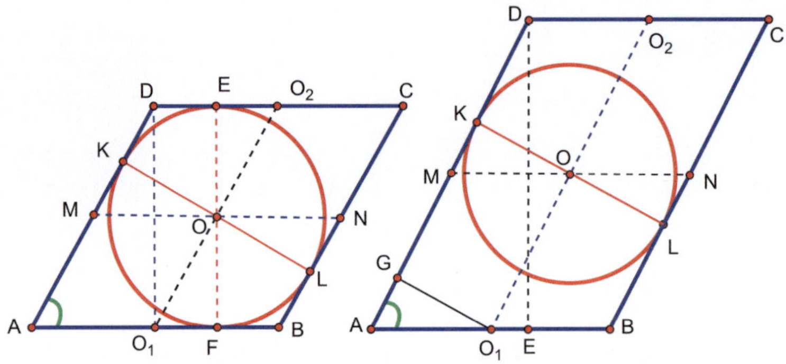

Fig. 4.52 A ball placed in two different oblique cylinders

The center of the ball that can be placed in such a cylinder must be the midpoint of O_1O_2, and the radius of the ball must be either less than or equal to half of the height of the oblique cylinder or it must be less than or equal to the radius of the base of the oblique cylinder multiplied to the sine of angle α. Otherwise, the ball will never fit the cylinder.

We can see that in both axial sections we marked segment KL as the diameter of the ball. We emphasize that $KL \perp O_1O_2$. Let us drop a perpendicular from the center of the base of the second cylinder O_1 to the segment AD. Triangle O_1GA is a right triangle and $GO_1 = r \sin\alpha$. We can see that a sphere with radius R can be placed in an oblique cylinder if either

$$R \leq GO_1 = r \sin\alpha$$

or

$$R \leq \frac{a \sin\alpha}{2}.$$

If we have α, R, r, and a are given, then we need to evaluate both $r \sin\alpha$ and $\frac{a \sin\alpha}{2}$, then make sure that the radius of the sphere is less than or equal to the smallest value of two. This can be written

$$R \leq \min\left(\frac{a \sin\alpha}{2}; r \sin\alpha\right) \tag{4.49}$$

Answer. $R \leq \min\left(\frac{a \sin\alpha}{2}; r \sin\alpha\right)$

Problem 206 Can a sphere of radius 1 be placed in an oblique cylinder whose base radius is 1 and whose side edge is 2 and if it is inclined to the plane of the base at angle of 60°?

Solution No, since the height of the cylinder is $h = 2 \cdot \sin 60° = \sqrt{3}$, which is less than the diameter of the sphere, 2. The maximum radius R of the sphere that can be placed inside the given oblique cylinder is equal to $\frac{\sqrt{3}}{2}$.

Answer. No. It is impossible.

4.6.2.2 Sphere Inscribed in a Cone

A sphere is said to be inscribed in a cone if it is tangent to the base of the cone and the lateral surface of the cone. A sphere can always be inscribed in a right circular cone (Fig. 4.53).

Fig. 4.53 Sphere inscribed in a cone

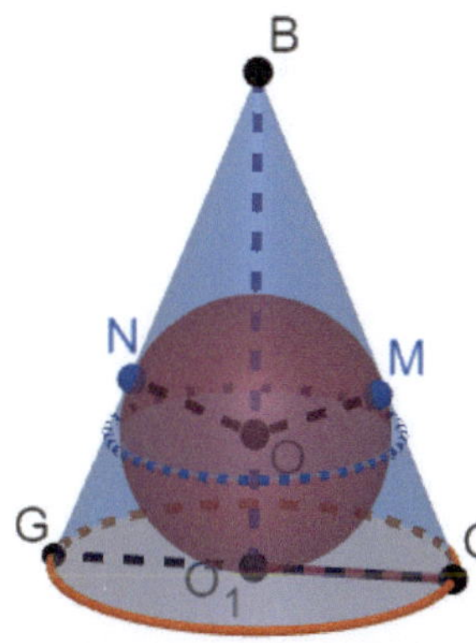

Problem 207 (*Kaganov*) The height of the cone is twice the diameter of the ball inscribed in it. Prove that the ratio of volumes and surfaces of the figures is equal to $2 \div 1$.

Proof Let us sketch a sphere inscribed into a cone as it is shown in Fig. 4.54. Let $OK = OD = R$ and $OA = r$ be the radii of the cone and sphere, respectively. Denote the angle $OBD = \alpha$. Then we can find the lengths of other segments in terms of R. By the condition of the problem the height of the cone B is twice as long as the diameter of the sphere, then $BK = 4R$ and $BO = 3R$ as can be seen in Fig. 4.54. From the right triangle BDO we have that, $\sin\alpha = \frac{OD}{OB} = \frac{R}{3R} = \frac{1}{3}$. Triangles BDO and BKA are similar by two angles (they are right triangles that share the angle α.) The following is valid for the ratio of the corresponding sides, $\frac{AK}{AB} = \frac{1}{3}$. The radius of the cone is $AK = r$ so $AB = 3r$. Applying the Pythagorean Theorem to the right triangle AKB, we obtain $AB^2 - AK^2 = BK^2 \Rightarrow 9r^2 - r^2 = 16R^2 \Rightarrow r = 3R\sqrt{2}$. Next, we will evaluate the volumes of the sphere and cone and then their full surface areas. The volume of the cone is $V_{\text{cone}} = \frac{1}{3}\pi AK^2 \cdot BK = \frac{1}{3}\pi 2R^2 \cdot 4R = 2[\cdot\frac{4}{3}\pi R^3] = 2V_{\text{sphere}}$. The expression inside the brackets is the volume of the sphere of radius R.

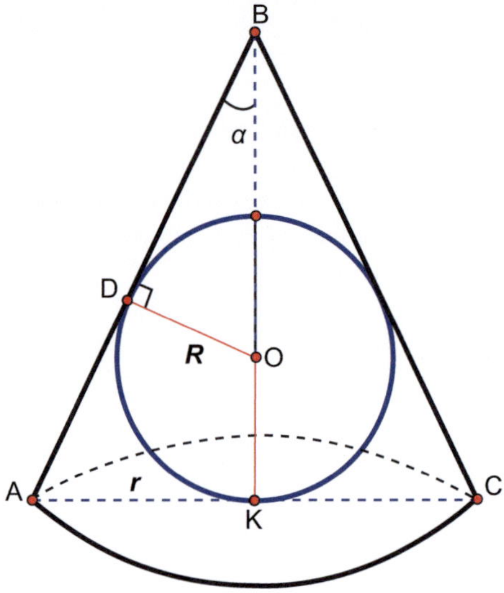

Fig. 4.54 Ball inscribed in a cone

While the surface area of the sphere of radius R is $4\pi R^2$, the full surface area of the cone equals the area of the base circle A_1 plus the lateral area A_2. $A_{\text{cone}} = A_1 + A_2$ where $A_1 = \pi r^2 = 2\pi R^2$. As we know, the lateral surface of a cone is π times the product of its side AB and the radius of the base AK, $A_2 = \pi AK \cdot AB = \pi R\sqrt{2} \cdot 3R\sqrt{2} = 6\pi\sqrt{2}R^2$. Hence, the cone's total surface area is, $A_{\text{cone}} = 8\pi \cdot R^2 = 2 \cdot [4\pi R^2] = 2A_{\text{sphere}}$.

Therefore the ratio of the volumes and surfaces of the figures is equal to $2 \div 1$. □

4.6.3 *Spheres Inscribed in Regular Polyhedra*

In the following problems we will consider a sphere inscribed in a regular octahedron and in an icosahedron, and derive a formula for the radius of a sphere inscribed in an icosahedron with side a, and then derive a formula for the volume of an icosahedron using the radius of the sphere inscribed in it.

Problem 208 Find the radius of a ball inscribed in a regular octahedron with side a.

Solution Let us make a sketch as shown in Fig. 4.55. Because the sphere is tangent to all faces of the octahedron, its radius $r = OG$ is equal to the radius of the circle inscribed in the rhombus $MENF$. Here $OG \perp ME$. Each side of the rhombus is equal to the height of an equilateral triangle with side a, i.e., $ME = EN = NF = FM = \frac{a\sqrt{3}}{2}$. One of the rhombus diagonals, $MN = a$, the side of the octahedron. Another diagonal $EF \neq MN$. It is half the length of the segment EO which can be found from triangle EOM and it is equal to $\frac{a\sqrt{2}}{2}$. Next, the radius of the sphere can be found from the right triangle EOM, if we write its area in two different ways as follows:

$$\text{Area}(\triangle \text{EOM}) = \frac{1}{2}\text{EO} \cdot \text{MO} = \frac{1}{2}\text{GO} \cdot \text{ME},$$

which can be rewritten as

$$\frac{a\sqrt{2}}{2} \cdot \frac{a}{2} = r \cdot \frac{a\sqrt{3}}{2}.$$

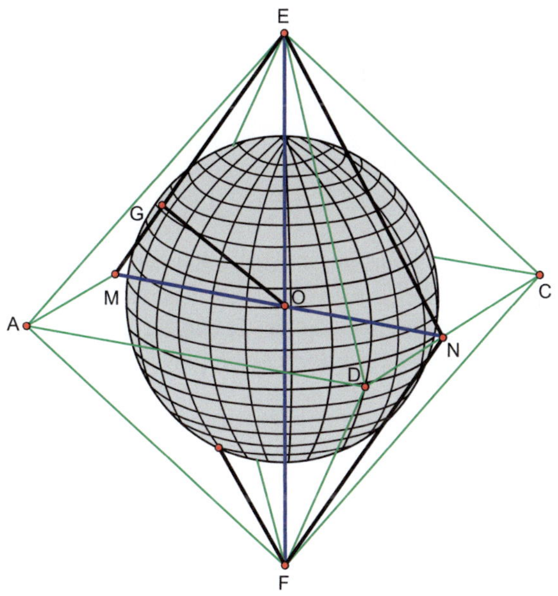

Fig. 4.55 Sphere inscribed in an octahedron

Solving this equation for r and after simplification, we obtain the radius of the sphere inscribed in the octahedron with side a, $r = \frac{a\sqrt{6}}{6}$.

Answer. The radius of the sphere is $\frac{a\sqrt{6}}{6}$.

Problem 209 Find the radius of a sphere inscribed in a regular icosahedron with side $a = 1$.

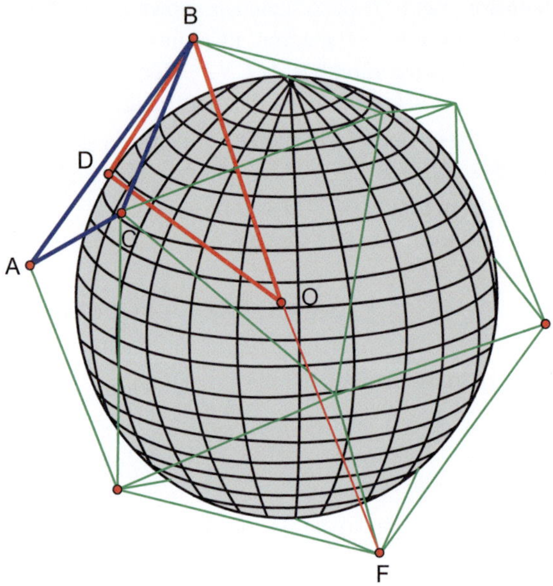

Fig. 4.56 Ball inscribed in an icosahedron

Solution To solve this problem we can think of a regular icosahedron with side a both inscribed in a sphere of radius R (see (4.41)) and simultaneously circumscribed about another sphere of radius r. In Fig. 4.56 consider the equilateral triangle ABC with side a. In this triangle $DB = \rho$ the radius of the circumscribed circle around this triangle. Segment BF connecting two vertices of the icosahedron is the diameter of the sphere circumscribed about it so half of this distance is its radius, $R = OB$. Because an inscribed sphere is tangent to all faces of the polyhedron, its radius $r = OD$ must be perpendicular to the corresponding tangent plane. Hence,

$$OD \perp (ABC) \Rightarrow OD \perp DB.$$

Therefore triangle BDO is the right triangle and we can find the inscribed sphere radius applying Pythagorean theorem:

$$DO^2 = OB^2 - DB^2,$$

which can be written as

$$r^2 = R^2 - \rho^2. \tag{4.50}$$

In Eq. (4.50) we do not know ρ yet. We can find it easily as

$$\rho = \frac{a}{2} \div \cos 30^\circ = \frac{a}{\sqrt{3}}. \tag{4.51}$$

Substituting the values of R and ρ from (4.41) and (4.51), respectively, and taking $a = 1$, we obtain

$$r^2 = \left(\frac{\sqrt{10 + 2\sqrt{5}}}{4}\right)^2 - \frac{1}{3} = \frac{7 + 3\sqrt{5}}{24}.$$

Taking the square root, we have the formula for the radius of the sphere inscribed in the icosahedron,

$$r = \frac{1}{2}\sqrt{\frac{7 + 3\sqrt{5}}{6}}. \tag{4.52}$$

Answer. $r = \frac{1}{2}\sqrt{\frac{7+3\sqrt{5}}{6}}$

Problem 210 Find the volume of an icosahedron with side a.

Solution I guess you're wondering why I provided Table 2.4 with all the formulas for regular polyhedra in Chap. 2, but ask you to derive some of them here in another chapter? Maybe we can use some ideas about a sphere inscribed in a regular polyhedron? Please think about this. We have just derived the formula (4.52) for the radius of an inscribed sphere. Can that relation be applied here?

We understand that if there is a sphere inscribed in the icosahedron, then it must be tangent to each of its 20 triangular faces. However, this means that for each of the twenty faces of the icosahedron there is one perpendicular erected from the center of the sphere. If we connect the center of the inscribed sphere with all three vertices of one of the faces, we see a regular triangular pyramid with base a and height h equal to the radius of the inscribed sphere, r. (For example, using Fig. 4.56, we can mentally connect the vertex A of triangle ABC and the center of the sphere O and observe one of such pyramids $ABCO$ with base ABC and height $OD = r$.) If we repeat this procedure for each of the twenty faces, we will form exactly 20 equal pyramids. Together they make up the entire volume of our icosahedron.

Assume that V_p is the volume of one pyramid and V is the volume of the icosahedron with side a. The following is true:

$$V = 20V_p = 20 \cdot \frac{1}{3}r \cdot \frac{a^2\sqrt{3}}{4}.$$

Substituting (4.52) into this formula, we obtain,

$$V = 5a \cdot \frac{1}{2}\sqrt{\frac{7+3\sqrt{5}}{6}} \cdot a^2\sqrt{3}.$$

For the volume of icosahedron we have,

$$V = \frac{5a^3\sqrt{3}}{2\cdot 3}\sqrt{\frac{7+3\sqrt{5}}{6}} = \frac{5a^3}{2\cdot\sqrt{3}} \cdot \sqrt{\frac{7+3\sqrt{5}}{6}} = \frac{5a^3}{2}\sqrt{\frac{7+3\sqrt{5}}{3\cdot 3\cdot 2}} = \frac{5a^3}{6}\sqrt{\frac{7+3\sqrt{5}}{2}}.$$

We can continue simplifying the radicals. First, we can multiply the numerator and denominator of the radicand by 2 and then factor 2 out from the radical. We have,

$$V = \frac{5a^3}{6}\sqrt{\frac{2\left(7+3\sqrt{5}\right)}{4}} = \frac{5a^3}{12} \cdot \sqrt{14+6\sqrt{5}}.$$

This formula looks "better," but maybe we can reduce the number of the radicals? Let us try to complete the square in the new radicand,

$$14+6\sqrt{5} = 9+2\cdot 3\cdot\sqrt{5}+5 = \left(3+\sqrt{5}\right)^2.$$

Luckily, it can be done. Finally, the volume of the icosahedron with side a is equal to

$$V = \frac{5a^3\left(3+\sqrt{5}\right)}{12}. \tag{4.53}$$

It is a perfect match to the entry in Table 2.4.

Answer. $V = \frac{5a^3\left(3+\sqrt{5}\right)}{12}$.

4.7 Different Circumscribed and Inscribed Figures

Let us focus in this section only on the combinations of right circular cylinders, cones, prisms, and pyramids.

4.7.1 Circumscribed Cones

What can we inscribe into a right circular cone? Consider Fig. 4.57.

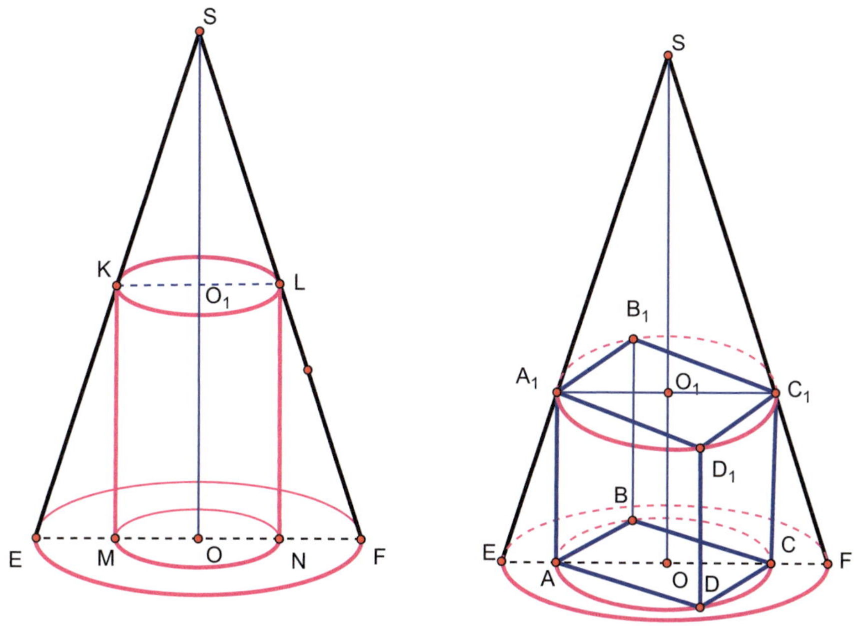

Fig. 4.57 A cylinder and a prism inscribed in a cone

1. A cylinder with the radius of the base less than the radius of the base of the cone, i.e., $r = OM < R = EO$, and the height less than the height of the cone ($h = OO_1 < H = SO$) can be inscribed in a right circular cone (see Fig. 4.57 on the left).) The axes of both the cone and cylinder must be on the same line.
2. You can similarly inscribe a right prism into a cone, if a circle can be circumscribed about the polygon of its base. For example, in Fig. 4.57 on the right, a right prism is inscribed into a cone with the axial section of ESF shown. The base quadrilateral $ABCD$ must be cyclic such that the radius of the circumscribed circle $r = AO$ is less that the radius of the base of the cone $R = EO$. Also, the height of the prism $h = OO_1$ must be shorter that the height of the cone $H = OS$.
3. In order to decide if a cylinder with a height h and radius of the base r or if a prism with a given radius of the circumscribed circle around the base of a prism can be inscribed into a given cone with radius R and height H, the four measurements of two solid figures must satisfy the relationship,

$$\frac{r}{R} = \frac{H - h}{H}, \quad h < H, \quad r < R. \tag{4.54}$$

This follows from consideration of the similarity of triangles SOE and SO_1K in case of a cylinder and triangles; SOE and SO_1A_1 in case of the right prism shown in Fig. 4.57 on the left and right, respectively.

Let us apply formula (4.54) right away.

Problem 211

(a) Can a right cylinder with radius of 1 and height of 1 be inscribed in a cone with radius of the base of 2 and height of 2?

(b) Can a cube with edge 1 be inscribed into a cone with radius of the base 2 and height of 2?

Solution

(a) If a cylinder is inscribed into a cone, its upper base is inscribed into one of the cross-sectional circles obtained by a plane parallel to the base of the cone. And since (4.54) is satisfied, i.e., $\frac{1}{2} = \frac{2-1}{2}$, the answer is yes and such a cylinder can be inscribed in the given cone.

(b) If a cube is inscribed in a cone, then its base is inscribed in the circle of radius $r = \sqrt{2}$. Use the sketch on the right of Fig. 4.57, so $A_1O_1 = \sqrt{2}/2$, and $SO_1 = 2 - 1 = 1 = AA_1$. If $\Delta SO_1A_1 \sim \Delta SOA$, then $\frac{1}{2} = \frac{\frac{\sqrt{2}}{2}}{2}$, which is not true. So a cube with edge 1 cannot be inscribed in a cone with height and radius of 2.

Remark. Using the left sketch in Fig. 4.57 we can derive the formula (4.7) for **the volume of truncated cone**. Consider such a cone $EKLF$ with the base radii $R = OF$ and $r = O_1L$, and height $h = OO_1$. This truncated cone is obtained by removing the small cone SO_1KL with height $h_1 = SO_1$ from the big cone $SEOF$ with height $H = SO$. The volume of the truncated cone is $V_{\text{truncated}} = \frac{1}{3}\pi\left(R^2H - r^2h_1\right)$. If follows for the similar triangles SOF and SO_1L that $\frac{h_1}{H} = \frac{r}{R}$, and so, $h_1 = \frac{r}{R}H$. Substituting this in the volume's formula above, we obtain

$$V_{\text{truncated}} = \frac{1}{3}\pi H \cdot \left(\frac{R^3 - r^3}{R}\right),$$

applying the difference of cubes formula to which, we can simplify it as follows

$$V_{\text{truncated}} = \frac{\pi H}{3}\frac{(R-r)(R^2 + Rr + r^2)}{R}$$

Since triangles SOF and LNF are similar, then $\frac{LN}{SO} = \frac{NF}{OF} \Rightarrow \frac{h}{H} = \frac{R-r}{R}$, substituting which in the formula for the truncated cone, we immediately obtain

$$V_{\text{truncated}} = \frac{\pi h}{3} \cdot \left(R^2 + Rr + r^2\right)$$

4.7.2 Circumscribed and Inscribed Cylinders

Consider a cylinder with radius $R = OM$ and height $H = OO_1$. We understand that a right circular cone is inscribed in a cylinder if its vertex is the center of one of the bases of the cylinder and the base of the cone coincides with the other base of the cylinder. For example, such one cone is shown in Fig. 4.58 on the left.

A right pyramid is inscribed in a cylinder if its vertex is the center of the one of the bases of the cylinder and if the base polygon of the pyramid can be inscribed in the circle of the other base of the cylinder (see quadrilateral pyramid inscribed in a cylinder on the right of Fig. 4.58.) Since a triangle can always be inscribed in a circle, then obviously, a triangular pyramid can always be inscribed in a cylinder.

Fig. 4.58 Cone and pyramid inscribed in a cylinder

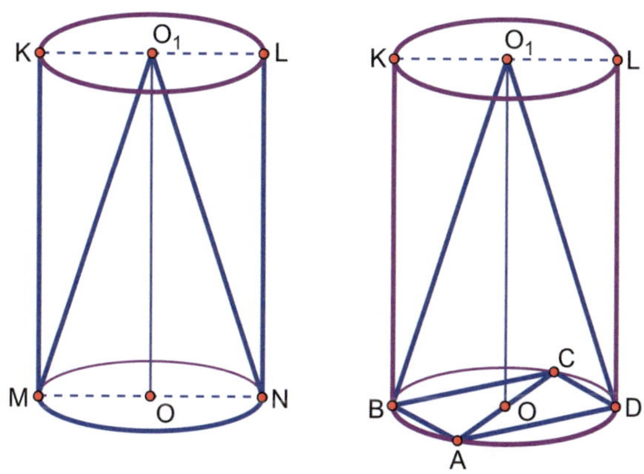

For a right circular cylinder with radius R and height H the following is true:

- A circular cone (right or oblique) with radius of the base R and height H can be inscribed in this cylinder, if the vertex of the cone lies inside the circle of one of the bases of the cylinder.
- A pyramid with height H is inscribed in a cylinder with height H and radius of the base R if the vertex of the pyramid lies on the one of the bases of the cylinder, and the circle of the other base of the cylinder with radius R is circumscribed about the polygon of the base of the pyramid.
- A prism is called inscribed in a cylinder if its bases are inscribed in the bases of the cylinder. The radius of the base of the cylinder is equal to the radius of the circle circumscribed about the base of the prism. The height of the cylinder is equal to the height of the prism.

A cylinder circumscribed about a straight prism is sketched in Fig. 4.59. A cylinder inscribed in a prism is a cylinder whose base circles are inscribed in polygons that are

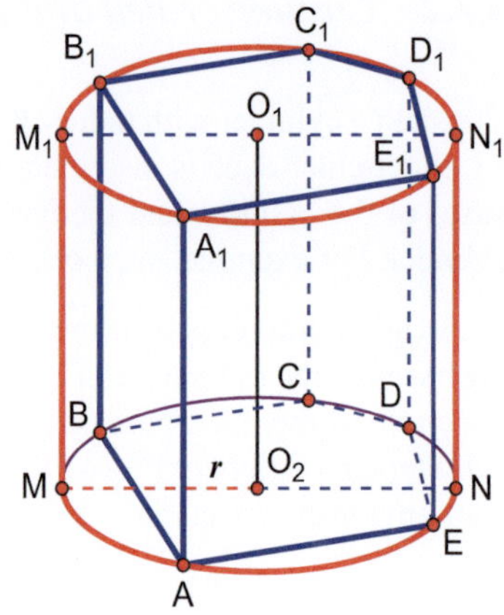

Fig. 4.59 A cylinder circumscribed about a prism

the bases of the prism, and the planes of the lateral faces of the prism are tangent the cylinder along the generatrices of the cylinder (Fig. 4.60). Note, that spatial figures can be inscribed into each other in different ways. Usually all the vertices of an inscribed polyhedron must belong to the surface of another figure, and sometimes to its boundaries.

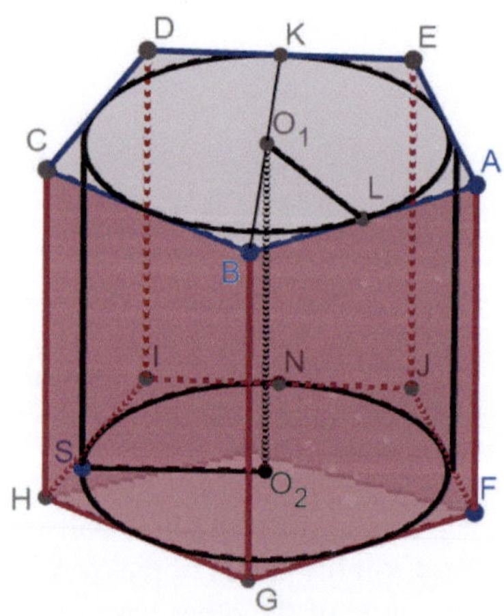

Fig. 4.60 A cylinder inscribed in a pentagonal right prism

Problem 212 A cylinder is circumscribed about a regular tetrahedron with side a, so that the vertices of the tetrahedron belong to the circumferences of the bases of the cylinder. Find the radius of the base and the height of the cylinder.

Solution Figure 4.61 shows two different views of a tetrahedron inscribed in a cylinder so that all its vertices are on the bases of the cylinder. Consider the left sketch. Because a regular tetrahedron has all edges of the same length, edge DC must go through the center of the top circle, O_1 and AB must be the diameter of the bottom circle with center O. Otherwise, DC and AB would not be equal so $R = \frac{a}{2}$.

Next, consider the equilateral triangle ADB with side a. Segment DO is the height of this triangle, $DO = \frac{a\sqrt{3}}{2}$. Because segment OO_1 is the height of the cylinder, then $OO' \perp CD$. In fact, OO_1 is the common perpendicular of two skew edges, DC and AB, of the regular tetrahedron. Then triangle OO_1D is a right triangle. Using the Pythagorean theorem, we can find the height of the cylinder as

$$h^2 = DO^2 - DO'^2 = \left(\frac{a\sqrt{3}}{2}\right)^2 - \left(\frac{a}{2}\right)^2$$

or $h = \frac{a\sqrt{2}}{2}$.

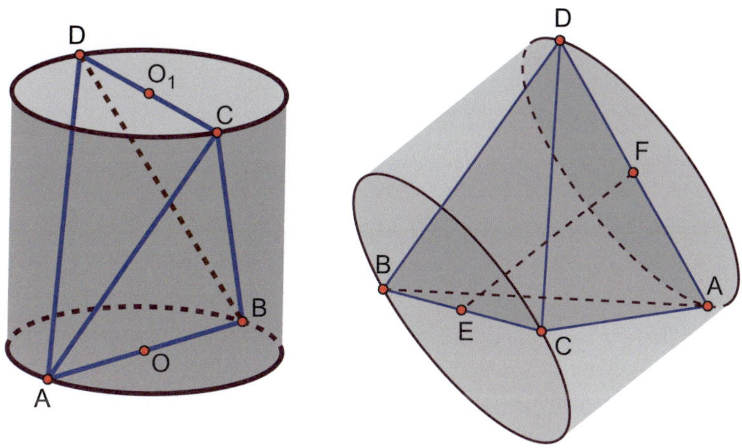

Fig. 4.61 Tetrahedra inscribed in cylinders

Answer. $R = \frac{a}{2}$, $h = \frac{a\sqrt{2}}{2}$.

Problem 213 A cylinder is circumscribed about a regular octahedron with side a so that two opposite vertices of the octahedron are located at the centers of the bases of the cylinder with the remaining vertices on the lateral surface of the cylinder. Find the radius of the base and the height of the cylinder.

Solution Consider a regular octahedron $ABCISS_1$ inscribed in a cylinder (Fig. 4.62). Segment SS_1 is the height h of the circumscribed cylinder. Its radius is that of the circle circumscribed about the square $ABCI$. If the octahedron has side a, the radius is $R = \frac{a\sqrt{2}}{2}$. The height can be found from the right isosceles triangle SOA, in which $AS = a$ and $OS = AO = \frac{a\sqrt{2}}{2}$, so $h = 2 \cdot OS = a\sqrt{2}$.

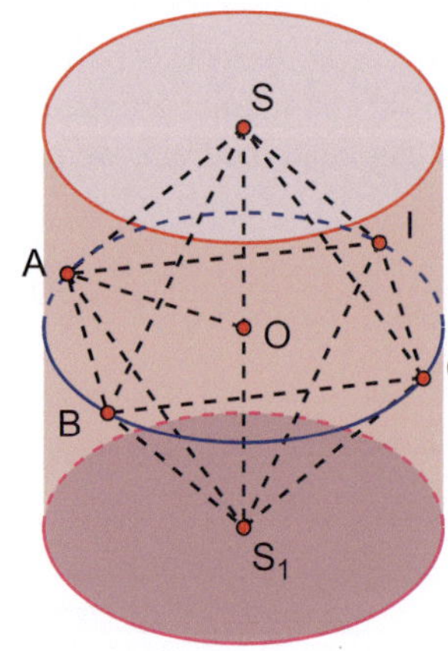

Fig. 4.62 A cylinder circumscribed about a regular octahedron

Answer. $R = \frac{a\sqrt{2}}{2}$ and $h = a\sqrt{2}$.

4.8 Miscellaneous Challenging Problems

Problem 214 (*Rivkin*) A cube is inscribed in a cone so that one of its edges lies on the diameter of the base of the cone. The vertices of the cube that do not belong to this edge lie on the lateral surface of the cone and the center of the cube lies at the height of the cone. Find the ratio of the volume of the cone to the volume of the cube.

Solution Perhaps one of the most difficult in this problem is to draw or imagine how a cube can be inscribed in a cone in accordance with the condition of the problem. Here, of course, it is clear that you can start from one of the edges of the cube, which is part of the diameter, and in due to the fact that the center of the cube lies on the axis of the cone, it is clear that the edge of this cube must be symmetrical about the axis of the cone. That is, we can draw this edge not just lying on the diameter at random, but symmetrically about the center of the base of the cone.

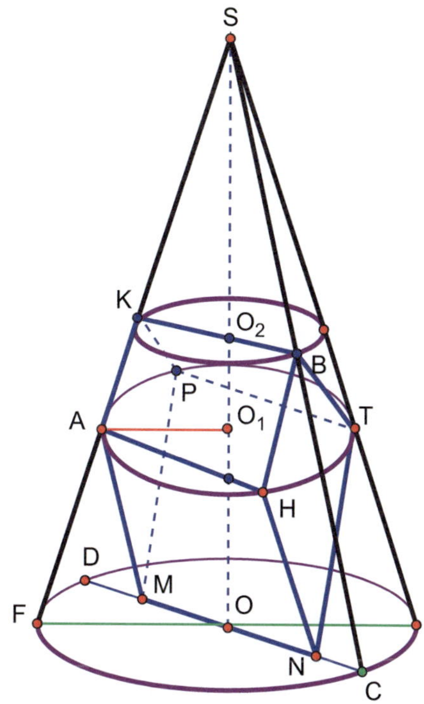

Fig. 4.63 A cube inscribed in a right cone in an unusual way

In Fig. 4.63 the edge MN of cube $NMAHTPKB$ lies on the base of the cone and symmetric with respect to the center of the circle O. Maybe our sketch is not perfect and not in scale, but by the condition of the problem O_1 is the center of the cube that must lie on the cone's axis SO. Denote $OC = R$, $MN = a$, and $SO = H$.

Triangles FSO and SOC are congruent. Because $\triangle AO_1S \sim \triangle FOS$, then $\triangle AO_1S \sim \triangle SOC$ and we can write that

$$\frac{OC}{O_1A} = \frac{SO}{SO_1}, \tag{4.55}$$

where $OC = R$, AO_1 is half diagonal of a cube with edge of a, then $AO_1 = \frac{a\sqrt{3}}{2}$, additionally $OO_1 = \frac{a\sqrt{2}}{2}$. Now (4.55) can be rewritten as $\frac{R}{\frac{a\sqrt{3}}{2}} = \frac{H}{H-\frac{a\sqrt{2}}{2}}$. After simplification we obtain that

$$R = \frac{a\sqrt{3}H}{2H - a\sqrt{2}}.$$

Note that for the same reason that the first two triangles are similar, the triangles SO_1A and SO_2B are also similar.

$$\frac{SO_1}{SO_2} = \frac{AO_1}{BO_2}$$

that can be simplified as

$$\frac{H - \frac{a\sqrt{2}}{2}}{H - a\sqrt{2}} = \frac{\frac{a\sqrt{3}}{2}}{\frac{a}{2}} = \sqrt{3}$$

and solved for H in terms of the edge of cube a,

$$H = \frac{a\sqrt{2}(2\sqrt{3} - 1)}{2\sqrt{3} - 1}.$$

From the first relationship we can obtain the radius R in terms of a as

$$R = \frac{a(2\sqrt{3} - 1)}{2}.$$

The ratio of the volumes of the cone and cube can be expressed as

$$V_1 \div V_2 = \frac{1}{3} \cdot \frac{\pi R^2 H}{a^3}.$$

In this formula, we can rationalize the denominator by using the difference of squares formula and also by simplifying the square roots. Finally, the requested ratio is

$$\frac{V_1}{V_2} = \frac{\pi\sqrt{2}(2\sqrt{3} - 1)^3 \cdot (\sqrt{3} + 1)}{48}.$$

Answer. $\frac{V_1}{V_2} = \frac{\pi\sqrt{2}(2\sqrt{3}-1)^3 \cdot (\sqrt{3}+1)}{48}$.

Problem 215 (*Moscow State University Entrance Exam*) The sphere of radius $\sqrt{41}$ passes through vertices $B,\ C,\ C_1$ and through the midpoint of the edge A_1D_1 of cube $ABCDA_1B_1C_1D_1$. ($AA_1 \| BB_1 \| CC_1 \| DD_1$) Find the surface area of the cube.

Solution First, we will draw a cube and decide on how the sphere mentioned in the condition of the problem would intersect the cube (Fig. 4.64).

Fig. 4.64 A sphere of radius $\sqrt{41}$ and a cube

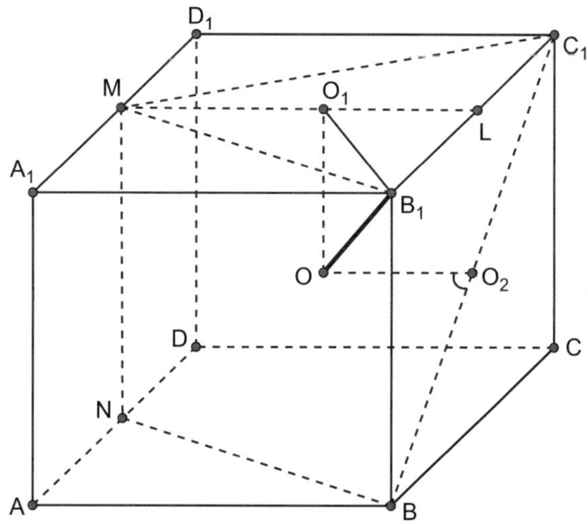

1. Let M be the midpoint of A_1D_1 and O_1 the center of the circle circumscribed about triangle MC_1B_1. Let O_2 be the center of face BB_1C_1C and L be the midpoint of side B_1C_1.
2. Since the sphere goes through points B, C, and C_1, then it must go through vertex B_1 as well because BB_1C_1C is a square.
3. Let us drop perpendicular from point M to face $ADCB$. Let N be the foot of the perpendicular. The sphere is such that the prism NMC_1B_1BC is inscribed into it. The center of the sphere O is the intersection of two perpendiculars O_1O and O_2O, where O_1O is perpendicular to $A_1D_1C_1B_1$ and O_2O is perpendicular to B_1C_1CB.
4. Let a be the side of the given cube and $b = MB_1$. From the isosceles triangle B_1MC_1 inscribed into circle of radius R and having center at O_1, we can find its area in two different ways: (a) using the radius of the circumscribed circle R (see formula (4.45)) and (b) by evaluating the area as the half product of height $ML = a$ and the base $B_1C_1 = a$, we have the following relationships:

$$\frac{b^2a}{4R} = \frac{a \cdot a}{2} \tag{4.56}$$

5. Sides a and b of the triangle B_1MC can be found from the right triangle B_1ML using the Pythagorean theorem,

$$b^2 = \left(\frac{a}{2}\right)^2 + a^2 = \frac{5}{4}a^2 \tag{4.57}$$

Substituting (4.57) into (4.56) and after simplification we obtain that

$$R = \frac{5a}{8} \tag{4.58}$$

6. From the right triangle OO_1B_1 and using (4.58), we obtain

$$\begin{aligned} &\left(\tfrac{a}{2}\right)^2 + \left(\tfrac{5a}{8}\right)^2 = 41 \\ &\tfrac{41a^2}{64} = 41 \\ &a^2 = 64. \end{aligned}$$

7. Finally, the surface area of the cube with side a is

$$6a^2 = 6 \cdot 64 = 384\,\text{units}^2.$$

Answer. $A = 384$ units^2.

Problem 216 (*Moscow State University Entrance Exam*) Rectangular parallelepiped $ABCDA_1B_1C_1D_1$ is given. ($ABCD$ and $A_1B_1C_1D_1$ are the bases, $AA_1 \| BB_1 \| CC_1 \| DD_1$, the segments M_1N_1, M_2N_2, M_3N_3 are the common perpendiculars to the pairs of segments A_1C_1 and AB_1, BC_1 and AC, DC_1 and AD_1, respectively. The volume of the parallelepiped is V, the radius of circumscribed sphere is R, and the sum of the lengths of sides AA_1, AB and AD equals m. Find the sum of the volumes of pyramids $AA_1M_1N_1$, ABM_2N_2, and ADM_3N_3.

Solution Assume that the parallelepiped shown in Fig. 4.65 is inscribed into the sphere of radius R.

Fig. 4.65 Parallelepiped

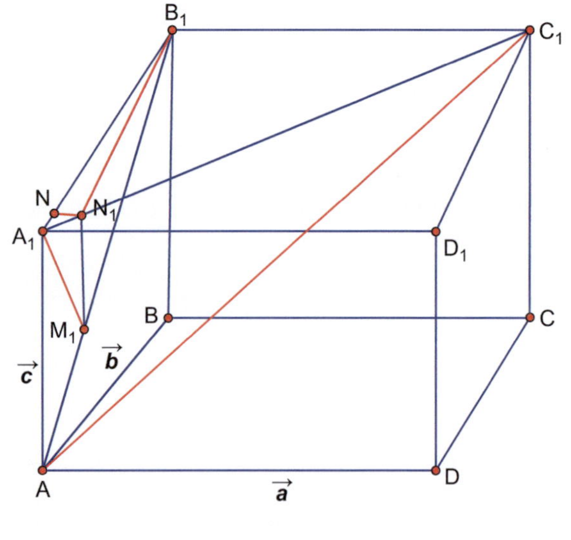

1. Introduce three vectors:

$$\mathbf{a} = AD, \quad \mathbf{b} = AB, \quad \mathbf{c} = AA_1.$$

These vectors are orthogonal, so the following scalar products are zero:

$$(\mathbf{a}, \mathbf{b}) = (\mathbf{b}, \mathbf{c}) = (\mathbf{a}, \mathbf{c}) = 0.$$

And also the following is true:

$$(\mathbf{a}, \mathbf{a}) = a^2, \quad (\mathbf{b}, \mathbf{b}) = b^2, \quad (\mathbf{c}, \mathbf{c}) = c^2.$$

2. Since all the vertices of the parallelepiped are on the sphere of radius R, then we have the vector equality:

$$\mathbf{a} + \mathbf{b} + \mathbf{c} = AC_1 = 2\mathbf{R}$$

3. We obtain the scalar system,

$$\begin{cases} a + b + c = m \\ a^2 + b^2 + c^2 = 4R^2 \\ abc = V \end{cases} \tag{4.59}$$

4. Let $\alpha = \frac{AM_1}{AB_1}$, $\beta = \frac{AN_1}{A_1C_1}$. Pyramids $AA_1M_1N_1$ and $AA_1B_1N_1$ have the same height dropped from the vertex N_1 to the bases. The areas of the bases also have the same height dropped from the common vertex A_1, so the ratio of their areas equals the ratio of AM_1 and AB_1 to which this height is dropped. Therefore $V_{AA_1M_1N_1} = \alpha V_{AA_1B_1N_1}$. Next, we can express the volume of pyramid $AA_1B_1N_1$ in terms of the volume of pyramid $AA_1B_1C_1$. These pyramids have the common base AA_1B_1, but different heights that are in the same ratio as A_1N_1 to A_1C_1, that is β. Further we can express the volume of pyramid $AA_1B_1C_1$ in terms of the volume of parallelepiped V. Thus,

$$V_{AA_1B_1C_1} = \frac{1}{3} \cdot B_1C_1 \cdot [AA_1B_1] = \frac{1}{3} \cdot a \cdot \frac{1}{2}bc = \frac{1}{6} \cdot V.$$

Now we can denote by V_1 the volume of the first pyramid, $AA_1M_1N_1$:

$$V_1 = V_{AA_1M_1N_1} = \alpha V_{AA_1B_1N_1} = \alpha\beta V_{AA_1B_1C_1} = \frac{\alpha\beta}{6} \cdot V.$$

5. We have the following vector expressions:

$$A_1M_1 = AM_1 - AA_1 = \alpha AB_1 - AA_1$$

$$A_1N_1 = \beta \cdot A_1C_1 = \beta \cdot (\mathbf{a} + \mathbf{b})$$

So $M_1N_1 = A_1N_1 - A_1M_1 = \beta \cdot \mathbf{a} + (\beta - \alpha) \cdot \mathbf{b} + (1 - \alpha) \cdot \mathbf{c}$.

6. Since $M_1N_1 \perp A_1C_1,\ M_1N_1 \perp AB_1 \Rightarrow (M_1N_1, A_1C_1) = (M_1N_1, AB_1) = 0$. This can be written as the following system,

$$\begin{cases} (\beta \cdot \mathbf{a} + (\beta - \alpha) \cdot \mathbf{b} + (1 - \alpha) \cdot \mathbf{c},\ \mathbf{a} + \mathbf{b}) = 0 \\ (\beta \cdot \mathbf{a} + (\beta - \alpha) \cdot \mathbf{b} + (1 - \alpha) \cdot \mathbf{c},\ \mathbf{b} + \mathbf{c}) = 0 \end{cases}$$

Or in algebraic form,

$$\begin{cases} \beta a^2 + (\beta - \alpha) \cdot b^2 = 0 \\ (\beta - \alpha) \cdot b^2 + (1 - \alpha) \cdot c^2 = 0 \end{cases}$$

From which we obtain that $\alpha = \frac{c^2 \cdot (a^2 + b^2)}{Q}, \quad \beta = \frac{b^2c^2}{Q}$, where

$$Q = a^2b^2 + a^2c^2 + b^2c^2.$$

7. Therefore, for the first pyramid we obtain

$$V_1 = \frac{\alpha\beta}{6} \cdot V = \frac{b^2c^4 \cdot (a^2 + b^2)}{6Q^2} \cdot V = \frac{1}{6} \cdot \frac{abc}{Q} \cdot (c^4a^2b^2 + c^4b^4)$$

8. The volumes of two other pyramids can be similarly evaluated. For the sum of the volumes of three pyramids denoted by $V_\Sigma = V_1 + V_2 + V_3$ we get

$$V_\Sigma = \frac{1}{6} \cdot \frac{abc}{Q^2} \cdot \left(a^2b^2c^2 \cdot (a^2 + b^2 + c^2) + a^4b^4 + a^4c^4 + c^4b^4\right) \tag{4.60}$$

9. Next, let us express $a^2b^2 + a^2c^2 + c^2b^2$ and $a^4b^4 + a^4c^4 + c^4b^4$ in terms of $m,\ R^2,\ V$. If we square both sides of the first equation of system (4.59) we obtain

$$(a + b + c)^2 = m^2$$
$$a^2 + b^2 + c^2 + 2(ab + bc + ac) = m^2$$

10. Substituting in the last equation the second equation from system (4.59), we have that

$$ab + bc + ac = \frac{m^2 - 4R^2}{2},$$

then

$$Q = \left(\frac{m^2 - 4R^2}{2}\right)^2 - 2V \cdot m$$

and

$$a^4b^4 + a^4c^4 + c^4b^4 = Q^2 - 8V^2 \cdot R^2.$$

11. Finally, with the use of formula (4.60), the sum of the volumes of three indicated pyramids equals

$$V_\Sigma = \frac{1}{6}V - \frac{2}{3}V^3 \cdot R^3 \cdot \left(\left(\frac{m^2 - 4R^2}{2}\right)^2 - 2Vm\right)^{-2}.$$

Answer. $\frac{1}{6}V - \frac{2}{3}V^3 \cdot R^3 \cdot \left(\left(\frac{m^2-4R^2}{2}\right)^2 - 2Vm\right)^{-2}$.

Problem 217 Find the largest quadrangular pyramid that can be inscribed in a sphere of radius R.

Solution Let us inscribe a regular quadrangular pyramid $ABCDF$ into a sphere of radius R. Another pyramid with five vertices would be possible to inscribe as well but we can prove that such a pyramid would not have the volume bigger than a regular pyramid. Suppose that the base of the pyramid is inscribed not in the big circle of radius $R = OF = OC$, but in another, smaller circle of radius $r = EC$, where E is the intersection point of the diagonals of the square $ABCD$ (see Fig. 4.66).

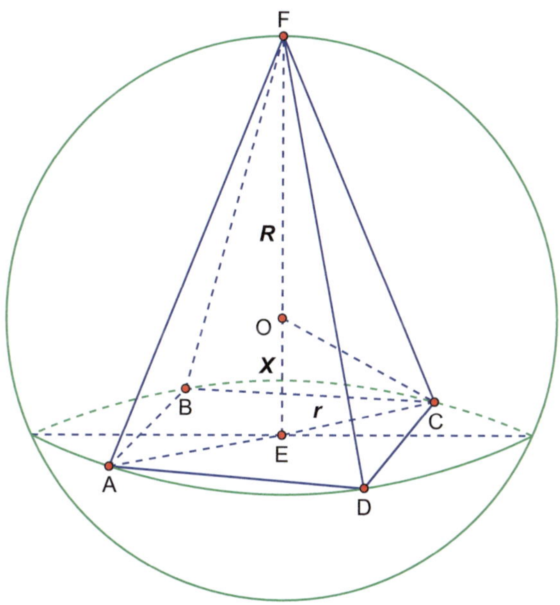

Fig. 4.66 A quadrangular pyramid inscribed in a sphere

The height of the pyramid $FE = h = R + x, \quad x = OE$. This x can be found using the Pythagorean theorem for the right triangle OEC,

$$x^2 = R^2 - r^2 \tag{4.61}$$

The side of the base of the pyramid $a = AB$ is related to the small radius,

$$a = r\sqrt{2} \tag{4.62}$$

Next, the volume of the pyramid is

$$V = \frac{2}{3}(R + x)\left(R^2 - x^2\right) \tag{4.63}$$

Let us show that the maximum will be at $x = \frac{R}{3}$, not at $x = 0$ when the base of the pyramid is inscribed into the major circle. Taking derivative of (4.63) with respect to x, we obtain

$$\frac{dV}{dx} = -\frac{2}{3}\left(3x^2 + 2Rx - R^2\right) \tag{4.64}$$

Equating the expression inside parentheses of (4.64) to zero, we find zeros of the first derivative,

$$x = \frac{-R\pm\sqrt{R^2+3R^2}}{3} = \frac{-R\pm 2R}{3}$$
$$x = -R, \quad x = \frac{R}{3}$$

for which the value,

$$x = \frac{R}{3} \tag{4.65}$$

maximizes the volume according to (4.63). Substituting (4.65) into (4.63) we obtain the formula for the maximum volume of the regular quadrilateral pyramid inscribed into a sphere of the given radius R,

$$\begin{aligned} V_{\max} &= \tfrac{2}{3}\left(R + \tfrac{R}{3}\right)\left(R^2 - \left(\tfrac{R}{3}\right)^2\right) \\ &= \tfrac{64R^3}{81} \end{aligned}$$

This value is larger than that at $x = 0, \quad V(0) = \frac{2R^3}{3}$ because $\frac{64}{81} > \frac{2}{3} = \frac{2\cdot 27}{3\cdot 27} = \frac{54}{81}$. Using (4.65) we can evaluate the corresponding value of x, and then using (4.62) and (4.61) to find the side of the basic square, $a = \frac{4R}{3}$ that is equal to the height of the pyramid, h.

Now we will show that no other quadrangular pyramid inscribed in a sphere of radius R can have a larger volume than the pyramid from (4.65). Suppose that the base of the pyramid is a cyclic quadrilateral $ABCD$ (see [2], p. 126). One such quadrilaterals is shown in Fig. 4.67 where the radius r is marked in red.

The area of a cyclic quadrilateral, A (i.e., for a quadrilateral $ABCD$) can be found as the sum of the areas of four isosceles triangles (AOB, BOC, COD, and AOD) that can be written as

$$\begin{aligned} A &= \frac{1}{2}r^2 \cdot (\sin\angle AOB + \sin\angle BOC + \sin\angle COD + \sin\angle AOD) \\ &= \frac{1}{2}r^2 \cdot (\sin\alpha + \sin\beta + \sin\angle\gamma + \sin\angle\delta) \\ &\le \frac{1}{2}r^2. \end{aligned} \tag{4.66}$$

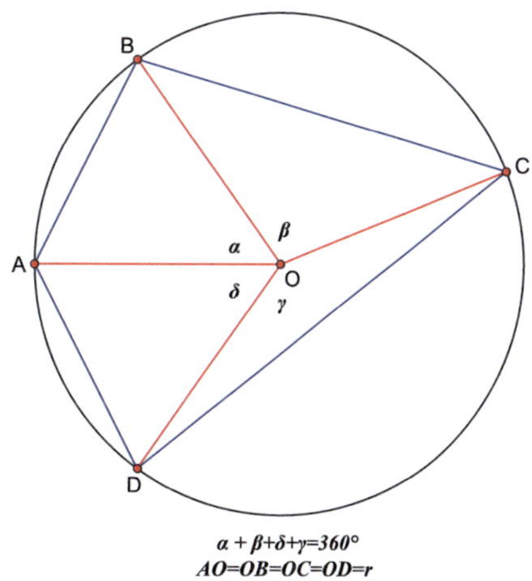

Fig. 4.67 Cyclic quadrilateral

Obviously, the maximum of (4.66) happens at $\alpha = \beta = \gamma = \delta = 90°$ which happens if $ABCD$ is a square. This proves that among all quadrilaterals inscribed into a circle, the square has the maximum area. Finally, if a quadrilateral pyramid is inscribed into a sphere of radius R, then the maximum volume will have a regular square pyramid with equal base and height, i.e., $a = h = \frac{4R}{3}$

$$V_{\max} = \frac{64R^3}{81}. \tag{4.67}$$

Answer. $V_{\max} = \frac{64R^3}{81}$.

Problem 218 A convex polyhedron with five vertices is inscribed in a sphere of radius R. Find the largest such polyhedron.

Solution A convex polyhedron with 5 vertices can be either a quadrangular pyramid or a bi-pyramid that is the union of two tetrahedra (triangular pyramids) with a common base. Let us focus on the bi-pyramid case since the first case was solved in the previous problem.

If AB, AD, AC and AE are four consecutive edges of a tetrahedral angle with vertex A, then our polyhedron is the union of tetrahedra $ABCD$ and $ABCE$ with a common base ABC, a sketch of which is shown in Fig. 4.68. Let us evaluate the volume of the bi-pyramid,

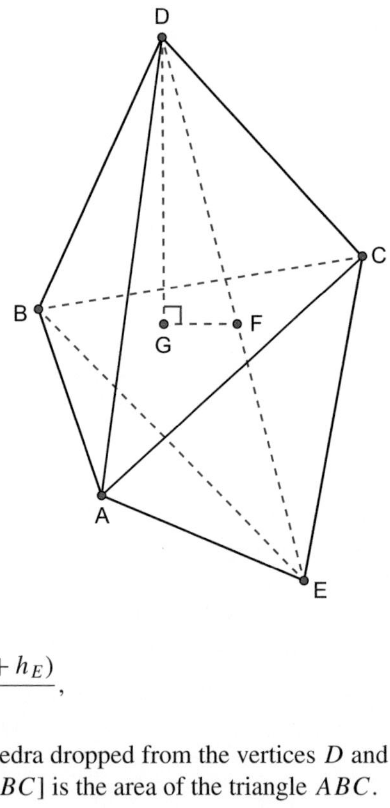

Fig. 4.68 A bi-pyramid $ABCDE$

$$V = \frac{B \cdot (h_D + h_E)}{3},$$

where h_D and h_E are the heights of the tetrahedra dropped from the vertices D and E to the base ABC, respectively and $B = [ABC]$ is the area of the triangle ABC.

Let F be the point of intersection of DE with plane ABC and $h_D = DG \perp ABC$. It follows from the right triangle DGF that $h_D = DG < DF$. Similarly, we can drop the height from vertex E and show that $h_E < EF$. Hence,

$$h_D + h_E \leq DF + FE = DE \leq 2R,$$

where we used the fact that the distance between any two points on a sphere is not greater than its diameter. The volume of the bi-pyramid will be bounded,

$$V \leq \frac{2}{3}[ABC] \cdot R.$$

Let r be the radius of the circle circumscribed about the triangle ABC with sides $a,\ b,\ c$ (that is, the section of the sphere by the plane ABC). Then the area of the triangle can be expressed in terms of its sides and the radius of the circumscribed circle as

$$[ABC] = B = \frac{abc}{4r} \leq \frac{a^3}{4r}.$$

Since for an equilateral triangle, the radius of the circumscribed circle is $2/3$ of its height (median or bisector), we can write that $r = \frac{2}{3} \cdot \frac{a\sqrt{3}}{2} = \frac{a}{\sqrt{3}}$, so $a = r\sqrt{3}$. Then

$$[ABC] \leq \frac{a^3}{4r} = \frac{r^2 3\sqrt{3}}{4} \leq \frac{R^2 \cdot 3\sqrt{3}}{4} = \frac{3\sqrt{3}R^2}{4}.$$

Because the radius r of any cross section of a sphere is at most the radius of the sphere R, the volume of the bi-pyramid inscribed in this sphere is $V \leq \frac{\sqrt{3}R^3}{2}$. Therefore, the maximum volume is $V = \frac{\sqrt{3}R^3}{2}$ in the case where ABC is a regular triangle inscribed in the largest circle of the sphere, the "equator," and D and E are the North and South poles of the sphere. For such a bi-pyramid, point F is the center of the sphere where ED is its diameter, and the center of the circle circumscribed about triangle ABC. The maximum volume of a bi-pyramid inscribed into a sphere of radius R is

$$V_{\max} = \frac{\sqrt{3}R^3}{2}. \tag{4.68}$$

Since a quadrangular regular pyramid is also a five-vertex convex polyhedron, we need to find out which is greater—the volume of a quadrangular pyramid or a bi-pyramid? So, we need to compare the formulas (4.67) and (4.68) for the same sphere radius R.

Which is greater: $\frac{64}{81}$ or $\frac{\sqrt{3}}{2}$?

We can compare the squares of two quantities, which gives us $\frac{4096}{6561} < \frac{3}{4}$. Therefore, a five-vertex convex polyhedron of maximum volume that can be inscribed in a sphere of the same radius is a bi-pyramid. In this bi-pyramid $ABCDE$, $\triangle ABC$

is an equilateral triangle inscribed in the great circle of the sphere. The distance between two opposite vertices D and E is the diameter of the sphere. In this case, the midpoint of the segment ED is simultaneously the center of the sphere and of the circumscribed triangle ABC.

Answer. A bi-pyramid with triangular base is five-vertex polyhedron with maximum volume $V_{\max} = \frac{\sqrt{3}R^3}{2}$.

Problem 219 Prove that the ratio of the volumes of a sphere and a truncated cone circumscribed about the sphere is equal to the ratio of their surface areas.

Proof Let us sketch the problem (see Fig. 4.69). The sphere is tangent to the top and bottom of the truncated cone. Denote the radius of the sphere by ρ, then its volume and surface area are $V(\text{sphere}) = \frac{4}{3}\pi\rho^3$, $\text{A(sphere)} = 4\pi\rho^2$, respectively. The truncated cone has radii: $r = MC$, $R = ND$. Using (4.7), the volume of the truncated cone with the height $MN = 2\rho$ is equal to $V_{\text{cone}} = \frac{1}{3}2\rho\pi \cdot (R^2 + Rr + r^2)$. Its surface area is equal to $A_{\text{cone}} = \pi\left(R^2 + r^2 + L \cdot (R + r)\right)$, where $L = CD$ is the lateral edge of the cone.

Fig. 4.69 Ball inscribed into a truncated cone

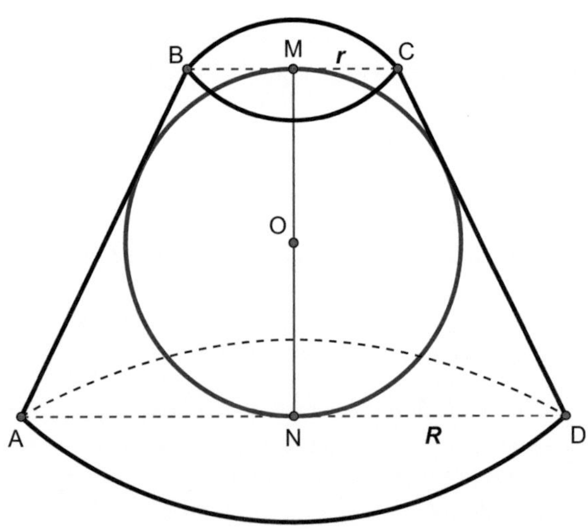

In order to find CD of the truncated cone, we can use the fact that the main circle of the sphere is inscribed into trapezoid $ABCD$. Then by the property (see [2], p. 154) of the circle inscribed into a quadrilateral, we can state that the sum of the opposite sides of the quadrilateral is the same,

$$AB + CD = BC + AD,$$

which can be written as $2L = 2R + 2r$ or $L = R + r$. Now we can rewrite the surface area of the truncated cone as

$$A_{\text{cone}} = \pi(R^2 + r^2 + (R + r)^2) = 2\pi(R^2 + r^2 + Rr)$$

Next, we can write down the ratio of the corresponding volumes

$$\frac{V(\text{sphere})}{V(\text{cone})} = \frac{4\pi\rho^3 \cdot 3}{3 \cdot 2\pi\rho(R^2 + r^2 + Rr)} = \frac{2\rho^2}{R^2 + r^2 + Rr}.$$

Then we can write down the ratio of the areas,

$$\frac{S(\text{sphere})}{S(\text{cone})} = \frac{4\pi\rho^2}{2\pi(R^2 + r^2 + Rr)} = \frac{2\rho^2}{R^2 + r^2 + Rr}.$$

We see that both ratios are equal. □

Problem 220 An equilateral cone stands on its top and a ball is placed inside it. Then water is poured into the cone so that it covers the top of the ball. The ball is taken out. To what level will the water level drop inside the cone?

Solution Consider a cone with its base upside down and with an equilateral triangle in an axial section. A ball of radius r is placed in such a cone (Fig. 4.70). Water is poured in the cone so that it is level with the top of the ball. Water with and without the ball both form equilateral cones. We call a cone equilateral if its slant height equals the diameter of the base. The level of water before the ball was removed is shown by the mark AK, the height of the cone. When the ball is removed, the level of water will drop. In this new situation it will take the new mark and the height of the cone will be less than AK. We need to find where the level of water in the cone will drop, so we need to find the new height of the new cone formed by the remaining water.

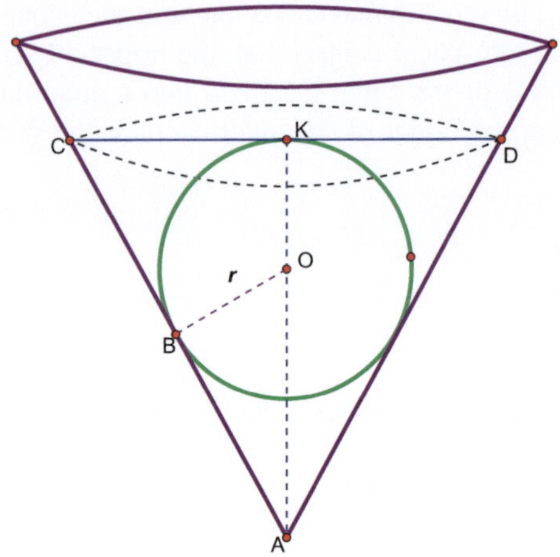

Fig. 4.70 Water in the cone when the ball is inside

Let the radius of the ball OB be r, then $OA = 2r$, and $AK = 3r$,

$$CK = AK \tan 30^{\circ} = 3r \cdot \frac{1}{\sqrt{3}} = r\sqrt{3}.$$

Now we can evaluate the volume of the cone

$$\begin{aligned} V_{cone} &= \tfrac{1}{3}\pi\ |CK|^2 \cdot |AK| = \tfrac{\pi}{3} 3r^2 \cdot 3r \\ &= 3\pi r^3. \end{aligned} \tag{4.69}$$

The volume of the ball is

$$V_{\text{ball}} = \frac{4}{3}\pi r^3. \tag{4.70}$$

When we remove the ball, the remaining water will take the volume of the difference of (4.69) and (4.70)

$$V_{\text{water}} = 3\pi r^3 - \frac{4}{3}\pi r^3 = \frac{5}{3}\pi r^3. \tag{4.71}$$

This water will take the shape of a new cone with different height, h, and radius, R, of the base. It follows from the right triangle CKA that they will be also connected by the formula,

$$R = \frac{h}{\sqrt{3}}. \tag{4.72}$$

Now we can evaluate the volume of the new cone,

$$V_1 = \frac{1}{3}\pi R^2 \cdot h = \frac{1}{3} \cdot \pi \frac{h^2}{3} \cdot h = \pi \frac{h^3}{9}. \tag{4.73}$$

We need to equate formulas (4.71) and (4.73), $\frac{5}{3}\pi r^3 = \pi \frac{h^3}{9}$, from which we obtain that $h^3 = 15r^3$ that can be solved for h, $h = r\sqrt[3]{15}$. This formula gives us the new level of the water or where the level of the water will drop when the ball is taken out. In Fig. 4.70, AK denotes the water level with the ball inside. The initial water level with the ball inside, shown by the dotted line, was $AK = 3r$, now it is $Ak = r\sqrt[3]{15}$, that is, the water level has decreased by a factor of $\sqrt[3]{\frac{9}{5}}$.

Answer. The new water level is $h = r\sqrt[3]{15}$.

Problem 221 Two equilateral cones are inserted vertically into each other. Find the volume and surface area of the resulting body.

Solution A spatial geometric figure consists of two identical truncated cones. Please see the sketch in Fig. 4.71.

Let the slant height of one cone be a, then $AB = a$, and $R = BC = \frac{a}{2}$, $r = ED = \frac{a}{4}$, $CA = \frac{a\sqrt{3}}{2}$. The height of the truncated cone is half of segment CA. Let its volume be V_1. The volume of two truncated cones is equal to

$$V = 2V_1 = 2 \cdot \frac{1}{3} \cdot \frac{\sqrt{3}}{2}\pi\left(R^2 + Rr + r^2\right) = \frac{\pi a\sqrt{3}}{3 \cdot 2} \cdot \left(\frac{a^2}{4} + \frac{a^2}{8} + \frac{a^2}{16}\right) = \frac{7\sqrt{3}\pi a^3}{96}.$$

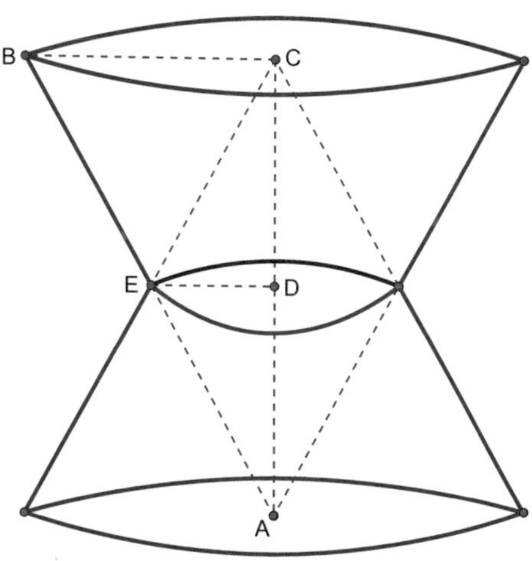

Fig. 4.71 Two truncated cones

The surface area is equal to two surface areas of the truncated cones.

$$A = 2A_1 = 2\left(\pi R^2 + \pi (r + R) \cdot l\right) = 2\pi \left(\frac{a^2}{4} + \frac{3a^2}{8}\right) = \frac{5\pi a^2}{4}$$

where l is the slant height of the cone.

Answer. $A = \frac{5\pi a^2}{4}$.

Problem 222 (*Moscow State University Entrance Exam 2019*) In a rectangular parallelepiped, a plane π passing through its three nonadjacent vertices cuts off a tetrahedron. Inside the sphere circumscribed around the parallelepiped, there are two balls of maximum radii on opposite sides of the plane. Find the ratio of the radii of these balls if the edges of the parallelepiped are equal to 1, $\sqrt{5}$ and 5.

Solution Let's make a sketch. Without loss of generality, we can assume that the plane passes through the vertices A, D_1, and C (Fig. 4.72). The center of the sphere lies at the intersection point of the diagonals of the parallelepiped, at the point O. By the condition, the balls are located on opposite sides of the π plane, are inscribed in the sphere, and have maximum radii. Therefore, the balls are tangent to the sphere and the π plane and the centers of the balls lie on a straight line passing through the center of the sphere and perpendicular to the π plane.

Referring to Fig. 4.73, the segment KL is the diameter of the sphere perpendicular to the plane. The point O_1 is the point of intersection of the diameter KL and the plane π. With a different arrangement of balls, they would have a smaller radius. The ratio

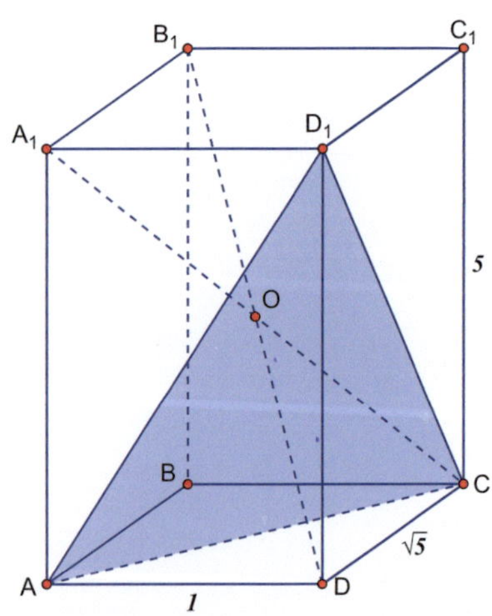

Fig. 4.72 Parallelepiped and the cutting plane π

of these radii equals $KO_1 \div O_1L$. Let us use the vector-coordinate method. Introduce a coordinate system with point A as its center as it is demonstrated in Fig. 4.74. Our solution plan follows:

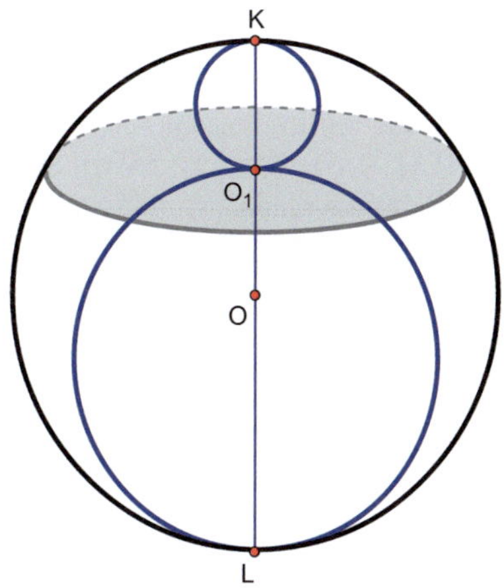

Fig. 4.73 Two tangent spheres

1. Find the coordinates of the point O and the radius of the sphere.
2. Find the equation of the plane π.
3. Find the distance from the point O to the plane π, that is, the length of the segment OO_1.
4. Find the ratio $KO_1 \div O_1L$.

Let's proceed.

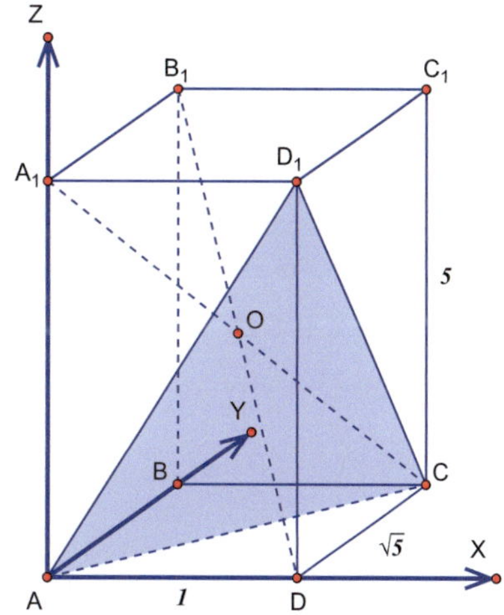

Fig. 4.74 Parallelepiped in a coordinate system

1. Since the point O is the point of intersection of the diagonals of the parallelepiped, that is, it is the midpoint of the diagonals, the coordinates of the point are $O(\frac{1}{2}, \frac{\sqrt{5}}{2}, \frac{5}{2})$. The radius of a sphere is half the length of the diagonal of the parallelepiped, that is $R = \frac{\sqrt{AD^2+DC^2=+DD_1^2}}{2} = \frac{\sqrt{1+5+25}}{2} = \frac{\sqrt{31}}{2}$.
2. We are looking for the equation of the plane π. Since the plane passes through the origin, its general equation is $ax + by + cz = 0$ where (a, b, c) are the coordinates of the normal vector to the plane. The plane passes through the points $D_1(1, 0, 5)$ and $C(1, \sqrt{5}, 0)$. Substitute the coordinates of the points in the equation of the plane,

$$D_1(1, 0, 5): \quad a + 5c = 0 \Rightarrow c = -\frac{a}{5}$$

$$C(1, \sqrt{5}, 0): \quad a + \sqrt{5}b = 0 \Rightarrow b = -\frac{a}{\sqrt{5}}$$

Obtain the equation of the plane, $ax - \frac{a}{\sqrt{5}}y - \frac{a}{5}z = 0$. Divide both sides by $a \neq 0$ and obtain the equation of the plane, $x - \frac{y}{\sqrt{5}} - \frac{z}{5} = 0$.

3. Let us find the length of segment OO_1. Recall the formula for finding the distance from a point to a plane from Chap. 1 (1.79)

$$OO_1 = \frac{\left|\frac{1}{2} - \frac{1}{\sqrt{5}} \cdot \frac{\sqrt{5}}{2} - \frac{1}{5} \cdot \frac{5}{2}\right|}{\sqrt{1^2 + \left(\frac{1}{\sqrt{5}}\right)^2 + \left(\frac{1}{5}\right)^2}} = \frac{5}{2\sqrt{31}}$$

Thus, the radius of the sphere, $R = \frac{\sqrt{31}}{2}$.

4. Finally, let us find the ratio of two diameters $KO_1 \div O_1L$ that is equal to the ratio of the corresponding radii. Since $KO_1 = R - OO_1$ and $O_1L = R + OO_1$, i.e.,

$$R - OO_1 = \frac{\sqrt{31}}{2} - \frac{5}{2\sqrt{31}} = \frac{1}{2\sqrt{31}} \cdot (31 - 5) = 26 \cdot \frac{1}{2\sqrt{31}},$$

and

$$R + OO_1 = \frac{\sqrt{31}}{2} + \frac{5}{2\sqrt{31}} = \frac{1}{2\sqrt{31}} \cdot (31 + 5) = 36 \cdot \frac{1}{2\sqrt{31}}.$$

Dividing two quantities we obtain the answer,

$$KO_1 \div O_1L = \frac{\frac{\sqrt{31}}{2} - \frac{5}{2\sqrt{31}}}{\frac{\sqrt{31}}{2} + \frac{5}{2\sqrt{31}}} = \frac{13}{18}$$

Answer. $13 \div 18$.

Problem 223 Four balls of radius r are placed on a table tangent to each other (Fig. 4.75). A fifth ball of the same radius is placed in the hollow formed by the four other tangent balls. Find the distance from the top point of the fifth ball to the plane of the table.

Solution Let us experiment a little bit and arrange real five balls of equal radius as described in the problem (see Fig. 4.75).

Let S be the center of the top ball and A, B, C, and D the centers of other four tangent balls. We can look from the top at this construction and from that position the composition of the balls will look like one shown in Fig. 4.76.

By the condition of the problem the following is true

$$AD = DC = BC = AB = 2r$$

and

$$AS = SD = SC = SB = 2r,$$

where $SABCD$ is a regular square pyramid formed by the centers of all five balls, with base side of $2r$ and height SO, $SO = \sqrt{AS^2 - AO^2}$, $AO = r\sqrt{2}$. Next, $SO = \sqrt{4r^2 - 2r^2} = r\sqrt{2}$. Now we can find the distance (d) to the plane on which four balls are located as

$$d = 2r + SO = 2r + r\sqrt{2} = r\left(2 + \sqrt{2}\right).$$

Answer. $d = r\left(2 + \sqrt{2}\right)$.

Fig. 4.75 Photo of five tangent balls

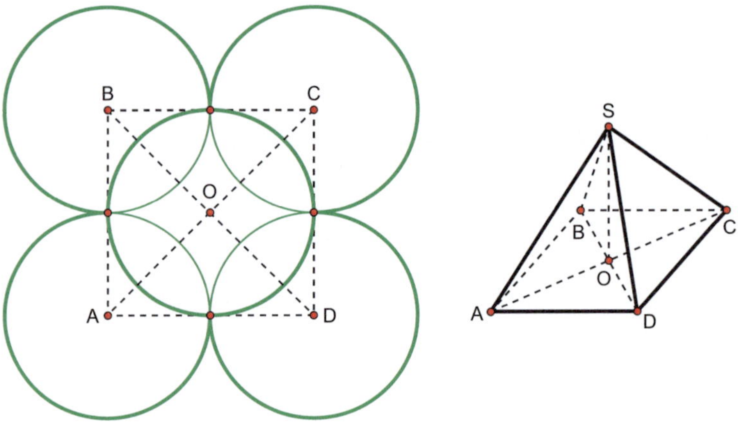

Fig. 4.76 Five tangent balls model

Problem 224 An equilateral cone is inscribed in a hemisphere. Prove that their surfaces are the same and that the line of intersection is half the size of the circumference of the great circle.

Proof Let us sketch an equilateral cone inscribed into hemisphere (see Fig. 4.77). By the condition of the problem, $AS = 2AO = 2r$. The lateral surface of the equilateral cone equals,

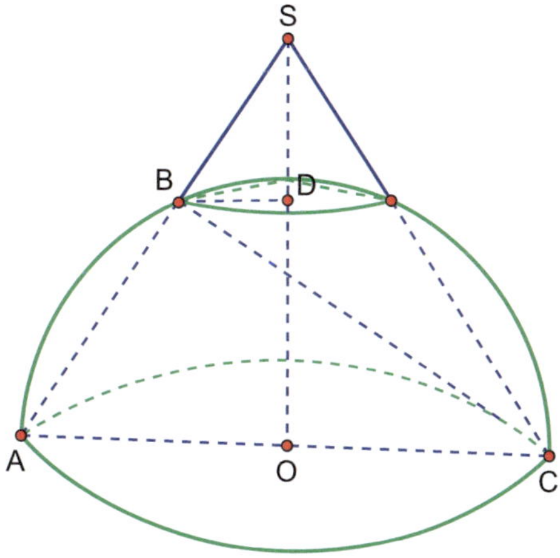

Fig. 4.77 An equilateral cone inscribed into a hemisphere

$$A = \pi r L = \pi r \cdot 2r = 2\pi r^2,$$

which is exactly the surface area of the hemisphere. Since AC is the diameter, the angle ABC is equal to 90°. Then the angle ACB is equal to 30°. The triangle ABC is the right $90 - 30 - 60$ triangle and so the leg AB is equal to the radius of the hemisphere, r. If $AB = r \Rightarrow BS = r,\ \angle BSD = 30°,\ BD = \frac{r}{2}$. Since the cone and the hemisphere intersect by circumference with radius BD, then this circumference is equal to

$$C_1 = 2\pi \cdot \frac{r}{2} = \frac{1}{2} \cdot 2\pi r = \frac{C}{2},$$

where C is circumference of the base circle. □

Problem 225 (*Lomonosov Tournament 2002*) It is known that a polyhedron is inscribed in a sphere if all ends of its edges lie on the sphere. So, a polyhedron is inscribed in a sphere. Could it be that this polyhedron is not convex?

Solution Let us give an example of a non-convex polyhedron inscribed in a sphere. **Method 1**. Take a regular octahedron and circumscribe a sphere around it. Now let's take two of its adjacent faces ABC and BCD and delete them together with edge BC, but instead add edge AD and faces ABD and ACD. We have obtained a new polyhedron such that the segment BC no longer belongs to the figure, so this figure is not convex, but the vertices of this polyhedron are the same as those of a regular octahedron. Therefore this polyhedron is inscribed into a sphere. In Fig. 4.78 we did not show a circumscribed sphere but sketched a regular octahedron that we assumed to be inscribed into a sphere. Next to it on the right is a non-convex polyhedron that is obtained from the octahedron inscribed into the same hypothetical sphere. This non-convex figure has all its vertices the same and hence they are on the sphere. Obviously, if we connect any different points taken on the faces colored in yellow, we will obtain a segment that will be outside the figure. This is by definition a non-convex solid figure.

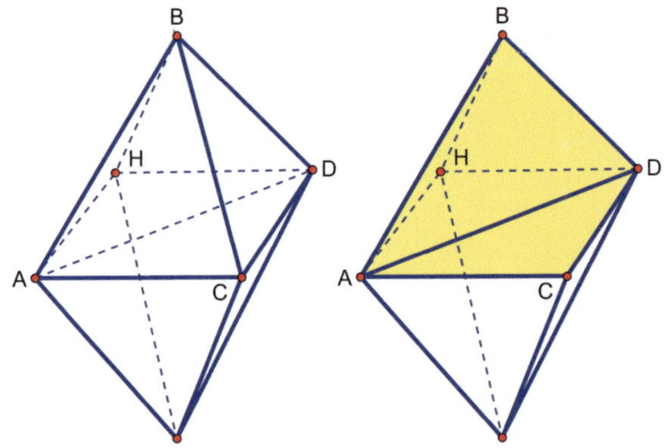

Fig. 4.78 Nonconvex polyhedron created from an octahedron and inscribed in a sphere

Method 2. We could act as follows: take a cube and make a "hole" in it in the form of a rectangular parallelepiped through two parallel faces of the cube. Then connect 8 points of the edges of the "hole" with edges to the nearest vertices of the cube and raise (lower) them to the sphere circumscribed around the cube. Please try to sketch this polyhedron on your own.

Problem 226 (*Kolmogorov*) A spatial quadrilateral $ABCD$ is circumscribed about a sphere. Prove that its four tangent points lie in the same plane.

Proof This problem was devised by academician Andrey Kolmogorov for gifted students of the Moscow State University boarding school.

Method 1. To visualize a spatial quadrilateral, we can look at the sketch on the right in Fig. 4.78, where the points A, B, C and D do not lie in the same plane. However, instead of the special case of a spatial quadrilateral shown in this figure, where $AB = BD = CD = AC$, consider the general case where the sides of the quadrilateral are not necessarily equal. Consider a spatial quadrilateral $ABCD$ (that is, a closed spatial polygonal line) circumscribed about a sphere. Suppose that the sphere is tangent to the sides AB, BD, CD, and CA at points M, N, K, and L, respectively (Fig. 4.79). By the property of equality of tangents drawn to a circle from a single point, it is easy to prove that in a spatial quadrilateral circumscribed about a sphere, the rule of equality of sums of opposite sides also holds. Let $AM = AL = a$, $BM = BN = b$, $DN = DK = c$ and $CK = CL = d$. It is clear that $AB + CD = a + b + c + d = AC + BD = a + d + b + c$.

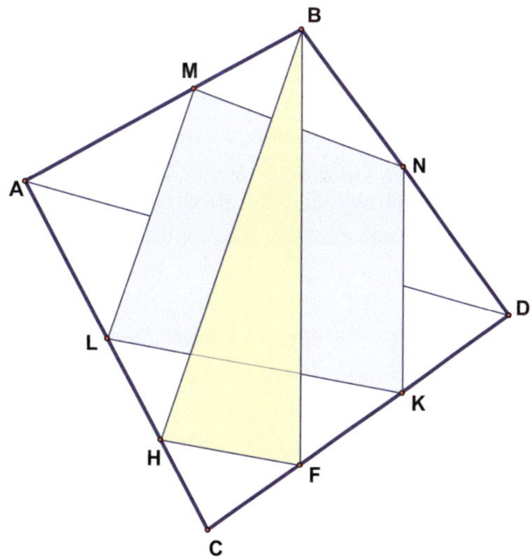

Fig. 4.79 Method 1: Spatial quadrilateral circumscribed about a sphere

Assume that $BD < CD$, then $AB < AC$ and let $F \in CD$, such that $BD = DF = b + c$. Connect points N and K, and then B and F. Triangles BDF and NDK are similar isosceles triangles with common vertex D that share the same angle, then $NK \| BF$. Similarly, let us place point H on side AC, such that $AB = AH = a + b$. Connect points M and L, then connect points B and H, then similarly $ML \| HB$. Connect points H and F. Denote the plane of $\triangle BHF$ by α. (It is shaded yellow in Fig. 4.79)

Let us connect the points L and K and show that $LK \| HF$. Triangle HCF is isosceles, since it has two equal sides ($HC = AC - AB = a + d - (a + b) = d - b$, as well as $CF = CD - BD = c + d - (b + c) = d - b$.) Since $\triangle LCK$ is also isosceles, in which $LC = CK = d$ and similar to $\triangle HCF$, then $LK \| HF$.

Consider plane $MLK = \beta$ that is formed by two intersecting lines ML and LK. Plane $BHF = \alpha$ contains interesting lines BH and HF. Plane $NKL = \gamma$ contains intersecting lines NK and LK. By Theorem 6 we know that if two intersecting lines of the same plane are respectively parallel to two intersecting lines of another plane, then these planes are parallel.

$$\left.\begin{array}{l} ML \cap LK = L \\ BH \| ML \\ HF \| LK \end{array}\right\} \Rightarrow \alpha = (HBF) \| \beta = (MLK).$$

Similarly,

$$\left.\begin{array}{l} NK \cap LK = K \\ NK \| BF \\ HF \| LK \end{array}\right\} \Rightarrow \alpha = (HBF) \| \gamma = (NKL).$$

Since $\alpha \| \beta$ and $\alpha \| \gamma$ and since $\gamma \cap \beta = LK$, then $\beta = \gamma$. Therefore, the tangent points N, K, L and M belong to the same plane. In the Fig. 4.79) this plane $MNKL$, shown in gray, is parallel to the plane HBF and is located behind it.

Method 2. I believe that each of us can always imagine a tetrahedron, $ABCD$. Let's use four vertices of the tetrahedron and make them four vertices of a spatial quadrangle. However, a tetrahedron has 6 edges, while a spatial quadrilateral has only 4. Then we can connect some pairs of four vertices of the tetrahedron by making an edge of the spatial quadrilateral and painting it, for example, blue, so that in each triangular face of the tetrahedron only two edges will be painted in the same color. Next, consider the union of four segments painted blue—we have an example of a spatial quadrangle. If we mentally connect points A and B, B and D, D and C and then C and A, we can imagine a spatial quadrangle $ABDCA$ made from a tetrahedron $ABCD$, in which we used only the edges AB, BD, CD and AC, as if a sphere inscribed in a spatial quadrilateral $ABDCA$ is tangent to only these four edges of the tetrahedron as was sketched in Fig. 4.79 used for Method 1. If we connect vertices A and B, B and C, C and D and then D and A, we will obtain a different spatial quadrilateral $ABCDA$ shown in Fig. 4.80.

Let the sphere be tangent to the sides of this spatial quadrilateral AB, BC, CD, and DA at points K, L, M and N, respectively (see our sketch in Fig. 4.80). Since two tangent segments drawn to the same circle from one external point are congruent, then $AN = AK$, $BK = BL$, $CL = CM$, $DM = DN$, as pairs of equal tangent segments to the sphere drawn from points A, B, C and D, respectively. The obvious true equality $\frac{AK}{AN} \cdot \frac{BL}{BK} \cdot \frac{CM}{CL} \cdot \frac{DN}{DM} = 1$ can be rewritten as

$$\frac{AK}{BK} \cdot \frac{BL}{CL} \cdot \frac{CM}{DM} \cdot \frac{DN}{AN} = 1, \tag{4.74}$$

which is also true.

We have to prove that the tangent points K, L, M, N are coplanar. Suppose contradiction, i.e., that point N does not belong to plane $\alpha = KLM$. Consider point N' at which the plane $KLM = \alpha$ intersects the straight line DA. ($\alpha \cap DA = N'$) Let's show that $N = N'$ by demonstrating that

$$\frac{AK}{BK} \cdot \frac{BL}{CL} \cdot \frac{CM}{DM} \cdot \frac{DN'}{AN'} = 1. \tag{4.75}$$

Let us construct a line a perpendicular to the plane α and let $a = PQ$, $\alpha \cap a = P$. Consider the projection of all points onto the line $a = PQ$. Recall that in three-

Fig. 4.80 Method 2: Spatial quadrilateral circumscribed about a sphere

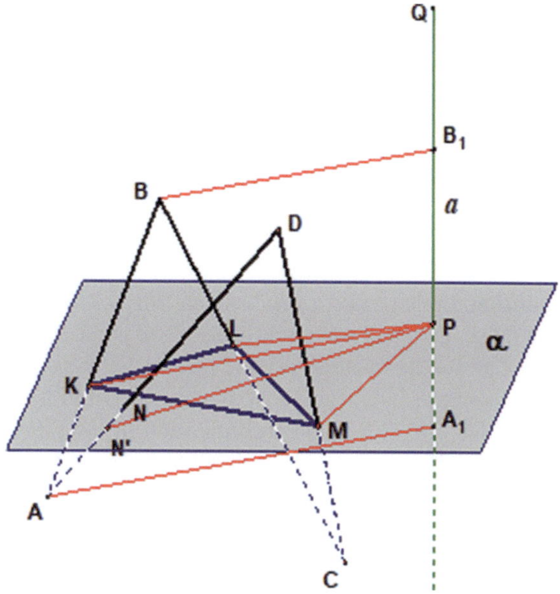

dimensional space, the projection of a point G onto a line a is a point $G_1 \in a$ such that GG_1 is perpendicular to the line a. If a point belongs to the plane α perpendicular to the line a, then the projection of this point onto the line a will be the intersection point of the line a and the plane α. Therefore coplanar points $K,\ L,\ M$ and N' must be projected onto the same point P.

Next, let $A_1,\ B_1,\ C_1,\ D_1$ be the projections of points $A,\ B,\ C$ and D onto the line $a = PQ$. According to Thales's theorem, the ratios of segments lying on the same line are preserved under orthogonal projection. For example, $\frac{AK}{BK} = \frac{A_1P}{B_1P}$, etc.

Let us replace the product of the ratios on the left of (4.75) by equal to them ratios of the corresponding projections,

$$\frac{AK}{BK} \cdot \frac{BL}{CL} \cdot \frac{CM}{DM} \cdot \frac{DN'}{AN'} = \frac{A_1P}{B_1P} \cdot \frac{B_1P}{C_1P} \cdot \frac{C_1P}{D_1P} \cdot \frac{D_1P}{A_1P} = 1.$$

Because the product is equal to 1, this means that (4.75) is correct.

It follows from the equalities (4.74) and (4.75) that $DN \div AN = DN' \div AN'$, then $N = N'$ (since both points N and N' lie on the same segment AD). We found that the tangent points $K,\ L,\ M$ and N are coplanar and therefore these points lie on the same circle of the inscribed sphere. □

4.9 Homework on This Chapter. Problems to Solve

1. A sphere is inscribed in a regular quadrangular pyramid, which is tangent to the base and all lateral faces of the pyramid. The sphere divides the height of the pyramid into ratio of one to three, counting from the top. Find the volume of the pyramid, if the apothem of the pyramid is a.

   Answer. $\frac{48}{125}a^3$

Solution Let us make a good sketch (Fig. 4.81). If a sphere is inscribed in pyramid $ABCDE$, then its great circle is inscribed in the isosceles triangle MEN, where EN and EM are the apothems of the faces EBC and ADE, respectively. The height of this triangle EH is simultaneously the height of the pyramid. Denote the height of the pyramid $h = EH$. By the condition of the problem $EK \div KH = 1 \div 3$, then $EK = \frac{1}{4}EH,\ \ KH = \frac{3}{4}EH$, then the radius of the sphere is

$$r = OH = \frac{3}{8}h. \tag{4.76}$$

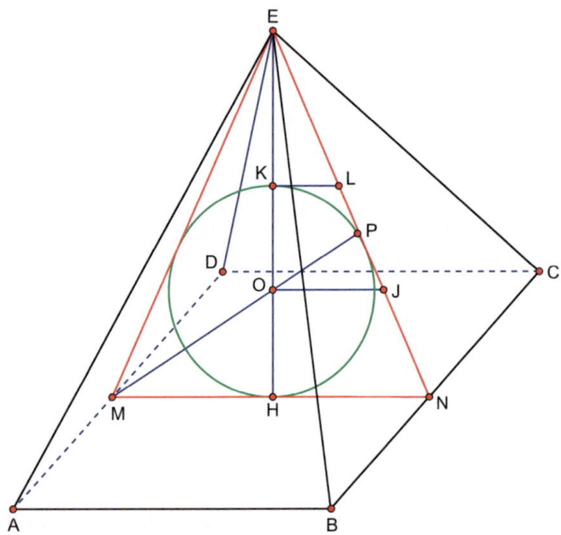

Fig. 4.81 Sphere inscribed in a regular square pyramid

Assume the base of a pyramid is $b = AB = 2HN$, where $HN = \sqrt{EH^2 - HN^2}$ which can be written as

$$h^2 = a^2 - \frac{b^2}{4}. \tag{4.77}$$

Since the circle is inscribed into triangle EMN, then we can write the area of triangle EMN in two different way as follows

$$\frac{1}{2}ah = \frac{1}{2}(a + 2b)\,r. \tag{4.78}$$

Substituting (4.76) in (4.78) we obtain that

$$b = \frac{6a}{5}. \tag{4.79}$$

Using (4.78) in (4.77) we have for the height of the pyramid:

$$h = \sqrt{a^2 - \frac{36a^2}{100}} = \frac{4a}{5}. \tag{4.80}$$

Next, using (4.80) and (4.79) the volume of the pyramid $ABCDE$ is

$$V = \frac{1}{3}b^2h = \frac{1}{3}\cdot\frac{36a^2}{25}\cdot\frac{4a}{5} = \frac{48a^3}{125}.$$

2. The lateral surface area of the cylinder is 10π, the base area is 2π. A regular hexagonal prism is inscribed into a cylinder. Find the distance from the center of one base of the cylinder to the side of the hexagon inscribed in the other base.

   Answer. $\sqrt{14}$

3. The diagonals of a rectangular parallelepiped form angles of 45° and 60° with the diagonals of its base. The distance between the diagonal of the parallelepiped and the diagonal of the base that does not intersect it is equal to L. Find the area of the sphere circumscribed about the parallelepiped.

   Answer. $24\pi L^2$

4. The base of the pyramid is a rhombus with an acute angle α. The lateral faces are inclined to the plane of the base at angle β. Determine the volume of the pyramid, if the radius of the circle inscribed in the rhombus is r.

   Answer. $\frac{4r^3\tan\beta}{3\sin\alpha}$

5. Two balls of radius r and one ball of radius $r/2$ lie inside the cylinder, so that each ball is tangent to the two others, the lower base of the cylinder and its lateral surface. Find the radius of the base of the cylinder.

   Answer. $\frac{9r}{4}$

6. A sphere of radius $1/\pi$ is inscribed into a cone. Find the volume of the cone, if the volume of the cone with the same base and with the vertex at the center of the inscribed sphere is 3.

   Answer. $\frac{54\pi^2}{9\pi^2-1}$

7. A pyramid inscribed in a sphere, has the base as is a rectangle with the diagonal of 10; the lateral edges of the pyramid are inclined to the base at an angle of $60°$. Find the surface area of the sphere.

   Answer. $S = \frac{400\pi}{3}$

8. The area of the axial section of a truncated cone is equal to the difference in the areas of the bases and the radii of the bases are 6 and 3, respectively. Find the volume of the cone.

   Answer. $V = 54\pi^2$

9. A ball with a radius of 2 units is inscribed in a truncated cone. The generatrix of the cone is inclined to the plane of the base at an angle $\alpha = \frac{\pi}{3}$. Find the volume of the truncated cone.

   Answer. $\frac{208\pi}{9}$

10. A cone is inscribed in a sphere, the volume of which is four times less than the volume of the sphere. What is the ratio of the height of the cone to the radius of the sphere?

    Answer. $1, \frac{1+\sqrt{5}}{2}$

11. Given a cube with side a and a ball whose center coincides with the center of the cube. Determine the radius of the sphere if the surface area of the part of the ball that protrudes from the cube is twice the area of the part of the cube that protrudes from the ball.

    Answer. $\frac{a}{8}\left(1 + \sqrt{1 + \frac{8(\pi+4)}{\pi}}\right)$

12. The sphere passes through the vertices of one face of the cube and touches the sides of the opposite face of the cube. Find the ratio of the volumes of the ball and the cube.

    Answer. $\frac{41\sqrt{41}}{284}$

13. The axial section of the cylinder is a square and the axial section of the cone is a regular triangle equal in area to this square. Find the ratio of the volumes of the cylinder and the cone.

    Answer. $\frac{\sqrt[4]{3}}{64}$

14. In a regular triangular pyramid $SABC$, the side of the base AB is p times the length of the side edge AC. Sphere centered at point O_1 is tangent to the planes SAB and SAC at points B and C. Sphere centered at O_2 is tangent to the planes SAC and SBC at points A and B. Find the ratio of the volumes of the pyramids SBO_1O_2 and $SABC$.

    Answer. $\frac{1}{3-p^2}$

15. Pyramid $ABCD$ is inscribed into a sphere. The second sphere of radius 1 is tangent to the first sphere internally at point D and also is tangent to plane ABC. It is known that $AD = 3$, $\cos\angle BAC = 4/5$, $\cos\angle BAD = \cos\angle CAD = \frac{1}{\sqrt{2}}$. Find the volume of pyramid $ABCD$.

    Answer. $V = 18/5$

16. Given a pyramid $ABCD$. A sphere is tangent to the planes DAB, DAC and DBC at points K, L, M respectively. Point K belongs to side AB, point L is on side AC, and point M is on side BC. The radius of the sphere equals 3. $\angle ADB = 90°$, $\angle BDC = 105°$, $\angle ADC = 75°$. Find the volume of the pyramid.

    Answer. $V = 48$

17. Eight equal spheres are inscribed into a large sphere S of radius R in such way that each small sphere is tangent to three other small spheres so as to the large sphere. Find the radius r of the small inscribed spheres if their centers lie in the vertices of a cube.

    Hint. The diameter of the large sphere is equal to the diagonal of the cube plus two small radii:
$$2R = a\sqrt{3} + 2r.$$

    The distance between the centers of the small spheres is equal to $2r = a$, where a is the edge of the cube.

    Answer. $r = \frac{R}{\sqrt{3}+1}$

18. Find the radius of a sphere circumscribed about a regular quadrangular pyramid with all edges equal to a.

    Answer. $R = \frac{a\sqrt{2}}{2}$

19. Find the radius of a sphere circumscribed about a regular tetrahedron with the edges equal to 1.

    Answer. $R = \frac{\sqrt{6}}{4}$

20. Two spheres are inscribed in a right cone with vertex C. One sphere of radius R_1 is tangent to the lateral surface and base of the cone. The second sphere with radius R_2 is tangent to the first sphere and to the lateral surface. Find the angle that generatrix of the cone forms with the base of the cone.

    Solution.

    Let us show the axial section of the solid figure described in the problem. In Fig. 4.82 it is triangle CAB with two tangent circles inscribed. Let O_1 and O_2 be the centers of the spheres and F and E the points at which the sphere tangent to the lateral surface of the cone. Let O_2E be perpendicular to O_1F. Drop a perpendicular segment CD from vertex to the base, $CD \perp AB$.

$$O_1G = R_2 - R_1, \quad O_1O_2 = R_2 + R_1$$

Because $\angle O_2O_1G = \angle CAB$, then $\Delta O_2O_1G = \Delta CAD$, and the following is true:

$$\cos \angle CAD = \cos \angle O_2O_1G = \frac{R_2 - R_1}{R_2 + R_1}$$

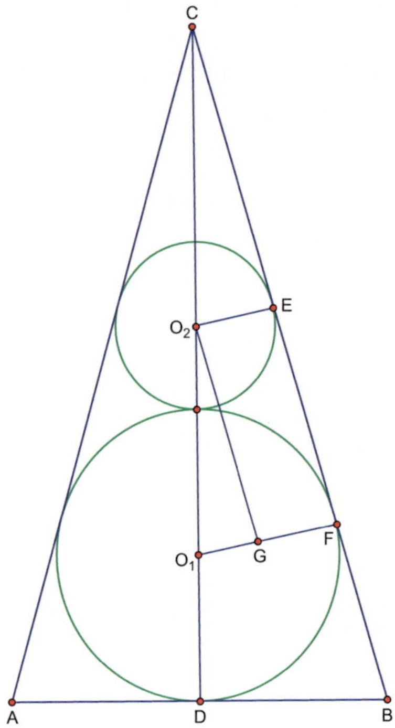

Fig. 4.82 Two balls inscribed in a right cone

Answer. $\arccos \frac{R_2-R_1}{R_2+R_1}$

21. Is it possible that a sphere circumscribed about a triangular pyramid to have its center outside the pyramid?

    Hint. Think about the location of the center of circles circumscribed about different triangles.

    Answer. The pyramid can be right with an obtuse triangle as the base. Or it can be a pyramid with all faces—obtuse triangles (see Fig. 4.83).

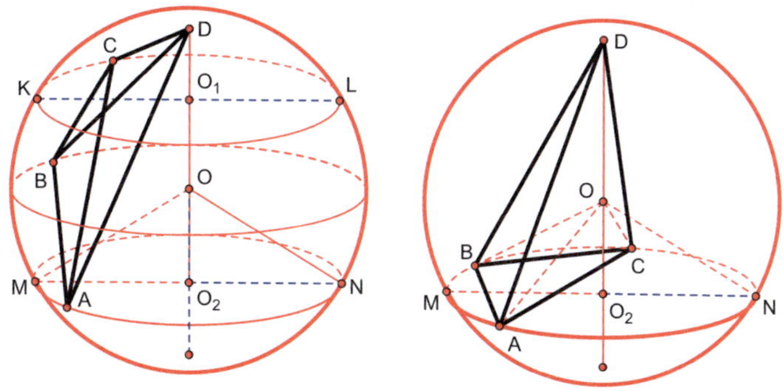

Fig. 4.83 The centers of the spheres are located outside the triangular pyramids inscribed in the spheres

22. Circular segment is rotated about diameter of the circle. Prove that the volume of the solid figure resulted as this rotation does not depend on the radius of the circle.
23. A hemisphere is inscribed in a regular tetrahedron, whose edge is equal to a, so that three faces of the tetrahedron touch its spherical surface, and the fourth serves as its diametral plane. Determine the radius of the hemisphere.
24. Answer the following questions:

    - Can a sphere be inscribed in a cuboid (rectangular parallelepiped) other than a cube?
    - Can a sphere be inscribed in an oblique parallelepiped all of whose faces are rhombuses?
    - Can a sphere be inscribed in an oblique triangular prism?

25. Find the volume of torus obtained by rotation of circle $x^2 + (y - b)^2 = a^2$ around the X-axis, if $b > a$.

    Hint. This problem is very similar to one (Problem 189) solved in this chapter. You also can simply use formula (4.26)

    Answer. $V = 2\pi^2 a^2 b$

26. Take a closer look at a standard soccer ball. It consists of 20 white hexagons and 12 black pentagons. Show that you can't sew a new soccer ball from just hexagons. Prove that a soccer ball must have exactly 12 pentagonal patches.

    Hint: The soccer ball problem is similar to the solution to Problem 129 of Chap. 2 about the fullerene molecule. Draw mentally a picture of a soccer ball on the sphere. It should consist of hexagonal and pentagonal patches. At any vertex exactly 3 edges must converge. Let x be the number of pentagons and y the number of hexagons. Use formula $V - E + F = 2$ and show that $x = 12$ is the only solution.

Solution. The total number of faces is $F = x + y$, of vertices is $V = \frac{5x+6y}{3}$ and of edges is $E = \frac{5x+6y}{2}$, then we can solve the system,

$$\begin{cases} 2x + 2y = V + 4 \\ 5x + 6y = 3V \end{cases}$$

and obtain $x = 12$, and so a soccer ball can only have 12 pentagonal sections, but in principle there can be any number of hexagonal sections. This is different from Problem 129. Since the number of vertices in a fullerene must be even, we had restrictions on the number of hexagons (y), but in the soccer ball problem there are no restrictions. If we want, we can make a soccer ball from 129 hexagonal and 12 pentagonal patches, or another ball from 40 hexagonal and 12 pentagonal patches. However, no matter how many times you try to sew a soccer ball from 10 or 21 pentagons, nothing will work. Also, no soccer ball can consist only of hexagonal patches. It is clear that pentagons give curvature to an object.

27. In a cube $ABCDA_1B_1C_1D_1$ with edge a, a plane P is drawn through a point A parallel to line BD, forming an angle equal to $\arcsin \frac{1}{2\sqrt{2}}$ with line AB. Find the radius of the sphere tangent to plane P and faces $ABCD$, BCC_1B_1 and DCC_1D_1 of the cube.

Answer. $R = \frac{a\sqrt{2}}{2+\sqrt{2}+\sqrt{3}}$.

References

1. Hartshorne R (2002) Geometry: euclid and beyond. Springer, Berlin, P 226
2. Grigorieva EV (2013) Methods of solving complex geometry problems. Birkhäuser. https://www.springer.com/us/book/9783319007045
3. Grigorieva EV (2015) Methods of solving nonstandard problems. Birkhäuser. https://www.springer.com/us/book/9783319198866
4. Grigorieva EV (2016) Methods of solving sequence and series problems. Birkhäuser. https://www.springer.com/us/book/9783319456850
5. Grigorieva EV (2018) Methods of solving number theory problems. Birkhäuser. https://www.springer.com/us/book/9783319909141
6. Kiselev AP, Givental A (2008) Kiselev's geometry book 2. Stereometry, Sumizdat. https://www.cut-the-knot.org/books/Reviews/KiselevsStereometry.shtml
7. David E, Twenty one proofs of Euler's formula. https://ics.uci.edu/~eppstein/junkyard/euler/
8. Imre L, Proofs and refutations. https://en.wikipedia.org/wiki/Proofs_and_Refutations
9. Smirnova IM, Smirnov VA (2008) Geometry for high school students. Education, Moscow, p 288. ISBN: 978-5-346-01106-4
10. Smirnova IM, Smirnov VA (2018) Geometry with geogebra. Stereometry, Moscow
11. Evan C (2024) Euclidean geometry in mathematical olympiads. An Imprint of the American Mathematical Society, MAA Press. https://bookstore.ams.org/prb-27
12. Friedman SH, DeCamp DL, Sijbesma RP et al (1993) Inhibition of the HIV-1 protease by fullerene derivatives: model building studies and experimental verification. J Am Chem Soc 115:6506–9
13. Pogorelov AV (2014) Textbook in school geometry. Prosveshenie, Moscow. 978-5-09-032026-9
14. Dalinger VA, Symogenkov SD (2013) Implementation of intersubject relationships when solving mathematical problems through cognitive-visual activities. Omsk University Press, Omsk, p 159. ISBN: 978-5-8268-1825-1
15. Dolbilin Y (2001) Three theorems on convex polyhedra Kvant 5:7–12
16. Gellert W, Gottwald S, Hellwich M, Kästner H, Küstner H (1975) The vnr concise encyclopedia of mathematics. Springer, Dordrecht
17. Alexandrov AD (2005) Convex polyhedra. Springer, Berlin (Monographs in Mathematics. Translation of the 1950 Russian edition by N. S. Dairbekov, S. S. Kutateladze, and A. B. Sossinsky.)
18. Cromwell P (1999) Polyhedra. Cambridge Univerity Press
19. Prasolov VV (2010) Problems in stereometry. Moscow, p 352

E. Grigorieva, *Methods of Solving Solid Geometry Problems*,
https://doi.org/10.1007/978-3-031-86969-3

20. Euler L (1758) Proof of Some Notable Properties with which Solids Enclosed by Plane Faces are Endowed, Demonstratio nonnullarum insignium proprieatatum, quibus solida hedris planis inclusa sunt praedita. Novi Commentarii Academiae Scientiarum Petropolitanae 4:72–93. http://eulerarchive.maa.org/docs/translations/E231.pdf
21. Archimedes (2009) On the sphere and cylinder by Archimedes. Book 1, 1665. Cambridge University Press. ISBN: 0521117984
22. Lidsky B, Ovsyannikov l, Tulaikov A, Shabunin M, (1973) Problems in elementary mathematics. MIR Publisher, Moscow (in Russian)
23. Popov G (1999) Historic problems in elementary mathematics. Moscow State University Publisher, Moscow (in Russian)
24. Budak V, Zolotareva N, Fedotov M (2012) Geometry course for high school. Moscow State University Publisher, Moscow (in Russian)
25. Leahy A, On Pappus–Guldin theorems. https://maa.org/book/export/html/117332
26. Sullenger D, Kennard CHL (1966) Boron crystals. Sci Am 215(1):96–107. Scientific American, a division of Nature America, Inc. Stable https://www.jstor.org/stable/10.2307/24930995
27. Eremin V (2016) Mathematics in chemistry, 2nd edn. Moscow State University, Moscow, p 64. ISBN 978-5-4439-0923-3
28. Moussa F, Living Longer—C_{60} Olive Oil Interview with Fathi Moussa https://www.youtube.com/watch?v=XLZS1WQl_1E
29. Andy L (2001) Hungarian problem Book III. Intermediate and challenging problems from old hungarian competitions. MIR, Moscow (1976), American Mathematical Society
30. Wikipedia, Timaeus dialogue. https://en.wikipedia.org/wiki/Timaeus_(dialogue)
31. John B, Who Discovered the Icosahedron? https://math.ucr.edu/home/baez/icosahedron/
32. Benno A (2006) The mathematical conquest of the third dimension. Bull Am Math Soc (N.S.) 43(2):231–234. https://www.ams.org/journals/bull/2006-43-02/S0273-0979-06-01111-6/S0273-0979-06-01111-6.pdf
33. Eves JH (1990) An introduction to the history of mathematics. Thomson Brooks/Cole, NY, USA
34. Bakry R, Vallant R, Najam-ul-Haq M, Rainer M, Szabo Z, Huck Ch, Bonn G (2007) Medicinal applications of fullerenes. Int J Nanomedicine 2(4):639–649
35. Kaganov ED (1997) 400 most interesting problems with solutions. MIR, Moscow (in Russian)
36. Rivkin AA (2003) Problem book for preparation for math entrance exams. MIR, Moscow (in Russian)
37. Ross DE (2009) Master math geometry. Cengage learning, 2nd edn. https://discoverbooks.com/products/master-math-geometry-1598639846
38. CORD (2014) CORD geometry: learning in context. CORD Communications, 4th edn. https://www.cordcommunications.com/geometry-4th.php
39. Prentice Hall (2020) Geometry, Texas Teacher's Edition, Prentice Hall, p 490
40. Pearson (2004) Essentials of Geometry for College Students, Pearson
41. Gary M, Lynn E, Mauer V (2008) College geometry: a problem-solving approach with applications, 2nd edn. Pearson
42. Alsina C, Nelsen R (2015) A mathematical space odyssey. Solid Geometry in the 21st Century, MAA Press
43. Pesado-Gómez C, Serrano-García JS, Amaya-Flórez A, Soto-Contreras A, Morales-Morales D, Colorado-Peral R (2024) Fullerenes: Historical background, novel biological activities versus possible health risks. Coordination chemistry reviews, vol 501 (2024). https://www.sciencedirect.com/science/article/pii/S0010854523005398 Books and Contest Problems for Further Reading

Books and Contest Problems for Further Reading

44. Gardiner A (2011) The mathematical olympiad handbook. Oxford University Press, NY
45. Shklarsky D, Chestnov N, Yaglom (1993) The USSR olympiad problem book: selected problems and theorems of elementary mathematics. Dover
46. Galperin G (2005) Moscow olympiads. Nauka, Moscow
47. Andreescu T, Feng Z (2000–2010) USA and international math olympiads. MAA
48. The William Lowell Putnam Mathematics Examination 1985–2000 by Kedlaya, Poonen, Vakil
49. Djukic D, Jankovic V, Matic I, Petrovic N (2011) The IMO compendium. a collection of problems suggested for the international mathematical olympiads: 1959–2009, 2nd edn. Springer, Berlin

Index

E. Grigorieva, *Methods of Solving Solid Geometry Problems*,
https://doi.org/10.1007/978-3-031-86969-3

D

Q

R

U

V

W

Z

MIX
Papier aus verantwortungsvollen Quellen
Paper from responsible sources
FSC® C105338

If you have any concerns about our products,
you can contact us on
ProductSafety@springernature.com

In case Publisher is established outside the EU,
the EU authorized representative is:
Springer Nature Customer Service Center GmbH
Europaplatz 3, 69115 Heidelberg, Germany

Printed by Libri Plureos GmbH
in Hamburg, Germany